Pathogenesis of Bacterial Infections in Animals

SECOND EDITION

Pathogenesis of Bacterial Infections in Animals

SECOND EDITION

EDITED BY

CARLTON L. GYLES

Professor of Veterinary Microbiology
Ontario Veterinary College
University of Guelph
Guelph, Ontario

AND

CHARLES O. THOEN

Professor of Veterinary Microbiology
College of Veterinary Medicine
Iowa State University
Ames, Iowa

WITH

TWENTY-EIGHT CONTRIBUTORS

IOWA STATE UNIVERSITY PRESS / AMES

♾ Printed on acid-free paper in the United States of America

First edition, 1986
Second printing, 1988
Second edition, 1993

Library of Congress Cataloging-in-Publication Data

Pathogenesis of bacterial infections in animals / edited by Carlton L. Gyles and Charles O. Thoen ; with twenty-eight contributors.—2nd ed.
p. cm.
Includes bibliographical references and index.
ISBN 0-8138-1339-5
1. Veterinary bacteriology. I. Gyles, C. L. (Carlton L.) II. Thoen, Charles O.
SF780.3.P37 1993
636.089′6014—dc20 93-1289

Contents

Authors

Numbers in parentheses beside each name refer to chapters.

Art A. Andersen (26)
National Animal Disease Center
Agricultural Research Service
Ames, IA 50010

David A. Bemis (17)
Department of Microbiology
College of Veterinary Medicine
University of Tennessee
Knoxville, TN 37916

Bernard Bizzini (9)
Unit of Molecular Toxinology
Division of Immunology
Pasteur Institute
28. Rue du Dr. Roux 75724
Paris, France

Brian W. Brooks (22)
Agriculture Canada Food Production
and Inspection Branch
Animal Diseases Research Institute
3851 Swallowfield Road
P.O. Box 11300, Station "H"
Nepean, Ontario, Canada
K2H 8P9

Eugene H. Burns, Jr. (17)
Department of Microbiology
College of Veterinary Medicine
University of Tennessee
Knoxville, TN 37916

Norm F. Cheville (20)
National Animal Diseases Centre
United States Department of Agriculture
Ames, IA 50010

R. J. Chiodini (4)
Mycobacterial Unit
Rhode Island Hospital
Providence, RI 020903

Robert C. Clarke (13)
Animal Pathology Laboratory
Agriculture Canada
Guelph, Ontario, Canada
N1G 1Y4

Jennifer A. Rice Conlon (18)
Langford Inc
131 Malcolm Rd
Guelph, Ontario, Canada

Charles J. Czuprynski (6)
Department of Pathobiological Sciences
School of Veterinary Medicine
University of Wisconsin Madison
Madison, WI 53706-1102

Fred Enright (20)
Department of Veterinary Science
Louisiana State University
Baton Rouge, LA 70803

John W. Ezzell, Jr. (3)
U.S. Army Medical Research Institute
of Infectious Diseases
Fort Dietrich
Frederick, MD 21701

M. M. Garcia (22)
Agriculture Canada Food Production
and Inspection Branch
Animal Diseases Research Institute
3851 Swallowfield Road
P.O. Box 11300, Station "H"
Nepean, Ontario, Canada
K2H 8P9

C. L. Gyles (7, 10, 12, 13, 14, 15, 19, 21)
Department of Veterinary Microbiology
and Immunology
Ontario Veterinary College
University of Guelph
Guelph, Ontario, Canada
N1G 2W1

Per Jonsson (20)
Department of Veterinary Medical Microbiology
Box 7073
S-750 07 Uppsala
Sweden

J. I. MacInnes (16)
Department of Veterinary Microbiology
and Immunology
University of Guelph
Guelph, Ontario, canada
N1G 2W1

L. Niilo (11)
825 - 12 A Street South
Lethbridge
Alberta
T1J 2T6

J. F. Prescott (5, 23, 24)
Department of Veterinary Microbiology
and Immunology
University of Guelph
Guelph, Ontario, canada
N1G 2W1

Tonie E. Rocke (8)
U.S. Department of the Interior
Fish and Wildlife Service
National Wildlife Health Research Center
6006 Schroeder Rd.
Madison WI 53711-6223

S. Rosendal (25)
Department of Veterinary Microbiology
and Immunology
University of Guelph
Guelph, Ontario, Canada
N1G 2W1

Patricia E. Shewen (18)
Department of Veterinary Microbiology
and Immunology
University of Guelph
Guelph, Ontario, Canada
N1G 2W1

Nonie L. Smart (16)
Veterinary Laboratory Services
Ontario Ministry of Agriculture and Foods
Guelph, Ontario, Canada

J. Glenn Songer (5)
Department of Veterinary Science
University of Arizona
Tucson Arizona 85721

Charles O. Thoen (4, 20)
Dept. of Veterinary Microbiology
and Preventive Medicine
College of Veterinary Medicine
Iowa State University
Ames, IA 50011

John F. Timoney (1)
Department of Veterinary Science
Gluck Equine Research Centre
University of Kentucky
Lexington, KY 40546-0099

Torkel Wadstrom (2)
Department of Medical Microbiology
Lund University
Solvegatan 23
S-223 62 Lund,
Sweden

Catherine L. Wilhelmsen (3)
Pathology Division
U.S. Army Medical Research Institute
of Infectious Diseases
Fort Dietrich
Frederick, MD 21701

R. L. Zuerner (24)
National Animal Diseases Centre
United States Department of Agriculture
Ames, IA 50010

Preface

Pathogenesis has always been a fascinating and exciting subject; in recent years the pace of discovery in this field has quickened and students of the subject have come to admire the ingenuity of brilliant researchers and the greater ingenuity of the many bacteria that have adapted to their hosts in a variety of intriguing ways. The second edition of Pathogenesis of Bacterial Infections in Animals retains the objective of bringing together the knowledge of experts on pathogenesis of animal diseases caused by various species or groups of bacteria. A new chapter on *Chlamydia* has been added and several new authors have provided contributions that enrich this edition. We have attempted to emphasize events in pathogenesis at the cellular and molecular level, and to place these developments in the context of the overall picture of disease. Accordingly, we expect that this edition will provide valuable information to graduate students and faculty and will also be useful as a reference source for undergraduate veterinary students.

There have been impressive advances in understanding the structure-function relationships for many important bacterial molecules that play critical roles in disease. These include molecules that make up structures such as pili and streptococcal M protein, and toxin molecules such as *Escherichia coli* verotoxins and enterotoxins. Combinations of molecular genetic approaches, mutagenesis, epitope analyses, and X-ray crystallography have provided new insights into the ways in which these molecules contribute to virulence. Tissue culture systems are being used increasingly to approximate the in vivo environment for pathogenic bacteria, and remarkable successes are being achieved through the use of such systems to study the association of certain pathogenic bacteria with cells. However, the limitations of these systems continue to be recognized: cells in culture are not "normal" cells and exist in isolation from important influences in the intact animal. Unfortunately, animals continue to be needed for some critical experiments.

Molecular genetic methods provide increased power for identification of virulence factors and for evaluation of the role of bacterial products in disease. These methods are very effective in creating isogenic pairs for the study of virulence factors and putative virulence factors, for identifying new genetic loci and bacterial products that contribute to virulence, and for allowing the creation of precise and defined mutants for structure-function studies and for vaccines.

A recent theme that is being "re-discovered" is that bacterial pathogens grown in vivo often produce a different array of surface structures and products compared with those produced during growth in vitro. Researchers are using differences between in vivo and in vitro bacterial products to identify substances that are important for growth in vivo and therefore for virulence. A recent report shows how methods of molecular genetics may be used to identify genetic loci that are turned on in vivo but not in vitro. It is expected that this approach will accelerate our ability to detect genes that are required for virulence.

Another dominant theme in pathogenesis is coordinate regulation of virulence attributes. It is becoming clear that pathogenic bacteria have well-developed systems for sensing various environments inside and outside the host and responding to changes by modifications of their metabolism and products. This method of regulation of response is complemented by random changes in bacterial metabolism, which enhance the chances that some segment of the bacterial population will be well prepared for whatever environment is encountered.

We gratefully acknowledge the efforts and contributions of the authors, who have gone to great lengths to provide interesting information that is on the cutting edge of present-day research.

Pathogenesis of Bacterial Infections in Animals

SECOND EDITION

1 / *Streptococcus*

BY J. F. TIMONEY

THE STREPTOCOCCI are gram-positive spherical bacteria less than 2 μm that typically grow by cell division in one plane so that nascent cells form a linear array which may consist of 50 or more attached cells. Most streptococci are facultatively anaerobic, catalase negative, non-spore-forming, and nonmotile. Their nutritional requirements are complex and variable, reflecting adaptation as commensals or parasites of a wide variety of vertebrates. Classification is based on a combination of characteristics including hemolytic properties, carbohydrate and protein antigen composition, fermentation and other biochemical reactions, growth characteristics, and DNA homology.

The majority of the pathogenic streptococci possess a dominant serologically active carbohydrate that is antigenically different from one species or group of species to another. These cell wall antigens are the basis of the Lancefield grouping system and are widely used by clinical laboratories in a variety of methodologies for serogrouping an unknown isolate. The group-specific antigens, designated A-H and K-V, are readily extracted from cell walls by autoclaving, formamide treatment, or by enzymic digestion. Groups B, C, D, E, G, L, and V contain the pyogenic animal streptococci which cause mastitis in cattle and other hosts; strangles in horses; meningoencephalitis, arthritis, and cervical lymphadenitis in swine; neonatal septicemias in kittens; and lymphadenitis in juvenile cats and laboratory rodents. Some pathogenic streptococci, notably *S. uberis, S. parauberis*, and *S. pneumoniae*, are not groupable in the Lancefield scheme and are identified by features such as fermentation behaviour, ability to grow at different temperatures, salt tolerance, DNA restriction fragment profile, optochin sensitivity, and bile solubility.

The virulence factors of the streptococci most frequently involved in animal disease are shown in Table 1.1. With the exception of *S. pneumoniae* and *S. suis*, the pathogens listed are often termed the "pyogenic streptococci" because of their association with pus and purulence. In general, virulence of the pathogenic streptococci is based on surface structures that directly or indirectly impede phagocytosis. The best understood streptococcal virulence factors are the hyaluronic acid capsule and the antiphagocytic M proteins. However, other molecules, which include streptolysins, leukocidal toxins, streptokinase, and possibly plasmin receptors found on the surface of the streptococcus or released into the surroundings, also contribute to lesion development. In addition, most pathogenic streptococci have the ability to bind components of the host's plasma, such as albumin, immunoglobulins, and fibrinogen. Organisms coated with one or more of these components may be able to evade host defenses either by escaping detection or by blocking deposition of opsonic components of complement.

TABLE 1.1. Pathogenic streptococci of animals

Species	Lancefield Group	Virulence Factors	Disease
S. agalactiae	B	Capsular polysaccharide, R proteins, CAMP factor	Mastitis in ruminants
S. dysgalactiae	C	Hyaluronidase, streptokinase, fibronectin-binding protein	Mastitis in ruminants
S. equi subsp. *equi*	C	Capsule, M protein, hemolysin, streptokinase, plasmin receptor, IgG-binding protein, leukocidal toxin	Strangles in *Equidae*
S. equi subsp. *zooepidemicus*	C	Capsule, M protein, hemolysin, streptokinase	Opportunist pyogen of many animal hosts
S. equisimilis	C	M protein	Lymphadenitis, metritis, placentitis in *Equidae*, porcine arthritis
S. canis	G	M protein, hemolysin	Canine and feline metritis and vaginitis, neonatal bacteremia of kittens, lymphadenitis of juvenile cats, guinea pigs, and rats
S. suis	D	Capsule, 136- and 100-kDa virulence proteins	Meningoencephalitis, septicemia and arthritis in young pigs
S. parauberis	NA[a]	Not identified	Bovine mastitis
S. porcinus	P, U, V, E	M protein	Porcine cervical lymphadenitis
S. pneumoniae	NA	Capsule, pneumolysin, IgA protease, neuraminidase	Respiratory disease of horses in training
S. uberis	NA	Capsule, hyaluronidase, antiphagocytic 65-kDa protein	Bovine mastitis

[a] Note: NA = not applicable.

STREPTOCOCCUS AGALACTIAE

Streptococcus agalactiae, the lone member of Lancefield group B, is an important cause of chronic, contagious bovine mastitis. In humans, it causes neonatal septicemia and meningitis as well as a variety of infections in immunocompromised patients. Although it has also been reported as a cause of mastitis and invasive disease in camels, and as an occasional cause of disease in dogs, cats, fish, and hamsters, it has not been clearly determined whether human and animal populations of *S. agalactiae* are identical. However, epidemiologic evidence does not support animal to human transmission, and differences in salicin and lactose fermentation and in bacteriocin and bacteriophage types suggest that the populations in each host are distinct (Finch and Martin 1984).

The polysaccharide capsule includes the specific antigenic types Ia, Ib, Ic, II, and III. Type Ic is often associated with a C protein antigen. Type Ia constitutes about 70% of bovine isolates in New York State but different types may be numerous in other geographic areas (Norcross and Oliver 1976). About 25% of *S. agalactiae* strains are untypable.

S. agalactiae is an obligate parasite of the epithelium and tissues of ruminant mammary glands, and eradication of the organism from herds is therefore possible by identification of animals with mammary infection and treatment, or culling, of these animals. Calves that feed on mastitic milk may transmit infection by suckling penmates' immature teats. In humans, *S. agalactiae* is found on the perineum, oropharynx, and gastrointestinal tract.

VIRULENCE FACTORS

Much of the information on potential virulence factors of *S. agalactiae* has been derived from studies on human isolates and therefore must be cautiously interpreted with respect to mastitis. The capsular polysaccharide, including the polysaccharide type-specific antigen, is antiphagocytic, and antibody to these type-specific antigens are protective in mice (Lancefield et al. 1975). Type-specific antibody is also important in resistance of human infants to group B streptococcal infection.

Type III capsular polysaccharide has a terminal sialic acid residue that inhibits activation of the alternate complement pathway and prevents deposition of C3 on the bacterial surface. The capsule increases the affinity of factor H for C3b bound to the surface of the cell wall and thereby reduces both the activity of C3 convertase and further deposition of C3b on the cell (Marques et al. 1992).

Bovine strains express less capsular polysaccharide on their surfaces and thus are capable of activating the alternate pathway. Moreover, bovine antibodies to the group B polysaccharide antigen, although not opsonic by themselves, are able to fix complement and C3 via the classical pathway, and opsonization is thereby effected (Rainard and Boulard 1992). The C protein found on all type lb, 60% of type II, and occasional type III strains has been shown to elicit protective, opsonophagocytic antibodies (Ferrieri 1988). The mode of action of this virulence factor is not known.

R antigens are proteins found on filamentous protrusions from the cell surface of *S. agalactiae* (Wagner et al. 1982). Strains expressing R proteins do not express C protein and vice versa. Clinical studies and mouse protection experiments with type III strains of *S. agalactiae* indicate that R protein contributes to virulence, possibly because it enhances colonization of epithelia, although not by promoting adhesion (Kurl et al. 1984).

Another protein antigen of about 100 kDa, termed X, that occurs on many untypeable bovine strains of *S. agalactiae* from cases of bovine mastitis, is of yet unknown significance in the pathogenesis of the disease (Rainard et al. 1991). This antigen is opsonic and apparently different from the cell surface Sas 97/104 protein (Wanger and Dunny 1987), which is immunodominant for the bovine and found on about 50% of bovine strains. Its presence or absence does not affect bacterial virulence in a guinea pig model.

The CAMP factor is a ceramide-binding protein of *S. agalactiae* that potentiates the action of staphylococcal sphingomyelinase. The lethal properties of this protein for rabbits and mice suggest that it may have a cytotoxic action for mammary tissue. The protein binds to the Fc region of IgM and IgG, and insertional inactivation of the gene that encodes this protein increases the mouse LD_{50} by a factor of 50 (Hollingshead et al. 1989). These mutants have not been tested for virulence in the mammary gland.

Other potential virulence factors of *S. agalactiae* for the mammary gland include neuraminidase, hemolysin, vasoactive extracellular toxin, and lipoteichoic acid. The roles of these factors, if any, in virulence for the mammary gland have not been determined.

S. agalactiae enters through the end of the teat, and colonization of the gland is possibly favored by adhesion to the epithelium of the gland sinuses (Frost et al. 1977). Backjetting of contaminated milk against the teat ends at milking time is an important factor in the introduction of infection past the teat sphincter. Keratin and associated bacteriostatic long-chain fatty acids of the teat canal are the first host barrier to physical penetration of the epithelial lining. Multiplication of the invading streptococcus is in part controlled by the lactoperoxidasethiocyanate-H_2O_2 system, by lysozyme, and by the flushing action of milk at milking. Adherence and multiplication of the organism on the epithelium of the teat and duct sinuses result in a slowly progressing inflammation and fibrosis. Although *S. agalactiae* rarely penetrates the epithelium, some cows may experience a transient invasion during the first few days in which the organism enters the lymphatics and travels to the supramammary lymph nodes. Release of chemoattractants from damaged host cells and *S. agalactiae* attracts polymorphonuclear leukocytes (PMNs), which then ingest and kill many of the invading streptococci. Opsonization is probably derived from C3 in the inflammatory exudate, which becomes fixed on the bacterial surface following activation of the alternative complement pathway. Normal milk has a very low complement content and thus cannot itself serve as a source of C3. Initial invasion is more likely to result in colonization in older cows and in mammary glands where there is delay in arrival of PMNs at the site of invasion. Death of PMNs and release of lysosomal enzymes cause further tissue damage and inflammation. Fibrin plug formation in the smaller milk ducts may lead to involution of secretory tissue and loss of milk-producing capacity ("agalactiae"). Without treatment, the organism persists in the face of the host's immune response and the infection and mastitis become chronic. The mechanism(s) by which *S. agalactiae* evades clearance and the host immune response is not understood.

IMMUNITY

Most of the protective activity of colostrum against *S. agalactiae* type Ia has been shown to be associated with IgA and IgM classes (Yokomizo and Norcross 1978). Serum antibodies appear to have little or no protective effect. The presence of agglutinins in milk of infected cows together with the failure of the infection to be cleared naturally from the udder of most infected cattle suggest that active immune responses are ineffective in clearance of infection. However, specific immunoglobulins against the group B polysaccharide, for example, may play a role in ameliorating the disease process. This conclusion was reached by Norcross et al. (1968) as a result of their observation that clinical signs were often absent in experimentally infected cows with preexisting circulating antibody. They postulated that this antibody neutralized extracellular products of *S. agalactiae* involved in the inflammatory response.

The lack of definitive information on the protective immune response, the antigen(s) involved, the role of hypersensitivity in the pathogenesis, and chronicity of *S. agalactiae* mastitis reflects the fact that the disease is readily eradicated from herds by hygiene, therapy, and culling.

STREPTOCOCCUS DYSGALACTIAE

Streptococci described as *S. dysgalactiae* belong to Lancefield groups C, G, and L, and are closely related to *S. equisimilis* of human origin (Schleifer and Kilpper-Biilz 1987). They have long been recognized as a cause of sporadic cases of acute bovine mastitis. *S. dysgalactiae* group C is carried in the mouth, vagina, and skin lesions of the udder, sites that are considered to be the source of opportunistic infections of the teat, often in synergy with *Actinomyces pyogenes*. Relatively small numbers of the organism are capable of producing mastitis following intramammary inoculation (Higgs et al. 1980). Fibronectin-binding protein, hyaluronidase, and streptokinase are produced, but the importance of these potential virulence factors has not been evaluated.

A group G streptococcal strain of human origin and phenotypically similar in many respects to *S. dysgalactiae* has caused a prolonged outbreak of mastitis in a dairy herd. The caretaker was a nasal carrier and suffered intermittent bouts of pharyngitis (Gonzalez and Timoney, unpublished data).

STREPTOCOCCUS EQUI SUBSP. *EQUI*

S. equi subsp. *equi* causes strangles, a highly contagious infection of the upper respiratory tract and associated lymph nodes of solipeds. Given its ubiquity, the subspecies is remarkably conserved and shows no antigenic or genetic variation (Galan and Timoney 1988). Differences in virulence between isolates appear to be related to the amount of antiphagocytic M protein and hyaluronic acid capsule produced. *S. equi* subsp. *equi* is closely related to *S. equi* subsp. *zooepidemicus*, which is much less hostadapted and shows great variation in the DNA restriction fragment profile.

The distribution of *S. equi* subsp. *equi* is closely correlated with the distribution of horse, donkey, and mule populations because survival in the environment is of relatively short duration. Under optimum conditions survival times of 7-9 weeks have been recorded on sterilized wood or glass.

The most important source is a horse that is shedding the organism in nasal discharges or from a draining abscess. Horses that are carriers are infrequent and only a very few instances of true carrier states have been recorded. However, horses occasionally develop deeply located, highly encapsulated abscesses that may carry the organism latently for many months. When the abscess eventually ruptures, organisms may be carried back to the nasopharyngeal tissues in the blood stream and shed from the nasopharynx. Such a horse could be the source of a strangles epizootic in a herd previously free of the disease.

The incubation period varies from 3 to 14 days after exposure and onset of the typical disease is marked by fever, lassitude, nasal discharge, slight cough, difficulty in swallowing, and swelling of the intermandibular areas, with tenderness and swelling of the mandibular lymph node. As the disease progresses, abscesses in the submandibular and/or retropharyngeal lymph nodes enlarge and become painful. Pressure of the enlarging retropharyngeal lymph nodes on the airway may cause respiratory difficulty and is the source of the common name of the disease. Metastasis of organisms may result in abscess formation in other locations such as the lungs, abdomen, or brain. In older animals with residual immunity, strangles may present as an atypical or catarrhal form of the disease. The clinical signs include a slight nasal discharge, cough, slight fever in some animals, and abscessation of lymph nodes in a minority of cases. The mild form of the disease is often seen in older animals with preexisting antibody and experiencing a second infection. Nonencapsulated strains of *S. equi* subsp. *equi* expressing bacteriophage-encoded hyaluronidase have also been associated with a clinically mild form of strangles.

Most animals recover quickly and uneventfully. Sequelae include myocarditis, anemia, purpura hemorrhagica, and acute leukocytoclastic vasculitis and glomerulonephritis. These latter two sequelae involve formation of circulating immune complexes.

VIRULENCE FACTORS

The virulence factors of *S. equi* subsp. *equi* include a nonantigenic hyaluronic acid capsule, hyaluronidase, streptolysin O, streptokinase, IgG Fc-receptor proteins, peptidoglycan, and the antiphagocytic M protein. There is also some circumstantial evidence for the production of a leukocidal toxin.

The hyaluronic acid capsule is a high-molecular weight polymer consisting of alternating residues of N-acetylglucosamine and glucuronic acid. Isolates of *S. equi* subsp. *equi* from cases of strangles are almost always highly encapsulated and produce very mucoid colonies. A non-encapsulated mutant of *S. equi* subsp. *equi* that was produced by nitrosoguanidine mutagenesis was avirulent for mice at doses of challenge organisms that, for the parent strain, were fatal for all mice inoculated (Timoney and Galan 1985). The nonencapsulated strain had a normal amount of M protein and stimulated protective antibodies in mice and horses. The phagocytic capsule greatly reduces the numbers of streptococci that become associated with the surface of neutrophils and are subsequently ingested and killed. There is up to a 100-fold reduction in virulence of *S. pyogenes* associated with loss of capsule (Wessels et al. 1991). The capsular hyaluronic acid increases the negative charge and hydrophilicity of the bacterial surface and produces a localized reducing environment in the immediate vicinity of the streptococcus that preserves the activity of oxygen-labile proteases or toxins such as streptolysin O. Synthesis of capsular hyaluronic acid in *S. pyogenes* is possibly coregulated with synthesis of M protein and C5a peptidase by the *vir*R gene (Simpson et al. 1990). However, the gene for hyaluronate synthase appears to be only distantly linked to the *vir*R gene.

Streptokinase released by *S. equi* subsp. *equi* interacts with the C-terminal serine protease domain of equine plasminogen to form active plasmin which hydrolyses fibrin. The role of plasmin in virulence has not been proven but its lytic action on fibrin may aid in spread and dispersion of the bacteria in tissue. Other possible roles include in situ activation of complement and production of low-molecular weight nitrogenous substrates for bacterial growth. A receptor for plasmin also occurs on the surface of *S. equi* subsp. *equi* but its significance is not known.

Streptolysin O is produced by groups A, C, and G streptococci and derives its name from its lability to oxygen by which it is reversibly inhibited. It belongs to the family of thiol-activated toxins and is a potent membrane-damaging enzyme that causes formation of hydrophilic channels in membranes rich in cholesterol (Bhakdi et al. 1985). In culture supernatants, streptolysin O occurs as a 53-kDa protein. Although highly antigenic, streptolysin O is also highly conserved and has a wide spectrum of lytic activity for erythrocytes of different hosts. It therefore differs from M proteins, which exhibit great heterogeneity within and among streptococcal species.

Although primarily recognized by its hemolytic activity on erythrocytes, streptolysin O is also toxic for other cells and cell components, including PMNs, myocardial cells, platelets, and lysosomes. Its role in virulence of *S. equi* subsp. *equi* is not well defined. Strains in which the gene was insertionally inactivated with Tn916 were less virulent for mice. No experiments with nonhemolytic mutants appear to have been done on horses.

A proteinaceous cytotoxic activity unrelated to streptolysin O has been detected in culture supernatant of *S. equi* subsp. *equi*. Equine PMNs incubated in the presence of culture supernatant showed signs of toxicity and became chemotactically unresponsive (Mukhtar and Timoney 1988). The action of the toxin appears to be on mitochondrial membranes because suspensions of equine PMNs exhibited intense respiratory activity shortly after exposure to culture supernatant, suggesting sudden release of respiratory enzymes. Nonhemolytic mutants of *S. equi* subsp. *equi* exhibited the same toxic effect, suggesting a toxic effect distinct from that of streptolysin O.

Peptidoglycan of *S. equi* subsp. *equi* is a potent activator of the alternative complement pathway, and chemotactic factors (C3a, C5a) released following incubation of peptidoglycan with plasma are strongly chemotactic for equine PMNs (Muhktar and Timoney 1988). It is this phenomenon that underlies the basic pathologic process in strangles — the outpouring of PMNs in the infected lymph nodes and onto the upper respiratory mucosa. Peptidoglycan is also a potent pyrogen by inducing release of pyrogenic cytokines such as interleukin-6 and tumor necrosis factor from leucocytes and thus accounting for the febrile response of the infected horse.

The surface of *S. equi* subsp. *equi* also carries receptors for albumin and for the Fc region of IgG. The Fc receptors have not been fully characterized. Binding of host plasma proteins to the surface of the whole organism could be an effective mode of concealment from host cellular recognition mechanisms. The bound proteins might also block access of C3 or specific antibody to target sites on the organism. The acid-extracted M protein fragments bind rabbit and equine IgG. The lack of similar binding by the native M molecule extracted with mutanolysin suggests that Fc binding to fragments may be an artefact and a result of conformational changes produced by hot-acid treatment.

M proteins are antiphagocytic, acid-resistant, dimeric fibrillar molecules that project from the cell wall surface. A typical M protein molecule is about 50-60 nm long, with a long, coiled central region flanked by a short, random, coiled sequence at the N-terminus and by a specialized, highly conserved arrangement of hydrophobic and charged amino acids at the C-terminus, that is anchored in the cell wall (Robinson and Kehoe 1992) (Fig. 1.1). The M protein of *S. equi* subsp. *equi* has a molecular mass of about 58 kDa and often occurs as a dimer or trimer. The main fragments in acid extracts have molecular masses of 46 kDa, 41 kDa, and 29-30 kDa.

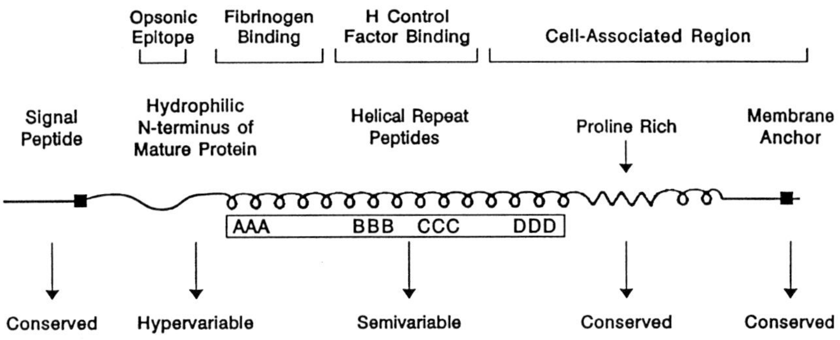

FIG. 1.1. Diagram showing the structural features of a typical M protein molecule.

The antiphagocytic action is apparently due to binding of fibrinogen and the complement control factor H to the M protein. These interactions mask C3b-binding sites on the bacterial surface and inhibit the alternative C3 and classical C5 convertases. Antibodies against specific epitopes on the N terminus override these effects and opsonize the streptococcus so that it is effectively phagocytosed. Unlike the M proteins of *S. pyogenes* or *S. equi* subsp. *zooepidemicus,* the M protein of *S. equi* subsp. *equi* is highly conserved and shows no variation in size or antigenicity. Variations in M proteins are probably due to deletions or additions of repeat regions within the structural gene that trigger changes in amino acid sequence.

Regulation of streptococcal M protein production in vivo is not well understood. Fresh isolates of *S. pyogenes* or *S. equi* subsp. *equi* are rich in M protein. Prolonged passage of these strains on agar media results in diminished M protein production, suggesting either that M protein synthesis is up-regulated in vivo and/or that there is a selection in vivo for clones with enhanced amounts of M proteins on the cell surface. M protein synthesis in *S. pyogenes* is not required for cell growth and has been shown to be coregulated with other virulence factors such as C5a peptidase and capsule (Simpson et al. 1990). Glucose is both a strong inducer and regulator of M protein synthesis by *S. pyogenes* (Pine and Reeves 1978), but the mechanism of regulation has not yet been determined.

Many bacterial and parasitic pathogens bind to host cells via the eukaryotic matrix protein fibronectin. The major fibronectin receptor protein on the group A streptococcus *S. pyogenes* is a 120-kDa protein designated protein F (Hanski and Caparon 1992). Strains of *S. pyogenes* in which the gene for this protein has been insertionally inactivated do not bind to epithelial cells in vitro, suggesting that other proteins, including M protein, are not important in adhesion.

S. equi subsp. *equi* enters via the mouth or nose and attaches to cells in the crypt of the tonsil and adjacent lymphoid nodules (Fig. 1.2). After a few hours, the organism is difficult to detect on the mucosal surface because it is translocated below the mucosa into the local lymphatics; it may be found in one or more of the lymph nodes that drain the pharyngeal/tonsillar region. Complement-derived chemotactic factors that are generated after interaction of C3 with bacterial peptidoglycan attract large numbers of PMNs. The lack of efficacy of these PMNs in phagocytosing and killing the streptococci appears to be due to a combination of the hyaluronic acid capsule, antiphagocytic M protein, and a leukocidal toxin released by the organism. The end result is many extracellular streptococci in the form of long chains in the presence of large numbers of degenerating PMNs. Final disposal of the organisms is dependent on lysis of the abscess capsule and evacuation of its contents.

Streptolysin O and streptokinase may also contribute to abscess development and lysis by damaging cell membranes and activating the proteolytic properties of plasminogen. Although strangles predominantly involves the upper airways and associated lymph nodes, metastasis to other locations may occur. Spread may be hematogenous or via lymphatic channels, which results in abscesses in lymph nodes and other organs of the thorax and abdomen. Metastasis to the brain has also been recorded. Evers (1968) reported bacteremia in horses inoculated intranasally with *S. equi* subsp. *equi* and in noninoculated contact horses that became infected. Blood cultures were more likely to be positive on days 6 to 12 following inoculation. These interesting findings have not been confirmed but clearly show the potential for localization of *S. equi* subsp. *equi* in body sites other than the lymph nodes of the head and neck and for the formation of circulating immune complexes.

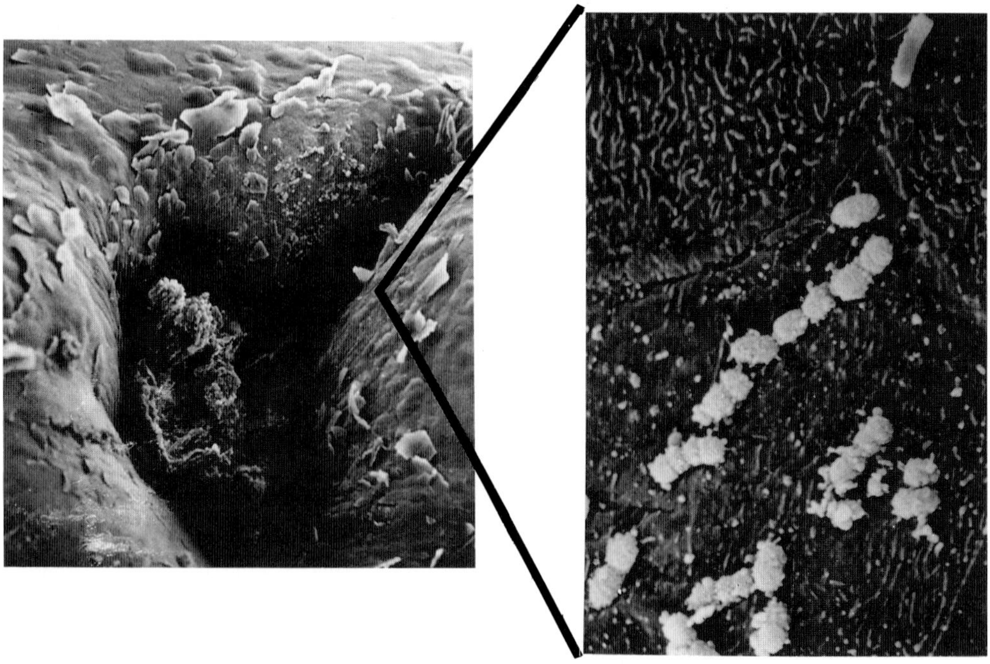

FIG. 1.2. Scanning electron micrographs showing a typical crypt in the palatine tonsil of a horse and (inset) short chains of *Streptococcus equi* subsp. *equi* adherent to a crypt epithelial cell.

Nasal shedding of *S. equi* subsp. *equi* usually begins after a latent period of 4-7 days and ceases between 3-6 weeks after the acute phase. However, some animals may continue to harbor infection for months after clinical recovery and be a source of contagion for other susceptible horses.

IMMUNITY

Approximately 75% of horses develop a solid, enduring immunity to strangles following recovery from the disease. The failure of 25% of recovered animals to develop protective immune responses effective against natural challenge has not been explained but probably represents a failure to produce the appropriate mucosal and systemic humoral antibodies. However, most animals that experience a second attack of the disease do subsequently develop a high level of protective immunity.

Protection against *S. equi* subsp. *equi* infection appears to be mediated by a combination of mucosal IgG and IgA antibodies that are locally produced in the nasopharynx together with opsonic IgG antibodies in the serum. Mucosal antibodies are usually produced earlier than serum opsonic antibodies, and horses recovering from strangles have been shown to be resistant to experimental reinfection weeks before serum opsonic antibody can be detected (Galan and Timoney 1985). The latter does not appear until late in convalescence and then only in a portion of horses (Timoney and Eggers 1985). The mucosal and systemic (serum) antibody responses to the M protein are independent and local mucosal responses appear to require local stimulation. Anamnestic mucosal responses are elicited following reinfection and contribute to protection against a second or further occurrence of disease. The mode of action of locally produced antibody is not understood. These antibodies may prevent adhesion to receptors on tonsillar cells or prevent interiorization of the organism after it adheres. Long-term studies on mucosal M protein antibody production and on maintenance of protective levels have not been reported.

Milk from mares that have recovered from strangles contains M protein antibody (IgG and IgA) similar to that found in nasopharyngeal mucus of convalescent horses. Suckling foals therefore benefit from the protective effects of this antibody until weaned. Colostral antibodies ingested during the first 24 hours of life have also been shown to recirculate to the nasopharyngeal mucosa, thus providing an additional source of protection to the foal during its first weeks. Foals that suckle immune mares are usually resistant to *S. equi* subsp. *equi* infection until weaning.

Serum antibodies to *S. equi* subsp. *equi* may be assayed by a variety of methods including the bactericidal and chain-length assays, radioimmunoassay, enzyme-linked immunosorbent assay (ELISA), mouse protection test, and gel diffusion precipitin test. Mouse protective and bactericidal antibodies occurring in horse sera appear to correlate with protection in the horse.

The antiphagocytic M protein is widely believed to be the major protective antigen of *S. equi* subsp. *equi*. This heat- and acid-resistant protein antigen is protective for mice (Timoney and Trachman 1985) and is an important immunogenic component of commercial strangles vaccines. A vaccine prepared in the author's laboratory from an M protein-negative strain that was otherwise identical to M protein-positive strains of *S. equi* subsp. *equi* did not stimulate protection in mice, indicating that other antigens on the organism do not contribute significantly to protection. However, there is some circumstantial evidence that a heat-labile component of *S. equi* does contribute to protection. Bazely (1943) found that heating of *S. equi* cultures for periods of greater than 12 minutes at temperatures of 56°C (133°F) or greater resulted in loss of protective immunogenicity for mice and horses. This temperature would not have been high enough to damage M protein, which is known to resist boiling, and suggests that another more heat-labile antigen is involved or that opsonophagocytic epitope(s) of the *S. equi* M protein may be more vulnerable to heat than the rest of the molecule.

Vaccines containing either the acid-extracted or enzymically extracted M protein are effective in stimulating serum opsonic antibodies when administered in a course of three inoculations but do not confer a high level of protective immunity against natural exposure. Studies on the efficacy of strangles vaccines in the field suggest that the clinical attack rate is reduced by 50% in vaccinates compared to control, nonvaccinated animals (Hoffman et al. 1991). Since the study was terminated 16 days after the third vaccination, no conclusions as to longer term protection can be made. However, the trial did prove that vaccination was of benefit during an epizootic. Substantial bactericidal (opsonic) activity has been noted in sera of vaccinated yearlings that developed strangles within a month or two of the date of collection of the sera, suggesting that opsonic activity by itself is not an adequate criterion of protection in the horse (Timoney and Eggers 1985). An experimental live vaccine capable of stimulating mucosal responses has been developed but is not available commercially at the time of writing (Timoney and Galan 1985).

STREPTOCOCCUS EQUI SUBSP. *ZOOEPIDEMICUS*

S. equi subsp. *zooepidemicus* is closely related to *S. equi* subsp. *equi*, and the subspecies show a very high degree of DNA homology. However, *S. equi* subsp. *zooepidemicus* is not host specific and produces disease as an opportunist in many species of animals. Unlike *S. equi* subsp. *equi, S. equi* subsp. *zooepidemicus* lives as a mucosal commensal in the horse and is apparently well tolerated by or evades the host immune response.

Diseases produced opportunistically in many species of animals include wound infections in horses, joint ill, lymphatic abscesses, nasal catarrh, and pneumonia of foals and weanlings. *S. equi* subsp. *zooepidemicus* is the most frequently isolated pathogen from the reproductive tract of the mare as well as causing mastitis in cattle and goats; pneumonia, septicemia, and wound infections in lambs, puppies, and greyhounds; septicemias in chickens; lymphadenitis in guinea pigs; and a variety of infections in other animals.

VIRULENCE FACTORS

S. equi subsp. *zooepidemicus* produces all of the virulence factors described for *S. equi* subsp. *equi*. However, hyaluronic acid production is much more variable and most strains isolated from normal mucosal sites or skin are not encapsulated. Isolates from invasive infections usually show a moderate degree of encapsulation, which is often quickly lost during culture on blood agar. Toxigenic strains have been implicated in outbreaks of septicemia in greyhounds (Sundberg et al. 1981).

Hyaluronidase is also produced by some strains. It is unclear whether this should be regarded as a virulence factor for it appears to render the organism more sensitive to phagocytosis by removing its capsule.

Unlike *S. equi* subsp. *equi*, the M proteins of *S. equi* subsp. *zooepidemicus* of equine origin show great variability in molecular mass and antigenic specificity (Timoney and Muhktar 1992). The N termini of M proteins of strains of equine origin show variations in amino acid sequence, which may account for observed variations in the opsonic epitopes.

IMMUNITY

Most animals carry antibodies to the M protein of *S. equi* subsp. *zooepidemicus*, but the significance of these in protection is unknown. Nasopharyngeal secretions of horses contain antibodies (IgA, IgG) to the M proteins of strains harbored in their tonsils, but these antibodies do not appear to be as effective in clearance of the organism from the nasopharynx as are the corresponding antibodies generated against *S. equi* subsp. *equi* M protein during recovery from strangles.

STREPTOCOCCUS EQUISIMILIS

S. equisimilis occurs infrequently in specimens from horses, cattle, and dogs. It is most frequently isolated from joints of piglets that have acquired infection from the sow. The piglet is invaded via wounds, the umbilicus, or tonsil and develops a suppurative arthritis. There appear to be at least four M types of *S. equisimilis* from swine, but the role of M proteins in virulence is unclear (Woods and Ross 1975).

S. equisimilis is occasionally isolated from abscessed lymph nodes from horses and from placentas of aborted foals.

STREPTOCOCCUS CANIS

Streptococcus canis belongs to Lancefield group G and is often included with the pyogenic streptococci such as *S. pyogenes, S. equi* subsp. *equi*, and *S. dysgalactiae*. The species *canis* is reserved for large colony beta-hemolytic streptococci from animals that react with group G typing antiserum and produce beta-galactosidase but not hyaluronidase. They are thereby distinguished from strains of *S. dysgalactiae* that react with group G antiserum. Such strains have been implicated in outbreaks of bovine mastitis in which the source of infection was the caretaker. However, group G streptococci from humans and animals represent genetically distinct and separate populations (Simpson et al. 1987).

The anal mucosa is the main carrier site of *S. canis* in the dog. The organism is also found on the genitals, groin, chin, mouth, and nose of dogs and cats. Invasive infections in the animals are usually sporadic and opportunistic. Thus, *S. canis* is found in wound infections, abscesses, proctitis, mastitis, prostatitis, maxillary adenitis, otitis, pyometra, and pyoderma. Outbreaks of abortion, neonatal septicemia, polyarthritis, and infertility in dogs were reported in the late 1930s in England, but not since then, suggesting that the virulent clone involved did not survive.

In cats, *S. canis* is the most commonly isolated bacterium from skin and lymph node abscesses, cases of mastitis, conjunctivitis, metritis, and septicemia in kittens (Swindle et al. 1981). Epizootics of contagious streptococcal lymphadenitis have been reported in colonies of laboratory cats and rats (Coming et al. 1991).

VIRULENCE FACTORS

Very little information has been reported on the virulence factors of *S. canis*. Neither hyaluronidase, streptokinase nor hyaluronic acid are formed. Streptolysin O and M protein are produced in large quantity by strains recently isolated from lesions or clinical specimens. Strains isolated from normal mucosal sites express very little detectable M protein. The M protein of a virulent *S. canis* clone from an epizootic of neonatal septicemia and mandibular lymphadenitis of juvenile cats had a molecular mass of about 75 kDa. Antiserum against this protein was strongly opsonic for the parent strain and for other clinical isolates of *S. canis* from disease outbreaks. However, isolates from opportunistic infections and from normal healthy cats were phagocytosed equally well by preimmune and hyperimmune serum to the M protein, suggesting either that an antigenically different M protein was present or that the M protein was absent or present in only low amounts on noninvasive, nonepizootic strains (Timoney and Blanchard, unpublished data).

The epizootic form of feline streptococcosis has been observed mainly in experimental colonies and occasionally in catteries (Blanchard 1987). Persian cats appear to have an enhanced susceptibility. Neonatal kittens are infected from vaginal secretions or from the mouth of the queen when she bites off the umbilical cord close to the abdominal wall. The organism enters the umbilical vein, often causing formation of a small local abscess, and then it is carried in the bloodstream to a variety of body sites. Bacterial thrombi form in the liver, spleen, lungs, kidneys and musculature, resulting in gross and microscopic abscess formation and death in a week, or so, after birth. The neonatal disease is seen only in litters from young queens. Older queens that have developed antibodies pass these antibodies to their kittens, which then become resistant to infection.

Juvenile streptococcosis occurs in young cats during the postweaning period from 2 to 4 months of age when maternal antibody is waning. *S. canis* acquired earlier during feeding invades the tonsil and local lymphatics of the head and neck, causing a purulent lymphadenitis of the mandibular lymph nodes. Lymphatic obstruction due to infection by *Brugia malayi.,* a filarial parasite, predisposes cats to *S. canis* lymphangitis. Lymphostasis apparently reduces the bacterial clearance capacity of the lymph glands.

IMMUNITY

Only one M type of *S. canis* has been isolated from outbreaks of feline streptococcosis. Opsonic antibodies that neutralize the antiphagocytic effects of the M protein are protective. These antibodies develop as a result of exposure to *S. canis* and thus older cats tend to have higher opsonic titers. Older queens also show much lower vaginal colonization rates and those queens that are colonized have fewer organisms. The role of locally produced mucosal antibodies in this phenomenon has not been studied.

STREPTOCOCCUS SUIS

The principal carrier site for *S. suis* in pigs is the palatine tonsil, and transmission is by the respiratory and oral routes. In some geographic regions, nearly 100% of pigs are tonsil carriers. However, most carriers do not develop clinical disease. Infection of piglets with *S. suis* type 2 may occur at birth and both types 1 and 2 are present in the noses of all piglets of infected herds by the time the piglets are 4-6 weeks old.

Both types 1 and 2 are associated with septicemia, meningitis, arthritis, bronchopneumonia and other lesions typical of generalized septicemia. Occurrence of disease is stress related.

VIRULENCE FACTORS

Sialic acid-rich, capsular polysaccharide antigens designated 1-8 are similar to those of *S. agalactiae* from cases of human meningitis and may function in part by preventing deposition of C3 and activation of the alternative complement pathway in the same manner as described for *S. agalactiae.* Type 2 capsular polysaccharide has a molecular mass of at least 100 kDa and elicits opsonizing responses, an indication of an antiphagocytic function. Nonencapsulated isolates are readily phagocytosed in the absence of type-specific antibody and complement, whereas both are required for efficient phagocytosis of encapsulated isolates.

A 100-kDa protein released into culture supernatant has been shown to be necessary for full expression of virulence. Strains that synthesized the protein produced the typical meningoencephalitis, polyserositis, and polyarthritis, but strains that did not produce this protein produced only a mild disease (Vecht et al. 1992). Another 136-kDa protein with an anchor sequence similar to that of the group A M proteins also appears to contribute to virulence, possibly by binding to fibronectin (Smith et al. 1992). However, the mechanisms of action of this or the 100-kDa proteins are not known.

The events that culminate in bacteremia are not understood but are often associated with stress. The tonsil appears to be the site of primary multiplication, with subsequent invasion of the regional lymphatics where the organism remains localized in the mandibular lymph node or escapes and multiplies in the bloodstream. A sustained, high-level bacteremia is necessary for the development of meningoencephalitis. The essential aspect of the host/parasite interaction relative to disease production is the organism's ability to be phagocytosed by, and survive in a circulating mononuclear cell (Williams 1990). *S. suis* does not survive in PMNs but does survive in circulating mononuclear cells in the absence of antibody and is carried to the joints and meninges. If the ependyma is breached, *S. suis* may enter the central nervous system via the choroid plexus and is spread in the cerebrospinal fluid. Hematogenously delivered organisms may be found beneath the surface of brain tissue.

IMMUNITY

The self-limiting nature of disease due to *S. suis* in contained pig populations strongly implicates the emergence of an acquired, protective immune response. Repeated inoculation of pigs with live and formalin-killed cultures of *S. suis* type 2 results in a strong protective response characterized by the appearance of both IgM and IgG antibodies to surface components (Holt et al. 1988). Protective sera also mediate killing of the organism in vitro. A 94-kDa antigen band, reactive with protective pig sera, stimulates antibodies that protect mice and are active in bactericidal tests. This antigen may be part of the cell envelope 135-kDa protein that binds fibronectin (Smith et al. 1992).

The disappointing results in field trials of formalin-killed vaccines may be due to a lack of protective antigen or destruction of protective epitopes during vaccine preparation.

STREPTOCOCCUS PORCINUS

Streptococcus porcinus, a group E streptococcus, is the cause of cervical lymphadenitis, a contagious disease of young 8- to 10-week-old swine. Although *S. porcinus* has been isolated from a variety of opportunist lesions in horses and cats, the pig is the normal host of the organism that is carried in the tonsils and transmitted by nose contact and via drinking water and feces. Antimicrobial supplementation of feed and changes in swine management have greatly reduced the incidence of disease in the United States.

In experimental infections, enlargement of the mandibular, parotid, and retropharyngeal lymph nodes is evident about 2 weeks after infection. Abscessed nodes drain during the following week or become encapsulated.

VIRULENCE FACTORS

S. porcinus is encapsulated and produces streptokinase specific for porcine plasminogen (Ellis and Armstrong 1971). An antiphagocytic factor similar to the M proteins of *S. pyogenes* is necessary for virulence (Deynes and Armstrong 1973). Antibody to this protein can be detected by opsonic and long-chain activities, both of which are specific for M protein antibody. However, complete characterization and purification of the antiphagocytic factor has not been reported. The phagocytic resistance of the organism is increased during growth in 10% porcine serum or 2% bovine serum albumin.

S. porcinus appears to enter the regional lymphatics of the head and neck via the palatine tonsils. The organism can be detected in tonsillar tissue within a few hours of oral exposure. Foci of infiltrated neutrophils and extracellular bacteria are visible in 48 hours. The developing abscess becomes necrotic and encapsulated and may eventually rupture to the exterior. The antiphagocytic factor resembling M protein is probably crucial in the hostparasite interaction. However, studies to define this interaction have not been reported.

IMMUNITY

Antibodies are formed in about 7 days in response to infection. Opsonic antibodies that neutralize the M proteinlike antiphagocytic factor persist for at least 20 weeks. An oral avirulent vaccine gives much better protection than bacterin-type vaccines, possibly by stimulating mucosal immunity in the nasopharynx. Bacterin-type vaccines effect only a partial reduction in the number and size of abscesses.

STREPTOCOCCUS PNEUMONIAE

Streptococcus pneumoniae type 3 has been associated with lower respiratory tract disease in young thoroughbreds in training (MacKintosh et al. 1988). Type 3 strains are among the most virulent for humans, the normal host of *S. pneumoniae.* It is as yet unclear whether humans are the source of infection for horses.

Disease in young thoroughbreds is characterized by febrile episodes, a highly cellular exudate in the bronchi and trachea in association with large numbers of *S. equi* subsp. *zooepidemicus,* poor racing performance, coughing, and nasal and ocular discharges.

VIRULENCE FACTORS

Adhesion of *S. pneumoniae* to epithelium of the tonsil and soft palate of ponies has been noted following experimental infection, suggesting that the organism possesses adhesins. Resistance to phagocytosis is mediated by a complex polysaccharide capsule that forms a hydrophilic gel on the surface of the organism. This gel shields the bacterium from antibodies and complement proteins. In addition, capsular sialic acid contributes to the antiphagocytic effect by inhibiting complement amplification and alternative pathway activation. Intrinsic complement inactivation mechanisms, which degrade C3b bound to the bacterial surface and prevent further C3 deposition, are also facilitated by capsular sialic acid. Capsular material has, however, been noted in the alveolar macrophages of ponies experimentally infected with *S. pneumoniae,* indicating that successful phagocytosis does take place. It is unclear how this relates to the clinically mild self-limiting nature of the naturally occurring respiratory disease of horses in training. Alveolar necrosis has also been observed in experimentally produced lesions in ponies, suggesting the elaboration of toxins. Toxins, including pneumolysin, purpura-producing principle, and autolysins, may play a role in this phase of pathogenesis (Boulnois et al. 1991). Toxin involvement in pneumococcal pneumonia in humans is suggested by the acute fulminating and toxic clinical character of the disease. Neuraminidase may act both to decrease the viscosity of mucus and to alter oligosaccharides of mucosal cells by removing N-acetyl neuraminic acid residues and thus expose receptors for bacterial attachment.

Stress associated with race training and occurrence of lower respiratory tract inflammatory disease in which *S. pneumoniae* is present in large numbers suggests that the host/parasite interaction is opportunistic. The increased respiratory rate during intense exercise may result in aspiration of *S. pneumoniae* from the tonsil and soft palate. At the same time, impairment of the mucociliary escalator mechanism results in failure to clear aspirated organisms. Fluid accumulation in the lower airway also predisposes to bacterial multiplication. The highly cellular exudate is in part due to the potent inflammatory effect of streptococcal cell wall products as well as the potent membrane-damaging effect of pneumolysin, a thiol-activated toxin similar to streptolysin O. The relative significance in lesion development of the large numbers of *S. equi* subsp. *zooepidemicus* that are often found with *S. pneumoniae* in tracheal aspirates is unknown. It is possible that IgA protease produced by *S. pneumoniae* may destroy protective antibodies against the M proteins of *S. equi* subsp. *zooepidemicus*. Pneumolysin also slows the ciliar beat and thus reduces bacterial clearance.

IMMUNITY

Type-specific capsular antibody is opsonizing and protective and is produced during convalescence. Binding of serum beta-globulin (C reactive protein) to C polysaccharide in the cell wall during the acute-phase disease, activates complement and mediates phagocytosis and thus promotes bacterial clearance.

STREPTOCOCCUS UBERIS AND *STREPTOCOCCUS PARAUBERIS*

Streptococcus uberis and *S. parauberis* are mucosal and epithelial commensals of cattle that opportunistically invade the mammary gland under conditions of poor hygiene. Between 10 and 20% of cases of clinical mastitis in North American herds are caused by these organisms. The genetically distinct, but phenotypically similar, *S. uberis* and *S. parauberis* are distinguished by the ability of the psychrophilic *parauberis* to grow well at 10°C and by immunoblot fingerprints (Groschup et al. 1991). Comparative epidemiologic and pathogenesis studies on the two species have not yet been reported.

VIRULENCE FACTORS

A hyaluronic acid capsule is present on many *S. uberis* strains. However, decapsulation of strains does not significantly affect their survival in bactericidal assays with bovine neutrophils, possibly because new capsular material is synthesized during the assay. Resistance to phagocytosis is also acquired by some strains grown in a chemically defined medium supplemented with casein, but the molecular basis of this increased virulence has not been elucidated (Leigh and Field 1991). Other potential virulence factors of *S. uberis* include hyaluronidase and an *uberis* factor analogous to the CAMP factor of *S. agalactiae*. A trypsin-resistant R-like antigen of 64 kDa that stimulates opsonic antibody has also been described. The role of the antigen in virulence or its response to casein supplementation of the growth medium is unknown.

The interaction of *S. uberis* with cells and tissues of the mammary gland is poorly understood. Environmental contamination and soiling of the udder are important predisposing factors. It is unclear, however, whether *S. uberis* enters opportunistically because udder defenses are nonspecifically overwhelmed by excessive microbial challenge or whether infection is a result of excessive numbers of *S. uberis* on the udder surface.

IMMUNITY

Specific immune mechanisms have been shown to be involved in clearance of *S. uberis* from the udder of cows experimentally challenged following recovery from a previous infection (Hill 1988). Sera from infected cows react strongly with *S. uberis* antigens of molecular masses 40-41, 59-65, and 118-122 kDa (Groschup et al. 1991). Milk becomes opsonic for *S. uberis* after mammary infection. Although the specificity of the opsonin in milk has not been determined, a trypsin-resistant protein antigen of 65 kDa has been shown to stimulate opsonic antibodies in guinea pigs that function well with bovine neutrophils (Groschup and Timoney 1992).

REFERENCES

Bazely, P. L. 1943. Studies with equine streptococci. Aust Vet J 19:62-85.

Bhakdi, S. J.; Tranum-Jensen, J.; and Sziegoleit, A. 1985. Mechanism of damage by streptolysin O. Infect Immun 47:5-60.

Blanchard, P. C. 1987. Group G streptococcal infections in kittens: Pathogenesis, Immune Response and Maternal Carrier State. PhD diss. University of California, Davis.

Boulnois, G. J.; Mitchell, T. J.; Saunders, K.; Owen, R.; Canvin, J.; Shepherd, A.; Camera, M.; Wilson, R.; Feldman, C.; Steinfort, C.; Bashford, C.; Pasternak, C.; and Andrew, P. W. 1991. In Genetics and Molecular Biology of Streptococci, Lactococci and Enterococci, Ed. G. M. Dunny, P. P. Cleary and L. C. McKay, pp. 83-87. Washington, D.C.: American Society for Microbiology.

Coming, B. F.; Murphy, J. C.; and Fox. J. G. 1991. Group G streptococcal lymphadenitis in rats. J Clin Microbiol 29:2720-23.

Deynes, R. A., and Armstrong, C. H. 1973. An antiphagocytic factor associated with group E streptococci. Infect Immun 7:298-304.

Ellis, R. P., and Armstrong, C. H. 1971. Production of capsules, streptokinase and streptodornase by streptococcus group E. Am J Vet Res 32:349-56.

Evers, W. O. 1968. Effect of furaltadone on strangles in horses. J Am Vet Med Assoc 152:1394-98.

Ferrieri, P. 1988. Surface-localized protein antigens of group B streptococci. Rev Infect Dis 10 (Suppl 2):S363-66.

Finch, L. and Martin, D. R. 1984. Human and bovine group B streptococci — two distinct populations. J Appl Bacteriol 57:273-78.

Frost, A. J.; Wamasinge, D. D.; and Woolcock, J. B. 1977. Some factors affecting selective adherence of microorganisms in the bovine mammary gland. Infect Immun 15:245-53.

Galan, J. E., and Timoney, J. F. 1985. Mucosal nasopharyngeal immune response of the horse to protein antigens of *Streptococcus equi*. Infect Immun 47:623-28.

_________. 1988. Immunologic and genetic comparison of *Streptococcus equi* isolates from the United States and Europe. J Clin Microbiol 26:1142-46.

Gonzalez, R. N., and Timoney, J. F. 1991. School of Veterinary Medicine, Cornell University, Ithaca, New York. Unpublished data.

Groschup, M. H., and Timoney, J. F. 1992. An R-like protein of *Streptococcus uberis* stimulates opsonic antibodies. Res Vet Sci 54:124-26.

Groschup, M. H.; Hahn, G.; and Timoney, J. F. 1991. Antigenic and genetic homogeneity of *Streptococcus uberis* strains from the bovine udder. Epidemiol Infect 107:297-310.

Hanski, E., and Caparon, M. 1992. Protein F, a fibronectin-binding protein is an adhesin of the group A streptococcus *Streptococcus pyogenes*. Proc Natl Acad Sci USA 82:6172-76.

Higgs, T. M.; Neane, F. K.; and Bramley, A. J. 1980. Differences in intramammary pathogenicity of four strains of *Streptococcus dysgalactiae*. J Med Microbiol 12:393-99.

Hill A. W. 1988. Protective effect of previous intramammary infection with *Streptococcus uberis* against subsequent clinical mastitis in the cow. Res Vet Sci 44:386-87.

Hoffman, A. M.; Staempfli, H. R.; Prescott, J. F.; and Viel, L. 1991. Field evaluation of a commercial M protein vaccine against *Streptococcus equi* infection in foals. Am J Vet Res 52:589-95.

Hollingshead, S. K.; Canfield, P. W.; and Pritchard, D. G. 1989. Insertional mutagenesis of CAMP factor in group B streptococci. Am Soc Microbiol Annu Meet, Abstr B-28.

Holt, M.; Enright, M. R.; and Alexander, T. J. L. 1988. Immunization of pigs with live cultures of *Streptococcus suis* type 2. Res Vet Sci 45:349-52.

Kurl, D. N.; Christensen, K. K.; and Christensen, P. 1984. Colonization of the upper respiratory tract of mice with group B streptococci type III with reference to the R protein. J Med Microbiol 17:347-51.

Lancefield, R. C.; McCarty, M.; and Everly, W. M. 1975. Multiple mouse protective antibodies directed against group B streptococci. J Exp Med 142:165-79.

Leigh, I. A., and Field, T. R. 1991. Killing of *Streptococcus uberis* by bovine neutrophils following growth in chemically defined media. Vet Res Commun 15:1-6.

MacKintosh, M. E.; Grant, S. T.; and Buffell, M. H. 1988. Evidence for *Streptococcus pneunoniae* as a cause of respiratory disease in young thoroughbred horses in training. In Equine Infectious Diseases V, Ed. D. G. Powell, pp. 41-46. Lexington: University of Kentucky Press.

Marques, M. B.; Kasper, D. L.; Pangbum, M. K.; and Wessels, M. R. 1992. Presentation of C3 deposition by capsular polysaccharide is a virulence mechanism of type III streptococci. Infect Immun 62:3986-93.

Mukhtar, M. M., and Timoney, J. F. 1988. Chemotactic response of equine polymorphonuclear leucocytes to *Streptococcus equi*. Res Vet Sci 45:225-29.

Norcross, N. L., and Oliver, N. 1976. The distribution and characterization of group B streptococci in New York State. Cornell Vet 66:240-48.

Norcross, N. L.; Dodd, K.; and Stark, D. M. 1968. Use of the mouse protection test to assay streptococcal antibodies in bovine serum and milk. Am J Vet Res 29:1201-5.

Pine, L., and Reeves, M. W. 1978. Regulation of the synthesis of M protein by sugars, Todd Hewitt broth, and horse serum in growing cells of *Streptococcus pyogenes*. Microbios 21:185-212.

Rainard, P., and Boulard, C. 1992. Opsonization of *Streptococcus agalactiae* of bovine origin by complement and antibodies against group B polysaccharide. Infect Immun 60:4801-8.

Rainard, P.; Lautrou, Y.; Sarradin, P.; and Poutrel, B. 1991. Protein X of *Streptococcus agalactiae* induces opsonic antibodies in cows. J Clin Microbiol 29:1842-46.

Robinson, J. H., and Kehoe, M. A. 1992. Group A streptococcal M proteins: Virulence factors and protective antigens. Immunol Today 13:362-67.

Schleifer, K. H., and Kilpper-Balz, R. 1987. Molecular and chemotaxonomic approaches to the classification of streptococci, enterococci, and lactococci: A review. Syst Appl Microbiol 10:1-19.

Simpson, W. J.; Robbins, J. C.; and Cleary, P. P. 1987. Evidence for group A-related M protein genes in human but not animal-associated group G streptococcal pathogens. Microb Pathogen 3:339-50.

Simpson, W. J.; LaPenta, D.; Chen, C.; and Cleary, P. 1990. Coregulation of type 12 M protein and streptococcal C5a peptidase genes in group A streptococci: Evidence for a virulence regulon controlled by the *vir*R locus. J Bacteriol 172:696-700.

Smith, H. E.; Vecht, U.; Gielkens, A. L. J.; and Smits, M. A. 1992. Cloning and nucleotide sequence of the gene encoding the 136-kilodalton surface protein (muramidase-released protein) of *Streptococcus suis* type 2. Infect Immun 60:2361-67.

Sundberg, J. P.; Hill, D.; Wyand, D. S.; Ryan, M. I.; and Baldwin, C. H. 1981. *Streptococcus zooepidemicus* as the cause of septicemia in racing Greyhounds. Vet Med Small Anim Clin 177:839-42.

Swindle, M. M.; Narayan, O.; Luzarraga, M.; and Bobbie, D. L. 1981. Contagious streptococcal lymphadenitis in cats. J Am Vet Med Assoc 177:829-30.

Timoney, J. F., and Blanchard, P. C. 1986. School of Veterinary Medicine, University of California, Davis. Unpublished data.

Timoney, J. F., and Eggers, D. E. 1985. Serum bactericidal responses to *Streptococcus equi* of horses following infection or vaccination. Equine Vet J 17:306-10.

Timoney, J. F., and Galan, J. E. 1985. The immune response of the horse to an avirulent strain of *Streptococcus equi*. In Recent Advances in Streptococci and Streptococcal Diseases, Ed. Y. Kimura, S. Kotami, and Y. Shiokawa, pp. 294-95. Bracknell, UK: Reedbooks.

Timoney, J. F., and Mukhtar, M. M. 1992. Variability in the M proteins of equine strains of *Streptococcus equi* subsp. *zooepidemicus*. In Equine Infectious Diseases Vol. 6. Ed. W. Plowright, P. D. Rossdale, and J. F. Wade, pp. 15-20. Newmarket, U.K.: R and W Publications (Newmarket) Ltd.

Timoney, J. F., and Trachman, J. 1985. Immunologically reactive proteins of *Streptococcus equi*. Infect Immun 48:29-34.

Vecht, U.; Wisselink, H. J.; van Dijk, J. E.; and Smith, H. E. 1992. Virulence of *Streptococcus suis* Type 2 strains in newbom germfree pigs depends on phenotype. Infect Immun 60:550-56.

Wagner, M.; Kubin, V.; Wagner, M.; and Gunther, E. 1982. Immunoelectron microscopic analysis of polysaccharide and protein surface antigens in wild strains of group B streptococci. Zentralbl Bakt Hyg Immunol (Abt 1 Orig) 253:331-43.

Wanger, A. R., and Dunny, G. M. 1987. Identification of a *Streptococcus agalactiae* protein antigen associated with bovine mastitis isolates. Infect Immun 55:1170-75.

Wessels, M. R.; Moses, A. E.; Goldberg, J. B.; and DiCessare, T. J. 1991. Hyaluronic acid capsule is a virulence factor for mucoid group A streptococci. Proc Natl Acad Sci USA 88:8317-21.

Williams, A. E. 1990. Relationship between intracellular survival in macrophages and pathogenicity of *Streptococcus suis* type 2 isolates. Microb Pathog 8:189-96.

Woods, R. D., and Ross, R. F. 1975. Purification and serological characterization of type-specific antigen of *Streptococcus equisimilis*. Infect Immun 12:881-87.

Yokomizo, Y., and Norcross, N. L. 1978. Bovine antibody against *Streptococcus agalactiae*, type 1A, produced by preparturient intramammary and systemic vaccination. Am J Vet Res 39:511-16.

2 / *Staphylococcus*

BY P. JONSSON AND T. WADSTROM

STAPHYLOCOCCI have been associated with disease ever since the Scottish surgeon Sir Alexander Ogston (1880) discovered round microorganisms in infected tissues. He introduced the name "staphylococcus" (Greek *staphyle*, bunch of grapes; *kokkos*, berry), referring to the microscopic picture of stained organisms. The importance of these microorganisms as pathogens for animals has been known for more than 100 years. In 1887, Nocard isolated staphylococci from mastitis in sheep and, in 1890, Guillebeau suggested that these organisms were responsible for mastitis in cattle.

Staphylococcus aureus is the main pathogen within the genus, which currently includes 29 different species. The ability to produce the enzyme coagulase divides staphylococci into two main groups: coagulase-positive staphylococci, consisting of the species *S. aureus, S. intermedius, S. schleiferi* subsp. *coagulans*, and *S. delphini* (Table 2.1); and coagulase-negative staphylococci (CNS), consisting of 25 different species. The species *S. hyicus* is variably coagulase-positive but most often is included in the group of CNS.

Strains of *S. aureus* seem to be generally species-specific and a scheme for recognition of five biotypes of *S. aureus* has been proposed (Hajek and Marsalek 1971; Devriese and Oeding 1976). There is an association between biotype and host species; thus, biotype A consists of strains of human origin; biotype B, strains from pigs and poultry; biotype C, strains from cattle and sheep; biotype D, strains from hares; and biotype E, strains from dogs, horses, and pigeons (Hajek 1976). There are deviations from these associations of biotypes of the organism with particular hosts; pathogenic strains from one animal species are not always non-pathogenic in other species. It is known that human strains can cause mastitis in cows; thus, strains of biotype A can be isolated from bovine mastitis. Similarly, strains of biotypes A and C can be isolated from horses and dogs. These observations demonstrate that *S. aureus* strains are not strictly species specific in regard to pathogenicity. It is probable that *S. aureus* of canine origin described in earlier literature was *S. intermedius*. Although showing little DNA homology, *S. aureus* and *S. intermedius* are, phenotypically, very similar.

Species of CNS associated with humans can also be isolated from animals, but recently, CNS occurring solely in animals have been reported. The most important diseases caused by CNS in animals are mastitis in dairy cows (several species of *Staphylococcus*) and exudative

TABLE 2.1. Association of coagulase-positive *Staphylococcus* with animal diseases

Coagulase-Positive Species	Host	Major Diseases
S. aureus	Bovine	Mastitis, suppurative lesions
	Ovine	Mastitis, tick pyemia
	Porcine	Suppurative lesions
	Equine	Botryomycosis, arthritis, mastitis
	Canine	Pyoderma, urinary tract infection, discospondylitis
	Avian	Bumblefoot, skin lesions, arthritis
S. intermedius	Canine	Pyoderma, suppurative lesions, otitis externa
	Equine	Suppurative lesions
	Avian	Abscesses, arthritis
S. schleiferi subsp. *coagulans*	Canine	Otitis externa
S. delphini	Dolphin	Suppurative skin lesions

epidermitis in pigs (*S. hyicus*). Coagulase-negative staphylococci (Table 2.2) are predominantly skin pathogens of pigs, cattle, and horses, but there are reports suggesting CNS as the causative agent of arthritis and osteomyelitis in pigs, cattle, and poultry (Devriese 1986). It has not been common practice for veterinary or medical microbiologists to identify CNS to the species level, but this is beginning to change because commercial identification systems for CNS from humans are now available. The reliability of these identification systems for animal strains has yet to be proven.

Staphylococci are normal inhabitants of the skin and mucous membranes, particularly at junctions with skin. Resistance of staphylococci to dry conditions and high salt and lipid concentrations make them well-suited to this ecological niche.

VIRULENCE FACTORS

Staphylococcus aureus and other coagulase-positive species, such as *S. intermedius*, produce a large number of cell-associated and extracellular proteins. A number of these proteins are important for colonization and growth in various body tissues. The presence of this variety of surface and extracellular products probably explains why these pathogens are very successful in establishing an infection in most organs of the body of many animal species (Table 2.1). Various cytolytic toxins (often called hemolysins) and enzymes such as proteases, lipases, hyaluronate lyase, and hyaluronidase act together to create tissue degradation that provides low-molecular weight nutrients, which the microbe can assimilate and use for rapid growth. Disease results from a complex interplay among various bacterial surface proteins involved in colonization of cells and extracellular matrix (ECM) and enzymes and toxins that may cause damage to tissues (Table 2.3).

No single virulence factor is preeminent in overcoming host resistance, and only in the case of exfoliative toxin and toxic shock syndrome toxin (TSST-1) is there a bacterial product clearly primarily responsible for the symptoms of disease. However, it appears that certain cytolytic exotoxins and enzymes contribute to the disease process. Furthermore, mutants lacking more than one virulence factor (such as alpha toxin and coagulase) in combined and pleiotropic mutants show a drastic decline in virulence when tested in animal models such as a mouse mastitis model (Jonsson et al. 1985).

TABLE 2.2. Association of coagulase-negative *Staphylococcus* with various animal hosts

Species	Host	Major Disease
S. hyicus	Porcine	Exudative epidermitis
	Bovine	Mastitis
	Avian	Suppurative lesions
S. chromogenes	Bovine	Mastitis
	Porcine, avian	NA
S. simulans	Bovine	Mastitis
	Porcine	NA
S. xylosus	Bovine	Mastitis
	Porcine, equine, avian, caprine	NA
S. epidermidis	Bovine	Mastitis
S. cohnii	Bovine, avian	Bovine mastitis, skin lesions
S. sciuri	Bovine, ovine, caprine, equine	Bovine mastitis, skin lesions
	Avian	NA
S. gallinarum	Avian, bovine	Suppurative skin lesions
S. lentus	Porcine, ovine, caprine, avian	Suppurative skin lesions
S. equorum	Equine	Genital tract infections
S. haemolyticus	Bovine	Mastitis
S. warneri	Bovine	Mastitis
S. felis	Feline	Otitis externa

Note: NA = Not associated with any major clinical syndrome.

Exotoxins

Alpha-Toxin. The best characterized exotoxin is alpha-toxin (alpha-hemolysin), a 34-kDa protein that is produced by most strains of *S. aureus*. Known initially as a hemolysin, it lyses red blood cells of a number of animal species. Alpha-toxin binds in hexamers to cell membranes of target cells, such as platelets and mast cells, and induces pore formation in the membrane. The cells then leak ions rapidly and lyse. Killing of sensitive target cells in various organs by this pore-forming toxin causes a release of prostaglandins and other inflammatory mediators. There is evidence that alpha-toxin also impairs macrophage function and promotes local tissue damage in abscesses. Interestingly, alpha-toxin is also dermonecrotic upon injection into the skin of rabbits and other animals, and it has a powerful action on vascular smooth muscle contraction. Injection of small quantities of alpha-toxin into the udder of goats and sheep results in tissue necrosis; detoxified alpha-toxin (alpha-toxoid) induces partial protection against *S. aureus* mastitis in goats and sheep and in a rabbit mastitis model. The critical importance of alpha-toxin is indicated by the reduced ability of alpha-toxin (Hla^-) mutants to kill mice, following intraperitoneal or intramammary injection.

Beta-Toxin. Beta-toxin induces classical hot-cold hemolysis of sheep red blood cells by specific degradation of sphingomyelin in the membrane. Beta-toxin is a toxic enzyme (a sphingomyelinase). Susceptibility of various cells of different species depends on the content of sphingomyelin in the cell membrane. This toxic enzyme appears to act synergistically with a number of other toxins and membrane-damaging enzymes and has an important role in tissue degradation under certain circumstances. Moreover, experimental beta-toxoid vaccines give some protection against experimental mastitis in rabbits.

TABLE 2.3. Virulence determinants of *Staphylococcus aureus*

Determinant	Activity in Host Tissue
Capsular polysaccharide	Antiphagocytic
Cell Wall Components	
Peptidoglycan	Pyrogenic, chemoattractant
Teichoic acid	Released teichoic acid may protect against complement
Cell Surface Proteins	
Protein A	Interacts with Fc region of IgG
Fibrinogen-binding protein	Binds fibrinogen
Fibronectin-binding protein	Binds fibronectin
Laminin-binding protein	Binds laminin
Collagen-binding protein	Binds collagen
Vitronectin-binding protein	Binds vitronectin
Extracellular Toxins and Enzymes	
Alpha-toxin, beta-toxin, delta-toxin, gamma-toxin	Cytotoxic for tissue cells and leukocytes
P-V leukocidin	Destroys leukocytes
Toxic shock syndrome toxin	Binds major histocompatibility complex (class II) molecules, induces synthesis of cytokines, causes multiple organ dysfunction
Enterotoxins	Emetic, diarrheagenic
Epidermolytic toxins	Lyse attachment among cells of the stratum granulosum
Coagulase	Catalyzes conversion of fibrinogen to fibrin
Lipase	Breaks down lipids on skin and at other sites
Fatty acid modifying enzyme	Modifies fatty acids released by lipid degradation; contributes to abscess formation
Proteases	Degrade a variety of proteins including a number involved in host defense
Phospholipases	Degrade phospholipids
Staphylokinase	Converts plasminogen to fibrinolytic plasmin
Hyaluronidase	Degrades hyaluronic acid
Nuclease	Cleaves both DNA and RNA

Most bovine isolates of *S. aureus* produce beta-toxin, but only the occasional human isolate does. It is therefore interesting that beta-toxin seems to be of particular significance in mastitis; the toxin enhances bacterial growth in the mammary gland (Foster 1992).

Gamma-Toxin. Gamma-toxin has only recently been studied in detail, and the possible role of this toxin in animal infection has not been explored. The toxin was shown to comprise two proteins, one of which was a protease, that may uncover specific cell membrane-binding structures for the second component.

Delta-Toxin. Delta-toxin behaves like a detergent and inflicts damage on all types of cells in all host species. However, skin lipids can rapidly bind and neutralize delta toxin on local injection in the skin of young mice and rabbits.

Leukocidin. A low percentage of strains of *S. aureus* produce a leukocidin (Panton-Valentine leukocidin), which is composed of two proteins, called F (fast) and S (slow), on the basis of their electrophoretic mobility in agarose. The toxin binds to human polymorphonuclear leukocytes (PMNs) and kills them after rapid degranulation. There is some evidence to suggest that bovine strains of *S. aureus* produce a leukocidin that is toxic for bovine PMNs but much less toxic for human PMNs.

Toxic Shock Syndrome Toxin (TSST-1). Certain strains of *S. aureus* produce a 22-kDa protein toxin, called TSST-1, that can be absorbed into the circulation and, probably together with endotoxin from gram-negative organisms absorbed from mucosal lesions, cause a toxin-mediated disease. TSST-1 belongs to the same superfamily of toxins as staphylococcal enterotoxin and erythrogenic toxins of group A streptococci; they can act as immunomodulators or superantigens. These toxins bind to conserved regions on major histocompatibility complex (MHC) class II antigens on host cells and induce excessive lymphokine production, resulting in tissue damage. These toxins also can cause release of tumor necrosis factor (TNF) and other cytokines and may induce cardiovascular shock, associated with microthrombus formation in capillaries. Furthermore, the cells to which the toxin binds are destroyed by certain cytotoxic T cells.

Our understanding of the role of these toxins or superantigens in animal infections, such as bovine mastitis, is very limited. Whereas bovine strains of *S. aureus* produce TSST-1 that is identical to TSST-1 produced by human strains, isolates from sheep and goats produce a variant that is similar in molecular mass and antigenicity but different in isoelectric point.

It seems likely that these immunotoxins may down-regulate the humoral and cell-mediated immune responses in local and systemic staphylococcal infections in animals and humans. The immunotoxins may be responsible for the poor immune response in severe staphylococcal infections, often reported in fulminant infections in humans, but also in experimental animal infections with potent TSST-1-producing strains of *S. aureus*.

Enterotoxins. Enterotoxins are produced by approximately 30% of strains of *S. aureus* isolated from various human infections, and from foods contaminated with this organism. There are six antigenic types of enterotoxin (A, B, C_1, C_2, D, and E). The toxins are all similar in protein structure and are heat-stable, withstanding exposure to 100°C for several minutes. When ingested as preformed toxins in contaminated foods, microgram amounts of toxin can induce nausea, vomiting, and diarrhea (all symptoms of staphylococcal food poisoning) in humans. Interestingly, most animal species, except the very young kitten and monkeys, are highly resistant to the toxins. Vomiting is believed to be due to the effect of the toxin on nerve centers in the stomach and intestine, that send a message to the vomiting center in the brain; the basis for the diarrheal effect is not known.

Enterotoxins, like TSST-1, are potent mitogens that induce release of immunomediators such as interleukin (Il-1) from lymphocytes. However, the effect on lymphocytes of various subclasses of different animal species have not been studied systematically. Immunological effects previously attributed to other staphylococcal proteins, such as protein A, have been shown to be due to contamination by extremely small amounts of staphylococcal enterotoxins, the most potent protein mitogens so far described.

Exfoliative Toxins (Epidermolytic Toxins). Two kinds of exfoliative toxins (ETA and ETB) commonly produced by *S. aureus* strains, which belong to phage group II, cause blistering in the skin of young infants and adult humans with underlying immunological disorders. ETA is the product of a chromosomal gene and ETB is the product of a plasmid gene. Both toxins are proteins of approximately 30 kDa, but they differ antigenically and in their composition. Both toxins disrupt the attachment of one cell to another in the stratum granulosum. Injection of toxin A or B into the skin of neonatal mice can induce blistering similar to that seen in pemphigus neonatorum or impetigo in humans. Outbreaks of disease involving a high percentage of infants have occurred in nurseries. The possible role of these toxins in animal diseases has not been studied.

Other Toxins. Many strains from human and animal infections, as well as CNS, produce a membrane-damaging polypeptide (MW 12 kDa), with amphiphilic properties similar to mellitin found in bee venom.

Exoenzymes

Coagulases. *S. aureus* produces two forms of coagulase: cell-bound coagulase, or clumping factor; and extracellular, or free, coagulase. Interestingly, fibrinogen-binding protein on the staphylococcal cell surface, with no known enzymatic action, can directly convert fibrinogen to insoluble fibrin and cause staphylococci to clump. On the other hand, the exoenzyme, cell-free coagulase, causes specific activation of plasma thrombin, leading to conversion of fibrinogen to fibrin. Formation of a fibrin layer around a staphylococcal tissue infection is believed to localize the process and protect the microbe from phagocytosis. The role of coagulase in virulence is not clear; when the gene for coagulase production was replaced with a defective gene, the coagulase-negative organism showed no reduction in virulence in a number of mouse models (Foster 1992).

Lipases, Esterases, and Fatty Acid-Modifying Enzyme. Certain strains of *S. aureus* that are able to multiply rapidly in skin tissues produce lipid-degrading enzymes (lipases and esterases) and an enzyme called FAME (or fatty acid-modifying enzyme). Antibacterial fatty acids released by lipid degradation are chemically modified by FAME so that they are no longer antibacterial. Thus, FAME is an important enzyme in strains that are able to form abscesses efficiently. CNS lack the ability to produce FAME, and this may be a major reason why these organisms are generally poor inducers of tissue infections with abscess formation.

Catalase. All staphylococci produce catalase, which catalyzes the conversion of hydrogen peroxide to water and oxygen. Catalase protects the pathogen from the toxic hydrogen peroxide that accumulates during bacterial metabolism and is released upon phagocytosis.

Proteases. *S. aureus* produces at least four proteases, and a specific lesion-degrading enzyme has recently been reported. This protease, like other hydrolytic exoenzymes such as hyaluronidase and nuclease (DNA- or RNA-degrading enzymes), is important in tissue degradation in early infection, as well as in degradation of pus in an abscess to release nutrients for young cells to continue to multiply in an "old infection." However, the local abscess with a fibrin barrier and with organisms with a low staphylokinase and protease production often will eventually become sterile.

Cell Surface Structures as Possible Virulence Determinants

Peptidoglycan. The major cell wall polymer of *S. aureus* and CNS is the peptidoglycan (glycopeptide or murein), which contains the unique cross-linking pentaglycine peptide in *S. aureus* (and with a few amino acid substitutions in various CNS). The polymer is released in great amounts in local infections, such as skin and muscle abscesses and joint infections. The peptidoglycan stimulates production of endogenous pyrogen and is a potent leukocyte chemo-attractant.

Polysaccharide Capsule. Capsules of *S. aureus* have been divided into 12 immunotypes. Production of a true capsule that can be visualized in the microscope after India ink staining is rare among *S. aureus* of both human and animal origin. Macrocapsules produced by strains of serotype 1 impair opsonophagocytosis and are associated with high virulence. A thin polysaccharide microcapsule is produced by most mastitic isolates and human clinical isolates of *S. aureus*. The majority of these strains produce a serotype 5 or 8 microcapsule. Those strains that do produce such a capsule tend to resist phagocytosis in in vitro studies. The role of these capsules in virulence appears to vary with the type of infection investigated. Mutants of *S. aureus* strains that lose the ability to produce capsule show a decrease in virulence in bovine mastitis. On the other hand, when mutants that are defective in the microcapsule were investigated for virulence in mice, there was no difference in virulence of the capsule-negative mutants compared with their capsule-positive parent (Albus et al. 1991). In a rat model of catheter-induced experimental endocarditis, mutants that lacked the capsule had a significantly lower 50% infective dose than did the capsule-positive parent (Badour et al. 1992).

Teichoic Acids. These acids are polyribitol or polyglycerol phosphates that are either associated with peptidoglycan or found extracellularly. Antibody to teichoic acids are often present in normal serum; extracellular teichoic acids are believed to protect the organism from opsonization, by activating complement and reducing the complement components available for interaction with the bacterium.

Protein A. Protein A is a major immunoglobulin (Ig)-binding surface protein found on over 98% of *S. aureus* of human and animal origin. Protein A binds to the Fc portion of IgG1, IgG2, and IgG4 subclasses, and inhibits opsonization and phagocytosis. Studies of protein A-deficient organisms in mice showed that loss of protein A resulted in a reduction in virulence in subcutaneous and peritoneal infections but no difference in mastitis (Foster 1992).

Fibronectin (Fn)-binding Proteins (FnBPs). Like protein A, FnBPs are present on the surface of the vast majority of *S. aureus* isolates. Interestingly, studies of fibronectin binding to *S. aureus* and fibrinogen binding to *Streptococcus pyogenes* showed that binding of these serum and tissue proteins also induces a host protein that inhibits phagocytosis. We have recently shown that pretreating *S. aureus* with specific antibody to staphylococcal FnBP enhances phagocytosis. Furthermore, experimental studies in murine mastitis and peritonitis models have shown that these antibodies may be protective by promoting phagocytosis and, probably, also by blocking tissue adherence mediated by specific binding of staphylococci to extracellular matrix (ECM) proteins in damaged tissues.

Fibrinogen-binding Proteins. Recently, several studies have shown that more than 95% of both human and animal strains of *S. aureus* (as well as *S. intermedius* and possibly *S. hyicus*) agglutinate or clump when purified fibrinogen from human or various animal species is mixed with the bacteria. At least three cell surface proteins have been identified, and it is likely that the family of fibrinogen-binding proteins will soon expand. It is probable that strains of animal and human origin possess different fibrinogen-binding proteins, since differences have been observed in proteins from *S. aureus* from various host species. It now seems logical to define staphylococcal enzymes that cause blood clotting as coagulases and those that activate dissolution of blood clots as staphylokinases.

Other Binding Proteins. Collagen-binding protein and a vitronectin-binding protein are two other staphylococcal ECM-binding proteins that have recently been identified, purified, and characterized (Wadström 1993). These binding proteins have been detected in *S. aureus* and various CNS, which are pathogenic for humans and the bovine udder. Moreover, surface structures that bind tissue plasminogen, as well as thrombospondin and laminin, are also being characterized in different laboratories, but their possible roles in tissue adhesion have not yet been investigated.

Other Less Well Defined Virulence Factors

Slime Produced by Coagulase-Negative Staphylococci. Strains of CNS commonly isolated from biomaterial-associated infections are often found to be *S. epidermidis*. Upon growth in carbohydrate-rich media, these strains produce large amounts of slime that is predominantly extracellular polysaccharides that are poorly defined chemically, but are known to include various sugar constituents such as manuronic acid and galacuronic acids. This slime of CNS that grows on surfaces, such as intramammary devices, milking machine equipment, and intravascular catheters (or catheters for the urinary bladder) becomes mixed with host proteins to form an amorphous biofilm. It is now well established that biofilm and biomaterial-associated infections are very difficult to treat due to poor penetration of antibiotics, which often bind to biofilm polymers. Moreover, phagocytes cannot penetrate a biofilm efficiently so that the infective agent is well protected. Furthermore, the slime material shows immunosuppressive properties when injected into animals with various standard antigens.

Iron Regulation and In Vivo Growth

Recent studies in several laboratories have shown that both *S. aureus* and *S. epidermidis* and other CNS growing in in vivo models, such as tissue cages in the mouse or rabbit peritoneum, show a great difference in surface proteins compared to the same strain grown in a conventional laboratory medium. Moreover, growing staphylococci in vitro in media supplemented with peritoneal fluid, bovine milk, or various milk fractions such as lactoferrin clearly suggests that ferrous (Fe^{++}) ions and possibly other divalent cations may be involved in gene regulation of various cell surface proteins as well as exo- and intracellular proteins.

PATHOGENESIS

The virulence of both coagulase-positive and coagulase-negative species cannot be explained in terms of a single virulence determinant. Complex interactions that occur among a variety of adhesion mechanisms, extracellular toxins, surface proteins, and enzymes are responsible for staphylococcal virulence.

Typically, *S. aureus* requires some breach in the primary host defense, commonly the skin or mucosal barrier, in order to cause disease. Early steps in infection are facilitated by tissue damage in open skin wounds and in small lesions on various mucosal surfaces, such as the bovine udder epithelium and tampon-related microlesions in the human female genital tract. Such lesions allow the pathogen to establish an infection locally and spread from microcolonies into surrounding tissues. The presence of a foreign body, such as suture material, facilitates establishment of staphylococci. *S. aureus* is, however, able to cause disease in the unbroken skin: the organism initiates infection in the root of the hair shaft and in sebaceous glands.

A hallmark of the most common staphylococcal lesion in animals is that it is a localized, suppurative lesion, as in mastitis and skin infections. Only occasionally, as in tick pyemia in lambs, do the organisms become invasive. Peptidoglycan and proteases of organisms introduced into an area of damaged tissue will attract neutrophils from the blood, will activate the alternate pathway of complement in the nonimmune host, and thereby generate chemotactic compounds. In the immune host, or later on in infection, specific antibody will react with the bacterium and activate the classical complement pathway. Typically, staphylococci are ingested and killed by neutrophils; the oxygen-dependent antibacterial mechanisms appear to be most significant, but low pH, lactoferrin, lysozyme, and granular cationic proteins also contribute to killing of the bacteria. Some neutrophils are destroyed by toxic bacterial products. Because staphylococci are readily killed following phagocytosis, mechanisms for prevention of phagocytosis are important for survival of the bacteria in tissue. The lesion tends to be walled off, possibly due to the action of coagulase, which causes deposition of fibrin. Pathogenesis of a number of important infections in animals is discussed below.

Bovine Mastitis

Mastitis is an inflammatory reaction of the mammary gland. The most important mastitis pathogen, worldwide, is *S. aureus*. CNS are usually regarded as minor pathogens, but they can sometimes give rise to severe mastitis. There are differences in the CNS species reported from abnormal bovine milk in various countries. However, *S. epidermidis, S. chromogenes, S. simulans, S. xylosus, S. hyicus,* and *S. haemolyticus* seem to be the most prevalent species that are implicated in bovine mastitis (Hogan et al. 1987; Harmon and Langlois 1989; Jarp 1991; Matthews et al. 1991; Perrin-Couilloud 1992; Birgersson et al. 1992). Differences in isolation rates reported by various groups may be due to their use of different methods for identification of staphylococcal species, inconsistencies in definitions, and differences in milk-collecting procedures, as well as differences in geographical areas.

Staphylococcal mastitis may result in clinical signs or it may be subclinical. In the clinical form, the disease may be peracute, acute, or chronic. The peracute form, which is rare, is caused by *S. aureus*, often shows a gangrenous udder, and is frequently fatal. Usually, staphylococcal mastitis begins with an acute episode, including some or all of the cardinal signs of inflammation in the udder: *tumor, rubor, calor, dolor* and *functio laesa*. Systemic signs of disease are not always seen. If not successfully treated, the acute form becomes chronic, with few clots in the milk, or with no clinical signs; the disease subsequently passes into a subclinical, chronic form. This form of mastitis can only be detected by an elevated somatic cell count and isolation of bacteria from the milk.

The route of infection in staphylococcal mastitis is via the teat, and the milking machine is considered to be the most important predisposing factor. However, staphylococcal mastitis does occur in hand-milked cows, in suckled cows, and during the dry period (Hunter and Jeffrey 1975). The bacteria pass the teat duct into the teat and udder cisterns, and subsequently into the milk ducts (Frost et al. 1977). They may subsequently establish in an area of secretory tissue. *S. aureus* and other mastitis pathogens, at 37°C, are able to rise with the milk fat. The phenomenon seems to involve IgA antibodies associated with membranes of fat globules in previously infected animals. This phenomenon may aid the bacterium to invade upper parts of the milk ducts and udder.

The mechanism of penetration via the teat duct is not fully understood, but it is suggested that the organisms gain entrance to the teat cistern by propulsion and/or by bacterial growth. The staphylococci colonize the tip of the teat, especially in areas of teat lesions or erosions. Colonization often precedes an intramammary infection but does not always result in mastitis. Staphylococci can colonize the teat duct for a prolonged period of time (Davidson 1961; Forbes 1968; du Preez 1985). In the intact teat, the organisms probably exist as distinct microcolonies in the stratum corneum, as they do in other areas of skin. When contact is made between the staphylococci and milk, then a medium is provided in which the staphylococci can grow into the gland. The nearer the organisms are to the milk in the teat cistern, the greater is the risk for mammary infection to occur (Anderson 1983). Once the organisms pass the barriers of the teat duct, adherence to epithelial cells is thought (but still not proven) to be the next critical step in pathogenesis. Adherence is believed to involve a combination of cation-bridging, hydrophobic interactions, and bacterial adhesin-specific epithelial surface receptors (Lindahl et al. 1990). Olmsted and Norcross (1992) have shown that the ability of *S. aureus* to bind to epithelial cells of the ductules and alveoli in the bovine mammary gland is an important virulence factor, and that antibodies against whole cells inhibit the adherence.

Once established on the surface, nucleases and phosphatases degrade pus into substances that can be utilized as nutrients for the bacteria. Although several potential virulence determinants have been purified, or partly purified, and tested for toxicity in different experimental systems, their relative importance as virulence determinants is not known. Jonsson et al. (1985) isolated mutants defective in alpha-toxin, coagulase, and protein A and studied them in a mouse mammary gland model of mastitis. The researchers concluded that both alpha-toxin and coagulase contributed to virulence, which was drastically reduced in a strain with a double mutation but was regained by a wild-type recombinant strain. Clear evidence for a role of protein A in virulence was lacking. Later, Bramley et al. (1989) used a mouse model to study strains in which alpha-toxin genes, beta-toxin genes, or both, were inactivated by site-specific mutagenesis. The acutely lethal effect of alpha-toxin was confirmed, since alpha- and beta-toxin-producing parent strains often killed mice, while strains that lacked alpha-toxin did not. Comparisons of parent strains with double mutants that lacked alpha- and beta-toxins showed that both toxins contributed to a significantly higher recovery of *S. aureus* from the gland, 48 hours postinfection. Histopathological examination of mammary glands showed that phagocytosis of bacteria occurred irrespective of toxigenicity.

During the course of staphylococcal multiplication, chemotactic substances are liberated. The chemotactic substances cause an infiltration of polymorphonuclear leukocytes (PMNs) into the mammary gland. The role of these neutrophil granulocytes is to ingest and kill the bacteria. The outcome of an infection depends on the interaction between invading staphylococci and the number and activity of PMNs in the mammary gland. In some cases, PMNs overcome the staphylococci, resulting in a self-cure; but, usually, the outcome is a chronic mastitis. A suggested explanation for the development of a chronic infection is that the capacity of the PMNs is reduced by engulfment of fat droplets and casein (Paape et al. 1975; Russel et al. 1976). It has also been suggested that milk PMNs have reduced glycogen reserves and a decreased phagocytic capacity, in comparison with blood PMNs.

Hallén Sandgren et al. (1991) reported a decrease in the phagocytic activities of PMNs from bovine milk in the first lactation, compared with the activities of blood neutrophils isolated from the same animal at the same time. They found a significantly lower luminol-dependent chemiluminescence, most prominent in the seventh week of lactation, towards opsonized yeast by milk neutrophils compared with blood neutrophils. When fat and casein are phagocytosed, the cells presumably release some of their primary granule component, myeloperoxidase, so that milk neutrophils will have fewer primary granules compared with their blood precursors. The expected outcome of the loss of this enzyme would be the observed lowered luminol-dependent chemiluminescence on stimulation of the cells; as this reaction is generally considered to reflect both oxygen-radical production and the release of myeloperoxidase by the neutrophil (Edwards 1987).

Lucigenin-dependent chemiluminescence, reflecting only oxygen-radical production aimed at C3-opsonized particles, was impaired in milk neutrophils compared with blood neutrophils, although not as severely in cells isolated in early and late lactation. Oxygen-radical formation thus appeared somewhat less affected than the degranulation process in the milk cells

(Aniansson et al. 1984). Together, these findings indicate a sequential decline in functional activities of neutrophils once they arrive in the milk, the myeloperoxidase-dependent luminol-enhanced chemiluminescence being the most sensitive marker for such functional alterations.

Despite the depressed milk neutrophil activity described above, milk neutrophils show some signs of activation, as indicated by a markedly enhanced recognition of unopsonized yeast. In other species, this recognition has been attributed to an increased expression of the complement receptor CR3 and has been experimentally induced by incubating blood neutrophils with cytokines (Tsujimoto et al. 1986). The putative upregulation of the CR3 receptor on the surface of milk neutrophils is of interest, for this receptor has been associated with the recognition of a number of different bacteria, including *S. aureus*. Such a recognition mechanism may diminish the relative importance of opsonins in milk. The rapid onset of both phagocytosis and chemiluminescence by milk neutrophils also indicated that the cells are activated in some respects.

Mamo et al. (1991a,b) showed that *S. aureus* strains grown in milk whey were more virulent in two mouse models than were the homologous strains grown in trypticase soy broth (TSB). In addition, the strains grown in whey resisted phagocytosis more effectively than the strains grown in TSB, as measured in three different models for evaluation of phagocytic capacity. The substance produced in whey under in vivo-like conditions is probably a polysaccharide microcapsule (Watson 1989; Hallén Sandgren et al. 1991). Thus, it appears that the microcapsule produced in vivo may be a major factor in the ability of *S. aureus* to resist uptake by phagocytic cells.

Mastitis in Other Animal Species

Staphylococcal mastitis is undoubtedly of greatest economic importance in dairy cows, but this disease is also of considerable concern in sheep and goats. In these animal species, the mastitis is predominantly of the gangrenous type. Staphylococcal mastitis also occurs in sows, rabbits, horses, dogs and cats.

Exudative Epidermitis

Exudative epidermitis (greasy pig disease) is an acute, generalized infection of the skin of suckling and weaned piglets. The disease is characterized by excess sebaceous secretion, exfoliation, and exudation. These changes result in loss of skin function, most often affecting the entire skin surface. Pigs become depressed and lose their appetite; pigs affected at a very young age may die.

Exudative epidermitis is caused by *S. hyicus*, a bacterium that reacts variably in the tube coagulase test but is most often considered along with the true coagulase-negative species. In a Danish investigation, *S. hyicus* was recovered from the vagina of 92% of prepubertal gilts in four herds. The vaginal carrier rates varied between 32% and 74%, and were lowest immediately prior to farrowing. *S. hyicus* was isolated from skin samples collected from 61% of the offspring within 24 hours after farrowing, increasing to 74% 3 weeks after farrowing (Wegener and Skov-Jensen 1992). *S. hyicus* can also be isolated from nares, external ears, and skin of healthy pigs.

Outbreaks of exudative epidermitis are frequently associated with litters from gilts, but may spread to other litters in the herd. Mortality may be as high as 90% in severely infected herds, but a low mortality is more common.

Very little is known about pathogenesis of the disease, but the organisms appear to enter the body through lacerations on the skin. The disease can be produced by subcutaneous inoculation of the bacterium (Mebus et al. 1968; Amtsberg et al. 1973), and subcutaneous inoculation of an exfoliative toxin preparation from *S. hyicus* induces clinical signs of exudative epidermitis (Amtsberg 1979). It is likely that the disease is due to the effects of a single protein product of the bacterium (Sato et al. 1991; Andresen et al. 1992).

Pyoderma in Dogs and Cats

Pyoderma is one of the most common skin diseases of dogs, but it is relatively rare in cats. The normal cutaneous microbial flora comprises both resident and transient bacteria, with the resident flora being fairly consistent within a limited area of the animal's body. The resident flora consists of harmless bacteria (commensals) that may inhibit the growth of pathogenic bacteria. Maintenance of the resident bacterial flora is dependent on pH, moisture, sebum production, and the condition of the stratum corneum. Destruction of the normal bacterial flora can predispose to pyoderma by bacteria of the transient or resident bacterial flora.

Coagulase-positive *Staphylococcus* is the most commonly isolated organism from pyoderma in dogs. Prior to 1982, the organism was called *S. aureus*, but subsequently it has been identified as *S. intermedius* (Phillips and Kloos 1981; Cox et al. 1984; deBoer 1990). It is suggested that changes in sebum production provide a milieu in which *S. intermedius* may multiply. Alterations to the skin surface could possibly create conditions in which the organism may adhere. Factors that predispose to development of staphylococcal pyoderma include immunodeficiency, primary dermatosis caused by parasites or other forms of skin damage, and allergy. Thus, staphylococcal pyoderma often represents secondary involvement of the organism in favorable milieu for its proliferation caused by one of these conditions.

Hypersensitivity reactions are proposed to be a major factor in the development of recurrent staphylococcal hypoderma. Repeated exposure to staphylococci has been shown to induce delayed hypersensitivity to the cell wall peptidoglycan. The explanation offered by deBoer (1990) is as follows. High levels of IgE against staphylococcal surface antigens develop in serum, and IgE becomes associated with mast cells in the skin. When the staphylococcal antigens are present, the mast cells release their granules, resulting in an intense local inflammatory reaction. The inflammatory mediators are responsible for pruritis, erythema, and inhibition of leukocyte function, thereby promoting persistence of the bacteria.

Tick Pyemia of Lambs

Tick pyemia is a perinatal staphylococcal infection of lambs which occurs in areas infested with the ticks *Ixodes ricinus*. Staphylococcal abscesses develop particularly in the joints but may affect other tissues. *S. aureus* is present on the skin of the lambs and enter the body through the wounds caused by the ticks. No specific virulence determinants have been found in the strains of *S. aureus* recovered from these animals and the spread of the organism is likely due to an inability of the young host to localize the bacteria at the site of inoculation.

STAPHYLOCOCCAL VACCINES

The first staphylococcal vaccines were chemically or heat-detoxified crude exoproteins that were assessed in animal models before the golden era of antibiotics started in the 1940s and 1950s. Several such vaccines contained partly purified, detoxified, alpha-toxin, beta-toxin, coagulase, as well as whole cells and cell lysates. Many studies showed partial protection (up to 40-60%) in various experimental models such as mastitis in mice. It seems likely that the rapid increase in our understanding today of how various virulence factors interplay will give us a new generation of virulence factor-based vaccines. A gene-fusion FnBP has recently been evaluated in clinical trials for immunoprophylaxis against bovine mastitis; staphylococci grown in vivo have also been tested. A staphylococcal vaccine could be combined with an *E. coli* J5-based bacterin to induce protection against two important agents of bovine mastitis.

FUTURE TRENDS

The availability of systems for genetic manipulation and for creation of pairs of isogenic mutants has lead to rapid advances in our understanding of aspects of pathogenesis of diseases due to *E. coli* and to *Salmonella*. Such systems are now becoming available for *Staphylococcus* and can be expected to aid in resolving some of the uncertainties about the roles of various structures and products in disease. It is now well-established that specific gene loci, such as *agr* (accessory regulatory gene complex), regulate the synthesis of a number of extracellular proteins and at least one surface protein, protein A, in *S. aureus*. Further studies of regulation of virulence genes will also contribute to understanding of the response of the bacterium to existence in a variety of environments.

The vast differences between bacteria grown in vitro compared with those grown in vivo are being recognized. Further studies in model systems are now necessary to define virulence factors of in vivo-grown cells in models that simulate natural staphylococcal infections in animal hosts. Defining conditions for experimental infections are also important in order to enhance the opportunities for evaluation of experimental immunoprophylactic and treatment regimens for animal and human infections.

REFERENCES

Albus, A.; Arbeit, R. D.; and Lee, J. C. 1991. Virulence of *Staphylococcus aureus* mutants altered in type 5 capsule production. Infect Immun 59:1008-14.

Amtsberg, G. 1979. Demonstration of exfoliation-producing substances in cultures of *Staphylococcus hyicus* of pigs and *Staphylococcus epidermidis* biotype 2 of cattle. Zentralbl Veterinaermed [B] 26:257-72.

Amtsberg, G.; Bollwahn, W.; Hazem, S.; Jordan, B.; and Schmidt, U. 1973. Bacteriological, serological, and experimental studies on the aetiological significance of *Staphylococcus hyicus* in moist eczema of pigs. Dtsch Tieraerztl Wochenschr 80:493-99; 521-23.

Anderson, J. C. 1983. Veterinary aspects of staphylococci. In Staphylococci and Staphylococcal Infections. Ed. C. S. F. Easmon and C. Adlam, pp. 193-241. London: Academic Press.

Andresen, L. O.; Wegener, H. C.; and Bille-Hansen, V. 1992. Skin reactions in piglets caused by extracellular products of *Staphylococcus hyicus*. In 7th Int Symp Staphylococci and Staphylococcal Infect. Stockholm. Abstr. 91.

Aniansson, H.; Stendahl, O.; and Dahlgren, C. 1984. Comparison between luminol- and lucigenin-dependent chemiluminescence of polymorphonuclear leukocytes. Acta Path Microbiol Immunol Scand Sect C 92:357-61.

Badour, L. M.; Lowrance, C.; Albus, A.; Lowrance, J. H.; Anderson, S. K.; and Lee, J. C. 1992. *Staphylococcus aureus* microcapsule expression attenuates bacterial virulence in a rat model of experimental endocarditis. J Infect Dis 165:749-53.

Birgersson, A.; Jonsson, P.; and Holmberg, O. 1992. Species identification and some characteristics of coagulase negative staphylococci isolated from bovine udders. Vet Microbiol 31:181-89.

Bramley, A. J.; Patel, A. H.; O'Reilly, M.; Foster, R.; and Foster, T. J. 1989. Roles of alpha-toxin and beta-toxin in virulence of *Staphylococcus aureus* for the mouse mammary gland. Infect Immun 57:2489-94.

Cox, H. N.; Newman, S. S.; Roy, A. F.; and Hoskins, J. D. 1984. Species of Staphylococcus isolated from animal infections. Cornell Vet 74:124-35.

Davidson, I. 1961. Observations on pathogenic staphylococci in a dairy herd during a period of six years. Res Vet Sci 2:22-40.

deBoer, D. J. 1990. Canine staphylococcal pyoderma: Newer knowledge and therapeutic advances. Vet Med Rep 2:254-66.

Devriese, L. A. 1977. Isolation and identification of *Staphylococcus hyicus*. Am J Vet Res 38:787-92.

__________. 1986. Coagulase-negative staphylococci in animals. In Coagulase-negative Staphylococci. Ed. P. -A. Mardh and U. H. Schleifer, pp 51-57. Stockholm: Almqvist and Wiksell.

Devriese, L. A., and Oeding, P. 1976. Characteristics of *Staphylococcus aureus* strains isolated from different animal species. Res Vet Sci 21:284-91.

du Preez, J. H. 1985. Teat canal infections. Kieler Milchwirtsch Forschungsber 37:267-73.

Edwards, S. W. 1987. Luminol- and lucigenin-dependent chemiluminescence of neutrophils: Role of degranulation. J Clin Lab Immunol 22:35-39.

Forbes, D. 1968. The passage of staphylococci through the bovine teat canal. J Dairy Res 35:399-406.

Foster, T. J. 1992. The use of mutants for defining the role of virulence factors in vivo. In Molecular Biology of Bacterial Infection: Current Status and Future Perspectives. Ed C. E. Hormaeche, C. W. Penn, and C. J. Smyth, pp.173-191. Cambridge: Cambridge University Press.

Frost, A. J.; Wanasinghe, D. D.; and Woolcock, J. B. 1977. Some factors affecting selective adherence of microorganisms in the bovine mammary gland. Infect Immun 15:245-53.

Guillebeau, A. 1890. Studien über Milchfehler und Eutenenzündungen bei Rindern und Ziegen. Landwirthsch Jahrb Schweiz 4:27-39.

Hajek, V. 1976. *Staphylococcus intermedius*, a new species isolated from animals. Int J Syst Bact 26:401-8.

Hajek, V., and Marsalek, E. 1969. A study of staphylococci of bovine origin *Staphylococcus aureus* var bovis. Zentralbl Bakteriol ParasitKde (Abt I Orig) 209:154-60.

__________. 1971. The differentiation of pathogenic staphylococci and a suggestion for their taxonomic classification. Zentralbl Bakteriol ParasitKde (Abt I Orig) 217a:176-82.

Hallén Sandgren, C.; Mamo, W.; Larsson, I.; Lindahl, M.; and Björk, I. 1991. A periodate-sensitive anti-phagocytic surface structure induced by growth in milk whey, on *Staphylococcus aureus* isolated from bovine mastitis. Microb Pathog 11:211-20.

Harmon, J. R., and Langlois, B. E. 1989. Mastitis due to coagulase-negative *Staphylococcus* species. Agric Pract 10:29-34.

Hogan, J. S.; White, D. G.; and Pankey, J. W. 1987. Effects of teat dipping on intramammary infections by staphylococci other than *Staphylococcus aureus*. J Dairy Sci 70:873-79.

Hunter, A. C., and Jeffrey, D. C. 1975. Sublicinal mastitis in suckler cows. Vet Rec 96:442-47.

Jarp, J. 1991. Classification of coagulase-negative staphylococci isolated from bovine clinical and subclinical mastitis. Vet Microbiol 27:151-58.

Jonsson, P.; Lindberg, M.; Haraldsson, I.; and Wadström, T. 1985. Virulence of *Staphylococcus aureus* in a mouse mastitis model: Studies of alpha-hemolysin, coagulase, and protein A as possible virulence determinants with protoplast fusion and gene cloning. Infect Immun 49:765-69.

Lindahl, M.; Holmberg, O.; and Jonsson, P. 1990. Adhesive proteins of haemagglutinating *Staphylococcus aureus* isolated from bovine mastitis. J Gen Microbiol 136:935-39.

Mamo, W.; Lindahl, M.; and Jonsson, P. 1991a. Enhanced virulence of *Staphylococcus aureus* from bovine mastitis induced by growth in milk whey. Vet Microbiol 27:371-84.

Mamo, W.; Hallén Sandgren, C.; Lindahl, M.; and Jonsson, P. 1991b. Induction of anti-phagocytic surface properties of *Staphylococcus aureus* from bovine mastitis by growth in milk whey. J Vet Med [B] 38:401-10.

Matthews, K. R.; Harmon, R. J.; and Langlois, B. E. 1991. Effect of naturally occurring coagulase-negative staphylococci infections on new infections by mastitis pathogens in the bovine. J Dairy Sci 74:1855-59.

Mebus, C. A.; Underdahl, N. R.; and Twiehaus, M. J. 1968. Exudative epidermitis. Pathogenesis and pathology. Pathol Vet 5:146-63.

Nocard, E. 1887. Mammite gangreneuse des brébis. Ann Inst Pasteur (Paris) 1:417-28.

Olmsted, S. B., and Norcross, N. L. 1992. Effect of specific antibody on adherence of *Staphylococcus aureus* to bovine mammary epithelial cells. Infect Immun 60:249-56.

Ogston, A. 1880. Über Abscesse. Arch Klin Chir 25:588-600.

Paape, M. J.; Guidry, A. J.; Kirk, S. T.; and Bolt, D. J. 1975. Measurement of phagocytosis of ^{32}P-labeled *Staphylococcus aureus* by bovine leucocytes: Lysostaphin digestion and inhibitory effect of cream. Am J Vet Res 36:1737-43.

Perrin-Couilloud, I. 1992. Staphylocoques et mammites bovines: importance des espèches differentes de *S. aureus* problème des échecs thérapeutiques. Bull GTV 2:7-16.

Phillips, W. E. Jr., and Kloos, W. E. 1981. Identification of coagulase-positive *Staphylococcus intermedius* and *Staphylococcus hyicus* subsp. *hyicus* isolated from veterinary clinical specimens. J Clin Microbiol 14:671-73.

Russel, M. W.; Brooker, B. E.; and Reiter, B. 1976. Inhibition of bactericidal activity of bovine polymorphonuclear leucocytes and related systems by casein. Res Vet Sci 20:30-35.

Sato, H.; Tanabe, T.; Kuramoto, M.; Tanaka, K.; Hashimoto, T.; and Saito, H. 1991. Isolation of exfoliative toxin from *Staphylococcus hyicus* subsp. *hyicus* and its exfoliative activity in the piglet. Vet Microbiol 27:263-75.

Tsujimoto, M.; Yokota, S.; Vilaek, J.; and Weissman, G. 1986. Tumor necrosis factor provokes superoxide anion generation from neutrophils. Biochem Biophys Res Commun 137:1094-1100.

Wadstrom, T. 1993. Department of Medical Microbiology, Lund, Sweden. Unpublished data.

Watson, D. L. 1989. Expression of a pseudocapsule by *Staphylococcus aureus*: influence of cultural conditions and relevance to mastitis. Res Vet Sci 47:152-57.

Wegener, H.C., and Skov-Jensen, E. W. 1992. A longitudinal study of *Staphylococcus hyicus* colonization of vagina of gilts and transmission to piglets. Epidemiol Infect 109:433-44.

3 / *Bacillus anthracis*

BY **J. W. EZZELL, JR.** AND **C. L. WILHELMSEN**

***B**ACILLUS ANTHRACIS* is the etiologic agent of anthrax, a disease recognized for centuries as an economically important zoonotic infection. The disease is infrequent in humans where it usually manifests itself as cutaneous lesions following contact with infected animals, carcasses, or animal products (i.e., hides, wool, meat). The appearance of these lesions gives rise to the term "anthrax" (Greek, coal) which denotes the black coallike appearance of crusts or eschars and the dark, almost black appearance of the blood of animals that die of the disease. Anthrax is historically important in that through the work of Koch in the mid-1800s, it was the first of all infectious diseases of both humans and animals in which the agent was definitively demonstrated as a specific microorganism. Through the studies of Pasteur, Chamberland, and Roux, anthrax was also one of the first infectious diseases for which an efficacious bacterial vaccine was made. Each year anthrax is responsible for the deaths of thousands of domesticated and wild herbivorous animals in enzootic areas. Parts of Africa, Asia, southern Europe, Australia, and North and South America continue to be subject to repeated outbreaks of anthrax. Incidence of disease varies with climate, soil, and efforts made to control it through vaccination and animal husbandry. Conditions conducive to maintenance of the anthrax bacillus and survival of its spores are temperate and alkaline soils with much organic matter and poor drainage.

Over the past two decades, anthrax has been the single most important cause of large-scale mortality in some antelope species and zebra in southern Africa. The rare roan antelope is threatened by anthrax in the Kruger National Park; and in Etosha National Park, anthrax has caused a dramatic decrease in the blue wildebeest and zebra populations, prompting immunization programs. Deaths of over 4000 hippopotami were reported in Zambia in 1987 and 1988, with additional losses reported in elephants and cape buffalo (Turnbull et al. 1991). The largest affected area in the United States extends from North Dakota and Montana down to Texas and the Mississippi River delta region. However, over the past decades, anthrax has been so well controlled in livestock through vaccination and improvements in animal husbandry

---

The views expressed in this chapter do not purport to reflect the positions of the Department of the Army or the Department of Defense (Para. 4,3, AR360-5).

that the disease attracts little public attention in the western hemisphere. A survey of the incidence of anthrax in livestock in the United States, conducted for the 5-year span from 1984 to 1988 (Whitford 1990), found that anthrax was diagnosed in 11 states. Those states (number of isolations) are as follows: Arkansas (24), Texas (20), South Dakota (7) Mississippi (6), Colorado (4), North Dakota (2) Louisiana (2), Alabama (1), Iowa (1), Montana (1) and Nevada (1). The estimated total number of livestock lost by species during this 5-year period was at least 700 cattle, 555 sheep, 8 horses and mules, 3 llamas, 1 pig, 1 dog, and 1 goat. From 1986 to 1991, Texas lost thousands of white-tailed deer; Mississippi had major outbreaks in deer and livestock in 1991 and 1992; and Nevada reported the loss of 15 head of cattle in 1992.

CHARACTERISTICS OF THE BACTERIUM

Bacillus anthracis is a large, gram-positive, rod-shaped, endospore-forming bacterium approximately 1.0-1.3 μm in diameter and 3-10 μm in length. On routine agar cultures such as 5% sheep blood agar or nutrient agar, *B. anthracis* grows as long chains that form rough colonies. The colonies have slightly serrated edges and a ground glass appearance and may display protrusions from their periphery, resulting in the classically described "medusa-head" or "spiked" appearance. However, within the diseased host, the bacilli occur singly or in chains of two or three in blood smears and produce a capsule composed of poly-D-glutamic acid. Capsule formation is also induced in culture under 5-10% CO_2 on media containing 0.5% sodium bicarbonate or serum. The presence of capsule gives rise to the formation of smooth, mucoid, convex colonies with entire edges. *B. anthracis* is nonhemolytic on sheep blood agar, which distinguishes it from the closely related species *B. cereus* and *B. thuringiensis*, which are typically beta-hemolytic. The anthrax bacillus also produces a highly specific N-acetylglucosamine/galactose polysaccharide in its cell wall (Ezzell et al. 1990a). Monoclonal antibody-fluorescein conjugates to both the capsule and polysaccharide are used to identify *B. anthracis* vegetative forms.

Traits required for virulence of *B. anthracis* within the host or for survival in the environment are regulated by CO_2 (Bartkus 1989). Where CO_2 levels are elevated (as within the host), the virulence factors (capsule and toxins) are induced and sporulation is inhibited. Conversely, when the atmospheric levels of CO_2 are sufficiently low (as when the diseased carcasses are opened by scavengers), sporulation occurs, allowing survival of the organism in the environment. The spores (0.75 μm by 1.0 μm) are oval, light refractive under phase microscopy, and resistant to drying, cold, heat, and disinfectants; spores remain viable for many years in soil, water, and animal hides and products.

SOURCES OF THE ORGANISM

The modes of transmission and spread of anthrax within endemic areas are by grazing on contaminated pastures, feeding activities of blood-sucking flies and ticks, drinking from dirty stagnant pools of water, and contamination of pasture by migrating nomadic herds. Spores that are consumed by flies feeding on infected carcasses are passed intact in feces deposited on leaves and foliage; these spores infect other grazing and foraging animals. Animals may become infected following ingestion of vegetation contaminated through insect feces or spore-laden soil associated with roots. Infection is initiated by penetration of mucous membranes by the spores; this event is facilitated by defects in the epithelium resulting from scratches from tough, fibrous foods. Occasionally, domesticated or captive wild animals are infected by ingesting contaminated feedstuffs such as spore-contaminated bonemeal or meat contaminated with spores or vegetative forms. Wild carnivores consuming infected prey often die; however, some survive and become immune through repeated sublethal exposure. There is evidence that rainwater can carry anthrax spores from soil, contaminated by diseased animals and their remains left by scavengers, into water holes where they settle to the bottom. Vultures may also contaminate water holes by defecation and regurgitation into the water following consumption of infected carcasses. Vos (1990) postulated that during times of drought, anthrax spores, possibly ingested during consumption of contaminated water remaining at the bottom of pools, may lead to anthrax outbreaks, which may serve an ecological role by reducing animal populations during times of low rainfall. During heavy rainfall in poorly drained areas, spores may be leached from the soil or suspended in puddles and ingested by animals.

BACTERIAL VIRULENCE FACTORS

The anthrax bacillus possesses three primary, plasmid-encoded virulence factors: a poly-D-glutamic acid capsule, encoded by the 60-MDa pXO2 plasmid (Green et al. 1985; Uchida et al. 1985); and lethal and edema toxins, encoded by the 110-MDa pXO1 plasmid (Leppla et al. 1985; Mikesell et al. 1983; Uchida et al. 1986). Lethal toxin is composed of two proteins: an 82-kDa protein called lethal factor (LF) and an 83-kDa protein called protective antigen (PA). Edema toxin is composed of an 89-kDa edema factor (EF), a calmodulin-dependent adenylate cyclase, and PA (Beall et al. 1962; Fish and Lincoln 1967; Leppla 1984; Leppla et al. 1985). British workers refer to EF, PA, and LF as factors I, II, and III, respectively (Stanley and Smith 1963). An excellent survey of anthrax toxin and molecular studies up to 1990 has been published (Stepanov 1991).

Experimental intradermal injection of a PA-EF mixture (edema toxin) produces edema in guinea pigs and rabbits. Edema toxin is presumed to be responsible for the edema seen around cutaneous lesions and other sites of infection. Intravenous injection of a PA-LF combination (lethal toxin) causes death. The Fischer 344 rat is the most sensitive to lethal toxin (Beall et al. 1962; Beall and Dalldorf 1966). Neither LF nor EF is active without PA. On the molecular level, Leppla et al. (1988) reported that the 83-kDa PA (PA83) does not bind LF or EF. Upon binding to surface receptors of target cells in vitro, an unidentified, cell-associated protease cleaves the 83-kDa PA, releasing a 20-kDa fragment (O'Brien et al. 1985; Friedlander 1986, 1990). The activated 63-kDa (PA63) fragment retained at the cell surface binds LF or EF, and the complex enters the cell by endocytosis (Leppla et al. 1988). Ezzell and Abshire (1992) demonstrated that PA exists in the blood of infected animals primarily as a 63-kDa protein complexed with LF. Cleavage of PA from 83- to 63-kDa is catalyzed by a calcium-dependent, heat-labile serum protease activity. Other than being complexed to PA, LF appears to be unaltered during the course of anthrax infection. The ubiquitous protective antigen-cleaving protease has been identified in primates, horses, goats, sheep, dogs, cats, and rodents.

During the early to mid-1900s researchers reported that *B. anthracis* had aggressin activities, which enabled the organism to evade the host's immune system during infection, resulting in enormous numbers of bacilli throughout the body at death. The aggressin activities were due in part to the virulence factors. Crude mixtures of the three toxin proteins decreased host resistance to challenge with *B. anthracis* spores and inhibited phagocytosis in vitro (Keppie et al. 1963). Showing EF to be an inactive adenylate cyclase that is activated by eukaryotic host-cell calmodulin (Leppla 1982, 1984), explained the observations made by Keppie and others. Polymorphonuclear neutrophil phagocytosis and oxidative metabolism, measured by chemiluminescence, is inhibited by elevated 3′, 5′ cyclic-adenosine monophosphate, resulting from the combination of PA and EF but not by other combinations of the three toxin components (O'Brien et al. 1985).

Edema toxin is implicated in other pathophysiologic manifestations present during systemic anthrax infections. Observed progressive hyperglycemia associated with marked decreases in liver and muscle glycogen is followed by severe terminal hypoglycemia (Fish et al. 1968). Changes in glycogen metabolism may reflect alterations in cAMP concentrations in liver and possibly other tissues.

The less well understood biologic effect of lethal toxin appears to be directed in part toward inhibition of macrophage function (Friedlander 1986, 1990). Protective antigen-lethal factor mixtures are highly toxic to mouse macrophages in vitro. The precise biologic or enzymatic basis for the toxicity of LF is unknown. Fischer 344 rats, injected with lethal toxin components, exude froth from the nares and mouth immediately before death. Analysis of this froth shows the presence of high concentrations of serum proteins, indicative of pulmonary edema, thereby supporting a premise that lethal toxin may alter membrane permeability (Fish and Lincoln 1968). Flux of body fluids into the tissues and body cavities, such as the lungs and peritoneal cavity, explains in part the increased hematocrit (hemoconcentration) observed in animals with anthrax. Death is attributed to respiratory distress resulting from fluid flux into the lungs and to vascular impediment due to excessive pressure on thoracic blood vessels from mediastinal edema (Smith and Stoner 1967). Pulmonary edema, circulatory collapse, oxygen utilization by the massive numbers of circulating bacilli, and lysis of red blood cells (discussed later) all contribute to the extreme hypoxia observed in dying animals.

Although edema and lethal toxins impair the immune system by damaging cells, the major mechanism for evasion of the host immune defense is the *B. anthracis* capsule. The highly negatively charged poly-D-glutamate capsule physically inhibits phagocytosis and interferes with opsonization (Keppie et al. 1963). The role of the capsule was also shown in studies by O'Brien et al. (1985), in which bacteria of the nonencapsulated Sterne strain, but not of the encapsulated Vollum strain, were opsonized with guinea pig complement.

INTERACTION WITH THE HOST

Anthrax is separated into two clinical forms: cutaneous and septicemic (Lincoln and Fish 1970). The cutaneous form is characterized by an intensely dark, relatively painless edematous lesion, that usually forms a black eschar. The lesions rapidly become sterile after antibiotic therapy and require an extended period of time to resolve, even with treatment. The cutaneous form has been reported only in humans, rabbits, swine, and horses.

The septicemic form can arise from various initial sites of infection, including cutaneous, oropharyngeal, gastrointestinal, or pulmonary infections. The course of disease is peracute, characterized by pronounced bacteremia and toxemia. Death is sudden in most herbivorous animals and carcasses exude tarry blood from the body orifices. The course of the disease depends largely on the portal of entry and the susceptibility of the animal species. The incubation period varies from 1 to 14 days, usually being 2 to 3 days. Cattle and sheep are most susceptible, with horses and goats slightly less so. Humans are intermediate in their susceptibility to infection, whereas pigs and strict carnivores are relatively resistant. Carnivorous birds are highly resistant and poultry are almost totally resistant, a trait that has been attributed to their higher body temperature.

The natural occurrence of pulmonary, or inhalation, anthrax in animals is open to speculation. Pulmonary anthrax is known to occur only in humans. Until the mid-1900s, pulmonary anthrax (once termed "wool-sorter's disease") was most commonly associated with humans who worked in wool-processing mills. Animals could inhale spore-laden dust and contract the disease, but this hypothesis has not been proven in controlled experiments. To date, evidence from laboratory animals indicates that inhaled spores do not germinate within the lungs, and respiratory infection does not occur. Instead, inhalation anthrax is initiated by phagocytosis of spores in the lungs by alveolar macrophages. Spore-laden macrophages pass through lymphatic channels to the sinuses of regional lymph nodes of the mediastinum and thoracic cavity where the spores germinate within the macrophages, multiply as vegetative cells, overwhelm and escape from macrophages to invade the efferent lymphatics, and eventually enter the blood.

The spleen is the principal defense against circulating bacilli during systemic anthrax. Pronounced splenomegaly is a characteristic finding at necropsy of animals dead of anthrax, and impression smears of spleen reveal enormous numbers of bacilli. This splenic line of defense is quickly overcome and bacteria circulate in the bloodstream, followed by secondary sites of infection, massive bacteremia, toxemia, and death. Studies by the author and by Turnbull (1990) show that in some laboratory animals (guinea pigs and rhesus monkeys), the bacilli in the blood may reach counts as high as 5×10^9 colony-forming units /ml, with lethal toxin levels in excess of 50 μg/ml of blood. Studies by the author have demonstrated that lethal toxin and bacilli appear in the blood almost simultaneously. Failure of the blood to clot, hemorrhages of skin, hemorrhagic meningitis, and absence of rigor mortis are important necropsy findings.

Systemic infections may stem from cutaneous or gastrointestinal infections. Although rare, vegetative cells from cutaneous lesions may spread to local lymph nodes, with progression of the disease being much the same as described with pulmonary anthrax. Progression of gastrointestinal anthrax to systemic infection is somewhat complex in that foodstuff may contain various combinations of spores, vegetative cells, and toxins. Therefore, some of the effects of toxins and vegetative cells within the gastrointestinal tract may be more direct (i.e., edema, hemorrhage) rather than as a result of the events following spore germination in regional mesenteric lymph nodes. The estimated case fatality rate of untreated gastrointestinal infections is 25-35%; however, these figures are only speculative.

Cattle, sheep, and swine contract anthrax primarily through ingestion of contaminated feed. The clinical progression is acute in cattle, sheep, and horses, and subacute in swine. Following ingestion and an incubation period of 3-7 days in cattle and sheep, the disease often runs an acute course. Typically, the first indication of disease in a herd or flock is sudden death with others possibly showing weakness, inappetence, high elevated temperatures, and convulsions followed by death. Clinical symptoms of disease include septicemia and pockets of edema. After death, bloody discharges from the mouth, anus, and vulva are common. Carcasses of anthrax-stricken animals are characterized by delayed or incomplete rigor mortis and blood clotting; dark (unoxygenated) blood; dark, blood-stained urine; rapid decomposition; and bloating. In the author's experience, urine of guinea pigs that were experimentally infected by intramuscular injection and developed fatal anthrax contained high concentrations of hemin, no intact red blood cells, and few or no anthrax bacilli. Virtually all of the guinea pig's red blood cells had apparently undergone hemolysis as a terminal event, and hemin released from the lysed erythrocytes was removed by the kidneys and excreted in the urine.

An acute clinical course of anthrax usually develops in horses that ingest *B. anthracis*. However, when infection originates with an insect bite, the disease is then characterized by rapidly progressing subcutaneous edema of areas on the neck, thorax, abdomen, and mammary glands. Goats and humans may become infected through bites of flies and other insects that carry the organism subsequent to feeding on infected animals and carcasses; this is not a major source of infection (Van Ness 1971).

Anthrax in swine is more likely to be subacute. Edema of the throat and neck is sometimes a prominent feature, following consumption of contaminated feedstuffs. Swelling of the throat can produce suffocation, whereas the external edematous areas are hot and usually not painful when palpated. With pharyngeal involvement, blood-stained froth may be discharged from the mouth.

Anthrax in carnivores clinically resembles anthrax in swine. The disease usually results from eating infected meat of carcasses containing high levels of bacilli and toxin. In dogs, pharyngeal or oral anthrax produces swelling about the head and throat; alimentary involvement is typically manifested as severe gastroenteritis. Older dogs are usually less susceptible.

Birds, reptiles, and fish are not susceptible to anthrax under natural conditions (Stein 1960), partially because their normal body temperatures are outside the optimum range required by anthrax bacilli.

IMMUNITY

Besides its central role in binding the two toxins by target cells, PA also plays an important role in the host's protective immune response against anthrax (Klein et al. 1962; Puziss and Wright 1963; Ivins et al. 1986). No vaccine lacking PA has been found to be protective (Hambleton et al. 1984; Ivins et al. 1986). In British and American human anthrax vaccines, PA bound to aluminum salts is the principal antigen (Brachman et al. 1962; Puziss and Wright 1963). The molecular mass of PA in these vaccines is primarily 83 kDa plus some breakdown products of 36 and 47 kDa in some lots. However, veterinary vaccines are composed of viable spores of *B. anthracis* Sterne strain, a nonencapsulated toxigenic variant (Sterne 1939). Full protection against anthrax with the veterinary vaccine is afforded by a primary vaccination followed by annual booster vaccinations. Although protective immunity is associated with PA in both human and veterinary vaccines, studies with laboratory animals show no definitive correlation between antibody titers to PA and the degree of protection. Guinea pigs vaccinated with the human vaccine develop significantly higher titers than those that received the veterinary vaccine, yet they are less well protected against lethal challenge (Little and Knudson 1986; Broster and Hibbs 1990). The in vitro-produced PA bound to aluminum salts (human vaccine) may have differed from the in vivo-produced PA (veterinary vaccine) with regard to its structural form, antigen processing, and/or presentation to the host's immune system. The effect of these differences on the type of host immune response (i.e., cell-mediated vs. humoral immunity) is unknown.

Studies have also been carried out to determine if surface antigens of *B. anthracis* contribute to stimulation of protective immunity in guinea pigs (Ezzell and Abshire 1988). Although animals vaccinated with live-spore vaccine demonstrated that a surface protein was the dominant antigen to which animals responded, the antibody that was generated was not protective. Similarly, antibodies against the N-acetylglucosamine/galactose polysaccharide in the cell wall of the organism were not protective.

REFERENCES

Bartkus, J. M., and Leppla, S. H. 1989. Transcriptional regulation of the protective antigen of *Bacillus anthracis*. Infect Immun 57:2295-300.

Beall, F. A., and Dalldorf, F. G. 1966. The pathogenesis of the lethal effect of anthrax toxin in the rat. J Infect Dis 116:377-89.

Beall, F. A.; Taylor, N. J.; and Thorne, C. B. 1962. Rapid lethal effect in rats of a third component found upon fractionating the toxin of *Bacillus anthracis*. J Bacteriol 83:1274-80.

Brachman, P. S.; Gold, H.; Plotkin, S. A.; Fekety, F. R.; Werrin, M.; and Ingraham, N. R. 1962. Field evaluation of a human anthrax vaccine. Am J Public Health 52:632-45.

Broster, M. G., and Hibbs, S. E. 1990. Protective efficacy of anthrax vaccines against aerosol challenge. In Proc Int Workshop on Anthrax, Salisbury Med Bull No. 68. Ed. P. C. B. Turnbull, pp. 91-92. Wiltshire, UK: Salisbury Printing Co.

Ezzell, J. W., and Abshire T. G. 1988. Immunological analysis of cell-associated antigens of *Bacillus anthracis*. Infect Immun 56:349-56.

__________. 1992. Serum protease cleavage of *Bacillus anthracis* protective antigen. J Gen Microbiol 138:543-49.

Ezzell, J. W.; Abshire, T. G.; Little, S.; and Brown, C. 1990a. Identification of *Bacillus anthracis* by using monoclonal antibody to cell wall galactose/N-acetyl-glucosamine polysaccharide. J Clin Microbiol 28:223-31.

__________. 1990b. Analyses of *Bacillus anthracis* vegetative cell surface antigens and of serum protease cleavage of protective antigen. In Proc Int Workshop on Anthrax, Salisbury Med Bull No. 68. Ed. P. C. B. Turnbull, pp. 43-44. Wiltshire, UK: Salisbury Printing Co.

Fish, D. C., and Lincoln, R.E. 1967. Biochemical and biophysical characterization of anthrax toxin. Fed Proc 26:1534-38.

__________. 1968. In vivo-produced anthrax toxin. J Bacteriol 95:919-24.

Fish, D. C.; Klein, F.; Lincoln, R. E.; Walker, J. S.; and Dobbs, J. P. 1968. Pathophysiological changes in the rat associated with anthrax toxin. J Infect Dis 118:114-24.

Friedlander, A. M. 1986. Macrophages are sensitive to anthrax lethal toxin through acid- dependent process. J Biol Chem 261:7123-26.

__________. 1990. The anthrax toxins. In Trafficking of Bacterial Toxins. Ed. C. Saelinger, pp. 121-29. Boca Raton, Fla.: CRC Press.

Green, B. D.; Battisti, L.; Koehler, T. M.; Thorne, C. B.; and Ivins, B. E. 1985. Demonstration of a capsule plasmid in *Bacillus anthracis*. Infect Immun 49:291-97.

Hambleton, P.; Carman, J. A., and Melling, J. 1984. Anthrax: The disease in relation to vaccines. Vaccine 2:125-32.

Ivins, B. E.; Ezzell, J. W.; Jemski, J.; Hedlund, K. W.; Ristroph, J. D.; and Leppla, S. H. 1986. Immunization studies with attenuated strains of *Bacillus anthracis*. Infect Immun 52:454-58.

Keppie, J.; Harris-Smith, P. W.; and Smith, H. 1963. The chemical basis of the virulence of *Bacillus anthracis*. 9. Its aggressins and their mode of action. Brit J Exp Pathol 44:446-53.

Klein, F. I.; Dearmon, A.; Lincoln, R. E.; Mahlandt, B. G.; and Fernelius, A. L. 1962. Immunological studies of anthrax. 2. Levels of immunity against *Bacillus anthracis* obtained with protective antigen and live vaccine. J Immunol 88:15-19.

Leppla, S. H. 1982. Anthrax toxin edema factor: A bacterial adenylate cyclase that increases cyclic AMP concentrations in eukaryotic cells. Proc Natl Acad Sci USA 79:3162-66.

__________. 1984. *Bacillus anthracis* calmodulin-dependent adenylate cyclase: Chemical and enzymatic properties and interaction with eukaryotic cells. In Advances in Cyclic Nucleotide and Protein Phosphorylation Research, Vol. 17. Ed. P. Greengard and G. A. Robison, pp. 189-98. New York: Raven Press.

Leppla, S. H.; Ivins, B. E.; and Ezzell, J. W. 1985. Anthrax toxin. In Microbiology 1985. Ed. L. Leive, pp. 63-66. Washington, D.C.: American Society for Microbiology.

Leppla, S. H.; Friedlander, A. M.; and Cora, E. M. 1988. Proteolytic activation of anthrax toxin bound to cellular receptors. In Bacterial Protein Toxins. Ed. F. Fehrenbach, J. E. Alouf, P. Falmagne, W. Goebel, J. Jelajaszewicz, D. Jurgens, and R. Rappuoli, pp. 111-12. Stuttgart: Gustav Fischer.

Lincoln, R. E., and Fish, D. C. 1970. Anthrax toxin. In Microbial Toxins, Vol. 3. Ed. T. Monte, S. Kadis, and S. Aji, pp. 361-414. New York: Academic Press.

Little, S. F., and Knudson, G. B. 1986. Comparative efficacy of *Bacillus anthracis* live spore vaccine and protective antigen vaccine against anthrax in the guinea pig. Infect Immun 52:509-12.

Mikesell, P.; Ivins, B. E.; Ristroph, J. D.; and Dreier, T. M. 1983. Evidence for plasmid-mediated toxin production in *Bacillus anthracis*. Infect Immun 39:371-76.

O'Brien, J.; Friedlander, A.; Dreier, T.; Ezzell, J.; and Leppla, S. 1985. Effects of anthrax toxin components on human neutrophils. Infect Immun 47:306-10.

Puziss, M., and Wright G. G. 1963. Studies on immunity in anthrax. 10. Gel-adsorbed protective antigen for immunization of man. J Bacteriol 85:230-36.

Smith, H., and Stoner, H. B. 1967. Anthrax toxic complex. Fed Proc 26:1554-57.

Stanley, J. L., and Smith, H. 1963. The three factors of anthrax toxin: Their immunogenicity and lack of demonstrable enzymic activity. J Gen Microbiol 31:329-37.

Stein, C. D. 1960. Anthrax. US Dept Agric Farm Bull 1736.

Stepanov, A. S. 1991. The molecular nature of *Bacillus anthracis* toxin. Mol Gen Microbiol Virusol 1:1-9.

Sterne, M. 1939. The use of anthrax vaccines prepared from avirulent (unencapsulated) variants of *Bacillus anthracis*. Onderstepoort J Vet Sci Anim Ind 13:307-12.

__________. 1959. Anthrax. In Infectious Diseases of Animals: Diseases Due to Bacteria. Ed. A. W. Stableforth and A. Galloway, pp. 16-52. New York: Academic Press.

Turnbull, P. C. B. 1990. Terminal bacterial and toxin levels in the blood of guinea pigs dying of anthrax. In Proc Int Workshop on Anthrax, Salisbury Med Bull No. 68. Ed. P. C. B. Turnbull, pp. 53-55. Wiltshire, UK: Salisbury Printing Co.

Turnbull, P. C. B.; Bell, R. H.; Saigawa, K.; Munyenyembe, F. E. C.; Mulenga, C. K.; and Makala, L. H. C. 1991. Anthrax in wildlife in the Luangwa Valley, Zambia. Vet Rec 128:399-403.

Uchida, I.; Sekizaki, T.; Hashimoto, K.; and Terakado, N. 1985. Association of the encapsulation of *Bacillus anthracis* with a 60 megadalton plasmid. J Gen Microbiol 131:363-67.

Uchida, I.; Hashimoto, K.; and Terakado, N. 1986. Virulence and immunogenicity in experimental animals of *Bacillus anthracis* strains harbouring or lacking 110-MDa and 60-MDa plasmids. J Gen Microbiol 132:557-59.

Van Ness, G. B. 1971. Ecology of anthrax. Science 172:1303-7.

Vos, V. de. 1990. The ecology of anthrax in the Kruger National Park, South Africa. In Proc Int Workshop on Anthrax, Salisbury Med Bull No. 68. Ed. P. C. B. Turnbull, pp. 19-23. Wiltshire, UK: Salisbury Printing Co.

Whitford, H. W. 1990. Incidence of anthrax in the USA: 1945-1988. In Proc Int Workshop on Anthrax, Salisbury Med Bull No. 68. Ed. P. C. B. Turnbull, pp. 5-7. Wiltshire, UK: Salisbury Printing Co.

4 / *Mycobacterium*

BY C. O. THOEN AND R. CHIODINI

MYCOBACTERIA are rod-shaped, acid-fast bacilli with a high lipid content in their cell wall, which has been the focus of much attention. Although numerous cell wall components of virulent mycobacteria have been isolated and identified, there is no definitive information on the role of these components in the pathogenesis of disease due to mycobacteria (Rastogi 1991).

Pathogenic mycobacteria produce granulomatous lesions in tissues of human beings and a wide range of domestic and wild animal species (Thoen and Williams 1993). *Mycobacterium bovis*, a slow-growing nonphotochromogenic organism is the etiologic agent of bovine tuberculosis and causes disease in a wide range of domestic animals (Table 4.1). Unexplained differences in susceptibility of different animals to various acid-fast bacilli occur. Guinea pigs are susceptible to *M. bovis* and *M. tuberculosis* and develop progressive lesions following experimental exposure; however, rabbits are susceptible to *M. bovis* and quite resistant to *M. tuberculosis* (Thoen and Karlson 1984). *M. tuberculosis*, the human tubercle bacillus, produces progressive generalized disease in nonhuman primates, dogs, swine, and certain exotic animals; cats and cattle are quite resistant. *M. tuberculosis* may also induce tuberculin skin sensitivity in cattle and other animals.

M. avium complex (previously referred to as *M. avium-M. intracellulare*) has the widest host range of mycobacteria. It has been suggested that glycolipids, peptidolipids, or glycopeptidolipids that accumulate on the periphery of the *M. avium* cell envelope contribute to pathogenicity (Barrow 1991). *M. avium* complex serovars 1, 2, and 3 are typical avian tubercle bacilli isolated from tuberculous lesions in avian species (Thoen and Karlson 1991), whereas strains of *M. avium* complex serovars 4-21 produce only minimal disease (microscopic foci in liver and spleen) in chickens inoculated intravenously or intraperitoneally. Interestingly, whereas birds are susceptible to *M. avium* complex serovars 1-3, they are very resistant to infection with *M. bovis* or *M. tuberculosis*. In birds, disease is usually progressive, with lesions in the liver and spleen; in nonhuman primates, cattle, and swine, infection due to *M. avium* complex is usually confined to lymph nodes associated with the intestinal tract. Rabbits are highly susceptible to experimental infection with *M. avium* serovars 1 and 2, but relatively resistant to other serovars of *M. avium* complex.

TABLE 4.1. Some animals susceptible to mycobacterial infection

Organism	Species
M. avium complex	Chickens, birds, swine, monkeys, great apes, cattle, sheep, goats, dogs, cats, reptiles, amphibians, fish, mink, horses
M. bovis	Cattle, goats, dogs, cats, monkeys, great apes, mink, swine, elephants, rhinoceri, foxes, kudus, parrots, deer, elk, tapirs, camels, bison, llamas, giraffes, nilghais, rabbits
M. tuberculosis	Monkeys, great apes, swine, elephants, dogs, oryxes, tapirs
M. paratuberculosis	Cattle, sheep, goats, deer, elk, auodads, mouflons, Siaga antelope, duikers, bighorn sheep, mountain goats
M. lepraemurium	Cats, rats, mice
M. leprae	Armadillos
M. fortuitum	Dogs, cattle, swine
M. marinum	Fish, toads
M. kansasii	Monkeys, cattle, swine
M. chelonei	Swine, manatees

Note: *M. xenopi, M. microti,* and other mycobacteria have been isolated from other animals.

M. paratuberculosis is the cause of a transmissible intestinal disorder of ruminants commonly known as paratuberculosis (Chiodini et al. 1984). Cattle, sheep, goats, and certain exotic ruminants are susceptible to *M. paratuberculosis*, but horses and swine fail to develop clinical disease following experimental exposure. A characteristic that is useful in differentiating this species is dependency on mycobactin, an iron-chelating agent, for in vitro growth. Mycobactin was initially extracted from *M. phlei*, but later, mycobactin J and certain extracellular iron-binding compounds were isolated from *M. paratuberculosis* (Barclay and Ratledge 1983). Mycobactin-dependence has also been reported for certain strains of *M. avium*; therefore, more definitive methods such as polymerase chain reaction and restriction endonuclease analysis have been developed for identifying *M. paratuberculosis* (Labidi and Thoen 1989; McFadden et al. 1990).

Other species of mycobacteria have been isolated from various animals. *M. fortuitum*, a rapid-growing, nonchromogenic organism has been isolated from dogs with lung lesions, cattle with mastitis, and from lymph nodes of slaughter cattle and swine. Granulomatous lesions in swine and cattle, which closely resemble lesions caused by *M. bovis*, have been reportedly caused by *M. kansasii*, a slow-growing, photochromogenic organism. *M. lepraemurium* has been isolated from leprosylike lesions in cats, rats, and mice, whereas *M. leprae*, the cause of leprosy in humans, has been isolated from armadillos in the United States. *M. marinum, M. intracellulare, M. nonchromigenicum, M. chelonei*, and certain other opportunistic mycobacteria have been isolated from granulomatous lesions in cold-blooded animals (Thoen and Schliesser 1984).

VIRULENCE

Development of mycobacterial disease in animals depends on the ability of mycobacteria to survive and multiply within macrophages of the host. Pathogenicity of mycobacteria is a multifactorial phenomenon, requiring the participation and cumulative effects of several components (Laneelle and Daffe 1991). The structure and biologic functions of the glycolipid-containing cell wall of mycobacteria have been investigated; however, an understanding of the functions of each cell well component in pathogenesis is unclear (McNeil and Brennan 1991). The core of the mycobacterial cell wall is composed of three covalently attached molecules, peptidoglycan, arabinogalactan, and mycolic acid. Glycolipids and lipid complexes in the cell wall of virulent tubercle bacilli, and other cellwall components (i.e., cord factor) of virulent and attenuated strains of mycobacteria, have been extensively examined to evaluate their significance in granuloma formation (Thoen and Himes 1986). Trehalose-6,6′dimycolate inhibits chemotaxis, induces disintegration of the rough endoplasmic reticulum and detachment of ribosomes in liver cells, and is leukotoxic. Sulfur-containing glycolipids (sulfatides) appear to promote the survival of virulent tubercle bacilli within macrophages by inhibiting phagolysosome formation and avoiding exposure to hydrolytic enzymes present in the lysosomes. More recently it has been reported that sulfolipids induce changes in phagocytic cell function that may be important in decreasing the ability of phagocytes to respond efficiently to *M. tuberculosis* (Zhang et al. 1988).

Tubercle bacilli and other mycobacteria contain several proteins and protein complexes (i.e., lipoproteins). Of particular interest are the secreted proteins in the antigen 85 complex, since they may play a role in development of cell-mediated responses and disease in the host (Andersen et al. 1991). Fibronectin binds to antigen 85 components, and the release of large amounts of this antigen could inhibit binding of fibronectin to tubercle bacilli (Abou-Zeid et al. 1988; Wiker and Harboe 1992). Although little or no evidence exists that fibronectin directly mediates phagocytosis of mycobacteria, there is some evidence that fibronectin enables monocytes to phagocytose C3b-sensitized cells (Pommier et al. 1983). This finding may be of importance, since complement receptors mediate phagocytosis of *M. tuberculosis* (Schlesinger et al. 1990).

Stress proteins produced by mycobacteria have been investigated for their role in protection against the host immune response (Kaufmann et al. 1990). Several stress proteins (heat shock proteins) have been identified as major immunodominant antigens of mycobacteria (Shinnick et al. 1987; Young et al. 1990). The elevated synthesis of stress proteins in response to changes in physiological conditions within the intracytoplasmic vacuole (phagolysosome) may protect mycobacteria from hydrolytic enzymes, reactive oxygen radicals (i.e., superoxide anion) and myeloperoxidase-killing mechanisms. Superoxide dismutase (SOD), produced and released by several mycobacterial pathogens, could protect the organisms from the toxic effects of reactive oxygen radicals generated during the oxidative burst by host macrophages (Andersen et al. 1991). The role of heat shock proteins; lipoproteins; and cytoplasmic, membrane-bound and secreted proteins are areas of intense research activity, which may provide more definitive information on the virulence of tubercle bacilli and other mycobacteria (Young and Elliot 1989).

IMMUNE RESPONSE

Cell-mediated immune (CMI) responses including specific T lymphocytes and activated mononuclear macrophages, are important in host resistance to virulent acid-fast bacilli (Rook 1988). Various cytokines (i.e., gamma interferon) have been shown to activate tuberculostatic macrophage functions and limit the replication of mycobacteria; however, the mechanism involved in killing of mycobacteria is unclear (Flesch and Kaufmann 1990; Denis 1991). The importance of active oxygen metabolites, such as superoxide anion or singlet oxygen, has been investigated, but convincing evidence is not available regarding the significance of these components in protection of the host. Recently, reactive nitrogen intermediates (i.e. nitrogen oxide) produced by peritoneal macrophages have been proposed to be important in killing *M. bovis* BCG and virulent *M. tuberculosis* (Flesch and Kaufmann 1991; Chan et al. 1992).

Numerous host mechanisms have been associated with susceptibility and development of disease in animals exposed to virulent mycobacteria. Phagocytosis is influenced by the presence of complement and complement receptors on monocytes (Schlesinger et al. 1990). Mature macrophages have receptors for both C3b and the cell-binding (Fc) portion of immunoglobulin G (IgG). Although neutrophils initially enter the infection site, activities of mononuclear macrophages are considered to be more important in protecting the host against pathogenic mycobacteria (Rook 1988). Macrophages may be activated after an encounter with a specific acid-fast organism or as a result of stimulation by cytokines, such as IL-2 and IFN gamma. Macrophages are also involved in processing of mycobacterial antigens and in presenting antigens to T lymphocytes, which are an important recognition unit in the immune response to mycobacteria. The interaction of lymphocytes with specific antigens stimulates the release of lymphokines that attract, activate, and increase the number of mononuclear cells at the infection site (Jeevan and Asherson 1988). Mononuclear macrophages are long lived and may reenter the circulation from a lesion, thereby acting as vehicles for the dissemination of ingested viable mycobacteria.

Intact mycobacteria are nontoxic, and the clinical symptoms and lesions that develop depend upon the types of immune responses that predominate in response to antigen. Mycobacteria have a wide repertoire of antigens, many of which are modifiers of the host responses or responsible for the pathologic manifestations of disease (Rastogi 1991). Since disease in mycobacterial infection is a direct result of the host's immunological response to mycobacterial antigens, an understanding of the pathogenesis and immunogenesis of mycobacterial infections is contingent upon comprehension of the architecture and function of mycobacterial antigens (Brennan 1989).

Immunity to mycobacterial infections is dependent on CMI responses; humoral immune factors are of little importance in protection of the host. Development of cell mediated responses to facultative intracellular organisms involves the cooperative action of T lymphocytes as specific inducers and macrophages as nonspecific effector cells. T lymphocytes recruit and assemble mononuclear phagocytes and release cytokines that activate macrophages for enhanced bactericidal activity (Denis 1990). Cytotoxic T cells may kill inactivated macrophages containing *M. tuberculosis* for subsequent phagocytosis (Orme et al. 1992). This activity appears independent of known cytokines produced and not limited to recognition of heat shock proteins of mycobacteria (Boom et al. 1991).

The key functional parameters determining the outcome of immune responses to infectious agents is the nature of the cytokines (cellular signals) produced locally by immune and other cell types. T-helper cells can be divided into two distinct populations, based on their cytokine profiles which dictate their functional role (Romagnani 1992). T-helper type-1 cells are involved in cellular immunity, while T-helper type-2 cells regulate antibody production. Type-1 and type-2 cells operate in a reciprocal fashion, whereby cellular and humoral immune responses are in opposition. Thus, while type-2 cells are associated with exacerbation and rapid lesion formation in several models of chronic infectious disease, type-1 cells are associated with resistance in the same model system (Salgame et al. 1991). Therefore, the existence or absence of an immune response does not predict resistance; this depends on the balance between the various types of immune responses of the host and the virulence of the bacterial strain for that host.

Mononuclear cell dysfunction has been reported in several animal species; it has been suggested that prostaglandins are mediators of abnormal immunoregulation and suppression of cell-mediated responsiveness. Specific immunosuppression mediated by T-suppressor lymphocytes and/or nonspecific immunosuppression mediated by monocytes or subpopulations of lymphocytes may play an important role in modifying the cellular response of the host to virulent mycobacteria (Rook 1988). Deficiencies of T-lymphocyte function and mononuclear dysfunction have been associated with opportunistic mycobacterial diseases. Lymphocyte responses to specific purified protein derivative (PPD) from *M. avium-M. intracellulare* increased in cultures containing indomethacin, suggesting dysfunction may be mediated by an imbalance of arachidonic acid metabolites (Hall and Thoen 1983). Recent reports reveal that *M. avium* can survive and grow in macrophages, and it has been suggested that SOD produced by *M. avium* is important in protection against host responses (Mayer and Falkingham 1986; Frehel et al. 1991). Studies utilizing human macrophages suggest that inhibition or killing of *M. avium* complex may be mediated by tumor necrosis factor and dependent on 1,25 dehydroxy vitamin D_3 (Bermudez et al. 1990).

Mononuclear macrophages activated by lymphokines exhibit marked changes, including an elevation of antimycobacterial mechanisms; however, it has been suggested that mycobacteria survive by inhibition of fusion of phagosomes and lysosomes, or by escape into the cytoplasm (Lowrie and Andrew 1988). Activated macrophages from rabbits inoculated with heat-killed *M. bovis* BCG, in oil, showed an increase in Krebs cycle activity and an increase in hexose monophosphate-shunt activity as compared to the activity observed in normal cells. Kato et al. (1984) reported that functional changes that occur in lung macrophages following exposure to *M. bovis* BCG varied for different strains of mice. Lung macrophages of high-responder mice receiving BCG had elevated acid phosphatase activity, superoxide anion production, and microbicidal activity to *M. bovis* as compared to macrophages of BCG low-responder mice. Macrophage functions such as chemotaxis, phagocytosis, enzyme secretion, and cytotoxicity are influenced by intracellular nucleotide levels (Verghese and Snyderman 1983). Studies on the activation of adenylate cyclase in isolated macrophage membranes suggest that certain nucleotides, such as cyclic guanosine-5′-monophosphate, present in the macrophage membrane, regulate adenylate cyclase activation. Moreover, prostaglandins can exert their effect on cyclic adenosine-5′-monophosphate (cAMP) production by stimulating membrane-bound adenylate cyclase. However, other agents such as arachidonic acid require additional intracellular components to elevate cAMP levels in macrophages. The importance of cAMP in phagolysosome formation in macrophages infected with *M. bovis* BCG and certain other mycobacteria has been investigated; an elevation of cAMP levels was found in macrophages ingesting *M. lepraemurium* (Lowrie and Andrew 1988). Granule fusion, assessed by electron microscopy, was found inversely related to the increase in cAMP. The ability of different mycobacteria to synthesize cAMP correlated with the increase in cAMP; therefore, survival of mycobacteria within cells may be related to inhibition of macrophage degranulation induced by increases in intracellular cAMP.

In addition to T-helper lymphocytes and mononuclear phagocytes, other lymphoid populations appear also to participate in immune responses to *M. paratuberculosis* and other mycobacterial infections. While classical T cells contain a T-cell receptor (TCR) composed of an alpha/beta heterodimer, a second type of TCR molecule has been identified that is composed of a gamma/delta heterodimer. It has recently been shown in mice that alpha/beta T-cell responses are not enough for resistance and that other T cell types, presumably gamma/delta T cells, are also required (Izzo and North 1992). It has been proposed that gamma/delta T populations may serve as a first line of defense to invasion by various intestinal pathogens and play a role in covering the interim between the phagocytic system and the highly evolved alpha/beta host defenses (Panchamoorthy et al. 1991).

Although much data on alpha/beta T cells are available, there is limited information on the antigen specificity, genetic restriction, functional activities, and physiologic role of the gamma/delta T cell. Nevertheless, it is increasingly clear that gamma/delta T cells appear to participate and contribute to resistance against *M. tuberculosis* and other mycobacteria (Banes et al. 1992). These cells increase in response to *M. tuberculosis* infection and produce a variety of important cytokines, including gamma-IFN, IL-2, and a yet undefined cytokine that synergizes with granulocyte-macrophage colony stimulating factor in macrophage aggregation (Follows et al. 1992).

GRANULOMATOUS LESION DEVELOPMENT

Aerosol exposure to acid-fast bacilli generally leads to involvement of pulmonary lymph nodes and lungs, while animals exposed by ingestion of contaminated food and water usually develop primary foci in lymph tissues associated with the gastrointestinal tract (Thoen 1988). The mucociliary clearance by mucus and epithelial cilia in the upper respiratory passages provides defense against infection by inhalation of mycobacteria. However, microorganisms on small particles such as dust and water droplets that do not impinge against the mucociliary layer can pass through terminal bronchioles, thus gaining access to alveolar spaces. The estimated size of terminal endings of bronchioles is about 20 μm as compared to 1-4 μm for an acid-fast bacillus. Mycobacteria multiply within macrophages, and after 10-14 days, CMI responses develop and host macrophages acquire an increased capacity to kill the intracellular bacilli. The CMI responses are mediated by lymphocytes, which release lymphokines (soluble substances) that attract, immobilize, and activate additional blood-borne mononuclear cells at the site where virulent mycobacteria or their products exist. The cellular hypersensitivity that develops contributes to cell death and tissue destruction (caseous necrosis). In some instances, liquefaction and cavity formation occur due to enzymatic action on proteins and lipids. Rupture of these cavities into the bronchi allows aerosol spread of bacilli (Fig. 4.1). Activated macrophages migrate to blind endings of lymphatic vessels and course to one or more of the thoracic lymph nodes, either bronchial or mediastinal. Engorged macrophages have been shown ultrastructurally to enlarge and to develop marked increases in the number of lysosomes, Golgi complexes, and vesicles.

Lymph nodes are more commonly infected than other tissues because fluids in an animal eventually pass through the nodes where the meshwork of trabeculae entraps the organisms. The enlargement and presence of macrophages in impenetrable passageways between reticular cell fibers of the lymph node provide an environment for mycobacterial growth and development of the granulomatous lesion in the node. Occasionally, some phagocytized mycobacteria remain in the lung, and both lung and thoracic nodes are affected (Fig. 4.2). Primary lesions often become localized in a node(s) and may become large and firm. Fibrous connective tissue development probably contributes to localization of the granulomatous lesions. Similar lesions have been observed in swine from which *M. tuberculosis*, *M. avium* complex, or *M. bovis* has been isolated.

Granuloma formation is an attempt by the host to localize the disease process and to allow inflammatory and immune mechanisms to destroy bacilli. A few lesions may appear to be regressing and becoming encapsulated by well-organized connective tissue; such lesions may contain viable bacilli. Typically, the microscopic appearance of a granuloma (a tubercle) is focal and has some caseous necrosis in a central area encircled by a zone of epithelioid cells, lymphocytes, and some granulocytes (Fig. 4.3). Mineralization may be present in necrotic centers; in more advanced lesions several foci of mineralization may coalesce. The zone near the necrotic area often contains multinucleated giant cells that contain several nuclei, often in a horseshoe or ring shape near the cytoplasmic border. An outer boundary of fibrous connective tissue is usually present between the lesions and normal tissue (Thoen and Himes 1986). Occasionally, fibrous tissue is not apparent and the lesion assumes a more diffuse appearance.

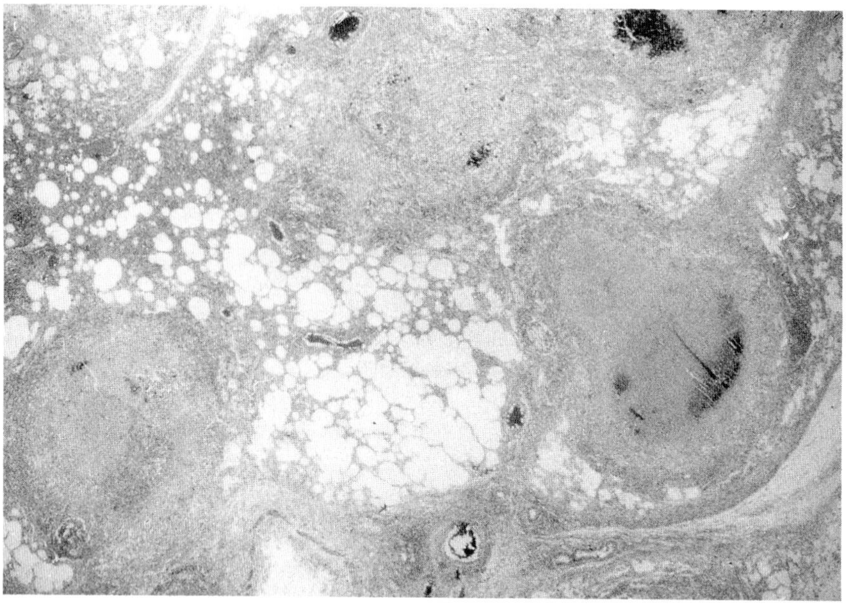

FIG. 4.1. Photomicrograph of lung of an elk from which *Mycobacterium bovis* was isolated. Note multiple granulomas, some caseation necrosis, and mineralization. One granuloma is adjacent to a bronchiole. H & E stain, x63. (Thoen et al. 1992.)

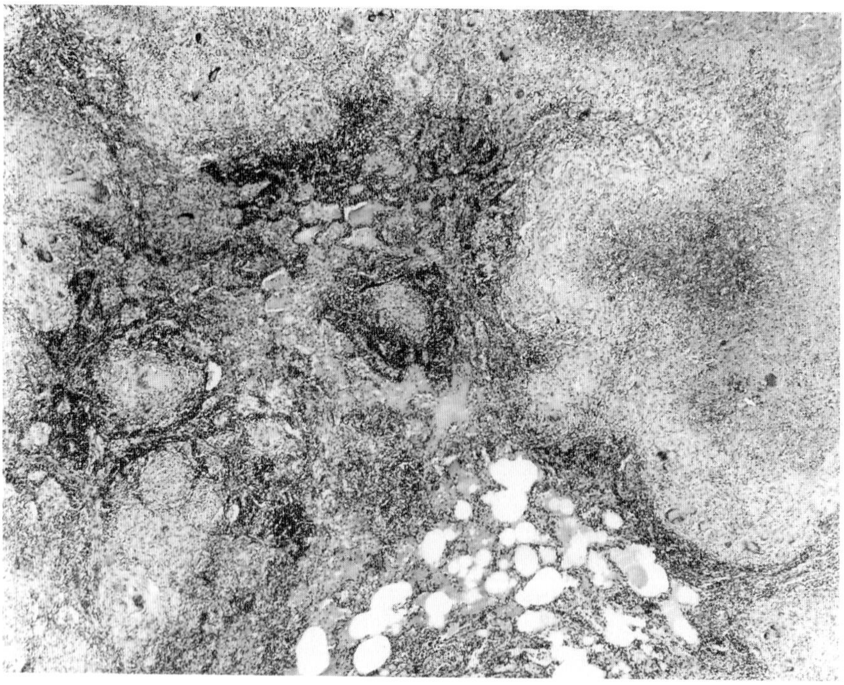

FIG. 4.2. Granulomas in lung of a nonhuman primate. Coalescing lesions have areas of caseous necrosis. Epithelioid cells, lymphocytes, and multinucleated giant cells are visible at higher magnifications. *Mycobacterium tuberculosis* was isolated. H & E stain, x50.

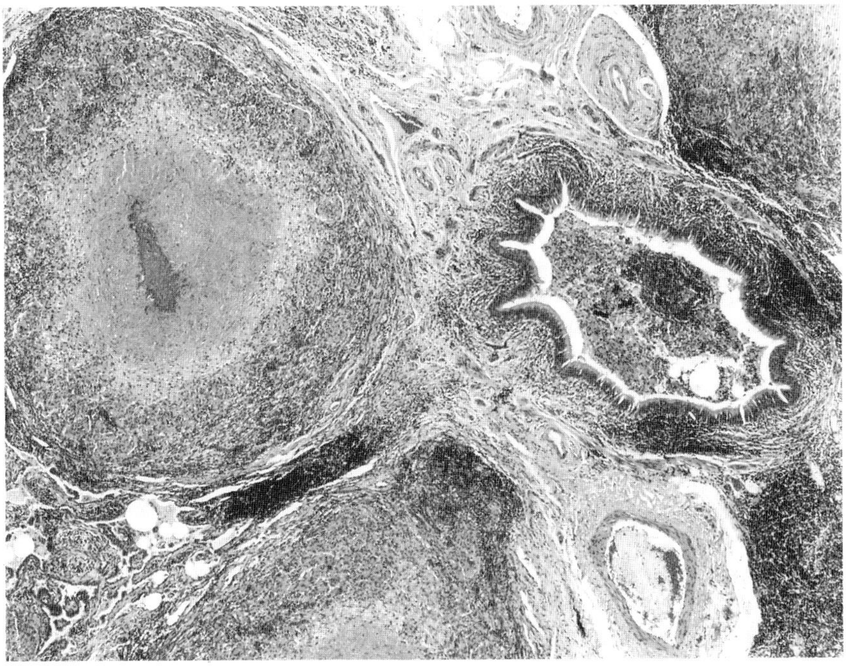

FIG 4.3. Tubercles in a cow. A central area of caseous necrosis has some mineralization. The outer zone has epithelioid cells, lymphocytes, and granulocytes, which can be distinguished at higher magnifications. *Mycobacterium bovis* was isolated. H & E stain, x40.

Lesions of lymph nodes associated with the gastrointestinal system, as with *M. avium* in cattle and swine, suggest infection by ingestion. Infection of the bovine tonsil is rarely seen, probably because the tonsil is located subepithelially and has no afferent lymph vessels. The medial retropharyngeal lymph nodes are a frequent site of *M. bovis* infection and are the most commonly infected site in the head. These nodes receive afferent lymph vessels from the floor of the mouth and adjacent parts. Other lymph nodes of the head (mandibular, parotid, and lateral retropharyngeal) are occasionally involved. The liver is only infrequently involved; hepatic nodes have afferent lymph vessels from the liver, duodenum, and abomasal lymph modes and are commonly involved. The greater part of the blood supply to the liver is derived from the portal vein, which drains the blood and lymph from the intestine; therefore, mycobacteria can pass directly to a hepatic node from the intestine, or through the portal vein to liver only, or subsequently into hepatic nodes. Occasionally, only mesenteric lymph nodes are found to be infected. Localized tubercles have not been reported in the mucous membrane of small intestine; mycobacteria are apparently able to diffuse into lymphatics of the lamina propria and be transported by phagocytes via the lymphatic vessels to mesenteric lymph nodes. In some animals lesions may develop in superficial cervical lymph nodes. Preputial lesions of skin and a local lymph node have been seen in a bull as the only involvement. Superficial iliac and popliteal lymph node lesions are seen only infrequently.

In paratuberculosis, disease is characterized by a granulomatous inflammation of the small and large bowel and regional lymph nodes. Information on the occurrence of cellular and humoral immune responses in cattle infected with *M. paratuberculosis* has been reviewed

(Thoen and Baum 1988). Clinical Johne's disease usually involves impaired intestinal function associated with chronic inflammatory responses. Tissue changes are accompanied by increased leakage of plasma proteins across the intestinal wall and malabsorption of amino acids from the gut lumen. Lesions in cattle are primarily in the intestinal wall and characterized by diffuse granulomatous changes with little or no evidence of necrosis. Accumulations of lymphocytes and epithelioid cells are present in the lamina propria. In advanced clinical cases inflammatory cells may be observed in the submucosa as a band of epithelioid cells along the muscularis mucosa (Fig. 4.4). Granulomas containing numerous acid-fast bacilli are usually present in lymph nodes associated with the intestinal tract in cattle. Caseous necrosis has been observed in such lesions in sheep and goats.

Infection with *M. paratuberculosis* occurs primarily in the young (<30 days of age) through the fecal-oral route; other routes that have been suggested include placental and transuterine. Adult-to-adult transmission and passage of organisms through infected milk have also been proposed. Classically, a newborn animal nurses on teats contaminated with feces. After an incubation period of 2-5 years (range 6 months to 15 years), animals develop a general unthriftiness, rough hair coat, chronic weight loss, and intermittent diarrhea. The clinical course usually lasts only a few months, terminating in severe diarrhea, emaciation, ventral edema, debilitation, and death.

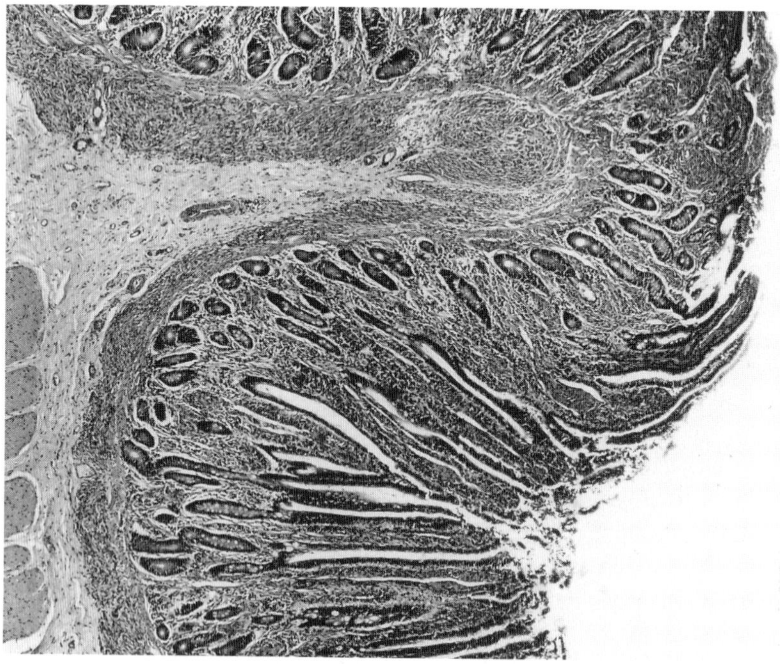

FIG 4.4. Accumulations of inflammatory cells in the mucosa and submucosa of small intestine of a cow. Lamina propria and part of submucosa are packed with epithelioid cells. *Mycobacterium paratuberculosis* was isolated. H & E stain, x40.

Mycobacterium paratuberculosis penetrates the intestinal epithelial layers through the follicle-associated epithelium (FAE) or M cells (Momotani et al. 1988). Recent studies on bovine paratuberculosis have shown that T-helper cell activity is hindered in animals with clinical disease (Chiodini and Davis 1993). It has also been shown that general cellular responses in infected cattle are depressed as compared to normal cattle (Kreeger and Snider 1992), which suggests a predominance of type-2 activity. This situation is similar to that recently described in human leprosy, in which disease appears to result from the presence of CDgamma^{+}T cells that accumulate in lesions and hinder T-helper activity (Salgame et al. 1991). A suppressed helper cell population implies an inability to properly activate mononuclear phagocytes, leading to bacterial proliferation. There is also a predominance of T-helper type-2 cells associated with progressive infection. On the other hand, in self-limiting tuberculoid leprosy lesions, type-1 cells predominate. Further efforts will be required to understand the mechanisms by which T-helper cells are impaired in paratuberculosis and to identify events that may lead to inactivation of these mechanisms.

The suggestion of a primary defense mechanism for gamma/delta T cells has important implications in mycobacterial infections, particularly paratuberculosis, since this disease is acquired by the newborn animal. Although clinical paratuberculosis does not usually occur until 2-5 years of age, ruminants appear to be infected with *M. paratuberculosis* within the first 30 days of life. A first-line defense by gamma/delta T cells with subsequent apoptosis is also supported by the high prevalence (80%) of gamma/delta T lymphocytes in the peripheral blood of young ruminants; prevalence gradually decreases as the animal matures (Chiodini and Davis 1993), presumably as alpha/beta T-cell responses become experienced. In view of these findings, gamma/delta T lymphocytes are likely to play a major role in the immunopathogenesis of *M. paratuberculosis* infection.

Although a great deal of new information and understanding has been acquired in recent years on the pathogenesis and immunogenesis of mycobacterial infections, definitive information on the mechanisms of these diseases remains obscure. Further research on tuberculosis and other mycobacterial infections, as well as other intracellular pathogens, will be required to fully understand and appreciate the complex immunoregulatory mechanisms of the host-parasite interrelationship that determine protective immunity, nonresponsiveness, tissue damage, and disease. The role of immunosuppressive agents and certain viruses on altering immune-cell functions in mycobacterial infections needs further investigation (Cox et al. 1989; Thoen and Waite 1990; Tomioka and Saito 1992).

REFERENCES

Abou-Zeid, C.; Ratcliff, T. L.; Wiker, H. G.; Harboe, M.; Bennedsen, J.; and Rook, G. A. W. 1988. Characterization of fibronectin-binding antigen released by *Mycobacterium tuberculosis* and *Mycobacterium bovis* BCG. Infect Immun 56:3046-51.

Andersen, P.; Askgaard, D.; Ljungquist, L.; Bennedsen, J.; and Heron, I. 1991. Proteins released from *Mycobacterium tuberculosis* during growth. Infect Immun 59:1905-10.

Barclay, R., and Ratledge, C. 1983. Iron-binding compounds of *M. avium*, *M. intracellulare*, *M. scrofulaceum*, and mycobactin-dependent *M. paratuberculosis* and *M. avium*. J Bacteriol 153:1138-46.

Barnes, P. F.; Grisso, C. L., Abrams, J. S.; Band, H.; Rea, T. H.; and Modlin, R. L. 1992. Gamma/delta T lymphocytes in human tuberculosis. J Infect Dis 165:506-12.

Barrow, W. W. 1991. Contributing factors of pathogenesis in the *Mycobacterium avium* complex. Res Microbiol 142:427-33.

Bermudez, L. E.; Young, L. S.; and Gupta, S. 1990. 1,25 dihydroxy vitamin D_3-dependant inhibition of growth or killing of *Mycobacteria avium* complex in human macrophages is mediated by TNF and GM-CSF. Cell Immunol 127:432-38.

Boom, W. H.; Wallis, R. S.; and Chervenak, K. A. 1991. Human *Mycobacterium tuberculosis*-reactive $CD4^+$ T cell clones: Heterogenicity in antigen recognition cytokine production, and cytotoxicity for mononuclear phagocytes. Infect Immun 59:2737-43.

Brennan, P. J. 1989. Structure of mycobacteria: Recent developments in defining cell wall carbyhydrates and proteins. Rev Infect Dis 11:S420-430.

Chan, J.; Xing, Y.; Maghiozzo, R. S.; and Bloom, B. R. 1992. Killing of virulent *Mycobacterium tuberculosis* by reactive nitrogen intermediates produced by activated murine macrophages. J Exp Med 175:1111-22.

Chiodini, R. J., and Davis, W. C. 1993. The cellular immunology of bovine paratuberculosis: The predominant response is mediated by cytotoxic gamma/delta T lymphocytes which prevent $CD4^+$ activity. Microb Pathog 16:In press.

Chiodini, R. J.; Van Kruiningen, H. J.; and Merkal, R. S. 1984. Ruminant paratuberculosis (Johne's disease): The current status and future prospects. Cornell Vet 74:218-62.

Cox, J. H.; Knight, B. C.; and Ivanyi, J. 1989. Mechanisms of recrudescense of *Mycobacterium bovis* BCG infection in mice. Infect Immun 57:1719-24.

Denis, M. 1991. Killing of *Mycobacterium tuberculosis* within human monocytes: activation by cytokines and calcitriol. Clin Expt Immunol 84:200-206.

Flesch, I. E. A., and Kaufmann, S. H. E. 1990. Activation of tuberculostatic macrophage functions by gamma Interferon, Interleukin 4, and tumor necrosis factor. Infect Immun 58:2675-77.

___________. 1991. Mechanisms involved in mycobacterial growth inhibition by gamma-interferon activated bone marrow macrophages: Role of reactive nitrogen intermediates. Infect Immun 59:3213-18.

Follows, G. A.; Munk, M. E.; Gatrill, A. J.; Conradt, P.; and Kaufmann, S. H. E. 1992. Gamma interferon and interleukin 2, but not interleukin 4, are detectable in gamma/delta T-cell cultures after activation with bacteria. Infect Immun 60:1229-31.

Frehel, C.; de Chasteller, C.; Offredo, C.; and Berche, P. 1991. Intramacrophage growth of *Mycobacterium avium* during infection in mice. Infect Immun 59:2207-14.

Hall, M. R., and Thoen, C. O. 1983. Lymphocyte immunostimulation responses following intravenous injection of *Mycobacterium bovis* PPD tuberculin in cattle experimentally exposed to *M. bovis*. Proc 26th Annu Meet Am Assoc Vet Lab Diagn, Las Vegas, Nev., 51-62.

Izzo, A. A., and North, R. J. 1992. Evidence for an alpha/beta T cell-independent mechanism of resistance to mycobacteria. Bacillus-Calmette-Guerin causes progressive disease in severe combined immunodeficient mice, but not in nude mice or mice depleted of $CD4^+$ and $CD8^+$ T cells. J Exp Med 176:581-86.

Jeevan, A., and Asherson, G. L. 1988. Recombinant Interleukin-2 limits the replication of *Mycobacterium lepraemuium* and *Mycobacterium bovis* BCG in mice. Infect Immun 56:660-64.

Kato, K.; Yamamoto, K.; Okuyama, H.; and Kimura, T. 1984. Microbicidal activity and morphological characteristics of lung macrophages in *Mycobacterium bovis* BCG wall-induced lung granuloma in mice. Infect Immun 45:325-31.

Kaufmann, S. H. E. 1990. Heat shock proteins and the immune response. Immunol Today 11:129-36.

Kaufmann, S. H. E.; Schoel, B.; Ward-Wurttenberger, A.; Steinhoff, U.; Munk, M. E.; and Koga, T. 1990. T-cells, stress proteins and pathogenesis of mycobacterial infections. Curr Top Microbiol and Immunol 155:125-31.

Kreeger, J. M., and Snider, T. G. 1992. Measurement of lymphoblast proliferative capacity of stimulated blood mononuclear cells from cattle with chronic paratuberculosis. Am J Vet Res 53:392-95.

Labidi, A., and Thoen, C. O. 1989. Genetic Relatedness of *Mycobacterium tuberculosis* and *M. bovis*. Acta Leprol 7:217-21.

Laneelle, G., and Daffe, M. 1991. Mycobacterial cell wall and pathogenicity: A lipodologist's view. Res Microbiol 142:433-37.

Lowrie, D. B., and Andrew, P. W. 1988. Macrophage antimycobacterial mechanisms. Brit Med Bull 44:624-29.

McFadden, J.; Kunze, Z.; and Seechurn, P. 1990. DNA probes for detection and identification. In Molecular Biology of Mycobacteria. Ed. J. McFadden, P. 139. London: Surrey University Press.

McNeil, M. R., and Brennan, P. J. 1991. Structure, function, and biogenesis of the cell envelope of mycobacteria in relation to bacterial physiology, pathogenesis, and drug resistance; some thoughts and possibilities arising from recent structural information. Res Microbiol 142:451-63.

Mayer, I. E. A., and Falkingham, J. O. 1986. Superoxide dismutase activity of *Mycobacterium avium, M. intracellulare*, and *M. scrofulaceum*. Infect Immun 53:631-35.

Momotani, E.; Whipple, D. L.; Thiermann, A. B.; and Cheville, N. F. 1988. Role of M cells and macrophages in the entrance of *Mycobacterium tuberculosis* into domes of ileal Peyer's patches in calves. Vet Pathol 25:131-37.

Orme, I. M.; Miller, E. S.; Roberts, A. D.; Furney, S. S. K.; Griffen, J. P.; Dobos, K. M.; Chi, D.; Rivoire, B.; and Brennan, P. J. 1992. T lymphcyte mediating protection and cellular cytosis during the course of *Mycobacterium tuberculosis* infection. J Immunol 148:189-96.

Panchamoorthy G.; McLean, J.; Modlin, R. L.; Morita, C. T.; Kshikawa, S.; Brenner, M. B.; and Band, H. 1991. A predominance of the T cell receptor V-gamma 2/V-delta 2 subset in human mycobacteria-responsive T cells suggests germline gene encoded recognition. J Immunol 147:3360-69.

Pommier, C. G.; Inada, S.; Fries, L. F.; Takahashi, T.; Frank, M. M.; and Brown, E. J. 1983. Plasma fibronectin enhances phagocytosis of opsonized particles by human peripheral blood monocytes. J Exp Med 157:1844-54.

Rastogi, N. 1991. Recent observations concerning structure and function relationships in the mycobacterial cell envelope: Elaboration of a model in terms of mycobacterial pathogenicity, virulence, and drug-resistance. Res Microbiol 142:464-76.

Rastogi, N., and David, H. L. 1988. Mechanisms of pathogenicity in mycobacteria. Biochimie 70:1101-20.

Romagnani, S. 1992. Induction of T_H1 and T_H2 responses: A key role for the "natural" immune response? Immunol Today 13:379-81.

Rook, G. A. W. 1988. Role of activated macrophages in the immunopathology of tuberculosis. Br Med Bull 44:611-23.

Salgame, P.; Abrams, J. S.; Clayberger, C.; Goldstein, H.; Convit, J.; Modlin, R. L.; and Bloom, B. R. 1991. Differing cytokine profiles of functional subsets of human CD4 and CD8 cell clones. Science 254:279-82.

Schlesinger, L. S.; Bellinger-Kawahara, C. G.; Payne N. R.; and Horwitz, M. A. 1990. Phagocytosis of *Mycobacterium tuberculosis* is mediated by human monocyte complement receptors and complement component C3. J Immunol 144:2771-80.

Shinnick, T. M.; Vodkin, M. H.; and Williams, J. L. 1987. The 65-kDa antigen is a heat shock protein which corresponds to common antigen and to the *E. coli* groEL protein. Infect Immun 56:446-51.

Thoen, C. O. 1988. Tuberculosis. J Am Vet Med Assoc 193:1045-47.

Thoen, C. O., and Baum, K. 1988. Current knowledge on paratuberculosis. J Am Vet Med Assoc 192:1609-11.

Thoen, C. O., and Himes, E. M. 1986. *Mycobacterium*. In Pathogensis of Bacterial Infections in Animals. Ed. C. L. Gyles and C. O. Thoen, pp.25-36. Ames: Iowa State University Press.

Thoen, C. O., and Karlson, A. G. 1984. Experimental tuberculosis in rabbits. In The Mycobacteria: A Sourcebook. Ed. G. P. Kubica and L. G. Wayne, pp. 978-89. New York: Marcel Dekker.

__________. 1991. Tuberculosis. In Diseases of Poultry, 8th ed. Ed. B. W. Calnek, H. J. Barnes, C. W. Beard, W. M. Reid, and H. W. Yoder, pp. 172-85. Ames: Iowa State University Press.

Thoen, C. O., and Schliesser, T. 1984. Mycobacterial infections in cold-blooded animals. In The Mycobacteria: A Sourcebook. Ed. G. P. Kubica and L. G. Wayne, pp. 1279-1311. New York: Marcel Dekker.

Thoen, C. O., and Waite, K. J. 1990. Some immune responses in cattle exposed to *Mycobacterium paratuberculosis* after injection with modified-live bovine viral diarrhea virus vaccine. J Vet Diagn Invest 2:176-79.

Thoen, C. O., and Williams, D. E. 1993. Handbook on Zoonoses. Ed. G.W. Beran. Boca Raton, Fla.: CRC Press. (In Press).

Thoen, C. O.; Throlsen, K. J.; Miller, L. D.; Himes, E. M.; and Morgan, R. L. 1988. Pathogenesis of *Mycobacterium bovis* infection in American bison. Am J Vet Res 49:1861-65.

Thoen, C. O.; Quinn, W. J.; Miller, L. D.; Stackhouse, L. L.; Newcomb, B. F.; and Ferrell, J. M. 1992. *Mycobacterium bovis* infection in North American elk (Cervus elaphus). J Vet Diagn Invest 4:423-27.

Tomioka, H., and Saito, H. 1992. Characterization of immunosuppressive functions of murine peritoneal macrophages induced with various agents. J Leuk Biol 51:24-31.

Verghese, M. W., and Snyderman, R. 1983. Hormonal activation of aderylate cyclase in macrophage membranes is regulated by guanine necleotides. J Immunol 130:869-73.

Young, D.; Garbe, T.; Lathigra, R.; and Abou-Zeid, C. 1990. Protein antigens: Structure, function and regulation. In Molecular Biology of the Mycobacteria. Ed. J. McFadden, p. 35. London: Surrey University Press.

Young, R. A., and Elliot, T. J. 1989. Stress proteins, infection and immune surveillance. Cell 59:5-14.

Wiker, H. G., and Harboe, M. 1992. The antigen 85 complex: A major secretion product of *Mycobacterium tuberculosis*. Microbiol Rev 56:648-61.

Zhang, L.; Goren, M. B.; Holzer, T. J.; and Anderson, B. R. 1988. Effect of *Mycobacterium tuberculosis*-derived sulfolipid I on human phagocytic cells. Infect Immun 56:2876-83.

5 / *Corynebacterium*

BY J. G. SONGER AND J. F. PRESCOTT

MEMBERS of the genus *Corynebacterium* comprise a diverse group, in spite of recent taxonomic work that has shifted several organisms into more appropriate genera (Table 5.1). In deference to these long-standing associations with the genus, several noncorynebacteria are discussed in this chapter.

TABLE 5.1. Taxonomy, disease production, and natural history of major pathogenic corynebacteria of humans and animals

Suggested Modern Classification	Type of Disease	Host		Habitat
		Main	Minor	
C. diphtheriae	Diphtheria	Humans	None	Throat
C. ulcerans	Pharyngitis (humans) Mastitis, abscesses (cows)	Humans	Cattle, primates	Throat (humans) ? Udder (cows)
C. pseudotuberculosis[a]	Lymphadenitis, abscesses	Sheep, goats, horses	Humans	Skin, mucosal surfaces ? Environment
C. renale *C. pilosum*[b] *C. cystitidis*[b]	Cystitis, pyelonephritis	Cattle	Sheep, horses, dogs, swine, humans	Lower urogenital tract
Actinomyces pyogenes[c]	Purulent disease processes	Cattle, sheep, swine	Other domestic animals	Mucosal surfaces
Rhodococcus equi[d]	Granulomatous pneumonia, lymphadenitis	Horses	Pigs, cattle, dogs, goats, sheep, humans	Soil (especially soils rich in feces of herbivores
Actinomyces suis[e]	Cystitis, pyelonephritis	Swine	None	Prepuce of boar
C. bovis	Commensal, mild mastitis	Cattle	Humans	Bovine teat canal
Arcanobacterium[f] *haemolyticum*	Pharyngitis, abscesses	Humans	None	? Mucosa, skin

[a]Previously *Corynebacterium ovis*. [b]Previously *Corynebacterium renale*. [c]Previously *Corynebacterium pyogenes* [d]Previously *Corynebacterium equi*. [e]Previously *Corynebacterium suis* and *Eubacterium suis*. [f]Previously *Corynebacterium haemolyticum*.

Corynebacterium is well-defined chemically and genetically, but the diversity within the genus makes it impossible to generalize regarding mechanisms of pathogenesis. In many cases no role for putative virulence attributes in pathogenesis has been defined. However, the application of molecular biological methods, and in particular the development of rudimentary systems for genetic manipulation, have brought promise of breakthroughs in defining mechanisms of pathogenesis.

CORYNEBACTERIUM RENALE

Corynebacterium renale, *C. pilosum*, and *C. cystitidis* are opportunistic pathogens that inhabit the urinary tract of cattle and other domestic animals, where they cause sporadic cases of cystitis and ascending pyelonephritis (Yanagawa 1986). The major factors that put an animal at risk are the shortness of the female urethra and the effects of pregnancy and parturition; thus, disease occurs most frequently in mature cows. *Corynebacterium pilosum*, a member of the normal flora of the lower urogenital tract, causes cystitis and, infrequently, pyelonephritis. *Corynebacterium cystitidis* is usually associated with chronic pyelonephritis and can cause the most severe cystitis of the three species in this group; isolation from normal animals is rare (Hiramune et al. 1971). *Corynebacterium renale* is also normal flora in the lower urogenital tract and is the most important of the three, based upon incidence of infection. The vulva may be an important portal of entry for *C. renale* into the urinary tract, in that this organism and *C. pilosum* adhere more readily to vulvar epithelial cells than to uroepithelial cells (Hayashi et al. 1985). The suggestion that *C. renale* is involved in the etiology of ulcerative posthitis in bulls (Schild et al. 1985) requires experimental confirmation.

The virulence factors of *C. renale* and their roles in pathogenesis have not been explored in depth, with the possible exception of pili. Renalin, a *C. renale* extracellular protein with a strong nonenzymatic affinity for ceramide (a product of the action of phospholipase C on sphingomyelin), may play a role in lysis of cell membranes (Bernheimer and Avigad 1982). Apparently, this protein is not produced by either *C. pilosum* or *C. cystitidis* (Hiramune et al. 1985). Iron regulation of the virulence of *C. renale* has been suggested (Henderson et al. 1978), but not confirmed.

Pili are produced by all three of these species, and only rare, apparently minor, antigenic cross reactions have been demonstrated between pili of *C. renale* and *C. pilosum* (Kudo et al. 1987). Piliated organisms are more resistant to phagocytosis by mouse neutrophils in vitro than nonpiliated organisms in the absence of opsonins, but in the presence of antipilus serum and complement, both are phagocytosed at an equal rate (Kubota and Yanagawa 1988). Phagocytosis of piliated bacteria by mouse peritoneal macrophages is likewise enhanced by complement and by antipilus polyclonal serum. Curiously, polyclonal antibodies prepared against nonpiliated bacteria also enhance phagocytosis of piliated bacteria, suggesting that nonpilus factors may play an antiphagocytic role (Kubota and Yanagawa 1987).

Loss of pili upon repeated in vitro passage in the presence of antipilus antibodies has been reported (Koga et al. 1988). In vivo selection for nonpiliated organisms in mice occurs and is antibody independent (Fukuoka and Yanagawa 1987b). Furthermore, there are no significant differences in mortality, number of culture positive mice, or numbers of bacteria recovered from the urinary tract in mice infected with piliated and nonpiliated bacteria, suggesting that pili are not obligate factors in virulence of *C. renale* or perhaps that the mouse is not a good model for bovine infection (Fukuoka and Yanagawa 1987a). Construction of isogenic pilus-minus mutants could be simplified by the availability of a gene for a 48-kDa pilus structural protein of *C. renale* (Abe et al. 1990); study of these mutants in cattle could provide useful information on the role of pili in pathogenesis.

The pathogenesis of infection begins when corynebacteria associate with the urethral epithelium and ascend to the bladder in the presence of diminished anatomical defenses related to pregnancy or damage at parturition. Bacteria grow readily in urine, producing cystitis, and ascend (through vesiculourethral reflux) to the kidney, where in most cases infection spreads chronically and relentlessly.

The role of humoral and cellular immunity in these conditions has not been investigated thoroughly. Circulating antibodies are present in cows that develop pyelonephritis rather than cystitis, but humoral antibodies in mice immunized with killed organisms apparently are not protective, since these mice develop pyelonephritis upon challenge. Data suggesting a selection for nonpiliated clones in mice may explain this lack of protection. In experimentally infected rats, IgG is apparently the major component of the humoral immune response to *C. renale* infection (de Buysscher et al. 1985).

CORYNEBACTERIUM PSEUDOTUBERCULOSIS

Corynebacterium pseudotuberculosis is a facultative intracellular parasite that causes caseous lymphadenitis (Fig. 5.1) in sheep and goats and ulcerative lymphangitis and ventral abscesses in horses. The prevalence of caseous lymphadenitis may be as high as 50% in adult animals, resulting in economic loss through trimming or condemnation of carcasses at slaughter and decreases in wool production, value of hides, body weight, milk production, and reproductive efficiency. About 20 human cases of *C. pseudotuberculosis* infection have been documented, but anecdotal reports of axillary lymphadenitis in sheep handlers and veterinarians suggest that human cases may be more common than is realized. *C. pseudotuberculosis*, like *C. ulcerans*, is capable of producing diphtheria toxin when lysogenized by corynephage β that bear the *tox* gene. Although there is considerable genetic relatedness between *C. ulcerans* and *C. pseudotuberculosis*, they are distinct species.

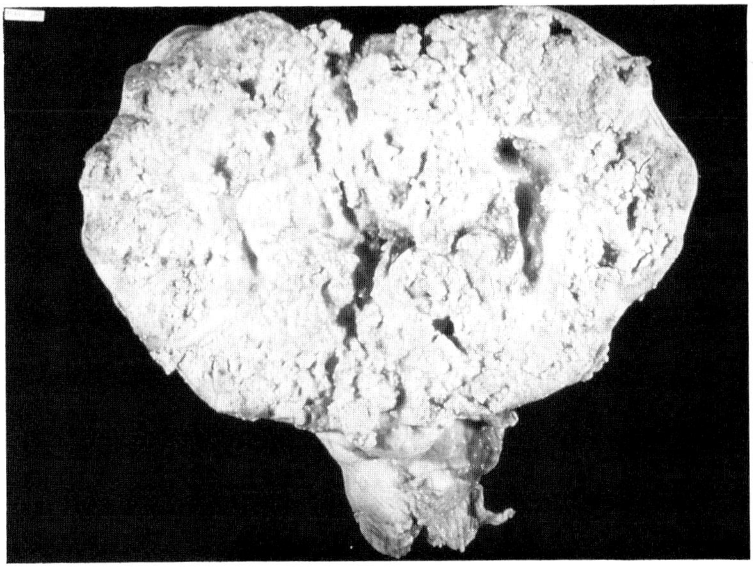

FIG. 5.1. Caseous necrosis in a lymph node caused by *Corynebacterium pseudotuberculosis*.

Intracellular survival by *C. pseudotuberculosis* is pivotal to eventual formation of abscesses and may be mediated by one or both of the organism's two major putative virulence attributes. Concurrent changes in bacterial viability and in the morphology of bacterial lipid external to the cell wall suggest that this lipid may facilitate survival in activated macrophages (Hard 1975). However, the lipid is not systemically toxic for guinea pigs, which are highly susceptible to *C. pseudotuberculosis* or its exotoxin. This suggests that the lipid may be similar to mycobacterial cord factor in its toxicity for cells, but different in its lack of systemic lethality (Carne et al. 1956). Its exact role in pathogenesis remains unclear.

The second putative virulence attribute is a toxic phospholipase D (PLD) (Fig. 5.2), which apparently is produced by all isolates of *C. pseudotuberculosis* and *C. ulcerans*. Production of PLD in the early phase of infection has profound effects on survival and multiplication of *C. pseudotuberculosis* in the host. This may be due to effects on phagocytic cells (Fig. 5.3) (inhibition of chemotaxis, degranulation, and lethality in PLD-treated neutrophils have been demonstrated) or complement depletion (PLD inactivates complement, reducing amounts available for opsonization of *C. pseudotuberculosis*). PLD-induced increases in vascular permeability may also play a role, since increased permeability increases extravascular circulation of fluid and facilitates spread of infection both locally and via the lymphatics. In fact, active or passive immunization of sheep against PLD limits movement of *C. pseudotuberculosis* from inoculation sites to regional lymph nodes in natural and experimental infections. If increased lymph node infection increases the incidence of lymph node caseation, then anti-PLD antibodies should have protective value against the disease in the field by neutralizing the permeability-promoting effect of PLD. Experimental studies generally support the idea that neutralization of the effects of PLD has value in prevention of lesion development in sheep and goats.

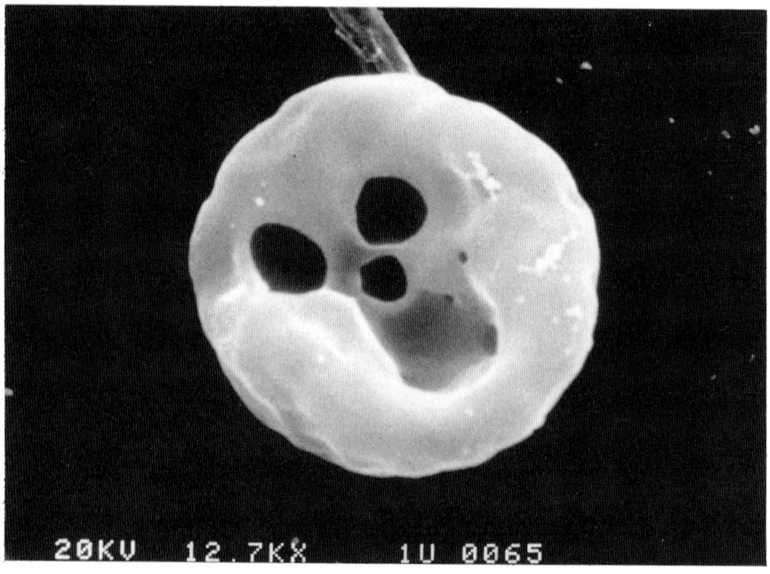

Fig. 5.2. Scanning electron photomicrograph of ovine erythrocyte following exposure to *Corynebacterium pseudotuberculosis* phospholipase D.

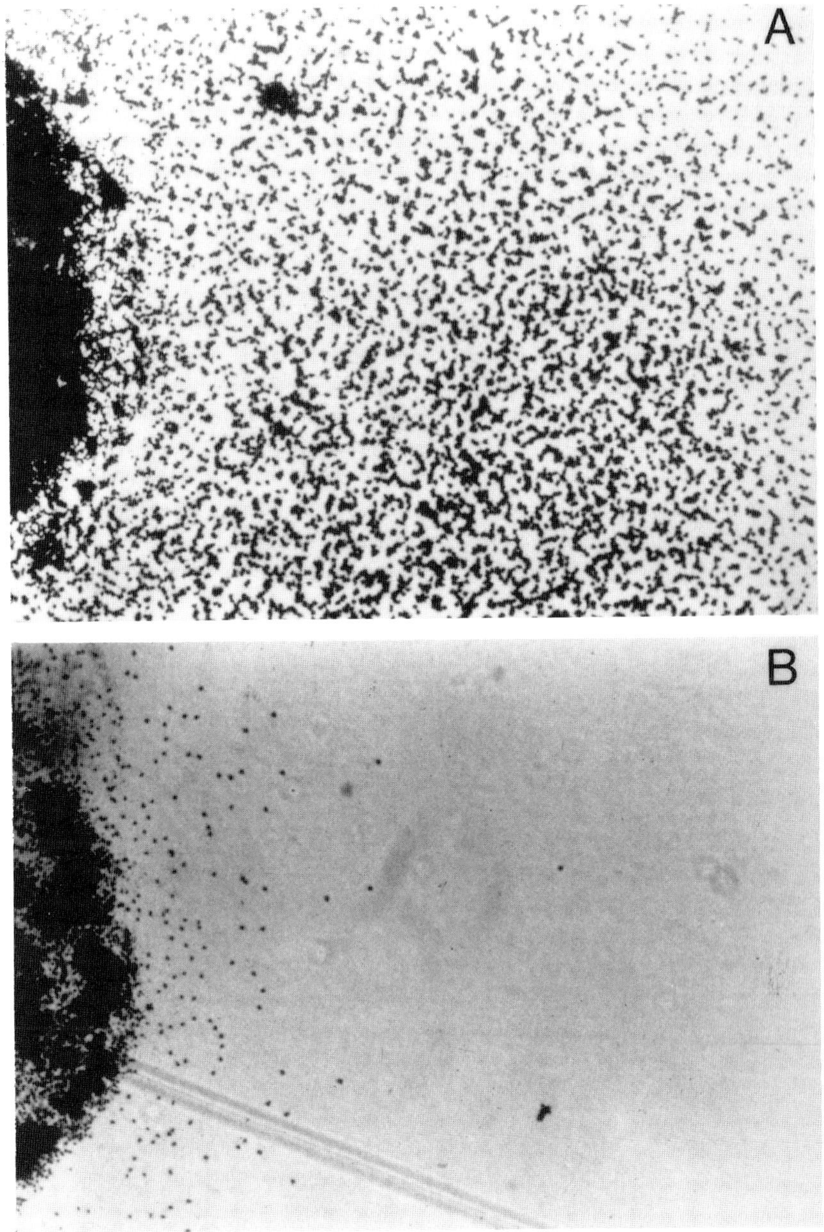

Fig. 5.3. Inhibition of ovine neutrophil chemotaxis by phospholipase D. **A.** Normal neutrophils migrate toward activated sheep serum. **B.** Phospholipase D-treated neutrophils lose both directed and random migration.

Cloning and characterizing the *pld* gene (Hodgson et al. 1990; Songer et al. 1990) and developing methods for genetic manipulation of *C. pseudotuberculosis* (Songer et al. 1991) have facilitated the construction of isogenic *pld*⁻ mutants that do not produce caseous lymphadenitis in sheep (Hodgson et al. 1992) or goats (McNamara and Songer, unpublished data). Thus, a direct and vital role for PLD in pathogenesis of *C. pseudotuberculosis* infections has been established; the exact nature of this role, particularly as it pertains to intracellular survival and multiplication, remains to be defined.

Pathogenesis of caseous lymphadenitis begins when bacteria enter the host via skin wounds, multiply, and are phagocytosed. Phagosome-lysosome fusion takes place, but *C. pseudotuberculosis* multiplies in the phagolysosome. If multiplication is sufficient, or if large numbers of organisms are engulfed, phagocytic cells die. Permeability of local blood vessels increases, encouraging the spread of infection from the initial site to other locations, often regional lymph nodes. Abscesses may develop at either primary or secondary sites, eventually rupturing and discharging a thick, caseous pus containing large numbers of viable bacteria (Pepin et al. 1991a,b); direct transmission via these discharges (through animal-to-animal contact or via shears) and indirect transmission through survival of the organism in the environment probably represent the primary means of spread of infection. Humoral and cell-mediated immune responses develop and immune macrophages engulf and kill *C. pseudotuberculosis*. The infection is usually controlled and sometimes eliminated, at least at the primary focus, but abscesses may recur at the same site. In some instances, lesions become metastatic and, as they increase in number, the "thin ewe syndrome" develops, resulting in progressive debilitation and death. Generally, the clinical and pathologic picture in caseous lymphadenitis illustrates the dual role of granulomatous lesions in chronic bacterial infections: dissemination is limited, but persistence of the infection leads to local tissue damage (Pepin et al. 1991b).

Although the immune response to *C. pseudotuberculosis* is understood in only the most incomplete fashion, vaccination is an increasingly accepted tool in preventing caseous lymphadenitis. Whole cell or cell wall preparations have been applied successfully (Menzies et al. 1991), but more work has been done with PLD toxoid vaccines. Degree of protection, measured by both incidence and extent of infection, is positively correlated with the amount of crude or purified toxoid administered to sheep, and potency of toxoid vaccines is not improved by the inclusion of cells or cell components.

CORYNEBACTERIUM BOVIS

Corynebacterium bovis is a commensal of the bovine udder where it adheres to the squamous epithelium of the teat duct; it is also found free in the teat cistern. Up to 20% of the udder quarters of cattle in most dairy herds carry *C. bovis*. Growth of *C. bovis* is apparently enhanced in milk during lactation, but growth is inhibited by mammary secretions during the nonlactating period. Inflammatory changes occur in the teat cisterns, Furstenberg's rosettes, and mammary parenchyma, but not consistently in teat canals. The basis of the negligible virulence of *C. bovis* has not been explored, but there is some interest in its ability to persist as a low-grade irritant in the udder, provoking a mild neutrophil response that may protect the udder against infection by more virulent bacteria. Evidence accumulated from studies of natural and experimental infections is inconsistent, both supporting (Pociecha 1989) and refuting (Honkanen-Buzalski et al. 1984) this viewpoint. Further, observed protection against mastitis pathogens is not universal; infection by *Staphylococcus aureus* has been inhibited, but animals have displayed increased susceptibility to streptococcal infection (Pankey et al. 1985).

CORYNEBACTERIUM ULCERANS

Corynebacterium ulcerans is best known as a cause of pharyngitis in humans (Coyle and Lipsky 1990), but is isolated with surprising frequency from cases of bovine mastitis. The organism is approximately 50% DNA:DNA homologous with *C. pseudotuberculosis*, and all strains examined by sensitive methods have been shown to produce a phospholipase D, which is antigenically similar to that produced by *C. pseudotuberculosis*. The genes for the two phospholipases are approximately 80% homologous. A matter of concern to public health officials is the capacity of *C. ulcerans* to produce diphtheria toxin when lysogenized by a phage that is carrying *tox*. Naturally occurring tox^+ isolates of *C. ulcerans* have been obtained from a variety of sources, including mastitic bovine milk. Infected cows may shed *C. ulcerans* for months or years in milk from a diseased quarter, and sporadic human cases of *C. ulcerans* infections associated with raw milk have also been reported (Barrett 1986). In a recent study, nearly 50% of isolates from milk samples produced diphtheria toxin. Clinical signs in affected cows are usually minor, but in severe cases, permanent loss of the affected quarter can follow the inflammation (Hedlund and Pohjanvirta 1989).

ACTINOMYCES PYOGENES

Actinomyces pyogenes is a commensal on the exposed mucosal surfaces of cattle, sheep, swine, and occasionally other domestic animals. It can disseminate from its normal habitat to cause a wide variety of nonspecific purulent infections that frequently involve visceral organs (particularly lungs) and sometimes result in abortion. *A. pyogenes* is often a sequel to earlier tissue injury, or to infection with other bacteria, and it is frequently isolated from abscesses in mixed culture with *Fusobacterium necrophorum* and *Peptostreptococcus indolicus*, for which it apparently provides factors stimulatory for growth (Smith et al. 1989). In spite of its versatility as a pathogen and its unquestioned importance as an agent of disease, studies of the virulence and even the basic biology of *A. pyogenes* have been extraordinarily neglected.

Actinomyces pyogenes produces a hemolytic exotoxin, which is dermonecrotic and lethal for laboratory animals (Lovell 1944); antitoxin is present in the serum of infected animals (Lovell 1939). Characterization of the hemolysin has been rudimentary at best, with estimates of molecular weight ranging from 36 to 900 kDa (Katsaras and Zeller 1978; Takeuchi et al. 1979a). *A. pyogenes* also produces a protease (Takeuchi et al. 1979b) and an extracellular neuraminidase, and it has hemagglutinating activity (Lammler et al. 1987; Schaufuss and Lammler 1989), but no role in pathogenesis has been suggested for any of these activities.

Experimentally, intrauterine infusion of *A. pyogenes* results in increased uterine antibody titers of IgG_2 and IgA, and the lack of such increases in vaginal secretions and serum suggests local uterine synthesis of antibody (Watson et al. 1990). Specific antibody against *A. pyogenes* can also be detected in milk whey from infected cows, and the presence of IgM and IgG_2 antibodies is positively correlated with the ability of neutrophils to kill whey-treated *A. pyogenes* (Watson 1989). On the other hand, bacterins or bacterin toxoids have generally failed to protect mice (Durner and Werner 1983); and mice surviving infection with the organism are not fully protected against subsequent challenge (Derbyshire and Matthews 1963). The reasons for these phenomena are unknown but point to the need for investigation of the interaction of *A. pyogenes* and its products with the immune system.

ACTINOMYCES SUIS

Actinomyces suis has been the subject of a taxonomic odyssey, beginning as *Corynebacterium suis* and being most recently classified as *Eubacterium suis*. Examination of

16S rRNAs has revealed that *A. suis* is a close relative of *A. pyogenes* (Ludwig et al. 1992), and it has now been suggested that the correct taxon should be *A. suis*.

Actinomyces suis causes cystitis and ascending pyelonephritis in swine, most often in sows. The preputial diverticulum and semen of boars as young as 6-12 weeks can yield isolates of *A. suis*. Transmission from these sites to the urogenital tract of the sow at coitus or at breeding by artificial insemination has not been demonstrated but seems a likely possibility (Dagnall 1987). Isolation of large numbers of *A. suis*-like bacteria from purulent vaginal discharges in swine, from aborted porcine fetuses, and from purulent abscesses in extragenital locations suggests possible involvement in abortion and infertility and perhaps other infections outside the urogenital tract. Hommez et al. (1991) found these organisms to be as common as *A. pyogenes* in pigs.

Disease is characterized initially by epithelial vacuolization, goblet cell differentiation, leukocytic infiltrations, and hemorrhages in the bladder. Antibody coating of *A. suis* in urine of affected sows is correlated with the presence of ascending pyelonephritis. When pyelonephritis and cystitis occur concurrently, anorexia; emaciation; anemia; subnormal body temperature; abortions; and erosive, ulcerative, hemorrhagic cystitis are also seen. In chronically affected animals, different stages of a mucopurulent, erosive-to-ulcerative cystitis can be seen; ureters and the urethra often show signs of a mild purulent inflammation. Severe fibrinopurulent and necrotizing pyelitis accompanied by bacterial invasion occurs in the renal pelvis, and severe, fibrosing, chronic, interstitial nephritis, leading to uremia and death, is observed in the terminal stages (Kaup et al. 1990; Langfeldt et al. 1990.)

RHODOCOCCUS EQUI

Rhodococcus equi is aerobic and gram-positive, with a characteristic mucoid appearance because of the possession of a capsule. Within the genus, it is the most pathogenic species for animals (Prescott 1991) and is an important pathogen of 1- to 3-month-old foals. *R. equi* is largely a soil organism, whose growth requirements appear to be met perfectly by herbivore manure and summer temperatures in temperate climates. Under ideal environmental conditions, which may include soil type, *R. equi* can multiply many thousands of times within days, thus considerably increasing exposure of foals in the area. Intestinal carriage of *R. equi* by adult herbivores is likely passive and represents acquisition from contaminated grass, but the organisms can multiply to large numbers in the intestine of foals up to about 3 months of age (Takai et al. 1986). *R. equi* pneumonia in foals is characteristically endemic on some farms, sporadic on others, and unrecognized on most. This reflects differences in environmental and management conditions, as well as differences in virulence of isolates (Takai et al. 1991a).

Rhodococcus equi causes subacute to chronic suppurative bronchopneumonia with extensive abscessation and associated suppurative lymphadenitis in foals, and mild cervical lymphadenitis in pigs. The foal is unique in its susceptibility to naturally occurring *R. equi* infection, probably the result of a natural deficiency or impairment of cellular immune mechanisms in the lungs of animals under the age of about 4 months. These conditions also make these foals susceptible to other lung infections in which cell-mediated immune mechanisms are important (e.g., *Pneumocystis carinii* and severe viral respiratory infections).

Reports of the isolation from other species are uncommon and are often associated with immunosuppression of some type. Recently, a significant number of immunosuppressed humans infected with human immunodeficiency virus (AIDS patients) have developed *R. equi* pneumonia with lesions similar to those in foals.

Rhodococcus equi is a facultative intracellular pathogen, surviving inside macrophages to cause granulomatous inflammation. Its ability to persist in, and eventually to destroy, alveolar macrophages is the basis of its pathogenicity. Organisms appear to evade killing by preventing phagosome-lysosome fusion. They may also cause nonspecific degranulation of lysosomes in vitro, which, if in vivo, would contribute to the neutrophil influx and tissue destruction characteristic of advanced lung lesions (Hietala and Ardans 1987).

Virulence in *R. equi* is associated with prominent 15- to 17-kDa antigens (Takai et al. 1991b), which are apparently encoded by a large 80- to 85-kb plasmid (Takai et al. 1991c; Tkachuk-Saad and Prescott 1991). The expression of these surface-exposed antigens is temperature regulated, being expressed in large amounts at temperatures of 34-41°C but not at lower temperatures (Takai et al. 1992). These recent discoveries are of exceptional importance, since they offer considerable potential in understanding the basis of virulence of the organism. In addition, the genetic similarities between the putative virulence plasmid of *R. equi* and that of pathogenic *Yersinia* species (intracellular pathogens with a similarly temperature-dependent expression of virulence plasmid genes) need to be explored. Plasmid-borne virulence genes in *Shigella* species are also characterized by temperature-dependent expression and, in *Yersinia*, expression of these plasmid-borne genes is controlled by low calcium concentrations, such as those occurring inside macrophages. Mycolic acid-containing glycolipids are also associated with virulence in *R. equi*, in that strains with longer chain mycolic acids are, experimentally, more lethal for mice and induced more granulomas than strains with shorter chain mycolic acids (Gotoh et al. 1991). Other unexplored candidates for possible virulence factors include capsular polysaccharides, cholesterol oxidase, choline phosphohydrolase, and phospholipase C exoenzymes (Machang'u and Prescott 1991).

The age of development of *R. equi* pneumonia coincides with the quantity of humoral antibody in foals detected by enzyme-linked immunosorbent assay (ELISA), reflecting the decline of maternally derived antibody by 5-10 weeks of age and then the steady rise as foal-derived antibody develops. The importance of opsonization by antibody in enhancing uptake and killing by macrophages has been demonstrated in vitro. Administration of specific immune plasma to foals prevents both experimentally produced and naturally occurring disease (Martens et al. 1990; Madigan et al. 1991). Surprisingly, however, foals born to vaccinated mares were not protected against experimental or natural infection (Martens et al. 1990; Madigan et al. 1991). The protective antigen has not yet been defined, and it is possible though unlikely that the protective effect of plasma is due to components other than antibody. Less is known about the critical role of cellular immunity in protecting foals against *R. equi* infection and the reasons for the peculiar susceptibility of foals to this infection (Prescott 1991). Recent studies in nude mice have, however, established the importance of $CD4^+$ and $CD8^+$ lymphocytes in clearing infection, as well as the dominant role of $CD8^+$ T cells (Nordmann et al. 1992).

Future research in the area of bacterial pathogenesis and immunity will address the fascinating role and nature of the virulence-associated plasmid of *R. equi,* its relationship to other virulence plasmids of facultative intracellular pathogens, the nature of the humoral protective antigen(s) of *R. equi*, the nature of cell-mediated immunity to this infection and the antigens of importance for cellular immunity, and the role of the temperature-controlled virulence-associated 15- to 17-kDa proteins in intracellular survival of the organism.

ARCANOBACTERIUM HAEMOLYTICUM

Arcanobacterium haemolyticum is associated with human pharyngitis, which can be severe enough to mimic that produced by *Streptococcus pyogenes* and *C. diphtheriae*. Isolates of the organism are also obtained from wound infections, chronic skin ulcers, brain abscesses, vertebral osteomyelitis, and bacteremia. Evidence supporting a role for this organism as an etiologic agent of pharyngitis includes its isolation as the sole or predominant species in the absence of other recognized bacterial pathogens, its absence from follow-up cultures of recovered patients, and the presence of high titers of specific antibody in clinical cases. *A. haemolyticum* produces a phospholipase D that is similar in molecular weight to those produced by *C. pseudotuberculosis* and *C. ulcerans*, and 65% DNA and amino acid sequence homologies have been demonstrated. Recent reports suggest that sequences with homology to streptococcal erythrogenic toxin may also be found in *A. haemolyticum* (Coyle and Lipsky 1990).

REFERENCES

Abe, S.; Saito, T.; Koga, T.; Ono, E.; Yanagawa, R.; Ito, T.; Kida, H.; and Shimizu, Y. 1990. Cloning and expression of a pilin gene of *Corynebacterium renale* in *Escherichia coli*. Jpn J Vet Sci 52:11-18.

Barrett, N. J. 1986. Communicable disease associated with milk and dairy products in England and Wales: 1983-1984. J Infect 12:265-72.

Bernheimer, A. W., and Avigad, L. S. 1982. Mechanism of hemolysis by renalin; a CAMP-like protein from *Corynebacterium renale*. Infect Immun 36:1253-56.

Carne, H. R.; Wickham, N.; and Kater, J. C. 1956. A toxic lipid from the surface of *Corynebacterium ovis*. Nature UK 178:701-2.

Coyle, M. B., and Lipsky, B. A. 1990. Coryneform bacteria in infectious diseases: Clinical and laboratory aspects. Clin Microbiol Rev 3:227-46.

Dagnall, G. J. R. 1987. An investigation of the bacterial flora of the preputial diverticulum and of the semen of boars. Index to Theses Accepted for Higher Degrees in the Universities of Great Britain and Ireland 36:380.

de Buysscher, E. V.; Appleton, J. A.; and Kadis, S. 1985. Appearance of immunoglobulin classes and complement (C3) during *Corynebacterium renale*-induced experimental pyelonephritis in the rat. Am J Vet Res 46:401-403.

Derbyshire, J. B., and Matthews, P. R. J. 1963. Immunological studies with *Corynebacterium pyogenes* in mice. Res Vet Sci 4:537-42.

Durner, K., and Werner, B. 1983. Untersuchungen zur Immunogenität und zu den Pathogenitäts-faktoren von *Corynebacterium pyogenes*. Arch Exp Vet Med Leipzig 37:541-47.

Fukuoka, T., and Yanagawa, R. 1987a. Population shift from piliated to non-piliated bacteria in kidneys, bladder, and urine or mice infected with *Corynebacterium renale* strain no. 115 piliated bacteria. Jpn J Vet Sci 49:1073-79.

__________. 1987b. Comparison of experimental infection in mice of *Corynebacterium renale* piliated and non-piliated clones. Jpn J Vet Res 35:79-86.

Gotoh, K; Mitsuyama, M.; Imaizumi, S.; Kawamura, I.; and Yano, I. 1991. Mycolic acid containing glycolipid as a possible virulence factor of *Rhodococcus equi* for mice. Microbiol Immunol 35:175-85.

Hard, G. C. 1975. Comparative toxic effect of the surface lipid of *Corynebacterium ovis* on peritoneal macrophages. Infect Immun 12:1439-49.

Hayashi, A.; Yanagawa, R.; and Kida, H. 1985. Adhesion of *Corynebacterium renale* and *Corynebacterium pilosum* to the epithelial cells of various parts of the bovine urinary tract from the renal pelvis to vulva. Vet Microbiol 10:287-92.

Hedlund, M., and Pohjanvirta, T. 1989. *Corynebacterium ulcerans* in dairy cattle — a potential risk for man. A review on *C. ulcerans* and a survey of its occurrence in mastitic milk in Eastern Finland. Suomen-Elainlaakarilehti 95:552-58.

Henderson, L. C; Kadis, S.; and Chapman, W. L. 1978. Influence of iron on *Corynebacterium renale*-induced pyelonephritis in a rat experimental model. Infect Immun 21:540-45.

Hietala, S. K., and Ardans, A. A. 1987. Interaction of *Rhodococcus equi* with phagocytic cells from *R. equi*-exposed and non-exposed foals. Vet Microbiol 14: 279-94.

Hiramune, T.; Inui, S.; Murase, N.; and Yanagawa, R. 1971. Virulence of three types of *Corynebacterium renale* in cows. Am J Vet Res 32:235-42.

Hiramune, T.; Kikuchi, N.; and Yanagawa, R. 1985. CAMP test for presumptive differentiation of *Corynebacterium renale* and other bovine urinary corynebacteria. Jpn J Vet Sci 47:295-96.

Hodgson, A. L. M.; Bird, P.; and Nisbet, I. T. 1990. Cloning, nucleotide sequence, and expression in *Escherichia coli* of the phospholipase D gene from *Corynebacterium pseudotuberculosis*. J Bacteriol 172:1256-61.

Hodgson, A. L. M.; Krywult, J.; Corner, L.A.; Rothel, J. S.; and Radford, A. J. 1992. Rational attenuation of *Corynebacterium pseudotuberculosis*: Potential cheesy gland vaccine and live delivery vehicle. Infect Immun 60:2900-2905.

Hommez, J.; Devriese, L. A.; Miry, C.; and Castryck, F. 1991. Characterization of 2 groups of *Actinomyces*-like bacteria isolated from purulent lesions in pigs. Zentralbl Veterinaermed [B] 38:575-80.

Honkanen-Buzalski, T.; Griffin, T. K.; and Dodd, F. H. 1984. Observations on *Corynebacterium bovis* infection of the bovine mammary gland. 1. Natural infection. J Dairy Res 51:371-78.

Katsaras, K., and Zeller, V. P. 1978. Studien über das Exotoxin des *Corynebacterium pyogenes* nach Reinigung mittels Gelfiltration und DEAE-Ionaustaucher. Zentralbl Veterinaermed [B] 25:596-604.

Kaup, F. J.; Liebhold, M.; Wendt, M.; and Drommer, W. 1990. *Corynebacterium suis* infections in swine. 2. Morphological findings in the urinary tract with special reference to the bladder. Tieraerztl-Prax 18:595-99.

Koga, T.; Ono, E.; and Yanagawa, R. 1988. Population shift between piliated and non-piliated bacteria in various culture conditions of *Corynebacterium renale*. Jpn J Vet Sci 50:1277-78.

Kubota, T., Yanagawa, R. 1987. Comparison of susceptibility to phagocytosis by mouse peritoneal macrophages between *Corynebacterium renale* piliated and non-piliated clones. Jpn J Vet Sci 49:663-72.

__________ . 1988. Phagocytosis of piliated and non-piliated *Corynebacterium renale* by murine polymorphonuclear leukocytes. Jpn J Vet Sci 50:199-207.

Kudo, Y.; Yanagawa, R.; and Hiramune, T. 1987. Isolation and characterization of monoclonal antibodies against pili of *Corynebacterium renale* and *Corynebacterium pilosum*. Vet Microbiol 13:75-85.

Lammler, C.; Niewerth, B.; and Blobel, H. 1987. Neuraminidase-enhanced hemagglutination of erythrocytes by cultures of *Actinomyces pyogenes*. Med Sci Res 15:1447-48.

Langfeldt, N.; Wendt, M.; and Amtberg, C. 1990. Vergleichende untersuchungen zum nachweis von *Corynebacterium suis* - infektion beim schwein mit hilfe der indirekten immunfluoreszenz und der kultur. Berl Munch Tieraerztl Wochenschr 103:273-276.

Lovell R. 1939. The *Corynebacterium pyogenes* antitoxin content of animal sera. J Pathol Bacteriol 45: 329-38.

__________ . 1944. Further studies on the toxin of *Corynebacterium pyogenes*. J Pathol Bacteriol 56:525-29.

Ludwig, W.; Kirchhof, G.; Weizenegger, M.; and Weiss, N. 1992. Phylogenetic evidence for the transfer of *Eubacterium suis* to the genus *Actinomyces* as *Actinomyces suis* comb. nov. Int J Syst Bacteriol 42:161-65.

Machang'u, R. S., and Prescott, J. F. 1991. Purification and properties of cholesterol oxidase and choline phosphhydrolase from *Rhodococcus equi*. Can J Vet Res 55:332-40.

McNamara, P. J., and Songer, J. G. 1991. Department of Veterinary Science, University of Arizona. Unpublished data.

Madigan, J. E.; Hietala, S; and Muller, N. 1991. Protection against naturally-acquired *Rhodococcus equi* pneumonia in foals by administration of hyperimmune plasma. J Reprod Fert (Suppl) 44:571-82.

Martens, R. J.; Martens, J. G.; and Fiske, R. A. 1990. *Rhodococcus equi* foal pneumonia: Pathogenesis and immunoprophylaxis. Proc Am Assoc Equine Pract 35:199-213.

Menzies, P. I.; Muckle, C. A.; Brogden, K. A.; and Robinson, L. 1991. A field trial to evaluate a whole cell vaccine for the prevention of caseous lymphadenitis in sheep and goat flocks. Can J Vet Res 55:362-66.

Nordmann, P.; Ronco, E.; and Nauciel, C. 1992. Role of T-lymphocyte subsets in *Rhodococcus equi* pneumonia. Infect Immun 60:2748-52.

Pankey, J. W.; Nickerson, S. C.; Boddie, R. L.; and Hogan, J. S. 1985. Effect of *Corynebacterium bovis* infection on susceptibility to major mastitis pathogens. J Dairy Sci 68:2684-93.

Pepin, M.; Fontaine, J. J.; Pardon, P.; Marly, J.; and Parodi, A. L. 1991a. Histopathology of the early phase during experimental *Corynebacterium pseudotuberculosis* infection in lambs. Vet Microbiol 29:123-34.

Pepin, M.; Pardon, P.; Lantier, F.; Marly, J.; Levieux, D.; and Lamand, M. 1991b. Experimental *Corynebacterium pseudotuberculosis* infection in lambs: Kinetics of bacterial dissemination and inflammation. Vet Microbiol 26:381-92.

Pociecha, J. Z. 1989. Influence of *Corynebacterium bovis* on constituents of milk and dynamics of mastitis. Vet Rec 125:628.

Prescott, J. F. 1991. *Rhodococcus equi*: An animal and human pathogen. Clin Microbiol Rev 4:20-34.

Schaufuss, P., and Lammler, C. 1989. Characterization of extracellular neuraminidase produced by *Actinomyces pyogenes*. Zentralbl Bakteriol 271:28-35.

Schild, A .L.; Riet-Correa, F.; del C Mendez, M.; Turnes, G. G.; Reyes, J. C.; Bermudez, J.; and del Carmen Mendez, M. 1985. Aetiological and epidemiological aspects of ulcerative posthitis of bulls. Pesqui Vet Brasil 5:41-46.

Smith, G. R.; Till, D.; Wallace, M.; and Noakes, D. E. 1989. Enhancement of the infectivity of *Fusobacterium necrophorum* by other bacteria. Epidemiol Infect 102:447-58.

Songer, J. G.; Hilwig, R. W.; Leeming, M. N.; Iandolo, J. J.; and Libby, S. J. 1991. Transformation of *Corynebacterium pseudotuberculosis* by electroporation. Am J Vet Res 52:1258-61.

Songer, J. G.; Libby, S. J.; Iandolo, J. J.; and Cuevas, W. A. 1990. Cloning and expression of the phospholipase D gene of *Corynebacterium pseudotuberculosis* in *Escherichia coli*. Infect Immun 58:131-36.

Takai, S.; Ohkura, Y.; Watanabe, Y.; and Tsubaki, S. 1986. Quantitative aspects of *Rhodococcus* (*Corynebacterium*) *equi* in foals. J Clin Microbiol 23:794-96.

Takai, S.; Ohbushi, S.; Koike, K.; Tsubaki, S.; Oishi, H.; and Kamada, M. 1991a. Prevalence of virulent *Rhodococcus equi* in isolates from soil and feces of horses from horse-breeding farms with and without endemic infection. J Clin Microbiol 29:2887-89.

Takai, S.; Koike, K.; Ohbushi, S.; Izumi, C.; and Tsubaki, S. 1991b. Identification of 15- to 17-kilodalton antigens associated with virulent *Rhodococcus equi*. J Clin Microbiol 29:439-43.

Takai, S.; Sekizaki, T.; Ozawa, T.; Sugawara, T.; Watanabe, Y.; and Tsubaki, S. 1991c. Association between a large plasmid and 15- to 17-kilodalton antigens in virulent *Rhodococcus equi*. Infect Immun 59:4056-60.

Takai, S.; Iie, M.; Watanabe, Y.; Tsubaki, S.; and Sekizaki, S. 1992. Virulence-associated 15- to 17-kilodalton antigens in *Rhodococcus equi*: Temperature-dependent expression and localization of the antigens. Infect Immun 60:2995-97.

Takeuchi, S.; Azuma, R.; and Suto, T. 1979a. Purification and some properties of hemolysin produced by *Corynebacterium pyogenes*. Jpn J Vet Sci 41:511-16.

Takeuchi, S.; Azuma, R.; Nakajima, Y.; and Suto, T. 1979b. Diagnosis of *Corynebacterium pyogenes* infection in pigs by immunodiffusion test with protease antigen. Natl Inst Anim Health Q 19:77-82.

Tkachuk-Saad, O., and Prescott, J. F. 1992. *Rhodococcus equi* plasmids: Isolation and partial characterization. J Clin Microbiol 29:2696-2700.

Watson, E. D. 1989. Specific antibody in milk whey and phagocytosis of *Actinomyces pyogenes* by neutrophils in vitro. Res Vet Sci 47:253-56.

Watson, E. D.; Diehl, N. K.; and Evans, J. F. 1990. Antibody response in the bovine genital tract to intrauterine infusion of *Actinomyces pyogenes*. Res Vet Sci 48:70-75.

Yanagawa, R. 1986. Causative agents of bovine pyelonephritis: *Corynebacterium renale, C. pilosum*, and *C. cystitidis*. In Progress in Veterinary Microbiology and Immunology, Vol. 2. Ed. R. Pandey, pp. 158-74. Basel: S. Karger.

6 / *Listeria*

BY C. J. CZUPRYNSKI

THE GENUS *Listeria* contains five species: *L. monocytogenes*, *L. ivanovii*, *L. innocua*, *L. seeligeri*, and *L. welshimeri*. Of these, *L. monocytogenes* is by far the chief pathogen of the genus, being the cause of septicemia, abortion, and central nervous system (CNS) infection in a wide range of animal species, including humans. *L. ivanovii* shares certain characteristics with *L. monocytogenes* (hemolysis) and is occasionally associated with abortion in ruminants, but it does not cause CNS infection and is not of pathogenic significance in other animal species or in humans. The latter three *Listeria* species are relatively avirulent and will not be considered further. From a practical veterinary medicine standpoint, *L. monocytogenes* is primarily of concern in cattle and sheep. In addition, listeriosis has received considerable attention during recent years as an opportunistic pathogen of humans, often with devastating results (Farber and Peterkin 1991; Schuchat et al. 1991). As such, the veterinarian must be aware of both the direct effects of *L. monocytogenes* on infected ruminants and of the public health concerns surrounding the entry of *Listeria*-contaminated meat and dairy products into the human food chain (Fig. 6.1).

CHARACTERISTICS OF THE BACTERIUM

Listeria monocytogenes has several characteristics that make it a unique and problematic pathogen. The organism is widely distributed in nature, having been isolated from soil, vegetation, water, feces, and tissues from a wide variety of vertebrate and invertebrate species (Gray and Killinger 1966). As a result, the exposure of animals to *L. monocytogenes* is unavoidable. As might be expected for a bacterium capable of free-living growth, *L. monocytogenes* can multiply in diverse environmental conditions. It is relatively resistant to high salt concentrations and can grow at a pH range of 5 to 9. Of particular note is its ability to grow at temperatures from 4° to 45°C. The ability to grow at low temperatures has been exploited by both the organism and the laboratory diagnostician. *L. monocytogenes* can multiply to high numbers in silage (Fenlon 1986) or in food products maintained at room or refrigeration temperature (Farber and Peterkin 1991). This poses a threat to the animal or person that ingests the contaminated material, but it also provides a means for the laboratory worker to select and enrich for listeriae in clinical or environmental specimens (i.e., cold enrichment).

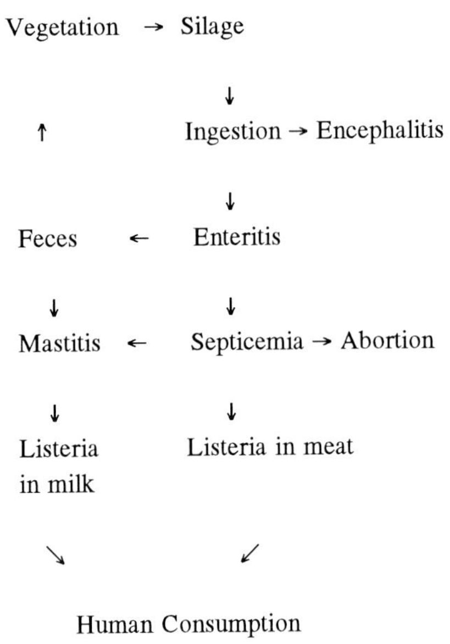

FIG. 6.1. Overview of transmission of *Listeria monocytogenes* in animals and humans.

In the laboratory, *L. monocytogenes* forms small translucent colonies in 24-48 hours when incubated aerobically or anaerobically at 37°C on common bacteriological media such as trypticase soy agar or brain-heart infusion agar. These colonies have a characteristic blue-green sheen when light is reflected obliquely at a 45° angle off their surface (Gray and Killinger 1966). *L. monocytogenes* is hemolytic on blood agar, but the zone of hemolysis can be rather narrow. *L. monocytogenes* produces a CAMP reaction with the hemolysin of *Staphylococcus aureus*. In contrast, *L. ivanovii* typically produces double-zone hemolysis and is negative in the CAMP reaction with *S. aureus* (Schuchat et al. 1991). Gram stain reveals thin, non-spore-forming, gram-positive rods that can vary in length. Glucose and rhamnose are fermented, whereas xylose and mannitol are not. Other diagnostic characteristics include being oxidase negative, catalase positive, hydrogen sulfide negative, and motile at room temperature but not at 37°C. This latter property reflects the production of peritrichous flagella at 25°C or lower, whereas flagella are rare when the organisms are incubated at body temperature (35°-39°C).

Although *L. monocytogenes* can often be isolated from infected tissues, sometimes specimens from suspected listeriosis cases do not yield detectable growth. In these situations, recovery of listeriae can be increased by incubating the tissue at 4°C in nonselective broth for 1 week or longer (Gray and Killinger 1966). Other strategies for the recovery of listeriae include the use of various selective media, such as McBride agar, followed by identification of *Listeria* colonies by the presence of a blue-green sheen when examined by oblique illumination. The emergence of *L. monocytogenes* as a foodborne pathogen of considerable notoriety has led diagnosticians to devise improved selective media, as well as novel methods such as antibody-based tests or DNA probes. For comprehensive reviews of these new procedures the reader is referred elsewhere (Farber and Peterkin 1991; Schuchat et al. 1991).

There are 11 recognized serotypes for the somatic and flagellar antigens of *L. monocytogenes*. Of these 11 serotypes, nearly all cases of animal and human infection are caused by serotypes 1/2a, 1/2b, and 4b (Gudding et al. 1989; Farber and Peterkin 1991; Schuchat et al. 1991). A phage typing system has also been described that has been useful in epidemiologic investigation of some outbreaks, but the proportion of untypable strains restricts its wider use.

TYPES OF DISEASE

Disease in Ruminants

Listeria monocytogenes is an infrequent, but serious, pathogen of both domestic and exotic ruminants (Gray and Killinger 1966). In the United States, greatest attention is usually placed on infections of dairy and beef cattle. In those areas of North America where sheep raising is an important industry and in Europe, listeriosis in sheep and goats assumes greater importance. In both cattle and sheep, listeriosis can manifest itself in one of four ways: (1) as a CNS infection (meningoencephalitis in adults and meningitis in the young); (2) as abortion; (3) as a generalized septicemia with involvement of the liver and other organs; and (4) as mastitis in dairy cattle.

Listeriosis of the CNS is a common presentation that must be included in the differential diagnosis of ruminants with neurologic changes. It is an unexplained curiosity that meningoencephalitis is seen only in adult ruminants; whereas, in monogastric species or young ruminants, before the rumen becomes functional, CNS infection typically presents as meningitis without brain involvement (Gray and Killinger 1966). The clinical presentation of meningoencephalitis in adult ruminants may begin with signs of depression and confusion (Low and Renton 1985; West and Obwolo 1989). The ears droop and the animal holds its head to one side. Protrusion of the tongue and salivation are common, and twitching or paralysis of the facial and throat muscles may occur. The animal may lean against fences or other objects for support. When the animal moves, it tends to be in a single direction, giving rise to the common name of "circling disease." In the terminal stages, the animal may fall and be unable to rise; tremors are common. Unlike meningitis in humans, *L. monocytogenes* is seldom recovered from the cerebrospinal fluid in ruminant meningoencephalitis. Spontaneous recovery from severe clinical disease is rare and may be accompanied by permanent neurologic damage.

The pathogenesis of meningoencephalitis in ruminants is only partially understood. There is a definite seasonal association, with disease being more common in winter or early spring when the animals are indoors rather than on pasture. There is also a strong association between listeriosis and animals fed silage rather than hay (Barlow and McGorum 1985; Low and Renton 1985; Gudding et al. 1989). The ability of *L. monocytogenes* to multiply in poor-quality silage is well documented (Fenlon 1986). It is thought that the infection is acquired when animals ingest *Listeria*-contaminated silage. It is not clear how *L. monocytogenes* breaches the barrier of the oral mucosal epithelium, although Barlow and McGorum (1985) have suggested it may penetrate through the dental pulp when sheep are cutting or losing teeth. There is persuasive evidence that once *L. monocytogenes* leaves the oral cavity, it invades the trigeminal nerves and travels centripetally along the axons to the brain (Asahi et al. 1957; Charlton and Garcia 1977). Observations of spontaneous disease in cattle (West and Obwolo 1985) are consistent with this route of invasion. In addition, morphologic evidence for a similar progression in the absence of septicemia has been presented for mice and rabbits inoculated in the lip with *L. monocytogenes* (Asahi et al. 1957). Grossly, the brains of ruminants suffering from encephalitis may appear normal. The typical histopathologic lesions observed are perivascular cuffing of mononuclear cells and occasional inflammatory foci (microabscesses) that contain both granulocytes and mononuclear cells. These lesions are most common in the midbrain, pons, and medulla oblongata (Asahi et al. 1957; Charlton and Garcia 1977). Gram-stained tissue sections may reveal listeriae in large lesions, but they are infrequently observed in the smaller perivascular cuffs. *L. monocytogenes* appears to have a predilection for invading the fetoplacental unit in a variety of animal species (Gray and Killinger 1966). This may or may not follow obvious septicemia. Experimental infections in mice suggest that T cells and macrophages do not reach the infected placenta in sufficient numbers to effect protective immunity (Redline et al. 1988). Abortion typically occurs in the third trimester of pregnancy and may not result in obvious clinical disease in the dam. The organism can be microscopically visualized or recovered by culture from the aborted fetus or placenta (Low and Renton 1985).

In addition to meningoencephalitis and abortion, *L. monocytogenes* has been associated with generalized septicemia and focal necrosis of the spleen and liver in various ruminant species (Gray and Killinger 1966; Low and Renton 1985). Considerable attention has been paid to the problem of *L. monocytogenes* mastitis in dairy cattle (Gitter 1980). This can range in severity from subclinical to severe suppurative infection. *L. monocytogenes* organisms can be found within neutrophils in mastitic milk (Doyle et al. 1987). Evidence was presented that this intracellular site protected some of the listeriae against inactivation by short-term pasteurization, but others have not found this to be the case (Bunning et al. 1988). Long-term infection of the udder may ensue, with shedding of listeriae in milk exacerbated by periodic episodes of immunosuppression (Wesley et al. 1989). This has practical implications, since successful treatment of *L. monocytogenes* mastitis is difficult to achieve (Gitter et al. 1980).

Listeriosis in Nonruminant Animal Species

Clinical listeriosis is rare in horses, pigs, dogs, and cats. Healthy pigs may excrete *L. monocytogenes* in their feces, but listeriosis in pigs is not of practical significance (Blenden 1986). Septicemic listeriosis in dogs and cats is extremely rare (Greene 1990). An outbreak of encephalitis in commercial broiler chickens has been reported; the source of the organism in that outbreak was not determined (Cooper 1989). Laboratory investigations of listeriosis rely heavily on the use of rodents, rabbits, and guinea pigs, all of which are quite susceptible to listeriosis. The prevalence of spontaneous listeriosis in these species kept as pets is not known, but practitioners whose practice includes pocket pets might keep listeriosis in mind as a possible diagnosis for unexplained septicemia or abortion.

Listeriosis in Humans

Listeria monocytogenes is considered a zoonotic agent. Infection may be transmitted directly from infected meat or milk, or indirectly by *L. monocytogenes*-containing manure that contaminates vegetables, which are then consumed by people. In humans, clinical listeriosis occurs most commonly in pregnant women (often resulting in abortion) and as septicemia and meningitis in immunodeficient adults (Farber and Peterkin 1991; Schuchat et al. 1991). There have been several large outbreaks of human listeriosis. The largest of these in the United States occurred in Los Angeles County, as a result of ingestion of *Listeria*-contaminated Mexican-style cheese made with unpasteurized milk. Of a total of 142 cases, 93 occurred in pregnant women and their fetuses. The case fatality rate was 63% for early neonatal or fetal infections and 37% for nonneonatal infections (Farber and Peterkin 1991; Schuchat et al. 1991). As in other outbreaks, most of the latter group of patients suffered from some underlying immunodeficiency. A recent Centers for Disease Control survey for the period 1988-1990 estimated that the overall annual incidence of listeriosis in the United States was 7.4 cases per 1 million people, with a fatality rate of 23% (Anderson et al. 1992). Foods incriminated with transmitting *L. monocytogenes* infection in this study included soft cheeses, delicatessen items, and undercooked chicken and hot dogs. Identifying the original source of *L. monocytogenes* contamination in foodborne infections can be difficult. It may well be that contamination occurs most frequently at the food-processing plant or at a food-preparation facility rather than on the farm (Farber and Peterkin 1991). Nonetheless, because listeriosis has achieved considerable visibility as a food safety concern among regulatory agencies and the public, veterinary practitioners and diagnosticians will be expected to assist food animal producers in reducing the likelihood of their raw product being contaminated with *L. monocytogenes*.

BACTERIAL VIRULENCE FACTORS

During the past several years, remarkable progress has been made in identifying virulence determinants of *L. monocytogenes*. Although this work has been done using tissue culture cells and mouse infection models, there is reason to believe that the general principles they have delineated are operative in *L. monocytogenes* infections of ruminants as well. A recent review of this subject provides a concise and lucid description of how various virulence factors allow *L. monocytogenes* to cause progressive infection (Portnoy et al. 1992). As illustrated in Figure 6.2, *L. monocytogenes* must invade a cell (both professional phagocytes and epithelial cells are susceptible) (A), escape from the phagosome (B), and enter the cytoplasm where it then multiplies with a doubling time of approximately 1 hour (C). It then directs the nucleation of host cell-derived actin filaments that, in an as yet unexplained manner, propel the organism to the cell's periphery (D). Projections of the infected cell then invaginate into adjacent cells (E), and transmit the listeriae, which then escape from the double membrane that encloses them (F), and the process of intracellular multiplication is repeated (G). As a result, *L. monocytogenes* is able to multiply and spread without direct exposure to the cells and soluble factors of the extracellular milieu.

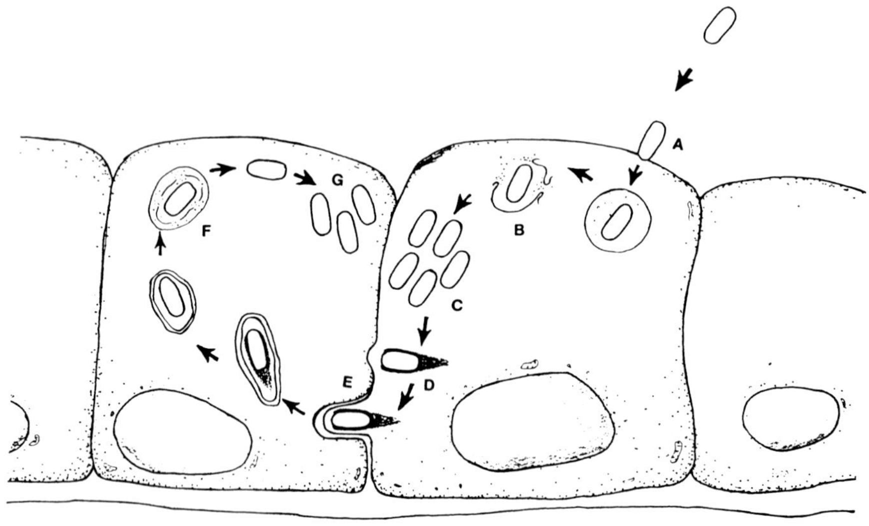

FIG. 6.2. Overview of events in the cycle of cell invasion and intracellular multiplication of *Listeria monocytogenes*.

Genes whose products are responsible for several of these intracellular events have been identified, cloned, and sequenced. Invasion (and perhaps adhesion) requires a unique 80-kDa surface protein known as internalin, which bears some structural similarities to streptococcal M proteins (Gaillard et al. 1991). Internalin binds to an as yet unidentified receptor on mammalian cells; the mechanism by which internalization of the listeriae occurs is also unknown.

Once inside the cell, *L. monocytogenes* relies on its cholesterol-binding hemolysin (listeriolysin O, related to streptolysin O) and perhaps a phosphatidylinositol-dependent phospholipase C (Portnoy et al. 1992) to lyse the phagosome membrane and escape into the cytoplasm. A 90-kDa surface protein (ActA) then directs the host-dependent arrangement of actin filaments that propel the listeriae about the cytoplasm (Kocks et al. 1992). Mutants that are unable to produce ActA can multiply intracellularly but are unable to invade adjacent cells. Other gene products that may play roles in cell-to-cell spread of *L. monocytogenes* include a lecithinase, which may aid in lysis of the double membrane the listeriae encounters after migrating from one cell to the next, and a temperature-regulated positive regulatory factor (prfA), which coordinates the expression of several virulence-associated genes (Portnoy et al. 1992). The studies that have provided the information described above represent an elegant marriage of contemporary molecular biology and traditional cell biology. What is particularly satisfying is the manner in which they provide a means of synthesizing and explaining earlier descriptive observations of the ability of *L. monocytogenes* to invade epithelial cells and neuronal cells in vivo (Racz et al. 1972; Charlton and Garcia 1977).

PATHOGENESIS

Being an intracellular pathogen of macrophages and epithelial cells, resistance to *L. monocytogenes* requires cellular immunity. Antibodies directed against *L. monocytogenes* antigens are not protective. Most of the available information on host defense against listeriosis is based on experimental studies in mice, which provide a popular and valuable model for assessing immunoregulation of antibacterial resistance to an intracellular pathogen. Despite the acknowledged role of *L. monocytogenes* as a pathogen of ruminants, there are very few published reports about host defense against listeriosis in cattle or sheep. The information given below, therefore, is based principally on murine studies, with relevant reports from cattle and sheep stated where available. For a concise recent review of anti-listeria resistance, the reader is referred to Portnoy (1992).

The pioneering studies of Mackaness and coworkers in the 1960's demonstrated the preeminent role of cellular immunity in host defense against listeriosis (Miki and Mackaness 1964). For many years, investigators adhered to a simple paradigm in which T-helper cells released IFN-γ which activated macrophages to kill intracellular listeriae. While macrophage activation certainly occurs in listeriosis, recent evidence indicates that the protective host response to listeriosis is considerably more complex. It has been demonstrated that considerable multiplication of *L. monocytogenes* occurs in hepatocytes and other nonphagocytic cells (Havell 1989; Conlan and North 1991; Gregory et al. 1992). This finding indicates that additional defense mechanisms must be operative to achieve the sterilizing immunity observed in experimentally infected mice.

Various events related to anti-listeria resistance have been described. The first cells to enter foci of infection are neutrophils. Neutrophils from several mammalian species (including bovine) can kill *L. monocytogenes* in vitro (Czuprynski et al. 1984, 1989). Although antibodies are not protective in vivo, these studies showed that antibodies and complement enhance the uptake and subsequent killing of *L. monocytogenes* by neutrophils and macrophages in vitro. Recently, evidence was presented that showed mice treated with a monoclonal antibody, which prevents influx of neutrophils were exquisitely sensitive to listeriosis (Conlan and North 1991). Thus, one can propose that neutrophils protect the host by both destroying infected cells that harbor the intracellular listeriae and by ingesting and killing the listeriae that are released from the lysed cells. Additional cells might also seek and destroy *L. monocytogenes*-infected target cells. $CD4^+$ and $CD8^+$ cytolytic T cells have been described that can lyse *L. monocytogenes*-infected macrophages (Kaufmann 1988). Perhaps similar T cells attack other types of cells infected with *L. monocytogenes*, as occurs in viral infections. Other investigators have presented evidence that natural killer cells (NK cells) release IFN-γ early in infection and that this is of considerable importance to host defense (Dunn and North 1991).

Lest one get the impression that macrophages have been relegated to a minor role, there is evidence that IFN-γ-treated activated macrophages restrict the ability of *L. monocytogenes* to escape from the phagolysosome (Portnoy et al. 1989). Thus, macrophage activation might prevent the intracellular multiplication of listeriae, as well as increase the ability of macrophages to inactivate ingested organisms. A report regarding the negative effects of increased intracellular concentrations of iron on macrophage anti-listeria activity is relevant in light of the known adverse effects of high iron levels on anti-listeria resistance in vivo (Alford et al. 1991). Although it has not been reexamined in light of contemporary concepts and reagents, earlier papers described the ability of sheep macrophages to ingest and inactivate *L. monocytogenes*; these studies provided evidence that serum factors both opsonized and activated the antibacterial activity of sheep macrophages (Njoku-Obi and Osebold 1961). Despite numerous studies of the interaction of *L. monocytogenes* with macrophages, a number of essential questions remain unanswered. Does opsonization influence the intracellular compartmentalization and fate of ingested listeriae? Which cytokines activate macrophage anti-listeria activity? What macrophage products (e.g., oxygen radicals, nitrogen intermediates, cationic peptides) are most important for intracellular anti-listeria activity? Do these exert their effects by killing the organism or by altering the intracellular environment in such a way that it is inhospitable for intracellular listerial multiplication and spread?

In addition to the cells described above, there is an abundant literature on the influence of various soluble mediators (cytokines) on anti-listeria resistance. Administration of a number of recombinant cytokines (e.g., IFN-γ, IL-1, IL-2, TNF-α, GM-CSF) has been shown to enhance resistance to experimental listeriosis in mice. Conversely, investigators have used cytokine-neutralizing monoclonal antibodies to study the roles of endogenous cytokines in anti-listeria resistance. In particular, endogenous IFN-γ and TNF-α have been shown to play critical roles; in vivo neutralization of either of these cytokines has dire consequences for the infected host (Buchmier and Schreiber 1985; Havell 1989). Conversely, the T cell-derived cytokine IL-4 may actually serve to restrict resistance, since mice treated with an anti-IL-4 MAb exhibited increased anti-listeria resistance (Haak-Frendscho et al. 1992). This latter observation is of interest in that it fits with the current working model that T-helper 1 cells (that produce IL-2 and IFN-γ) activate cellular immunity and T-helper 2 cells (that produce IL-4 and IL-5) down-regulate cellular immunity and select for antibody production.

IMMUNITY

Experimental studies with rodents indicate that clearance of a sublethal *L. monocytogenes* infection renders the host highly resistant to rechallenge (Miki and Mackaness 1964). These animals will also display a strong, delayed-type hypersensitivity response to listeriae antigens, although this skin response can be dissociated from protective immunity (Mielke et al. 1988). This heightened acquired cellular resistance is maximal for 2-4 weeks, and then it wanes over the ensuing several months. During the period of maximal resistance, protection can be transferred to naive recipient mice by immune spleen cells or purified T cells, but not with serum (Miki and Mackaness 1964). The level of resistance transferred in this way is substantial but less than that resulting from active immunization. It is generally agreed that active immunization requires the administration of viable *L. monocytogenes*; killed organisms do not generate protective cellular immunity. Immunization with live attenuated *L. monocytogenes* can protect sheep against naturally occurring or experimental listeriosis (Gudding et al. 1985; Vagsholm et al. 1991).

Listeriosis remains a sporadic disease problem for cattle and sheep producers in the United States. Most outbreaks of listeriosis have been traced to the use of poor-quality silage that contained increased numbers of *L. monocytogenes*. Ensuring use of good-quality, low-pH silage, keeping bulk-tank temperatures low, and practicing good hygiene to reduce fecal contamination should go a long way toward reducing *L. monocytogenes* contamination of milk, meat, and vegetables. Similar attention to scrupulous cleaning and hygiene must be maintained at dairy plants and food-production facilities. As a result, there would appear to be little economic justification for a listeriosis vaccination program in North America. In certain parts of Europe, however, vaccination with live attenuated strains of *L. monocytogenes* serotypes 1/2a, 1/2b, and 4b are a common practice that has resulted in cost-effective reductions in the incidence of listeriosis in sheep (Vagsholm et al. 1992).

Antibiotic treatment of food animal species with listeriosis may not be advisable considering the cost of therapy and losses associated with withdrawal times for the antibiotics used. Mastitis has proven refractory to treatment in several instances when it has been attempted. If latent udder infections occur, significant numbers of listeriae might be shed in the milk when the cow experiences transient immunosuppressive events, such as parturition (Wesley et al. 1989). In cases of abortion, the dam may clear the infection without treatment. In human listeriosis, the current treatment of choice is penicillin in combination with gentamicin. Sulfamethoxazole-trimethroprim, erythromycin, and tetracyclines have also proven effective, whereas cephalosporins are not recommended, even in combination with other antibiotics.

PROBLEMS

Basic research into the virulence factors of and host response to *L. monocytogenes* is in a dynamic state with new contributions being made at a rapid rate. Unfortunately, our understanding of the pathogenesis of listeriosis in ruminants has not kept pace. There is a paucity of published information on host defense mechanisms in cattle and sheep. Some of the observations made in these species (i.e., centripetal invasion of *L. monocytogenes* along the trigeminal nerve) offer exciting possibilities as models for further study of the roles of *L. monocytogenes* virulence factors. A better understanding of the site of bacterial invasion in the udder would also be helpful in evaluating the problem posed by mastitis.

REFERENCES

Alford, C. E.; King Jr., T. E.; and Campbell, P. A. 1991. Role of transferrin, transferrin receptors, and iron in macrophage listericidal activity. J Exp Med 174:459-66.

Anderson, G.; Mascola, L.; Rutherford, G.W.; Rados, M. S.; Hutcheson, R.; Archer, P.; Zenker, P.; Harvey, C.; and Smith, J. D.; 1992. Food borne Listeriosis — United States, 1988-1990. J Am Med Assoc 267:2446-47.

Asahi, O.; Hosoda, T.; and Akiyama, Y. 1957. Studies on the mechanism of infection of the brain with *Listeria monocytogenes*. Am J Vet Res 18:147-57.

Barlow, R. M.; and McGorum, B. 1985. Ovine listerial encephalitis: analysis, hypothesis and synthesis. Vet Rec 116:233-36.

Blenden, D. C. 1986. Listeriosis. In Diseases of Swine, 6th Ed. Ed. A. D. Leman, B. Straw, R. D. Glock, W. L. Mengeling, R. H. C. Penny, and E. Scholl, pp. 584-90. Ames: Iowa State University Press.

Buchmier, N. A., and Schreiber, R. D. 1985. Requirement of endogenous interferon-gamma for resolution of *Listeria monocytogenes* infection. Proc Natl Acad Sci USA 82:7404-8.

Bunning, V. K.; Donnelly, C. W.; Peeler, J. T.; Briggs, E. H.; Bradshaw, J. G.; Crawford, R. G.; Beliveau, C. M.; and Tierney, J. T. 1988. Thermal Inactivation of *Listeria monocytogenes* within bovine milk phagocytes. Appl Environ Microbiol 54:364-70.

Charlton, K. M., and Garcia, M. M. 1977. Spontaneous listeric encephalitis and neuritis in sheep. Vet Pathol 14:297-313.

Conlan, J. W., and North, R. J. 1991. Neutrophil-mediated dissolution of infected host cells as a defense strategy against a facultative intracellular bacterium. J Exp Med 174:741-44.

Cooper, G. L. 1989. Case Report — an encephalitic form of listeriosis in broiler chickens. Avian Dis 33:182-85.

Czuprynski, C. J.; Henson, P. M.; and Campbell, P. A. 1984. Killing of *Listeria monocytogenes* by inflammatory neutrophils and mononuclear phagocytes from immune and nonimmune mice. J Leuk Biol 35:193-208.

Czuprynski, C .J.; Noel, E. J.; Doyle, M. P.; and Schultz, R. D. 1989. Ingestion and killing of *Listeria monocytogenes* by blood and milk phagocytes from mastitic and normal cattle. J Clin Microbiol 27:812-17.

Doyle, M. P.; Glass, K. A.; Beery, J. T.; Garcia, G. A.; Pollard, D. J.; and Schultz, R. D. 1987. Survival of *Listeria monocytogenes* in milk during high-temperature, short-time pasteurization. Appl Environ Microbiol 53:1433-38.

Dunn, P. L., and North, R. J. 1991. Early gamma interferon production by natural killer cells is important in defense against murine listeriosis. Infect Immun 59:2892-2900.

Farber, J. M., and Peterkin, P. I. 1991. *Listeria monocytogenes*, a food-borne pathogen. Microbiol Rev 55:476-511.

Fenlon, D. R. 1986. Rapid quantitative assessment of the distribution of listeria in silage implicated in a suspected outbreak of listeriosis in calves. Vet Rec 118:240-42.

Gaillard, J. L.; Berche, P.; Frehel, C.; Gouin, E.; and Cossart, P. 1991. Entry of *L. monocytogenes* into cells is mediated by internalin, a repeat protein reminiscent of surface antigens from gram-positive cocci. Cell 65:1127-41.

Gitter, M.; Bradley, R.; and Blampied, P. H. 1980. *Listeria monocytogenes* infection in bovine mastitis. Vet Rec 107:390-93.

Gray, M. L., and Killinger, A. H. 1966. *Listeria monocytogenes* and listeric infections. Bacteriol Rev 30:309-82.

Greene, C. E. 1990. Listeriosis. In Infectious Diseases of the Dog and Cat. Ed. C. E. Greene, pp. 607-8. Philadelphia: W.B. Saunders Co.

Gregory, S. H.; Barczynski, L. K.; and Wing, E. J. 1992. Effector functions of hepatocytes and Kupffer's cells in the resolution of systemic bacterial infections. J Leuk Biol 51:421-24.

Gudding, R.; Nesse, L. L.; and Grønstøl, H. 1989. Immunization against infections caused by *Listeria monocytogenes* in sheep. Vet Rec 125:111-14.

Haak-Frendscho, M.; Brown, J. F.; Iizawa, Y.; Wagner, R. D.; and Czuprynski, C. J. 1992. Administration of anti-IL-4 monoclonal antibody 11B11 increases the resistance of mice to *Listeria monocytogenes* infection. J Immunol 148:3978-85.

Havell, E. A. 1989. Evidence that tumor necrosis factor has an important role in antibacterial resistance. J Immunol 143:2894-99.

Kaufmann, S. H. E. 1988. $CD8^+$ T lymphocytes in intracellular microbial infections. Immunol Today 9:168-74.

Kocks, C.; Gouin, E.; Tabouret, M.; Berche, P.; Ohayon, H.; and Cossart, P. 1992. *L. monocytogenes*-induced actin assembly requires the *act*A gene product, a surface protein. Cell 68:521-31.

Low, J. C., and Renton, C. P. 1985. Septicaemia, encephalitis and abortions in a housed flock of sheep caused by *Listeria monocytogenes* type 1/2. Vet Rec 116:147-50.

Mielke, M. E. A.; Ehlers, S.; and Hahn, H. 1988. T-cell subsets in delayed-type hypersensitivity, protection, and granuloma formation in primary and secondary *Listeria* infection in mice: Superior role of Lyt-2^+ cells in acquired immunity. Infect Immun 56:1920-25.

Miki, K., and Mackaness, G. B. 1964. The passive transfer of acquired resistance to *Listeria monocytogenes*. J Exp Med 129:93-108.

Njoku-Obi, A. N., and Osebold, J. W. 1961. Studies on mechanisms of immunity in listeriosis. 1. Interaction of peritoneal exudate cells from sheep with *Listeria monocytogenes* in vitro. J Immunol 89:187-94.

Portnoy, D. A. 1992. Innate immunity to a facultative intracellular bacterial pathogen. Current Opin Immun 4:20-24.

Portnoy, D. A.; Schreiber, R. D.; Connelly, P.; and Tilney, L. G. 1989. γInterferon limits access of *Listeria monocytogenes* to the macrophage cytoplasm. J Exp Med 170:2141-46.

Portnoy, D. A.; Chakraborty, T.; Goebel, W.; and Cossart, P. 1992. Molecular determinants of *Listeria monocytogenes* pathogenesis. Infect Immun 60:1263-67.

Rácz, P.; Tenner, K.; and Mérö, E. 1972. 1. An electron microscopic study of the epithelial phase in experimental listeria infection. Lab Invest 26:694-700.

Redline, R. W.; Shea, C. M.; Papaioannou, V. E.; and Lu, C. Y. 1988. Defective anti-listerial responses in deciduoma of pseudopregnant mice. Am J Pathol 133:485-97.

Schuchat, A.; Swaminathan, B.; and Broome, C. V. 1991. Epidemiology of human listeriosis. Clin Microbiol Rev 4:169-83.

Vagsholm, J.; Nesse, L. L; and Gudding, R. 1991. Economic analysis of vaccination applied to ovine listeriosis. Vet Rec 128:183-85.

Wesley, I. V.; Bryner, J. H.; Van Der Maaten, M. J.; and Kehrli, M. 1989. Effects of dexamethasone on shedding of *Listeria monocytogenes* in dairy cattle. Am J Vet Res 50:2009-13.

West, H. J., and Obwolo, M. 1987. Bilateral facial paralysis in a cow with listeriosis. Vet Rec 120:204-5.

7 / *Erysipelothrix rhusiopathiae*

BY C. L. GYLES

ERYSIPELAS is a disease caused by the bacterium *Erysipelothrix rhusiopathiae* and is of economic importance in turkeys and swine (Wood 1992). Immunization and effective antibiotic therapy have resulted in a reasonable degree of prevention and satisfactory treatment of the disease. The impetus for studies on pathogenesis has therefore been lacking, and very little is understood about features of the organism, host, and environment that are critical to the disease process.

E. rhusiopathiae is a slender, gram-positive, rod-shaped bacterium that may take the form of single bacteria, or short rods, or it may be filamentous. The organism is nonmotile, alpha-hemolytic, catalase-negative; produces H_2S on triple-sugar iron agar (TSI); and grows over a temperature range of 4-37°C. A characteristic reaction is test tube brush growth in gelatin incubated at 21°C (Jones 1986), and the ability to produce coagulase may be a useful feature in differentiating this organism from *Listeria* and *Corynebacterium* species (Tesh and Wood 1988).

E. rhusiopathiae is divided into 23 serotypes identified as 1-23 and those isolates that are nontypable are designated "N"; strains of serotype 1 are subdivided into 1a and 1b. Serotypes 1a, 1b, and 2 are most frequently implicated in disease in swine, but some strains of serovar 1a are avirulent (Takahashi et al. 1987b,c). Typing is based on heat-stable, soluble peptidoglycan extracted from the organisms and demonstrated in a precipitin test. Seven serovars (3, 7, 14, 20, 22, 23) of *E. rhusiopathiae* have been transferred to a new species of *Erysipelothrix*, named *E. tonsillarum* (Takahashi et al. 1987a; 1992); this realignment of the species was based on DNA-DNA hybridizations. Strains of *E. tonsillarum* are nonpathogenic or of low virulence for swine; however, some serotypes of *E. tonsillarum* are pathogenic for other species, a notable example being strains of serotype 7, which are pathogenic for dogs (Eriksen et al. 1987)

The bacterium is widespread in nature and has been recovered from a wide variety of wild and domestic animals including mammals, birds, reptiles, amphibians, and fish. The organism does not appear to survive for long in soil, and the major source of infection for swine and turkeys is carrier animals of the same species (Wood 1992). It is reported that 30-50% of pigs carry the bacterium in their tonsils and other lymphoid tissues; turkeys are known to carry the organisms in their cecal tonsils, liver, and other organs. *E. rhusiopathiae* is resistant to several chemicals including sodium azide, and to drying, pickling, salting, and smoking (Wood 1984).

DISEASES

Four forms of clinical disease in swine have been described: acute septicemia, urticarial or "diamond skin" lesions, vegetative endocarditis, and arthritis (Wood 1992). These may occur alone or in combinations. An essentially similar array of disease syndromes occurs in turkeys in which the most common form is an acute septicemia, but skin lesions, endocarditis, and arthritis are also seen. In sheep, *E. rhusiopathiae* causes polyarthritis ("joint-ill"), which occurs infrequently and is associated with wound infections in lambs. *E. rhusiopathiae* has been suggested as one of the causes of rheumatoid arthritis in dogs. The organism has been isolated from a limited number of cases in which antibody to the bacterium was demonstrated. It is an infrequent cause of septicemia and endocarditis in dogs. In humans, erysipeloid is a cutaneous infection by *E. rhusiopathiae* that typically occurs in veterinarians, fish handlers, meat-processing workers, and housewives; it is usually limited to the local site but occasionally can lead to bacteremia and infection of the heart valve or joints. Disease occurs in a wide variety of other species of mammals as well as in birds and fish.

Erysipelas in swine may run an acute, subacute, or chronic course. The acute disease is characterized by high temperature (40-42°C), inappetence, depression, a rapid course of illness, and death within 2-3 days in untreated animals. Some animals may show a stiff gait and reluctance to stand or move, and urticarial cutaneous lesions may develop. The diamond-shaped, raised skin lesions are pathognomonic. Pregnant sows may abort. Subacute disease is similar to the acute except that it is less severe and animals are likely to recover within 5-7 days. In the chronic disease, arthritis is the prominent feature; the hock, stifle, elbow, and carpal joints are most likely to be affected. The chronic course of the disease may be a sequel to acute, subacute, or unrecognized infection.

In turkeys, the disease is usually evident by the sudden death of birds in the flock. Toms are affected more often than are hens, but hens may die suddenly 4-5 days after artificial insemination. Cutaneous lesions, particularly those affecting the snoods, may be evident. A chronic course will develop in some birds, which may develop vegetative endocarditis or arthritis and become gradually emaciated. Intestinal damage and diarrhea are sometimes observed in infections in birds.

PATHOGENESIS

The course of disease depends markedly on the virulence of the strain and the resistance of the host. Individual strains of *E. rhusiopathiae* vary considerably in virulence, but the factors responsible for this variation have not been identified (Bohm and Suphasindhu 1980). Animals have varying degrees of resistance to the organism and at least a portion of this can be related to natural or artificial exposure to the bacterium. Exposure of a fully susceptible host to virulent strains is likely to lead to acute septicemia.

An overview of the steps in pathogenesis is shown in Figure 7.1. It is known that carrier animals are an important source of the organism, that the bacteria are able to adhere to epithelial cells, and that they invade the bloodstream and cause localizations. There is little information on detailed aspects of pathogenesis.

Entry of the organisms may be by the oral, cutaneous, or respiratory route. Ingestion of contaminated feed or water or contamination of abraded skin are the most common means of infection in swine. These routes, as well as insemination of hens with contaminated semen, are important in turkeys.

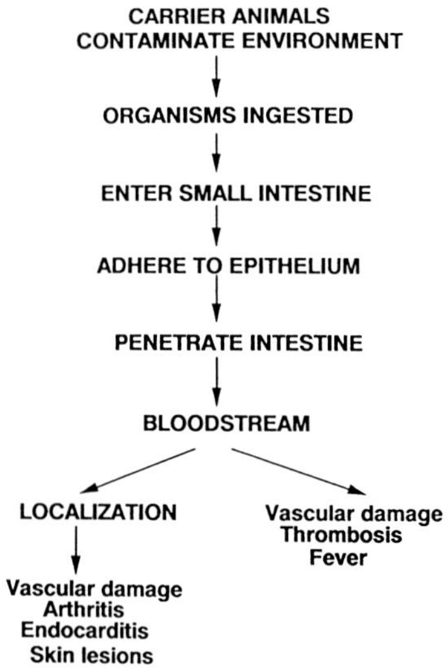

FIG. 7.1. Steps in the pathogenesis of disease due to *Erysipelothrix rhusiopathiae*.

Adherence may be an important factor in virulence, but there are few investigations of this phenomenon. Takahashi et al. (1987b) compared four strains that were virulent for mice and pigs with four strains that were avirulent, with respect to their ability to adhere to porcine kidney cells in culture. The number of adherent virulent strains per cell ranged from 0.1 to 1.4, whereas the corresponding numbers for the virulent strains were from 9.9 to 36.0. Thus, adherence may be related to virulence. The bacterial structure required for adherence was heat-labile and trypsin-sensitive. In an earlier study, Bratberg (1981) had shown that strains of *E. rhusiopathiae* recovered from heart valves or blood of swine showed a similar pattern of adherence to heart valves of swine in organ culture compared with isolates from other sources.

Intravenous or intraarticular inoculation of rabbits with live organisms, killed whole cells, or certain cell-free extracts of *E. rhusiopathiae* has resulted in chronic arthritis (Hermanns et al. 1981). In vitro studies have demonstrated that certain fractions of the bacterium are cytotoxic for cell cultures and other fractions are mitogenic (Rimpler et al. 1982). These substances have not been shown to play a role in disease.

Early lesions in the septicemic disease are dominated by vascular disturbances (Wood 1984). There is swelling of the endothelium of capillaries and venules, adherence of monocytes to the walls of blood vessels, and subsequent hyaline thrombosis. Within 4 days, there is progression to a fibrinous thrombosis, with diapedesis, penetration of bacteria into the vascular endothelium, and escape of fibrin into perivascular tissues. Hemolysis and ischemic necrosis may develop later in some animals. In the subacute and chronic forms of the disease, there is a predisposition of the organism for joints, heart valves, and arteries. It has been suggested that the predilection is associated with higher permeability of these tissues that are passively supplied by perfusion of blood plasma, difficulty in eliminating the antigen because of the absence of blood vessels, and persistence of fibrin depots because of insufficient plasminogen activators (Schulz et al. 1980). Development of microthrombi composed of fibrin and erysipelas bacteria have also been observed in intertubular capillaries in the kidneys of turkeys experimentally infected with *E. rhusiopathiae* (Tsangaris et al. 1980).

Arthritis is associated with initial infection of the joints and prolonged retention of bacterial antigens in the joints after viable bacteria can no longer be demonstrated (Muller 1980; Schulz et al. 1980; Drommer 1982). The early preimmune phase is characterized by proliferation of synovial tissue, leading to destruction of the cartilage. Later developments in the immune phase are due to reactions of antigen with antibody. The slowly progressive course and morphologic changes observed are similar to human rheumatoid arthritis (Schulz et al. 1980). Drommer (1982) has evidence that early in the course of infection the bacteria cause mild, direct damage to cartilage, but later damage is due to severe proliferation of synovial-lining cells and invasive pannus.

Early in the course of experimental *Erysipelothrix*-induced arthritis, the bacteria are observed on the endothelial surfaces of blood vessels and are associated with a marked increase in vascular permeability (Schulz et al. 1980; Drommer 1982). Whole, killed bacteria or cell-free extracts of *E. rhusiopathiae* administered by repeated intravenous or by intraarticular injection into rabbits result in a chronic purulent arthritis (Hermanns et al. 1982).

No toxins have been demonstrated for this organism, but the enzyme neuraminidase has been suggested as a product that may play a role in the virulence of strains. Neuraminidase is produced both in vitro and in vivo, and a correlation has been shown between the amount of neuraminidase produced and the virulence of strains. Muller (1980) proposed that neuraminidase plays a critical role in pathogenicity and suggested that the enzyme may enhance the penetration of cells by the microorganism and inflict damage on many different substrates in the host. Bacterial neuraminidases remove N-acetylneuraminic acid (NANA) from a number of substrates but exhibit varying specificities with respect to the bonds that they cleave. Removal of NANA from mucin, fibrinogen, and erythrocytes results in a reduction in protection afforded by mucin, formation of fibrinlike structures in blood vessels, increased viscosity of blood, and hemolysis. Because NANA is involved in control of the lifetime of glycoproteins, erythrocytes, leukocytes, and thrombocytes, its removal leads to enhanced catabolism resulting in anemia, leukopenia, and thrombocytopenia. Effects attributable to the removal of NANA from heart cells and immunoglobulin G (IgG) are disturbance in control of calcium exchange in the heart and deposition of IgG in the kidneys.

Infected animals produce antineuraminidase antibodies capable of neutralizing the enzyme activity. Some measure of protection has been demonstrated in mice following active immunization with purified neuraminidase or following passive immunization with specific antineuraminidase serum produced in rabbits.

Bohm and Suphasindhu (1980) reported that phagocytosis by neutrophilic leukocytes was not a major defence against *E. rhusiopathiae*. They found no correlation between virulence of strains of *E. rhusiopathiae* for 25 infected pigs and the ability of neutrophils from the pigs to phagocytose the organisms. However, Sawada and Takahashi (1987) investigated the effects of treatment of mice with carrageenin or cyclophosphamide on their susceptibility to infection by *E. rhusiopathiae* and concluded that polymorphonuclear leukocytes were important in defense and that ability to evade killing by phagocytes was a virulence attribute of *E. rhusiopathiae*. In a study of monocytes harvested from swine blood, Bohm et al. (1982) found that the average number of a virulent strain of *E. rhusiopathiae* that was ingested was 29, whereas the corresponding number for an avirulent strain was 96. Furthermore, the avirulent strain was digested and degraded more rapidly.

VACCINES

Vaccines have been used successfully in swine and poultry for many years. Two types of vaccines appear to be effective: a formalin-killed culture with aluminium hydroxide gel adjuvant produced from strains of serotype 2 *E. rhusiopathiae* that are efficient producers of the immunizing antigen and live attenuated or avirulent cultures (Bricker and Saif 1988; Wood 1992). Recent studies have identified an approximately 65-kDa extracted protein as a protective antigen (Galan and Timoney 1990; Groschup et al. 1991; Chin et al. 1992; Kobayashi et al. 1992), which elicits immunity that is protective against infection by a wide range of serotypes. Immunity tends to be short-lived and to confer protection against acute but not chronic disease (Wood 1992).

REFERENCES

Bohm, K. H., and Suphasindhu, V. 1980. Investigations on phagocytosis of erysipelas bacteria by neutrophil leucocytes with special reference to virulence. In Proceedings of the International Pig Veterinary Society. Ed. N. C. Nielsen, P. Hogh, and N. Bille, p. 197. Copenhagen: International Pig Veterinary Society.

Bohm, K. H.; Soliman, R.; and Leibold, W. 1982. Interactions between erysipelas bacteria and swine macrophages with special reference to virulence. In Proceedings of the International Pig Veterinary Society. Ed. R. R. Necoechea, C. Pijoan, A. Cesarin, and M. Guzman, p. 146. Mexico City: International Pig Veterinary Society.

Bratberg, A. M. 1981. Bacterial adherence: Adhesin-receptor interactions mediating attachment of bacteria to mucosal surfaces. Acta Vet Scand 22:39-45.

Bricker, J. M., and Saif, Y. M. 1988. Use of a live oral vaccine to immunize turkeys against erysipelas. Avian Dis 32:668-73.

Chin, J. C.; Turner, B.; and Eamens, G. J. 1992. Serological assay for swine erysipelas using nitrocellulose particles impregnated with an immunodominant 65 kDa antigen from *Erysipelothrix rhusiopathiae*. Vet Microbiol 31:169-80.

Drommer, W. 1982. Pathogenesis of joint alterations in experimental erysipelas polyarthritis. In Proceedings of the International Pig Veterinary Society. Ed. R. R. Necoechea, C. Pijoan, A. Cesarin, and M. Guzman, p. 159. Mexico City: International Pig Veterinary Society.

Eriksen, K.; Fossum, K.; Gamlem, H.; Grondalen, J.; Kucsera, G.; and Ulstein, T. 1987. Endocarditis in two dogs. J Small Anim Pract 28:117-23.

Galan, J. E., and Timoney, J. F. 1990. Cloning and expression in *Escherichia coli* of a protective antigen of *Erysipelothrix rhusiopathiae*. Infect Immun 58:3116-21.

Groschup, M. H.; Cussler, K.; Weiss, R.; and Timoney, J. F. 1991. Characterization of a protective protein antigen of *Erysipelothrix rhusiopathiae*. Epidemiol Infect 107:637-49.

Hermanns, W.; Kerlen, G.; Bohm, K. H.; Schulz, L.-Cl.; Winkler, F. C.; and Rimpler, M. 1981. Uber dic induktion einer chronischen polyarthritis mit bestandteilen von rotlaufbakterien (*Erysipelothrix rhusiopathiae*) 1. Mitteilung: Versuche zur arthritis-induktion bei kaninchen. Zentralbl Veterinaermed [B] 28:778-88.

Hermanns, W.; Jessen, H.; Schulz, L.-Cl.; Kerlen, G.; and Bohm, K. H. 1982. Uber die induktion einer chronischen polyarthritis mit bestandteilen von rotlaufbakterien (*Erysipelothrix rhusiopathiae*) 2. Mitteilung: Versuche zur arthritis-induktion bei ratten. Zentralbl Veterinaermed [B] 29:85-98.

Jones, D. 1986. Genus *Erysipelothrix* Rosenbach 1909. In Bergey's Manual of Systematic Bacteriology. Ed. P. H. A. Sneath. Baltimore: Williams and Wilkins.

Kobayashi, S.; Sato, H.; Hirose, K.; and Saito, H. 1992. Immunological characterization of protective antigens prepared by alkaline treatment of whole cells and from the culture filtrate of *Erysipelothrix rhusiopathiae*. Vet Microbiol 30:73-85.

Muller, H. E. 1980. Neuraminidase and other enzymes of *Erysipelothrix rhusiopathiae as* possible pathogenic factors. In Arthritis: Models and Mechanisms. Ed. H. Deicher and L. C. Schulz, pp. 58-67. New York: Springer-Verlag.

Rimpler, M.; Winkler, F. C.; Bohm, K. H.; Kerlen, G.; Leibold, W.; Hermanns, W.; and Mumme, J. 1982. Induction of chronic polyarthritis with cell components of erysipelas bacteria (*Erysipelothrix rhusiopathiae*). 3. Fractionation of erysipelas bacteria and testing in different bioassays. Zentralbl Veterinaermed [B] 29:426-33.

Sawada, T., and Takahashi, T. 1987. Cross protection of mice and swine given live-organism vaccine against challenge exposure with strains of *Erysipelothrix rhusiopathiae* representing ten serovars. Am J Vet Res. 48:81-84.

Schulz, L. C.; Ehard, H.; Hermanns, W.; Messow, C.; Drommer, W.; Langer, I.; Trautwein, G.; Winkelmann, J.; Leibold, W.; Bohm, K. H.; Rimpler, M.; Kirchoff, H.; Marquardt, K.; Burow, K.; and Rapp, K. 1980. The different phases of *Erysipelothrix* polyarthritis: Comparison with other microbial models. In Arthritis: Models and Mechanisms. Ed. H. Deischer and L. C. Schulz, pp. 12-23. New York: Springer-Verlag.

Takahashi, T.; Fujisawa, T.; Benno, Y.; Yamura, Y.; Sawada, T.; Suzuki, S.; Muramatsu, M.; and Mitsuoka, T. 1987a. *Erysipelothrix tonsillarum* sp. nov. isolated from tonsils of apparently healthy pigs. Int J Syst Bacteriol. 37:166-68.

Takahashi, T.; Hirayama, N.; Sawada, T.; Tamura, Y.; and Muramatsu, M. 1987b. Correlation between adherence of *Erysipelothrix rhusiopathiae* strains of serovar 1a to tissue culture cells originated from porcine kidney and their pathogenicity in mice and swine. Vet Microbiol 13:57-64.

Takahashi, T.; Sawada, T.; Muramatsu, M.; Tamura, Y.; Fujisawa, T.; Benno, Y.; and Mitsuoka, T. 1987c. Serotype, antimicrobial susceptibility, and pathogenicity of *Erysipelothrix rhusiopathiae* isolates from tonsils of apparently healthy slaughter pigs. J Clin Microbiol 25:536-39.

Takahashi, T.; Fujisawa, T.; Tamura, Y.; Suzuki, S.; Muramatsu, M.; Sawada, T.; Benno, Y.; and Mitsuoka, T. 1992. DNA relatedness among *Erysipelothrix rhusiopathiae* strains representing all twenty-three serovars and *Erysipelothrix tonsillarum*. Int J Syst Bacteriol 42:469-73.

Tesh, M. J., and Wood, R. L. 1988. Detection of coagulase activity in *Erysipelothrix rhusiopathiae*. J Clin Microbiol 26:1058-60.

Tsangaris, T. H.; Iliadis, N.; Kaldrymidou, E.; Lekkas, S. T.; Tsiroyannis, E. L.; and Artopios, E. 1980. Exerimentaler rotlauf der puten nach intravenoser infektion mit *Erysipelothrix insidiosa*. 1. Elektronmikroskopische befunde in den nieren. Zentralbl Veterinaermed [B] 27:705-13.

Wood, R. L. 1984. Swine Erysipelas: A review of prevalence and research. J Am Vet Med Assoc 184:944-49.

_________ . 1992. Erysipelas. In Diseases of Swine, 7th Ed. Ed A. D. Leman, B. E. Straw, W. L. Mengeling, S. D'Allaire, and D. J. Taylor, pp. 475-86. Ames: Iowa State University Press.

8 / *Clostridium botulinum*

BY T. E. ROCKE

CLOSTRIDIUM BOTULINUM denotes a group of bacteria that produce extremely potent neurotoxins. These toxins cause botulism, a disease characterized by flaccid paralysis in humans and many animals. Like other clostridia, *C. botulinum* is a strict anaerobe that can survive extreme environmental conditions by producing spores. *C. botulinum* spores are widely, but unevenly, distributed in soils and wetland sediments throughout the world. Spores can also be found in food items, invertebrates, animal tissues, and the feces of some animals. In both humans and animals, botulism is typically a food poisoning caused by ingestion of toxin-laden food items, but it can also result from toxico-infections when toxin-producing bacteria colonize the intestinal tract of an individual or secondarily infect a wound.

Bacteria classified as *C. botulinum* comprise a heterogenous group of strains that share a common characteristic: the production of toxin that blocks acetylcholine release from cholinergic nerve endings (Simpson 1981). At least seven different neurotoxins are produced by strains of *C. botulinum*; these have been designated types A, B, C_1, D, E, F, and G (Smith and Sugiyama 1988). Although they have similar pharmacologic actions, the different toxin types are serologically distinct, with a few exceptions. Common antigens are shared by some C_1 and D toxins (Oguma et al. 1984) and E and F toxins (Kozaki et al. 1986); these types may cross-react in serologic tests. Some strains of *C. botulinum* produce another toxin, designated C_2, which is not a neurotoxin, but a binary toxin with ADP-ribosylating activity; this toxin is highly lethal when injected into laboratory animals (Simpson 1989).

Strains of *C. botulinum* are typically denoted by the predominant toxin produced, but some strains produce more than one. Type C strains produce either C_1, the dominant toxin, combinations of C_1 and C_2, or C_2 alone (Eklund et al. 1987). Some type D strains also produce low titers of C_2 toxin. One strain of *C. botulinum* that produces both types A and F toxins (Sugiyama et al. 1972) has been designated A_F, with the subscript denoting the minor toxin. Other *C. botulinum* strains that produce more than one neurotoxin include types A_B, B_A, and B_F (Gimenez 1984; McCroskey and Hatheway 1984; Sakaguchi et al. 1986). In addition, bacteria other than *C. botulinum* can produce botulinum toxin. Isolates of two other clostridial species, *C. barati* and *C. butyricum*, were found to produce type F and type E botulinum toxin respectively (Hall et al. 1985; McCroskey et al. 1986); both were recovered from human infants with botulism.

Botulism occurs rarely in humans. When it does occur, types A, B, and E toxins are most frequently implicated; type F has been associated with only a few outbreaks. Type G toxin was demonstrated in autopsy specimens from several individuals (Sonnabend et al. 1981) but was not confirmed as a cause of death. Outbreaks of botulism occur more frequently in animals and are caused mostly by types C and D; however, types A, B, and E cause outbreaks as well. The predominant type varies between animal species and also varies geographically. Factors that affect the distribution of various types of *C. botulinum* include temperature, water activity, soil pH, and organic matter (Smith and Sugiyama 1988).

BOTULISM IN ANIMALS

Botulism in animals has been called a variety of names, including spinal typhus and shaker foal syndrome in horses; lamsiekte, loin disease, and contagious bulbar paralysis in cattle; and limberneck, alkali poisoning, and western duck sickness in waterfowl. Although the source of toxin and conditions that lead to botulism outbreaks differ among animal groups, the disease is caused primarily by inadvertent ingestion of toxin while feeding. The signs of botulism in most animals are similar and include incoordination, loss of motor control, paralysis of the hind quarters, labored breathing, anorexia, and excessive salivation. Death can result from cardiac and pulmonary failure. Botulism-intoxicated waterfowl first lose the ability to fly and then to swim or walk. A common sign is paralysis of the nictitating membrane, or inner eyelid. In advanced stages of the disease, birds can no longer hold their heads up (limberneck), and intoxicated waterfowl often drown. In chickens, diarrhea is a common sign, as is progressive paralysis. In horses with botulism, severe muscular tremors have been noted, and in cows and other mammals, paralysis of the eyes, tongue, and facial muscles are often evident.

Wild Birds

Of all animals, wild waterfowl, primarily ducks and shorebirds, incur the greatest mortality from botulism. Outbreaks of type C avian botulism occur yearly in waterfowl in the United States, with losses often reaching hundreds of thousands. The worst documented outbreak occurred near the Great Salt Lake in Utah in 1910, when "millions" of waterfowl were reported to have died from botulism. The disease has been diagnosed in wild birds on all continents except the Antarctic and in at least 117 species of birds (Jensen and Price 1987).

Despite many years of research on avian botulism, numerous questions remain unanswered regarding the epizootiology of this disease in waterfowl. Spores of *C. botulinum* type C are common in wetlands and are particularly abundant in areas where outbreaks of botulism have occurred previously (Wobeser et al. 1987). Spores can also be commonly found in marsh-associated invertebrates and in the tissues of healthy waterfowl. However, outbreaks of avian botulism in wetlands depend on many other factors in addition to the presence of spores. These factors include a suitable substrate for bacterial growth, favorable environmental conditions such as temperature and pH, and a mechanism for transfer of toxin to birds, presumably through invertebrate food items.

Several hypotheses have been posed to explain how, when, and where botulinum toxin is produced in a marsh environment, but only one has prevailed. According to this theory, called the microenvironment concept, *C. botulinum* type C grows and produces toxin in carcasses of aquatic invertebrates killed by receding waters or in terrestrial invertebrates killed by flooding (Bell et al. 1955). For many years, avian botulism had been attributed to shallow stagnant waters with low levels of dissolved oxygen, alkaline wetlands, and the flooding of mudflats during warm summer months. However, this theory and these associated conditions do not adequately characterize the timing and location of many outbreaks. Recent evidence shows that botulism occurs year round in certain areas and in a wide variety of habitats, including deep water with steep sides and even in rivers (National Wildlife Health Research Center, unpublished data). The ecological factors that precipitate avian botulism in waterfowl remain uncertain.

Once an outbreak begins, vertebrate carcasses can play a major role in perpetuating the disease. Decaying bird carcasses, regardless of the cause of death, provide a suitable substrate for the production of botulism toxin (Reed and Rocke 1992). Fly larvae and other invertebrates that feed on decaying carcasses ingest the toxin. Feeding waterfowl that consume these toxic maggots contract botulism and die, providing additional carcass substrates for toxin production, thus perpetuating the cycle. A similar carcass-maggot cycle of botulism has been reported in game-farm pheasants (Dohms 1987).

Type E botulism also occurs in wild birds, but outbreaks are sporadic and infrequent. In the United States, outbreaks are confined to the Great Lakes and primarily afflict fish-eating birds, such as common loons, gulls, and diving ducks. Little is known about the epizootiology of type E botulism in birds. Spores are widely distributed in the sediments of the Great Lakes and have also been found in the alimentary tracts of fish (Bott et al. 1966). Type E toxin has been detected in fish carcasses during loon die-offs in the Great Lakes, and toxin-laden fish have been found in the stomachs of sick and dead loons (Brand et al. 1988). However, it is unknown if toxin was produced postmortem in the fish carcasses or if the fish died from ingesting type E toxin. Large outbreaks of type E botulism have been reported in artificially reared salmon and trout (Eklund et al. 1984). The disease may also occur in natural populations of fish, but it has never been reported. The sporadic occurrence of type E botulism in birds suggests that conditions leading to outbreaks vary from year to year. It is also possible that outbreaks of type E botulism go unnoticed or unreported in remote areas.

Poultry

Botulism in poultry can range from small outbreaks in backyard flocks to devastating losses in commercial operations (Dohms 1987). Most outbreaks are from type C botulism, although type A has also been occasionally reported. Recent outbreaks in commercial operations have occurred primarily in intensively managed broiler chickens raised on deep litter; no outbreaks have been reported in caged layers. Cannibalism of decaying carcasses was found to be the cause of 2 outbreaks in broiler chickens, but in at least 35 outbreaks, the source of toxin could not be identified, despite thorough investigations (Dohms 1987).

Recent studies suggest that *C. botulinum* in the intestinal tract may constitute an important source of intoxication in broiler chickens. Several investigators postulated that spores of *C. botulinum* type C might proliferate in the intestine of chickens, resulting in toxico-infections (Dohms 1987), and that the cecum was the likely site of toxin production (Miyazaki and Sakaguchi 1978). However, chickens inoculated intracecally with spores of *C. botulinum* type C did not develop botulism, even though the spores germinated and produced toxin (Sakaguchi and Kaji 1980). It was later demonstrated that chickens fed type C spores produced toxin in fecal droppings, but only those chickens kept on a board floor, which allowed coprophagy, developed botulism and died (Huyn and Sakaguchi 1989). Chickens that were fed spores and maintained on a wire-net floor to prevent ingestion of fecal matter remained asymptomatic, even though toxin was detected in their feces for at least 3 days after spores were ingested. Presumably, the amount of toxin absorbed from the ceca was not sufficient to intoxicate the host chicken, but absorption of cell-bound toxin in fecal droppings ingested by chickens was sufficient to cause intoxication.

Cattle

Cattle can acquire botulinum toxin from diverse sources, such as animal carcasses, hay or silage, poultry litter, and brewers' grain (malt) (Smith and Sugiyama 1988). In some locations, cattle on range contract botulism by ingesting small-animal carcasses laden with toxin, usually type D. The disease has been reported in South Africa (lamsiekte), Australia (bulbar paralysis), southwestern United States (loin disease), and in other arid areas where the soil is markedly deficient in phosphorous. The cattle in these areas often become phosphorous deficient and develop pica, an abnormal appetite for material not usually eaten, in this case bone and flesh. Spores of *C. botulinum* type D are normal flora in the intestinal tract of animals where lamsiekte is common, and these spores can invade the tissues of dead animals, germinate, and produce toxin. Some carcasses may become quite toxic, and ingestion of even small amounts of flesh by cattle may be lethal. The lamsiekte form of botulism also occurs in protein-deficient sheep in South Africa and Australia.

Another source of botulinum toxin is hay or silage (forage poisoning) contaminated with the toxin-laden carcass of a small bird or mammal. Forage poisoning occurs sporadically worldwide in both cattle and horses and is usually caused by type C botulinum toxin. As in the case of lamsiekte, the toxin responsible for forage poisoning is produced after death in animal carcasses. Cat carcasses are considered to be an important source of toxin in forage poisoning, but other animals, such as wild rodents, are also probably involved. Toxin from the decomposing carcass diffuses into the surrounding feed, where it may persist for months.

Poultry waste has been recognized as a source of botulinum toxin for cattle only in the last several decades (Egyed 1987). In an effort to recycle animal waste, poultry litter, mostly from broiler operations, has been incorporated into animal feeds and bedding and spread on pastures as fertilizer. Poultry litter contains feces and carcass remnants, both of which may be a source of botulinum toxin. Outbreaks of botulism attributed to poultry waste are usually caused by type C or D toxin and have been reported in the United Kingdom (Smart et al. 1987), Israel (Egyed 1987), and the Netherlands.

Horses

Equine botulism occurs worldwide, but outbreaks are sporadic and infrequent (Smith and Sugiyama 1988). Like outbreaks in cattle, the disease in horses often results from forage poisoning and is usually caused by type C or D toxin. In North America, botulism due to type B toxin is the most common form of botulism, and is considered enzootic in horses and foals in eastern United States (Johnston and Whitlock 1987). The toxin appears to be produced in necrotic lesions of the gastrointestinal tract and disease is actually the result of a toxico-infection (Swerczek 1980). In foals, the disease is known as shaker foal syndrome, which usually affects animals between 3 and 8 weeks of age. The disease is more common in the foals of mares that are fed an excessively nutritious diet, supplemented with proteins, vitamins, and minerals, or those exposed to stress, such as inclement weather. It is thought that high levels of corticosteroids in mares' milk may play a role in the pathogenesis of the disease in foals (Swerczek 1980).

Other Mammals

Botulism has been a severe problem for commercially raised mink (Smith and Sugiyama 1988). Raw meat or fish, which are important components of their diet, is the usual source of toxin. Almost all outbreaks are caused by type C, although types A and E have also been reported. Type C botulism has been reported in other captive mammals, including dogs (Richmond et al. 1978), monkeys (Smart et al. 1980), and lions (Greenwood 1985). Chicken flesh is often the source of toxin in these cases. Botulism in free-ranging mammals is probably infrequent, with most cases occurring in scavenging animals, such as muskrats, raccoons, and weasels, that inhabit marsh environments where avian botulism is common.

TOXIN PRODUCTION

Much research has been devoted to the study of factors that influence the growth of *C. botulinum* and govern toxin production in human food items; less work has been done on factors that affect sources of toxin for animals. The basic requirements are an anaerobic environment, a suitable energy source, and a pH above 4.8; other physiologic and cultural characteristics vary between serotypes (Smith and Sugiyama 1988). Neurotoxin is synthesized intracellularly by *C. botulinum* and released when the bacterial cells undergo autolysis. In culture media, toxin titer is low during the logarithmic phase of growth but increases when cell growth ceases and cell membranes rupture (Boroff 1955). Interestingly, the toxin has no known role in the growth and physiology of the bacterium, and many naturally occurring isolates of *C. botulinum* do not produce toxin. Also, toxin production and bacterial growth have different nutritional requirements and can be controlled independently (Kindler and Mager 1955; Kindler et al. 1956).

For at least two types of *C. botulinum*, types C and D, production of neurotoxin depends on the presence of specific bacteriophages (Eklund et al. 1987). When type C and D strains are cured of bacteriophage infections by UV radiation or treatment with acridine orange, they lose the ability to produce C_1 or D neurotoxins, but not C_2 toxin. The cured strains resume production of neurotoxin only after they are reinfected with certain phages (TOX^+) derived from the toxigenic parent stock. Furthermore, TOX^+ phages isolated from type D strains can infect nontoxigenic type C strains and induce them to produce type D toxin, and visa versa (Eklund et al. 1987). Hybridization analysis has shown that the structural gene for neurotoxin is located on the genome of TOX^+ bacteriophages (Fujii et al. 1988). Because the relationship with type C and D host strains is unstable, TOX^+ phages have been described as pseudolysogenic (Eklund et al. 1987). The DNA of pseudolysogenic phages is not incorporated into the bacterial genome; however, like lysogenic phages, they can suppress the lysis of cells through several generations (Eklund et al. 1987). Replication of bacteria that contain pseudolysogenic phages can result in uninfected cells and cells that lyse and liberate phages, as well as intact cells that contain the phage. The stability of the host-phage relationship depends on the bacterial strain, environmental conditions such as temperature, and the growth phase of the bacteria (Eklund et al. 1987).

Despite numerous attempts, a similar mechanism for toxin production in types A, B, E, and F has not been discovered (Eklund and Poysky 1989). Toxin production in these types is much more stable than in types C and D, which are prone to lose toxigenicity after several passages. Bacteriophages have been detected in types A, B, E, and F; however, after curing with mitomycin C or acridine orange, they continue to produce neurotoxin. Also, no phages that can induce toxigenicity in nontoxigenic, phage-sensitive bacteria have been isolated from these types. For type G strains of *C. botulinum*, recent studies have provided strong evidence that toxigenicity is dependent on a plasmid (Eklund and Poysky 1989).

TOXIN STRUCTURE

The neurotoxins produced by *C. botulinum* share a common subunit structure (DasGupta 1988). Neurotoxin is initially produced by the bacteria as an inactive, single-chain protein with a molecular weight of 140-170 kDa. After lysis of the bacterial cell or secretion of the toxin from the cell, this single-chain protein is "nicked" by proteases (or trypsin) that are either endogenous or exogenous to the bacteria. The nicked protein is a dichain molecule, composed of a heavy (H) chain (85-100 kDa) and a light (L) chain (50-59 kDa) held together by non-covalent bonds and one or more disulfide bonds. The dichain molecule is much more toxic than its single-chain precursor, but it is not clear if nicking activates the neurotoxic properties of this protein. Some workers have suggested that the two events, nicking and activation, are separate and unrelated (Krysinski and Sugiyama 1981). It is possible that some other modification of the protein is necessary for activation, such as an alkaline pH-induced conformational change in the dichain molecule (DasGupta 1988).

Botulism neurotoxin in both natural substrates and in vitro cultures is generally associated with one or more nontoxin proteins (Sugiyama 1980). The toxin complexes are usually either bimolecular or trimolecular and vary in molecular weight from 230-500 kDa. Some of the nontoxin proteins have hemagglutinin activity, but their function is unknown. In general, the nontoxin proteins appear to protect the neurotoxin from inactivation by gastric enzymes in the gut. When administered perorally, toxin complexes are considerably more toxic than purified neurotoxin; trimolecular complexes are more toxic than bimolecular complexes (Ohishi 1984). In contrast, when administered parenterally, purified neurotoxin exhibits much greater toxicity than the toxin complexes (Sakaguchi et al. 1984).

MECHANISM OF TOXIN ACTION

Botulinum neurotoxin exerts its paralytic effects by blocking release of the neurotransmitter acetylcholine (Simpson 1981). The toxin acts on various sites, including the cholinergic neuromuscular junction, autonomic ganglia, postganglionic parasympathetic sites, postganglionic sympathetic nerves that release acetylcholine, and the adrenal glands (Simpson 1981). A three-step model has been proposed to account for the toxin's interference with neuromuscular transmission (Simpson 1981): (1) toxin binds rapidly and irreversibly to the presynaptic cell membrane (binding step); (2) the bound toxin penetrates the cell membrane and enters the cell (internalization step); and (3) intracellular toxin disables the mechanism for release of acetylcholine (lytic step). Although this hypothesis has been widely accepted, it was developed largely by analogy to other bacterial toxins (Simpson 1981). The exact sequence of events that leads to paralysis remains unknown; however, evidence supporting the proposed model is accumulating.

Binding Step

Binding of botulinum neurotoxin to cell membranes is the first critical step toward paralysis. Autoradiography studies have shown that neurotoxin accumulates selectively at the neuromuscular junction, and binding sites are confined to presynaptic nerve cell membranes (Dolly et al. 1984; Black and Dolly 1986a); however, the receptor for neurotoxin has not yet been identified. Cross-competition studies suggest that not all toxin types have the same receptor (Kozaki 1979). The region of the neurotoxin molecule that contains the binding site (binding domain) probably resides on the H chain (Bandyopadhyay et al. 1987). Isolated preparations of the H and L chains of neurotoxin are not paralytic when applied singly to tissues; both are required for neuromuscular blockage. In addition, the H chain must be applied before the L chain for paralysis to occur.

Internalization Step

Botulinum neurotoxin probably enters nerve cells by receptor-mediated endocytosis, similar to the mechanism of cell entry of exogenous substances (e.g., insulin), certain other toxins (e.g., diphtheria), and some viruses (e.g., Semliki Forest). This is a two-step process in which receptor-bound toxin becomes encapsulated into an endosome and then penetrates the endosomal membrane by a pH-dependent mechanism to enter the cytoplasm. Experimental findings support this theory, but do not provide direct evidence. Using electron-microscopic autoradiography studies, toxin has been observed inside cells in vesicles that resemble endosomes (Black and Dolly 1986b). This internalization process was impeded by low temperature and sodium azide, causing the toxin to remain longer at binding sites (Black and Dolly 1986 a,b). Acidotropic chemicals, such as chloroquine, ammonium chloride, and methylamine hydrochloride, that hinder receptor-mediated endocytosis are very effective antagonists of botulism neurotoxin (Simpson 1989). These drugs do not affect substances that are believed to act at the cell membrane. Also, microinjection of neurotoxin directly into adrenal medullary cells blocked calcium-induced exocytosis of catecholamines (Penner et al. 1986), which demonstrated that neurotoxin can act internally in cells to block the release of a chemical mediator similar to acetylcholine. Finally, pressure injection of botulism neurotoxin into cholinergic synapses of an invertebrate, *Aplysia californica*, caused a dose-dependent decrease in acetylcholine release (Poulain et. al. 1988). The neurotoxin probably crosses the endosomal membrane into the cytoplasm by creating pH-dependent channels (Simpson 1989); this function appears to be mediated by the aminoterminus of the H chain.

Lytic Step

It was postulated for many years that, once inside a nerve cell, neurotoxin inhibited the release of neurotransmitter through some enzymatic process (Simpson 1981), but the mechanism was unclear. In normal cells, an action potential at nerve endings triggers influx of calcium ions, which promote the release of acetylcholine from synaptic vesicles. Recent work provides evidence that the light chain of type B botulinum neurotoxin is a zinc endopeptidase that cleaves synaptobrevin, a protein located in the membrane of the synaptic vesicles (Schiavo et al. 1992). The function of synaptobrevin has not been determined but it is apparently critical for the release of acetylcholine from synaptic vesicles. Another clostridial neurotoxin, tetanus toxin, produced by *C. tetani* has an amino acid sequence similar to that of type B botulinum toxin and was also shown to cleave synaptobrevin. Types A and E neurotoxins, however, had no effect on this protein. The intercellular targets of types A and E botulinum toxin are apparently different from those of type B and tetanus toxin and remain unknown. Similar studies on the target protein have not been reported for types C and D botulinum toxins.

The structure and functional domains of botulinum neurotoxin are remarkably similar to those of other microbial toxins, including diphtheria toxin produced by *Corynebacterium diphtheriae* and tetanus toxin produced by *C. tetani* (Simpson 1986). All these toxins are initially produced as inactive single-chain polypeptides, which are further modified to produce the active structure, a dichain molecule, consisting of an L chain and an H chain in a 1:2 ratio. All possess a binding domain on the carboxyterminus of the H chain and a channel-forming domain on the aminoterminus of the H chain. The L chain of diphtheria toxin is an enzyme with ADP-ribosylating activity. Although the clostridial neurotoxins do not possess this same activity, the enzyme domain apparently resides on the L chain for these toxins as well (Simpson 1986).

C_2 TOXIN

C_2 toxin is not a neurotoxin and does not block neurotransmitter release or neuromuscular function (Simpson 1982). Instead, its effects in laboratory animals are characterized by increased movement of fluids across membranes, including increased vascular permeability, effusive secretions into the airway, pulmonary edema and bleeding, collection of fluids in the thoracic cavity, and extreme hypotension. C_2 is a binary toxin, consisting of two separate and independent polypeptides, an H and an L chain. The H chain mediates binding of the toxin to cell membranes (Ohishi 1983), and the L chain has been shown to possess ADP-ribosylating activity similar to toxins produced by *C. perfringens* and *C. spiriforme* (Simpson 1989).

IMMUNITY

In most animals, natural host immune defenses against neurotoxin probably do not play a significant role in either the pathogenesis of botulism or in prevention of the disease. The toxin is so poisonous that the amount required to immunize an animal is usually higher than the lethal dose. In one study, naturally occurring antibodies to neurotoxins (types A-F) were found in several carrion-eating species (Ohishi et al. 1979), including turkey vultures (*Cathartes aura*), crows (*Corvus brachyrhynchos*), and coyotes (*Canis latrans*). However, it is unknown if these antibodies were protective for the host, because the animals were not challenged. Botulism has not been reported in any of these species, and it is possible that they are innately resistant to the paralytic effects of neurotoxin. Turkey vultures have been shown to be highly resistant to both orally administered and injected botulinum toxin (Kalmbach 1939), which may be explained by failure of toxin to bind at presynaptic nerve endings (Cohen 1970). Another possibility is that the sero-positive animals developed antibody to toxin in response to a nonlethal toxico-infection in the gut (Ohishi et al. 1979).

Botulinum toxin can be inactivated with formaldehyde to produce a toxoid that is immunogenic. Toxoids have been used successfully as vaccines to prevent botulism in domestic and captive-reared animals. Vaccination of cattle with C_1 and D toxoid has been effective in preventing lamsiekte in South Africa and Australia (Jansen et al. 1976). In young horses, shaker-foal syndrome has been prevented by vaccinating pregnant mares with type B toxoid a few weeks prior to parturition (Johnston and Whitlock 1987). Type C toxoids have also been used to vaccinate commercially raised mink and game-farm pheasants (Smith and Sugiyama 1988).

PROBLEMS

In human medicine, the focus of research has shifted from botulism, the disease, to the use of botulinum toxin as a therapeutic agent. Capitalizing on its paralytic effect on the neuromuscular junction, low doses of botulinum neurotoxin have been used to treat many disorders, including strabismus (crossed-eye disorder), blepharospasm (spasmodic eye closure), hemifacial spasm, spasmodic torticollis (spasmodic head movements), spasmodic dysphonia (laryngeal muscle spasms), and finger dystonia (Scott 1989). The technique requires refinement, and many issues must be resolved, including the optimal dosages and injection schedules and the immunizing effects of injected toxin. For animals, prevention of botulism is still a primary concern. In most domestic and captive-reared animals, the disease can be prevented by vaccination, adequate storage of food items, and good sanitary management of their environment, including prompt removal of carcasses, litter, and feces. For wild birds, however, control of botulism is much more difficult. Carcass pickup after an outbreak begins can reduce losses from botulism (Reed and Rocke 1992); however, removal of every carcass is impossible. Mass vaccination of wild birds is not feasible at this time. Research is currently being conducted to identify environmental conditions that are associated with avian botulism outbreaks. With this information, methods of managing wetlands could possibly be developed to prevent the disease.

REFERENCES

Bandyopadhyay, S.; Clark, A. W.; DasGupta, B. R.; and Sathyamoorthy, V. 1987. Role of the heavy and light chains of botulinum neurotoxin in neuromuscular paralysis. J Biol Chem 262:2260-63.

Bell, F. G.; Sciple, G. W.; and Hubert, A. A. 1955. A microenvironment concept of the epizoology of avian botulism. J Wildl Manage 19:352-57.

Black, J. D., and Dolly, J. O. 1986a. Interaction of ^{125}I-labeled botulinum neurotoxins with nerve terminals. 1. Ultrastructural autoradiographic localization and quantitation of distinct membrane acceptors for types A and B on motor nerves. J Cell Biol 103:521-34.

__________. 1986b. Interaction of ^{125}I-labeled botulinum neurotoxins with nerve terminals. II. Autoradiographic evidence for its uptake into motor nerves by acceptor-mediated endocytosis. J Cell Biol 103:535-44.

Boroff, D. A. 1955. Study of toxins of *Clostridium botulinum*. III. Relation of autolysis to toxin production. J Bacteriol 70:363-67.

Bott, T. L.; Deffner, J. S.; McCoy, E.; and Foster, E. M. 1966. *Clostridium botulinum* type E in fish from the Great Lakes. J Bacteriol 91:919-24.

Brand, C. J.; Schmitt, S. M.; Duncan, R. M. and Cooley, T. M. 1988. An outbreak of type E botulism among common loons (*Gavia immer*) in Michigan's upper peninsula. J Wildl Dis 24:471-76.

Cohen, G. M. 1970. Studies on the resistance of roosters and vultures to type A botulinal toxin. PhD diss., Florida State University, Tallahassee.

DasGupta, B. R. 1988. The structure of botulinum neurotoxins. In Botulinum Neurotoxin and Tetanus Toxin. Ed. L. L. Simpson, pp. 296-314. New York: Academic Press.

Dohms, J. E. 1987. Laboratory investigation of botulism in poultry. In Avian Botulism: An International Perspective. Ed. M. W. Eklund and V. R. Dowell, Jr., pp. 296-314. Springfield, Ill.: Charles C. Thomas.

Dolly, J. O.; Black, J.; Williams, R. S.; and Melling, J. 1984. Acceptors for botulinum neurotoxin reside on motor nerve terminals and mediate its internalization. Nature 307: 457-60.

Egyed, M. N. 1987. Outbreaks of botulism in ruminants associated with ingestion of feed containing poultry waste. In Avian Botulism: An International Perspective, ed. M. W. Eklund and V. R. Dowell, Jr., pp. 317-380. Springfield, Ill.: Charles C. Thomas.

Eklund, M. W., and Poysky, F. T. 1989. Bacteriophages and plasmids in *Clostridium botulinum* and *Clostridium tetani* and their relationship to production of toxins. In Botulism Neurotoxin and Tetanus Toxin, ed. L. L. Simpson, pp.25-51. New York: Academic Press.

Eklund, M. W.; Poysky, F. T.; Peterson, M. E.; Peck., L. W.; and Brunson, W. D. 1984. Type E botulism in salmonids and conditions contributing to outbreaks. Aquaculture 41:293-09.

Eklund, M. W.; Poysky, F. T.; Oguma, K; Iida, H.; and Inoue, K. 1987. Relationship of bacteriophages to toxin and hemagglutinin production by *Clostridium botulinum* types C and D and its significance in avian botulism outbreaks. In Avian Botulism: An International Perspective, ed. M. W. Eklund and V. R. Dowell, Jr., pp. 191-222. Springfield, Ill.: Charles C. Thomas.

Fujii, N.; Oguma, K.; Yukosawa, N.; Kimura, K.; and Tsuzuki, K. 1988. Characterization of bacteriophage nucleic acids obtained from *Clostridium botulinum* types C and D. Appl Environ Microbiol 54:69-73.

Gimenez, D. F. 1984. *Clostridium botulinum* subtype B_a. Zentralbl Bakteriol Hyg A 257:68-72.

Greenwood, A. G. 1985. Diagnosis and treatment of botulism in lions. Vet Rec 117:58-69.

Hall, J. D.; McCroskey, L. M.; Pincomb, B. J.; and Hatheway, C. L. 1985. Isolation of an organism resembling *Clostridium barati* which produces type F botulinal toxin from an infant with botulism. J Clin Microbiol 21: 654-55.

Huyn, S., and Sakaguchi, G. 1989. Implication of coprophagy in pathogenesis of chicken botulism. Jpn J Vet Sci 51:582-86.

Jansen, B. C.; Knoetze, P. C.; and Visser, L. F. 1976. The antibody response of cattle to *Clostridium botulinum* type C and D toxoids. Onderstepoort J Vet Res 4:165-74.

Jensen, W. I., and Price, J. I. 1987. The global importance of type C botulism in wild birds. In Avian Botulism: An International Perspective. Ed. M. W. Eklund and V. R. Dowell, Jr., pp. 33-54. Springfield, Ill.: Charles C. Thomas.

Johnston, J., and Whitlock, R. H. 1987. Botulism. In Current Therapy in Equine Medicine - Two. Ed. N. E. Robinson, pp. 367-70. Philadelphia: W. B. Saunders Co.

Kalmbach, E. R. 1939. American vultures and the toxin of *Clostridium botulinum*. J Am Vet Med Assoc 94:187-91.

Kindler, S. H., and Mager, J. 1955. Production of toxin by resting cells of *Clostridium parabotulinum* type A. Science 122:926-27.

Kindler, S. H.; and Mager, J.; and Grossowicz, N. 1956. Toxin production of *Clostridium parabotulinum* type A. J Gen Microbiol 15:394-403.

Kozaki, S. 1979. Interaction of botulinum type A, B and E derivative toxins with synaptosomes of rat brain. Naunyn Schmiedebergs Arch Pharmacol 308:67-70.

Kozaki, S.; Kamata, Y.; Nagai, Y.; Ogasawara, J.; and Sakaguchi, G. 1986. The use of monoclonal antibodies to analyze the structure of *Clostridium botulinum* type E derivative toxin. Infect Immun 52:786-91.

Krysinski, E. P., and Sugiyama, H. 1981. Nature of intracellular type A botulinum neurotoxin. Appl Environ Microbiol 41:675-78.

McCroskey, L. M., and Hatheway, C. L. 1984. Atypical stains of *Clostridium botulinum* isolated from specimen in infant botulism cases. Annu Meet Am Soc Microbiol Abstr C159:263.

McCroskey, L. M.; Hatheway, C. L.; Fenicia, L.; Pasolini, B.; and Aureli, P. 1986. Characterization of an organism that produces type E botulinal toxin but which resembles *Clostridium butyricum* from the feces of an infant with type E botulism. J Clin Microbiol 23:201-2.

Miyazaki, S., and Sakaguchi, G. 1978. Experimental botulism in chickens: the cecum as the site of production and absorption of botulinum toxin. Jpn J Med Sci Biol 31:1-15.

National Wildlife Health Research Center, Madison, WI. 1992. Unpublished data.

Oguma, K.; Murayama, S.; Syuto, B.; Iida, H.; and Kubo, S. 1984. Analysis of antigenicity of *Clostridium botulinum* type C_1 and D toxins by polyclonal and monoclonal antibodies. Infect Immun 43:584-88.

Ohishi, I. 1983. Lethal and vascular permeability activities of botulinum C_2 toxin induced by separate injections of the two toxin components. Infect Immun 40:336-339.

__________. 1984. Oral toxicities of *Clostridium botulinum* type A and B toxins from different strains. Infect Immun 43:487-90.

Ohishi, I.; Sakaguchi, G.; Riemann, H.; Behymer, D.; and Hurvell, B. 1979. Antibodies to *Clostridium botulinum* toxins in free-living birds and mammals. J Wildl Dis 15:3-9.

Penner, R.; Neher, E.; and Dreyer, F. 1986. Intracellularly injected tetanus toxin inhibits exocytosis in bovine adrenal chromaffin cells. Nature 324:76-78.

Poulain, B.; Tauc, L.; Maisey, E. A.; Wadsworth, J. D. F.; Mohan, P. M.; and Dolly, J. O. 1988. Neurotransmitter release is blocked intracellularly by botulinum neurotoxin, and this requires uptake of both toxin polypeptides by a process mediated by the larger chain. Proc Natl Acad Sci 85:4090-94.

Reed, T. M., and Rocke, T. E. 1992. The role of avian carcasses in the epizootiology of avian botulism. Wildl Soc Bull 20:175-82.

Richmond, R. N.; Hatheway, C. L. and Kaufman, A. F. 1978. Type C botulism in a dog. J Am Vet Med Assoc 173:202-203.

Sakaguchi, G., and Kaji, R. 1980. Avian botulism. J Jpn Soc Poult Dis 16:37-49.

Sakaguchi, G.; Kozaki, S.; and Ohishi, I. 1984. Structure and function of botulinum toxins. In Bacterial Protein Toxins. Ed. J. E. Alouf, F. J. Fehrenbach, J. G. Freer, and J. Jeljiaszewicz, pp. 433-43. London: Academic Press.

Sakaguchi, G.; Sakaguchi, S., Kozaki, S.; and Takahashi, M. 1986. Purification and some properties of *Clostridium botulinum* type AB toxin. Fed Eur Microbiol Soc Microbiol Lett 33:23-29.

Schiavo, G.; Benfenati, F.; Poulain, B.; Rosetto, O.; Polverino de Laureto, P.; DasGupta, B. R.; and Montecucco, C. 1992. Tetanus and botulinum-B neurotoxins block neurotransmitter release by proteolytic cleavage of synaptobrevin. Nature 359:832-35.

Scott, A. B. 1989. Clostridial toxins as therapeutic agents. In Botulism Neurotoxin and Tetanus Toxin. Ed. L. L. Simpson, pp. 399-412. New York: Academic Press.

Simpson, L. L. 1981. The origin, structure, and pharmacological activity of botulinum toxin. Pharmacol Rev 33:155-88.

_________. 1982. A comparison of the pharmacological properties of *Clostridium botulinum* type C_1 and type C_2 toxins. J Pharmacol Exp Ther 223:695-701.

_________. 1986. Molecular pharmacology of botulinum toxin and tetanus toxin. In Annual Review of Pharmacology and Toxicology. Ed. R. George, R. Okun, and A. K. Cho, pp. 427-53. Palo Alto, Calif.: Annual Reviews, Inc.

_________. 1989. Peripheral actions of the botulinum toxins. In Botulinum Neurotoxin and Tetanus Toxin. Ed. L.L. Simpson, pp.153-78. New York: Academic Press.

Smart, J. L.; Jones, T. O.; Clegg, F. G.; and McMurtry, M. J. 1987. Poultry waste associated type C botulism in cattle. Epidemiol Inf 98:73-79.

Smart, J. L.; Roberts, T. A.; McCullagh, K. G.; Lucke, V. M.; and Pearson, H. 1980. An outbreak of type C botulism in captive monkeys. Vet Rec 107:445-46.

Smith, L. D., and Sugiyama, H. 1988. Botulism, the Organism, Its Toxins, the Disease. Springfield, Ill.: Charles C. Thomas.

Sonnabend, O.; Sonnabend, W.; Heinzle, R.; Sigrist, T.; Dirnhofer, R.; and Krech, U. 1981. Isolation of *Clostridium botulinum* type C and identification of type G botulinum toxin in humans: Report of five sudden unexpected deaths. J Infect Dis 143:22-27.

Sugiyama, H. 1980. *Clostridium botulinum* neurotoxin. Microbiol Rev 44:419-48.

Sugiyama, H.; Mizutani, D.; and Yang, K.W. 1972. Basis of type A and F toxicities of *Clostridium botulinum* strain 84. Proc Soc Exp Biol Med 191:1063-67.

Swerczek, T. W. 1980. Toxicoinfectious botulism in foals and adult horses. J Am Vet Med Assoc 176:217-20.

Wobeser, G.; Marsden, S.; and MacFarlane, R. J. 1987. Occurrence of toxigenic *Clostridium botulinum* type C in the soil of wetlands in Saskatchewan. J Wildl Dis 23:67-76.

9 / *Clostridium tetani*

BY **B. BIZZINI**

TETANUS has been known to humans since the beginning of the history of medicine. A description of this terrifying infection was reported by Hippocrates in the "Corpus Hyppocraticum" 24 centuries ago, but no significant progress in the knowledge of the disease was achieved until 1884. The Greek term "tetanos," which means contracture, has been taken from the Latin medicine "rigor" and the Arabian medicine "alcuzez."

In effect, it was not until 1884 that Carle and Rattone succeeded in transmitting the illness from a man who died from tetanus to rabbit, and from rabbit to rabbit, thus establishing that tetanus was an infectious disease. In the same year, Nicolaier produced tetanus in various animal species by inoculation of soil samples and identified the causative agent of the disease. He also assumed that the bacteria remained confined at the site of injection, ascribing the development of the disease to the diffusion into the blood of a strychninelike substance. In 1889, Kitasato isolated the etiologic agent in pure form, and in the next year, Faber showed that the pathogenic potency of the bacteria originated from a toxin capable of reproducing the entire clinical picture of tetanus. This finding has been a fundamental contribution of bacteriology to infectious disease.

Today, tetanus is still a major public health problem in most developing countries. Morbidity due to tetanus is, in fact, most prevalent in poor countries, in which sanitary conditions are unsatisfactory and the climate is warm and damp. The annual case fatality worldwide has been estimated to be close to 1 million individuals, 80% of whom are neonates.

Clostridium tetani is a strict anaerobe that grows as slender rods in young culture but develops long filaments in older cultures. As a strict anaerobe, *C. tetani* does not grow at oxidation-reduction potentials above +0.01 volts; germination of spores, growth of vegetative bacteria, and toxin production in a wound require anaerobic conditions. The organism is motile and numerous peritrichous flagella are present on the vegetative cells. The terminal, spherical spores, formed at the ends of vegetative cells, give rise to a characteristic tennis racket appearance. However, spores are not always in a terminal position. They can be oval and in a "subterminal position" or be present at each end of the bacillus, giving a dumbbell appearance. The gram-positive character of the organism fades rapidly with aging of cultures.

Soil is the usual source of *C. tetani*. As a sporulated anaerobe, *C. tetani* colonizes the soil, which explains the persistence of this microorganism. Although *C. tetani* spores exhibit a high resistance to a wide range of environmental conditions, the bacterium is not ubiquitous and is confined to certain tetaniferic regions, determined by the nature of the soil. In particular, an alkaline calcareous clay constitution of soil is favorable to anaerobiosis, and these soils contain tetanus spores. Acidic soils, such as volcanic soils and soils rich in granite, are free from tetanus spores. *C. tetani* is also present in soil rich in animal manure.

Another important source of *C. tetani* is house dust, which is a major factor in dissemination of spores, because it penetrates everywhere. *C. tetani* vegetative cells and spores are frequently detected in the digestive tract and feces of a number of domestic animals, such as cattle and horses, and other herbivores known to ingest soil regularly. Animals ingest fodder together with soil contaminated with *C. tetani* and disseminate *C. tetani* by passing the bacteria in their feces. *C. tetani* has also been found in human feces, but, in this case, the frequency of occurrence depends on the living circumstances. Nonetheless, the intestinal tract of animals should be considered as a secondary habitat of *C. tetani*, the primary one being the soil.

The spores of *C. tetani* will germinate both anaerobically and aerobically when put in a favorable environment (Shoesmith and Holland 1972). However, spores of *C. tetani* will not germinate in the intestine of germ-free rats, whereas vegetative cells of the organism will produce tetanus toxin in this environment and will induce formation of tetanus antitoxin by the host (Wells and Balish 1983).

DISEASE

Tetanus is characterized by muscular spasms resulting from the action of tetanus toxin on the central nervous system. All species of domestic animals appear to be susceptible, but there are remarkable differences in susceptibility among species. Horses and humans are most susceptible; cats and birds are comparatively resistant. On the basis of LD_{50} / kg body weight, birds are of the order of 10^5 times less susceptible than are horses. Disease occurs most commonly in horses, less frequently in other herbivores, and occasionally in pigs, dogs, and cats. The frequency of occurrence is related to susceptibility to tetanus toxin and to animal production practices that relate to opportunities for wound contamination, which is the usual mode of entry of the bacteria into the animal. It is noteworthy that the wound may not be evident at the time of clinical disease. It has been suggested that cattle may develop tetanus as a result of neurotoxin produced by *C. tetani* growing in the rumen (Smith 1975). Widespread vaccination of horses and sheep in many parts of the world has lead to drastic reductions in occurrence of the disease in these animals.

In many developing countries tetanus is the major cause of death in newborn infants. It is estimated that approximately 2 million people are affected each year by tetanus. If one considers that mortality is close to 50%, tetanus kills 1 million people each year, of which 80% are newborns. By contrast, in developed countries, systematic immunization and good sanitary conditions have resulted in marked reductions in mortality.

Four main forms of tetanus may be distinguished in humans: local, cephalic, neonatal, and generalized tetanus. Local tetanus is characterized by stiffness and rigidity of muscles around the site of injury. These signs are ascribed to a lack of inhibition by the spinal or medullary neurons innervating the group of muscles that are affected. Localized tetanus is usually mild, with a mortality rate as low as 1%. The symptoms may last for weeks to months, before resolving spontaneously or developing into generalized tetanus. Cephalic tetanus is a very rare occurrence. The incubation period is extremely short (1 or 2 days), and disease follows an injury or infection of the face or head. Facial or oculomotor palsy is the main clinical sign, and dysphagia is often observed. Usually, the prognosis is poor because the disease tends to progress to generalized tetanus. Neonatal tetanus develops most often as a result of contamination of the umbilical cord. Following a short incubation period, the newborn infant becomes unable to nurse. Thereafter, the body becomes stiff and is affected with generalized spasms. The stiffness of the jaw muscles results in a persistent "risus sardonicus." Spasms of the larynx result in inability to swallow and spasms of the respiratory muscles cause apneic episodes, resulting in death within 4-14 days.

Generalized tetanus is the most common and most severe form of tetanus, but mild, moderate, and severe forms have been described. In the severe form, the disease is usually fatal. The characteristic sign of opisthotonos appears as a consequence of the sudden contraction of muscles; the contraction of the muscles may be so powerful that only the head and the heels remain in contact with the bed. Spasms may be so intense as to cause vertebral fractures. Painful spasms also affect the larynx, diaphragm, and intercostal muscles, resulting in acute respiratory failure. Later, the autonomic nervous system becomes involved and causes tachycardia, hypertension, profuse sweating, and cardiac arrhythmias. Terminally, hypotension, hyperpyrexia, and pulmonary edema develop and cause death. Pneumonia is the most common cause of death. Generalized tetanus has been considered by Kryzhanovsky (1981) as multiple episodes of local tetanus.

The severity of tetanus depends on the quantity of toxin that reaches the central nervous system (CNS). The higher the amount of toxin reaching the CNS, the shorter the incubation period and the higher the mortality. The incubation period (the time elapsed from inoculation to the appearance of the first symptom) should be distinguished from the period of onset or the time elapsed from the first symptom to the first reflex spasm.

Tetanus in horses has often been described, but the essential features are similar in other animal species. Usually, the first sign of disease is impairment of muscle function in the head and neck. Difficulty in mastication and deglutition, a flared appearance to the nostrils, and stiffness of the ears in a vertical position may be observed. There may be rigidity in the muscles of the back and tail, extension of the body (orthotonos), or protrusion of the nictitating membrane. Hyperaesthesia is pronounced. As the disease progresses, respiration becomes impaired and respiratory failure may lead to death.

The course of disease may be acute with death within a few days or subacute to chronic. In the latter cases, recovery is more likely than in the acute disease.

PATHOGENESIS

Tetanus infection most often results from the contamination of a wound or the umbilicus by soil, but there are many routes of contamination. This has led to the use of descriptive terms such as puerperal, accidental, umbilical, gynecological, obstetrical, otogenic, surgical, and idiopathic tetanus. Slight penetrating wounds such as those produced by a thorn, rusty nail, or even a dirty abrasion are the types of lesions that most commonly result in tetanus. In animals, wounds created as a part of management procedures may be a portal of entry for *C. tetani*. Thus, docking and castration of lambs may provide wounds in which the organism may grow and produce toxin. Sometimes, the wound is so trivial as to remain unnoticed: minimal tissue necrosis suffices to create an environment appropriate for growth and toxin production by *C. tetani*.

The bacterium is noninvasive and symptoms resulting from the effect of the toxin alone will be manifest only after a certain incubation period. The length of the incubation period and severity of the disease will be determined by the amount of toxin formed in the primary lesion, which is dependent on the toxinogenicity of the infecting strain; the amount and rate of toxin reaching the neural pathways and the blood circulation; the toxin-transferring capacity of the neural pathways; the length of the neural pathways; and the susceptibility of receptors in the

CNS (which is species dependent). In humans, the incubation period often lasts 14 days (Kryzhanovsky 1981); the shorter this period is, the more severe tetanus is expected to be. In animals, the incubation period varies similarly in the range of 24 hours to 2 weeks or more (Kryzhanovsky 1981; Wellhöner 1982).

Toxin production depends on various factors, which are taken into consideration in culture production and must also exist in the wound. *C. tetani* produces two toxic substances; tetanus neurotoxin (or tetanospasmin) and tetanolysin, a hemolysin. Tetanolysin is produced in very low concentrations by pathogenic strains and is not considered to play a significant role in disease. The critical role of tetanus toxin in disease is indicated by the fact that strains that do not produce the toxin are nonpathogenic.

The toxinogenicity of *C. tetani* strains is variable. The factors that govern toxin production have as yet not been identified. Particular peptides might be necessary for toxin synthesis. Early studies had ruled out the involvement of phages in toxinogenicity, since identical phages were isolated from both mitomycin C-lysed toxinogenic and nontoxinogenic *C. tetani* strains. That toxinogenicity of strains was dependent on the presence of a plasmid was first suggested by Hara et al. (1977), and definitely confirmed by Eisel et al. (1986), who determined the whole sequence of the tetanus toxin gene, from which they deduced the complete amino acid sequence of the toxin. The toxin amino acid sequence was established independently and contemporaneously by Fairweather and Lyness (1986) in England. The *tox* gene is carried on a 75-kb plasmid.

Tetanus toxin is synthesized as a single-polypeptide chain with a molecular mass (Mr) of 150,700 Da. The toxin is not released from the bacterial cell as it is produced, but it can be extracted from the washed bacteria by the action of hypertonic solutions. This form of the toxin is called intracellular, cell, or extract toxin. Nicking of the intracellular toxin results in a three-fold increase of toxicity. When the culture is prolonged until bacteria lyse spontaneously, the toxin that is released is nicked by endogenous proteases. This form of the toxin is called extracellular, or filtrate toxin, and consists of a light chain (L-chain or fragment A) with a Mr of 52,288 Da held by a disulfide bond and noncovalent bonds to a heavy chain (H-chain or fragment BC) with a Mr of 98,300 Da. The two constituent chains may be isolated from the reduced toxin molecule under mild dissociating conditions. Either constituent toxin subchain itself is not toxic. A toxin molecule that was indistinguishable from native toxin and highly toxic was reconstituted from the separated chains.

Digestion of the toxin with papain cleaves the molecule into two fragments. One fragment, termed C, has a Mr of 51,562 Da and carries the binding site of the toxin for its receptors. It is nontoxic. The other fragment, called A-B, has a Mr of 99,138 Da and induces a flaccid, rather than a spastic, paralysis in experimental animals. Fragment A-B was shown to block not only inhibitory but also excitatory synaptic transmission to motor cells in the spinal cord when injected intraspinally in cats.

The purified, homogeneous tetanus toxin molecule is a highly toxic holoprotein that contains about 3×10^7 minimum mouse-lethal doses per mg protein. The high toxicity of tetanus toxin suggests that it should act through specific recognition sites that display very high affinity for a receptor. van Heyningen (1959) was the first to determine that the toxin-receptor compound in nervous tissue was a ganglioside. The toxin binds preferentially to gangliosides with a disialyl group attached to the inner galactose residue. The ganglioside-binding site is located in fragment C. However, recent studies have suggested the existence of a toxin-receptor substance that is different from gangliosides; a trypsin-sensitive-receptor component has been reported by Yavin and Nathan (1986). Under physiologic conditions, binding might involve membrane proteins capable of modulating the affinity for gangliosides. In this connection, Montecucco (1986) has proposed an attractive two-step model for toxin binding, in which the toxin first binds with low affinity to gangliosides. Subsequently, the resulting membrane-bound ganglioside-toxin complex moves laterally until it is attached with high affinity by the toxin-specific protein receptor. The binding step is followed by an internalization step, after which there is a retrograde axonal transport to the ventral spinal cord and brain stem.

Tetanus toxin, formed by *C. tetani* in an infected wound or injected into the body spreads to neighboring muscles, where it is bound to the presynaptic terminals of the motor axons. The toxin then enters the lymphatic and blood vascular systems to be distributed to all muscles and nerve endings. Distribution in the blood is characteristic for solipeds and humans. This mode of spread of the toxin results in generalized tetanus, also termed "descending tetanus." The blood-brain barrier and blood-peripheral nerve barrier exclude direct entry of the toxin into the nervous system. Transport of the toxin to the site of action in the CNS depends on binding to presynaptic membrane receptors (which are disialo- or trisialogangliosides or sialoglycoproteins abundant in synaptic vesicles) and internalization into the membrane and retroaxonal transport within the smooth endoplasmic reticulum carrier system. The toxin reaches the neuronal body, where it can be detected several hours before the appearance of the first symptoms of tetanus and where it can persist for a prolonged time.

Pathogenic action of the toxin is due to its unique ability to pass transsynaptically from the nerve body to the axon terminals, which synapse on the motor neuron, and its ability to selectively bind to the presynaptic membranes and to interfere with the release of neurotransmitters, glycine and gamma-aminobutyric acid (GABA). At later stages of intoxication, excitatory neurotransmission also is affected.

Localized and generalized tetanus have been produced experimentally and contribute to our understanding of the disease process. When a small dose of toxin is injected into the hind limb of a mouse, local (ascending) tetanus usually results. Muscle spasms first develop in the injected limb, then in the opposite limb. The toxin travels along the tissue spaces of the peripheral nerves to the anterior horns of the spinal cord and may spread anteriorly in the spinal cord (Kryzhanovsky 1981).

When the mouse is injected intravenously or given a large dose of toxin intramuscularly, the toxin travels by the lymphatic and blood vascular systems and is distributed to all motor nerves. Generalized (descending) tetanus results, possibly because the neural centers of the head and neck are more sensitive to toxin (Kryzhanovsky 1981).

Toxin produced in the wound by the bacteria, or experimentally injected into the body, reaches primarily the peripheral motor nerve endings and is transported intraaxonally retrograde to the soma of the neurons in the CNS. The toxin then passes transsynaptically to bind to presynaptic axonal terminals of neurons that synapse on the motor neurons. Initially, the toxin blocks the release of glycine and GABA, although at later stages of intoxication, excitatory transmission is also affected. Since excitatory transmission is allowed to continue in the absence of inhibition (disinhibition), a hyperactivity of motor neurons ensues and leads to manifestations of muscle spasticity.

An exception to the usual spastic effect of tetanus toxin is seen in the flaccid paralysis of the diaphragm occasionally observed in patients. In experimental animals, flaccid paralysis is readily induced with high doses of toxin injected subcutaneously or intramuscularly (Matsuda et al. 1982). High doses of toxin have a botulinum toxinlike effect at the neuromuscular junction; this effect can be reproduced with the A-B fragment of the toxin.

The neuromuscular blocking action of tetanus toxin was found to be about 1000 times lower than that of botulinum toxin A. Tetanus toxin can inhibit the K^+-stimulated release of GABA, glycine, acetylcholine, and norepinephrine. The complete blockade of Ca^{++}-stimulated catecholamine secretion was achieved by injecting either the whole tetanus toxin molecule or its A-B fragment into adrenal chromaffin cells (Penner et al. 1986). Therefore, tetanus toxin is likely to act on a process of exocytosis that is common to all vesicular and granular release systems and is considered to affect the release rather than the synthesis of the transmitter.

Tetanus toxin was shown to exert its paralytic effect on cholinergic nerve terminals in rabbits by interfering with cGMP metabolism (King et al. 1978). The toxin fragment A-B has the same effect. The toxin was also capable of inhibiting the depolarization-evoked cGMP accumulation in pheochromocytoma cells (PC12 cells). At the same time, acetylcholine release from these cells was blocked, indicating that cGMP plays a fundamental role in the mechanism of neurosecretion. A 180-kDa membrane guanylate cyclase, which is also an atrial natriuretic factor (ANF) receptor and is functionally coupled to the production of cGMP, was purified by Sharma et al. (1989). Tetanus toxin inhibited both the basal and ANF-stimulated generation of cGMP by this enzyme in the same manner as specific antibodies to the enzyme did.

Depending on the amount of toxin reaching the CNS, the spinal inhibitory reflexes are fully inhibited as early as 30-48 hours after toxin injection into laboratory animals, whereas inhibition of excitatory monosynaptic reflexes is observed only after about 26-144 hours. In contrast, fragment A-B blocks inhibitory and excitatory monosynaptic reflexes almost simultaneously.

Three steps have been characterized in the toxin action on isolated hemidiaphragm preparations: (1) the toxin binds to its fixation sites; this is temperature dependent and reversible; (2) the toxin molecule translocates within the membrane, which depends on transmitter release; and (3) paralysis results and is strictly temperature dependent and not related to transmitter release (Habermann et al. 1980).

It has also been postulated that contractile proteins of the presynaptic nerve endings might represent possible targets for the toxin action (Kryzhanovsky 1981). There is recent evidence that tetanus toxin is a zinc endopeptidase that targets and cleaves synaptobrevin, a membrane protein found in synaptic vesicles (Schiavo et al. 1992).

Tetanus toxin preferentially binds with high affinity to neuronal membranes. However, binding to plasma membranes of the thyroid, pancreatic, and islet cells, and of melanoma, renal medulla, and anterior pituitary, adrenal medulla, and thymic epithelial cells has also been demonstrated. The significance of this nonneuronal binding of the toxin in the pathogenesis of tetanus is still not defined.

Tetanus is accompanied by various changes in metabolism. Major changes occur in fluid, acid-base, and electrolyte balance, as well as in carbohydrate, lipid, protein, and nucleic acid metabolism. Variations in the levels of particular enzymes are also recorded (Kryzhanovsky 1981; Mellanby and Green 1981; Wellhöner 1982).

In severe tetanus, initial respiratory alkalosis is followed by acidosis, which is further increased as a result of lactic acid formation by muscular hyperactivity. Convulsive respiratory muscle spasms are accompanied by a dramatic decrease in pulmonary ventilation. Hypoxia develops, the severity of which varies with the degree of hypertonia of the respiratory muscles. Microcirculatory disturbances in pulmonary tissue occur early in the disease, and rather frequent complications are a major cause of mortality. However, toxin has been found to accumulate in the lungs. The accumulated, but not bound, material might represent the toxin-derived A-B fragment, which could account for the occurrence of respiratory paralysis that is not associated with the convulsion syndrome.

IMMUNITY

Some host factors are important in tetanus. For instance, increased resistance to the disease was observed during pregnancy, but there is no explanation for this phenomenon.

Similarly, susceptibility to tetanus was found to follow seasonal variations with a peak of incidence in late spring and summer. Another factor of susceptibility is the age of individuals. Increased susceptibility of elderly people and neonates probably results from some inadequacy of their homeostatic mechanisms. However, the main factor of resistance to tetanus toxin is represented by functional body barriers, the most important of which is the blood-brain barrier since the toxin moves along a neural pathway. The significance of this barrier is demonstrated by the observation that differences in susceptibility of various animals to the toxin disappear when the toxin is injected intraspinally. Other barriers are the uterine and intestinal barriers.

Macrophages represent a first line of defense against bacterial infections. The observation that tetanus toxin can interfere with the release of lysozyme from human macrophages (Ho and Klempner 1985) might explain a decrease in natural defences toward *C. tetani*.

In clinical tetanus, specific immunity does not develop in surviving patients for two reasons. First, the amount of toxin that induces the disease is too small to be immunogenic. Second, the toxin, as it is produced in situ by *C. tetani*, is readily bound by its receptors on the terminals of the nerves innervating the infected area.

For a long time, naturally-acquired immunity to tetanus has been thought not to occur. However, Veronesi et al. (1983) has provided epidemiological evidence for its existence. For instance, before the discovery of tetanus toxoid, one-third of the inhabitants in Peking were found to be carriers of tetanus spores and their sera were shown to contain significant levels of tetanus antibodies. More recently, naturally-acquired immunity to tetanus was demonstrated on a sound experimental basis: reliable and specific methods were used to identify high levels of tetanus antitoxin in the sera of individuals belonging to vaccination-free populations (Veronesi et al. 1983). Similar observations were made in animals. Development of natural antitetanus immunity is likely to be related to prolonged ingestion of tetanus spores, which are present in highly contaminated areas. In fact, production of tetanus antitoxin was elicited by protracted feeding of tetanus spores to mice.

Currently, antitetanus vaccination is performed with tetanus toxoid, the immunogenicity of which is enhanced by adsorption onto a mineral oil adjuvant. In humans, vaccination is aimed at inducing a protective level of antibodies, arbitrarily fixed at 0.01 IU / ml serum (Bizzini 1984). In animals, vaccination can be performed at all ages by injecting two doses of veterinary tetanus toxoid at 1-month intervals and a third dose 1 year later. In the case of valuable animals, booster injections are repeated every 5 years. It is also worth mentioning that vaccination of breeding animals will protect the newborn progeny via the colostrum.

The decision to vaccinate or not to vaccinate depends on the species of animal and the history of occurrence of tetanus in an area. Immunization with tetanus toxoid is generally carried out in horses and sheep. In immunized animals, a booster injection of toxoid is usually administered to animals at risk because of a wound. In unimmunized animals that are at risk, tetanus antitoxin may be administered to provide immediate passive protection at the same time that toxoid is given to stimulate an active antitoxic response.

PROBLEMS

The genetics of tetanus toxin production by *C. tetani* and the chemistry and pharmacology of the toxin are known in great detail. However, the actual mechanism of action of tetanus toxin has not as yet been elucidated. As a consequence, the main challenge to researchers is to determine the precise mechanism of action. It can safely be predicted that tetanus toxin is likely to exert its pathogenic action via multiple effects. Structure homology suggested that the toxin may be a zinc-protease and recent studies have demonstrated that it is a zinc-protease (Schiavo et al. 1992). Furthermore, the specific substrate for the toxin action in the nerve cell has been identified as synaptobrevin, a cell membrane protein. It is now important to determine the role of this protein in neurotransmitter release.

DNA recombinant techniques permit the generation of toxin fragments, such as the binding C-fragment, which has been used to successfully target a drug or a protein to the CNS. Such investigations should be developed further because they offer a way of exploring particular mechanisms in the CNS, as well as of increasing the efficacy of certain drugs by delivering them to their site of action.

In developing countries, complete vaccination programs are difficult to achieve; therefore, persisting efforts should aim at producing an effective oral vaccine.

For obvious reasons (blood contamination, insufficient supply of human blood), human antitetanus IgG preparations will have to be replaced by neutralizing antitetanus monoclonal antibodies. Preliminary reports have indicated that this is feasible.

REFERENCES

Bizzini, B. 1984. Tetanus. In Bacterial Vaccines, Vol. 1. Ed. R. Germanier, pp. 37-68. London: Academic Press.

__________. 1989. Axoplasmic transport and transynaptic movement of tetanus toxin. In Botulinum Neurotoxin and Tetanus Toxin. Ed. L. L. Simpson, pp. 203-29. London: Academic Press.

Eisel, V.; Jarausch, W.; Goretzki, K.; Henschen, A.; Engels, J.; Weller, U.; Hude, M.; Habermann, E.; and Niemann, H. 1986. Tetanus toxin: Primary structure, expression in *E. coli*, and homology with botulinum toxins. Eur Mol Biol Organ J 5:2495-502.

Fairweather, N. F., and Lyness, V. A. 1986. The complete nucleotide sequence of tetanus toxin. Nucleic Acids Res 14:7809-12.

Habermann, E.; Dreyer, F.; and Bigalke, H. 1980. Tetanus toxin blocks the neuromuscular transmission in vitro like botulinum A toxin. Naunyn Schmiedebergs Arch Pharmacol 311:33-40.

Hara, T.; Matsuda, M.; and Yoneda, M. 1977. Isolation and some properties of nontoxigenic derivatives of a strain of *Clostridium tetani*. Biken J. 20:105-15.

Ho, J. L., and Klempner, M. S. 1985. Tetanus toxin inhibits secretion of lysosomal contents from human macrophages. J Infect Dis 152:922-29.

King, Jr., L. E.; Fedinec, A. A.; and Lathami, W. C. 1978. Effects of cyclic nucleotides or tetanus toxin paralyzed rabbit sphincter pupillae muscles. Toxicon 16:625-31.

Kryzhanovsky, G. N. 1981. Pathophysiology. In Tetanus — Important New Concepts, Vol. 1. pp. 109-182. Amsterdam, Oxford, Princeton: Excerpta Medica.

Matsuda, M.; Sugimoto, N.; Ozutsumi, K.; and Toshiro, H. 1982. Acute botulinum toxin-like intoxication by tetanus neurotoxin in mice. Biochem Biophys Res Commun 104:799-805.

Mellanby, J., and Green, J. 1981. How does tetanus toxin act? Neuroscience 6:281-300.

Montecucco, C. 1986. How do tetanus and botulinum toxins bind to neuronal membranes? Trends Biochem Sci 11:314-17.

Penner, R.; Neher, E.; and Dreyer, G. 1986. Intracellularly injected tetanus toxin inhibits exocytosis in bovine adrenal chromaffin cells. Nature (London) 324:76-78.

Sharma, R. K.; Marala, R. B.; and Duda, T. 1989. Purification and characterization of 180-kDa membrane guanylate cyclase containing atrial natriuretic factor receptor from rat adrenal gland and its regulation by protein kinase C. Steroids: Nes Mem Issue 53:437-60.

Schiavo, G,; Benfenati, F.; Poulain, B.; Rossetto, O.; Polverino de Laureto, P.; DasGupta, B. R.; and Montecucco, C. 1992. Tetanus and botulinum-B neurotoxins block neurotransmitter release by proteolytic cleavage of synaptobrevin. Nature 359:832-35.

Shoesmith, J. G., and Holland, K. T. 1972. The germination of spores of *Clostridium tetani*. J Gen Microbiol. 70:253-61.
Smith, L. DS. 1975. *Clostridium tetani*. In the Pathogenic Anaerobic Bacteria, 2d Ed., pp. 177-201. Springfield, Ill.: Charles C. Thomas.
van Heyningen, W. E. 1961. The fixation of tetanus toxin by ganglioside. J Gen Microbiol 24:107-19.
Veronesi, R.; Bizzini, B.; Focaccia, R.; Coscina, A. A. L.; Mazza, C. C.; Focaccia, M. T.; Carraro, F.; and Honningman, M. N. 1983. Naturally acquired antibodies to tetanus toxin in humans and animals from the Galapagos Islands. J Infect Dis 147:308-11.
Wellhöner, H. H. 1982. Tetanus neurotoxin. In Reviews of Physiology, Biochemistry, and Pharmacology, Vol. 93. Ed. R. H. Adrian, pp. 1-68. Berlin, Heidelberg, New York: Springer Verlag.
Wells, C. L., and Balish, E. 1983. *Clostridium tetani* growth and toxin production in the intestines of germfree rats. Infect Immun 41:826-28.
Yavin, E. and Nathan, A. 1986. Tetanus toxin receptors on nerve cells contain a trypsin-sensitive component. Eur J Biochem 154:403-7.

10 / Histotoxic Clostridia

BY C. L. GYLES

CLOSTRIDIA that are discussed in this chapter are *Clostridium perfringens, C. chauvoei, C. septicum, C. novyi, C. haemolyticum*, and *C. sordellii*. These are all large, gram-positive spore-forming bacteria that produce at least one hemolysin. Their ability to form resistant spores allows them to persist in soils for long periods of time. These organisms typically are involved in infections of muscle or liver, but *C. perfringens* more commonly causes enterotoxemias (Chapter 11), and *C. septicum* sometimes causes infection of the stomach in sheep. There is a simple, consistent pattern to the diseases considered in this section (Fig. 10.1). Toxigenic clostridia, which are either normally present in tissue or deposited there as a result of wound contamination, multiply in traumatized tissue; toxins produced by the bacteria create a local lesion, and toxins absorbed from this site lead to toxemia (terminal bacteremia may develop).

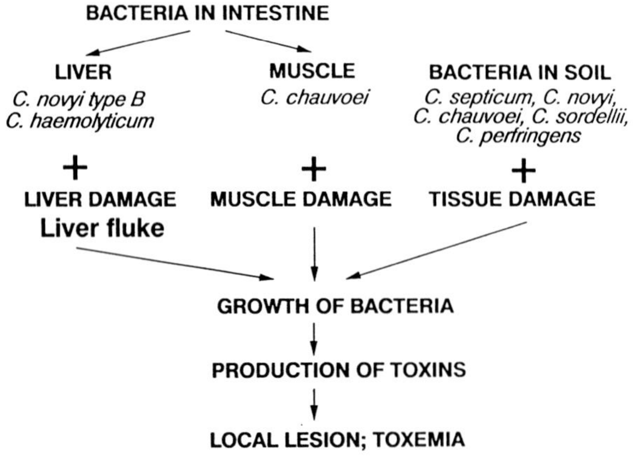

FIG. 10.1. Common features of pathogenesis of diseases due to the histotoxic clostridia.

Although there has been some new information about several toxins produced by these clostridia, there has been little new knowledge about the pathogenesis of infections due to the histotoxic clostridia since the excellent review by MacLennan (1962). Diseases are satisfactorily prevented by use of commercial, formalinized whole-culture vaccines in animals at greatest risk and now occur at low frequency (Rutter and Collee 1969; Stevenson and Stonger 1980).

Smith (1975a) describes three types of wound infection involving clostridia: simple wound contamination, anaerobic cellulitis, and gas gangrene, or clostridial myositis. The first two represent mild to moderate infections that usually resolve readily, but gas gangrene is severe and often fatal. Gas gangrene is initiated by deep trauma to large muscle masses. Low oxidation-reduction potential and a favorable supply of amino acids in the traumatized muscle lead to invasion of healthy muscle and an enlarged area of infection. The presence of ionizable calcium salts in the wound may contribute not only to tissue necrosis but also to the activity of bacterial toxins; lecithinases produced by *C. perfringens* and *C. novyi* all require calcium ions for their activity. The presence of other less-fastidious bacteria in the wound may also facilitate the growth of clostridia (MacLennan 1962).

CLOSTRIDIUM PERFRINGENS

Clostridium perfringens is nonmotile and encapsulated, and its spores are not usually formed during growth on ordinary laboratory media. The organism grows rapidly and, under anaerobic conditions, produces colonies on blood agar that have a characteristic double zone of hemolysis. The inner zone of complete hemolysis is due to theta-toxin, and the outer zone of incomplete hemolysis is due to alpha-toxin. *C. perfringens* grows over a temperature range of 20-50°C, but the optimal temperature for growth is 45°C, at which temperature generation time is as short as 8 minutes (Hatheway 1990). The species *C. perfringens* is divided into five types, A, B, C, D, and E, on the basis of four major lethal toxins produced by the organism. Alpha-toxin is produced by strains of all types but in largest quantities by strains of type A; beta-toxin, by strains of types B and C; epsilon-toxin, by strains of types B and D; and iota-toxin, by strains of type E (Smith 1975b, 1979).

Clostridium perfringens is part of the normal intestinal flora of humans and animals and is widespread in soil. Wound contamination with soil or feces may lead to gas gangrene. It is common to have more than one species of *Clostridium* involved in gas gangrene lesions, but the usual species is *C. perfringens* type A (Love et al. 1938; Smith 1975a; Miller et al. 1983). Other genera of bacteria may also be involved (Smith 1975a). *C. perfringens* is the most common species of *Clostridium* involved in wound infections (malignant edema) in horses. In many instances, the wound that becomes infected is caused by injections into muscle (Rebhun et al. 1985). Carnivores are not often affected, but there have been reports of gas gangrene occurring in dogs whose repaired fractures were contaminated with *C. perfringens*. Toxins produced by the organism are responsible for the destruction of tissue and phagocytes, and 16 toxins and enzymes that might play a role in disease have been described.

Alpha-toxin is the only lethal toxin produced by *C. perfringens* type A and it is considered to be the most important virulence factor in the initiation of muscle infections (Smith 1975b, 1979). The alpha-toxin content in culture fluids has been shown to be critical in experimental infections and alpha-antitoxin has been shown to be protective. The 54-kDa toxin is both a phospholipase C and a sphingomyelinase that requires zinc and free calcium ions for its enzymatic activities. It is an extremely potent toxin and has an LD_{50} in mice of only 0.75 hemolytic units (Rood and Cole 1991). Damage to tissues is mediated by hydrolysis of lecithin and sphingomyelin in cell membranes of erythrocytes, blood platelets, leukocytes, endothelial cells, and muscle cells (Elder and Miles 1957; Smith 1979). Tissue damage associated with gas gangrene can be partially reproduced by administration of alpha-toxin to guinea pigs. Local injury caused by alpha-toxin is almost identical to that observed with whole cultures of *C. perfringens*. The toxin attacks phospholipids, which are major components of cell membranes; microsomes; and mitochondria (Demello et al. 1974).

When a sufficient concentration of alpha-toxin is present in the blood, widespread damage to capillary endothelial cells, destruction of platelets, intravascular hemolysis, and liver toxicity are observed. In muscle, however, only damage to muscle cells and development of edema are evident. The edema is a result of capillary damage.

Nonlethal bacterial factors may also contribute to disease. Kappa-toxin, a collagenase produced by *C. perfringens,* is capable of destroying collagen in blood vessels. It is probably responsible for the pulpy muscle seen in gas gangrene (Smith 1975b, 1979). Amino acids that are released by the action of collagenase facilitate the growth of *C. perfringens* in the damaged tissue. Significant quantities of alpha-toxin do not appear to get into circulation during gas gangrene, and efforts have been made to determine the lethal factor in disease. There is evidence that vascular damage can occur in infected animals even when the actions of alpha- and other toxins are blocked by antitoxin. The ability of *C. perfringens* to take up large amounts of oxygen from its environment is probably responsible for vascular damage, resulting in hemoconcentration and shock due to hypovolemia (Smith 1979); this has been the basis for the use of hyperbaric oxygen in treating infected animals and humans.

Tissue necrosis probably is aided by gas formation during bacterial fermentation; pockets of gas exert sufficient pressure to interfere with venous return from the muscles. Cellulitis due to *C. perfringens* and accompanied by immune-mediated hemolytic anemia has been described in horses (Reef 1983). The autoimmune hemolytic anemia was considered to be a result of damage to the erythrocyte membrane by hemolysins and the subsequent formation or unmasking of new or altered antigen, recognized as foreign by the host. The injured erythrocytes were removed from circulation and sequestered in the spleen by complement-fixing immunoglobulin M (IgM). For a fuller discussion of *C. perfringens,* see Chapter 11.

CLOSTRIDIUM CHAUVOEI

Clostridium chauvoei is a highly motile organism, the principal habitat of which is the intestine and tissues of animals. However, survival of the organism in soil is likely to be a significant factor, since contaminated pasture appears to be a source of organisms. The organism causes a specific disease, blackleg of cattle, as well as wound infections in cattle, sheep, pigs, and occasionally horses. Blackleg is the result of the activation of latent spores of *C. chauvoei* deposited in muscle after transport from the liver or intestine in the circulation, possibly through macrophages. Young cattle (4 months to 3 years of age) on a high plane of nutrition are most susceptible. Bruising of the large muscle masses of these animals occurs readily and provides a favorable environment for spore germination and toxin production (Sterne and Batty 1975). Excessive exercise also appears to provide an environment in muscles that permits germination of *C. chauvoei* spores.

Experimental studies in the guinea pig demonstrate that tissue destruction and factors that stimulate sporulation are critical to development of the local muscle lesion (Princewill 1965), which has a characteristic dark appearance and often has a metallic sheen, a rancid odor, and gas bubbles. The dark appearance is attributed to staining of muscle with hemoglobin that is released from lysed erythrocytes; and the odor is due to formation of butyric acid as an end product of fermentation by the bacterium. Affected animals may show depression, elevated temperature, and lameness, but in most cases the course of disease is so acute that animals are not observed to be ill before they are found dead.

Castration, docking, parturition, and inoculation may provide opportunities for wound infection by *C. chauvoei*. Cattle and sheep are more commonly affected by myositis due to *C. chauvoei* than are swine.

Several toxic compounds are produced by *C. chauvoei* (Ramachandran 1969): alpha-toxin, a lethal, necrotizing, protein that causes hemolysis of sheep red blood cells; beta-toxin, a heat-stable deoxyribonuclease that may be responsible for the marked destruction of nuclei in muscle cells observed in the local lesion (Dixit and Khera 1968); gamma-toxin, a hyaluronidase; and delta-toxin, an oxygen-labile hemolysin.

CLOSTRIDIUM SEPTICUM

Clostridium septicum, found in soil and the intestine of animals, is an important cause of wound infection in cattle, sheep, and swine. The term malignant edema is used for wound infections that involve clostridial species, alone or in combination with other bacteria, and is typically a cellulitis, with minor involvement of muscle. There is often a foul odor associated with the lesion. *C. septicum* occasionally has been the only organism isolated from lesions characteristic of blackleg, but these are apt to be cases in which the causative agent, *C. chauvoei,* has been overgrown by *C. septicum,* a postmortem invader. Braxy is a specific infection of sheep (rare in North America), in which injured abomasum is invaded by *C. septicum.* The injury may be caused by an experience common to the flock (such as ingestion of frozen grass), and large numbers of animals may be affected.

The spectrum of toxins produced by *C. septicum* is similar to that of *C. chauvoei* (Bernheimer 1944a; Aikat and Dible 1960). Alpha-toxin is a lethal, necrotizing, hemolytic product that has a direct toxic effect on cardiac muscle (Kellaway et al. 1941) and is capable of causing capillary damage. Iron is required both for growth of the bacteria and for production of alpha-toxin (Bernheimer 1944b). Alpha-toxin has been purified and shown to be a 48 kDa basic protein widespread among strains of *C. septicum*; proteins immunologically related to this toxin were not found in the other histotoxic clostridia (Ballard et al. 1992). The LD_{50} for mice is 10 μg/kg body weight and immunization of mice with alpha-toxin resulted in partial protection against challenge with live *C. septicum* organisms. Little is know about the mechanism of action of alpha-toxin. It has been suggested that it may be a pore-forming hemolysin.

Beta-toxin is a deoxyribonyclease, delta-toxin is a hyaluronidase, and gamma-toxin is an oxygen-labile hemolysin. In the early literature, the alpha- and beta-toxins of *C. septicum* and *C. chauvoei* were reported to be related antigenically, but studies with the purified alpha-toxin of *C. septicum* indicate that no extracellular protein of *C. chauvoei* is antigenically related to it (Ballard et al. 1992). Neuraminidase produced by *C. septicum* (Gadalla and Collee 1968) is produced in vivo and may contribute to pathogenicity.

CLOSTRIDIUM NOVYI

The principal habitat of *C. novyi* is soil and the intestine of animals. Three types, A, B, and C, are distinguished on the basis of toxins produced. A fourth type, D, is considered to be a different species, *C. haemolyticum* (Smith 1975c). *C. novyi* synthesizes five toxins: alpha, beta, gamma, delta, and epsilon. Strains of type A, common in soil, produce all except beta-toxin. Strains of type B, rarely found in soil, produce alpha- and beta-toxins only. Strains of type B also produce zeta-, eta-, and theta-toxins. Strains of types A and B are recovered from the livers of normal animals (Niilo et al. 1969). Type C isolates are nontoxigenic and are not implicated in disease.

Wound infection involving *C. novyi* is characterized by extensive edema and occurs in cattle, sheep, and several other animal species. The edema is the result of vascular damage inflicted by the alpha-toxin. The condition known as "bighead" in rams develops when subcutaneous tissues are traumatized during fights and are subsequently invaded by *C. novyi* type A (Sterne and Batty 1975). Only a small number of clostridia are often detected in infections due to *C. novyi*; this has been attributed to the potency of the toxins produced by this species. Infectious necrotic hepatitis (black disease) is a special type of infection, in which spores of *C. novyi* type B organisms that normally reside in the liver of sheep, germinate and produce tissue necrosis. The damage is usually due to migration of immature liver fluke, *Fasciola hepatica*, and permits germination of spores, growth of the vegetative cells, and subsequent production of toxins, but any destruction of liver tissue could be the inciting factor. Alpha-toxin produced in the local area of necrosis in the liver is absorbed into the circulation and results in systemic effects. A similar condition occasionally occurs in cattle and horses.

Alpha-toxin is considered to be the major toxin influencing the course of infections. It is a protein of approximately 200-kDa molecular weight, whose LD_{50} for mice is approximately 200 ng/kg body weight. Alpha-toxin is lethal; necrotizing; causes increased capillary permeability (Elder and Miles 1957; Cotran 1967); and is toxic to several tissues including muscle, heart, and liver (Aikat and Dible 1960; Rutter and Collee 1969; Pemberton et al. 1971). On intravenous injection of mice, alpha-toxin causes formation of gaps between endothelial cells, edema, and hypovolemic shock, which is considered to be the cause of death (Bette et al. 1989). Similar changes in blood vessels are seen locally when the toxin is injected subcutaneously. Toxin appears to act by changing the microfilament system of the cell (Oksche et al. 1992); endothelial cells that are exposed to toxin undergo changes in microfilaments resulting in alterations in cell shape, accompanied by a partial loss of F-actin, and redistribution of vinculin, which usually marks the sites of attachment of microfilaments on membranes.

Beta-toxin is a lecithinase, which is produced in small quantity by strains of type B. Gamma-toxin is a necrotising phospholipase D. Delta-toxin is an oxygen-labile hemolysin, and epsilon-toxin is a lipolytic enzyme. Zeta-toxin is a poorly characterized hemolysin, and theta-toxin is a lipase produced in trace amounts by strains of *C. novyi* type B. Eta-toxin is a tropomyosinase, which degrades tropomyosin and myosin and may play a role in destruction of infected muscles.

CLOSTRIDIUM HAEMOLYTICUM

This organism was previously called *C. novyi* type D. The disease it produces is due to synthesis of large amounts of the lethal, necrotizing, and hemolytic beta-toxin (phospholipase C) of *C. novyi*. The organism also produces the eta and theta toxins of *C. novyi* (Smith 1975d).

C. haemolyticum causes bacillary hemoglobinuria of cattle. Pathogenesis of the disease is similar to that of infectious necrotic hepatitis of sheep except that the organism is *C. haemolyticum* and the dominant toxin is beta-toxin. Thus, a nidus of infection in the liver develops as a sequel to damage by liver flukes or other agents; local necrosis and intravascular hemolysis ensue. Experimental studies in rabbits (Van Kampen and Kennedy 1969) suggest the following sequence of events: migration of immature flukes through the liver causes damage to arterioles and subsequent necrosis of liver tissue; hypoxia and other changes in the necrotic liver induce germination of spores of *C. haemolyticum* that are present in Kupffer cells; vegetative cells of *C. haemolyticum* produce phospholipase C, which lyses the Kupffer cells and sets the bacteria free; and multiplication of the bacteria and further production of phospholipase C occur.

The strongly hemolytic beta-toxin can be shown to be responsible for all major developments in disease; locally, it causes lysis of hepatocytes and erythrocytes, leading to liver necrosis. Toxin produced in the local lesion is absorbed into the bloodstream and produces hemoglobinuria, as well as damage to capillary endothelium, which leads to loss of blood into tissues and serous cavities. The disease can be produced experimentally by intrahepatic inoculation of spores and calcium chloride (Erwin 1977).

CLOSTRIDIUM SORDELLII

Clostridium sordellii is a normal inhabitant of the intestinal flora of animals and is also found in soil. The organism is associated with sudden death in cattle and sheep, from which it can be recovered from lesions in muscle or liver (Coleman et al. 1975). Muscle lesions are characterized by myositis and considerable edema of the subcutaneous tissues and fascial planes surrounding the muscles. Subendocardial hemorrhages in the left ventricle and a terminal septicemia have been observed in experimentally infected cattle.

Toxic products of this organism have not been characterized to any great extent, but it is clear that toxic substances are formed that account for the signs of disease. An edema-producing factor is likely to be the most important toxin and is a lethal toxic factor (Arseculeratne et al. 1969). *C. sordellii* also produces a phospholipase C, a lysolecithinase, an oxygen-sensitive hemolysin, a neuraminidase, and a deoxyribonuclease. Deoxyribonuclease may play a role in disease by causing destruction of nuclei of cells in affected tissues, but the other substances are not considered to be of much significance in disease (Smith 1975e).

VACCINES

Because the course of disease is very rapid and often fatal, it is advisable to vaccinate animals that are at risk. Vaccines prepared from *C. septicum, C. chauvoei, C. novyi, C. hemolyticum*, and *C. sordellii* are available commercially. *C. perfringens* is only infrequently implicated in wound infections in animals, and vaccination is not designed to protect against occasional infection. The vaccines are prepared for administration of individual bacterial agents or combinations of the agents. For example, cattle might be immunized with *C. haemolyticum*; *C. haemolyticum* + *C. novyi* type B; *C. chauvoei* + *C. septicum*; combinations of *C. chauvoei* + *C. septicum* with *C. novyi* or *C. sordellii*; *C. chauvoei* + *C. septicum* + *C. novyi* + *C. sordellii*; or these four organisms + *C. haemolyticum*. The choice is determined by the agents that are prevalent in a particular area.

Typically, the vaccines consist of inactivated broth cultures of the bacteria, so that antibodies develop to both bacterial surface antigens and to toxic exoproducts. There is little understanding of the contribution to the vaccines made by specific bacterial structures or products, but there is some indication that antibodies to flagellin are opsonizing and induce protective immunity against *C. chauvoei*.

REFERENCES

Aikat, B. K., and Dible, J. H. 1960. The local and general effects of cultures and culture-filtrates of *Clostridium oedematiens, Cl. septicum, Cl. sporogenes* and *Cl. histolyticum*. J Pathol Bacteriol 79:227-41.

Arseculararatne, S. N.; Panabokke, R. G.; and Wijesundra, S. 1969. The toxins responsible for the lesions of *Clostridium sordellii* gas gangrene. J Med Microbiol 2:37-53.

Ballard, J.; Bryant, A.; Stevens, D.; and Tweten, R. K. 1992. Purification and characterization of the lethal toxin (alpha-toxin) of *Clostridium septicum*. Infect Immun 60:784-90.

Bernheimer, A. W. 1944a. Parallelism in the lethal and hemolytic activity of *Clostridium septicum*. J Exp Med 80:309-20.

_________ . 1944b. Nutritional requirements and factors affecting the production of toxin by *Clostridium septicum*. J Exp Med 80:321-31.

Bette, P.; Frevert, J.; Mauler, F.; Suttrop, N.; and Habermann, E. 1989. Pharmacological and biochemical studies of cytotoxicity of *Clostridium novyi* type A alpha-toxin. Infect Immun 57:2507-13.

Coleman, J. D.; Hill, J. S.; Bray, H. T.; Armstrong, D. A.; and Morgan, C. O. 1975. Prevention of sudden death caused by *Clostridium sordellii* in feedlot cattle. Vet Med Small Anim Clin 70:191-95.

Cotran, R. S. 1967. Studies on inflammation. Ultrastructure of the prolonged vascular response induced by *Clostridium oedamatiens* toxin. Lab Invest 17:39-60.

Demello, F. J.; Anderson, R.; Hitchcock, C. R.; and Haglin, J. J. 1974. Ultrastructure study of clostridial myositis. Arch Pathol 97:118-25.

Dixit, S. N., and Khera, S. S. 1968. Pathogenesis of *Clostridium chauvoei* infection in guinea pigs. Indian J Exp Biol 6:80-83.

Elder, J. M., and Miles, A. A. 1957. The action of the lethal,toxins of gas gangrene clostridia on capillary permeability. J Pathol Bacteriol 74:133-45.

Erwin, B. G. 1977. Experimental induction of bacillary hemoglobinuria in cattle. Am J Vet Res 38:1625-27.

Gadalla, M. S. A., and Collee, J. G. 1968. The relationship of the neuraminidase of *Clostridium septicum* to the haemagglutinin and other soluble products of the organism. J Pathol Bacteriol 96:169-85.

Hatheway, C. L. 1990. Toxigenic Clostridia. Clin Microbiol Rev 3:66-98.

Kellaway, C. H.; Reid, G.; and Trethewic, E. R. 1941. Circulatory and other effects of the toxin of *Cl. septicum*. Aust J Exp Biol Med Sci 11: 297-309.

Love, W. G.; Millar, J. A. S.; and Rawling, W. B. 1938. Gas gangrene (*C. welchii, B.perfringens*) infection in the dog. North Am Vet 19:51-59.

MacLennan, J. D. 1962. The histotoxic clostridial infections of man. Bacteriol Rev 26:177-276.

Miller, R. A.; McCain, C. S.; and Dixon, D. 1983. Canine clostridial myositis. Vet Med Small Anim Clin 78:1065-66.

Niilo. L,; Dorward, W. J.; and Avery, R. J. 1969. The role of *Clostridium novyi* in bovine disease in Alberta. Can Vet J 10:159-69.

Oksche, A.; Nakov, R.; and Habermann, E. 1992. Morphological and biochemical study of cytoskeletal changes in cultured cells after extracellular application of *Clostridium novyi* alpha-toxin. Infect Immun 60:3002-6.

Pemberton, J. R.; Matson, R. L.; Claus, K. D.; and Macheak, M. F. 1971. Changes in plasma enzyme levels of sheep infected with *Clostridium novyi* type B. Clin Chim Acta 34:431-36.

Princewill, T. J. T. 1965. Effect of calcium chloride on germination and pathogenicity of spores of *Clostridium chauvoei*. J Comp Pathol 75: 343-51.

Ramachandran, S. 1969. Haemolytic activities of *Cl. chauvoei*. Indian Vet J 46:754-68.

Rebhun, W. C.; Shin, S. J.; King, J. M.; Baum, K. H.; and Patten, V. 1985. Malignant edema of horses. J Am Vet Med Assoc 187:732-36.

Reef, V. B. 1983. *Clostridium perfringens* cellulitis and immune-mediated hemolytic anemia in a horse. J Am Vet Med Assoc 182:251-54.

Rood, J. I., and Cole, S. I. 1991. Molecular genetics and pathogenesis of *Clostridium perfringens*. Microbiol Rev 55:621-48.

Rutter, J. M., and Collee, J. G.. Studies on the soluble antigens of *Clostridium oedamatiens* (*Cl. novyi*). J Med Microbiol 2:395-417.

Smith, L. DS. 1975a. Clostridial Wound Infections. In The Pathogenic Anaerobic Bac-teria, 2d Ed. pp. 321-24. Springfield, Ill.: Charles C. Thomas.

_________ . 1975b. *Clostridium perfringens*. In The Pathogenic Anaerobic Bacteria, 2d Ed. pp. 115-76. Springfield, Ill.: Charles C. Thomas.

_________ . 1975c. *Clostridium novyi*. In The Pathogenic Anaerobic Bacteria, 2d Ed. pp. 257-70. Springfield, Ill.: Charles C. Thomas.

_________ . 1975d. *Clostridium haemolyticum*. In The Pathogenic Anaerobic Bacteria, 2d Ed. pp. 271-80. Springfield, Ill.: Charles C. Thomas.

_________ . 1975e. *Clostridium sordellii*. In The Pathogenic Anaerobic Bacteria, 2d Ed. pp. 291-98. Springfield, Ill.: Charles C. Thomas.

_________ . 1979. Virulence factors of *Clostridium perfringens*. Rev Infect Dis 1:254-60.

Sterne, M., and Batty, I. 1975. The role of clostridia in infections. In Pathogenic Clostridia. Ed. M. Sterne and I. Batty, pp. 18-32. Boston: Butterworths.

Stevenson, J. R., and Stonger, K. A. 1980. Protective cellular antigen of *Clostridium chauvoei*. Am J Vet Res 41:650-53.

Van Kampen, K. R., and Kennedy, P. C. 1969. Experimental bacillary hemoglobinuria. 11. Pathogenesis of the hepatic lesion in rabbit. Pathol Vet 6:59-75.

11 / Enterotoxemic *Clostridium perfringens*

BY L. NIILO

CLOSTRIDIUM PERFRINGENS is the causative organism of several enterotoxemic conditions of animals and humans. It is a spore-forming, anaerobic bacillus, but stringent anaerobic conditions are not required for growth. The various enterotoxemias caused by this bacillus differ in clinical manifestations and pathology, according to the toxigenic type of the organism involved and the particular toxins it produces. Strains of this species are divided into five types (A, B, C, D, and E), based on their ability to form four major lethal toxins: alpha, beta, epsilon, and iota (Brooks et al. 1957; Smith 1975; McDonel 1980). At least eight other toxins or antigenic substances may be produced, some of which have pathogenic significance or act as virulence factors. The most important of these is enterotoxin; followed by kappa (collagenase); mu (hyaluronidase); and two hemolysins, delta and theta. Based on the production of certain minor toxins, different varieties or subtypes have been identified within types A, B, and C (Brooks et al. 1957).

C. perfringens is normally present in large numbers in soil, sewage, feces, water, feeds, and the animal and human intestinal tracts. Type A strains of low toxigenicity are the most commonly found in normal intestinal microflora and in soil. Types B, C, D, and E appear to be restricted to the intestinal tract. However, types B, C, and D have been isolated from soil only in areas where these enterotoxemias are enzootic. Apparently, the strains of *C. perfringens* of intestinal origin die in soil within a few months. Survival of the organism for short periods is aided by the formation of spores; the spores of some strains are heat resistant.

DISEASES

C. perfringens is a versatile pathogen that produces a variety of diseases in a variety of hosts (Table 11.1). Terminology can be variable and confusing because of the number of toxins and pathological manifestations that are recognized. For example, necrotic enteritis and hemorrhagic enterotoxemia are often used interchangeably, and type D enterotoxemia is not caused by enterotoxin, but by epsilon toxin.

The most important enterotoxemia, the classical, rapidly fatal type D enterotoxemia in sheep (also known as sudden death, pulpy kidney, and overeating disease), is distributed throughout the world and affects sheep of all ages, except newborns. Isolated cases of type D enterotoxemia have been reported in goats, calves, and humans.

TABLE 11.1. Enterotoxemias, principal toxins, and related disease conditions caused by *C. perfringens*

C. perfringens Type	Toxins	Disease
A		
Classical	Alpha (phospholipase C) (+ minor toxins)	Gas gangrene in humans and animals; possibly colitis in horses; necrotic enteritis in fowl
Food poisoning	Enterotoxin	Human food poisoning
B		
2 varieties	Beta (+ epsilon and minor toxins)	Lamb dysentery; occasionally neonatal hemorrhagic enteritis in calves, foals; hemorrhagic enterotoxemia in adult sheep and goats in Iran
C		
5 varieties	Beta (+ minor toxins)	Neonatal hemorrhagic/necrotic enterotoxemia (enteritis) in lambs, calves, foals, piglets; acute enterotoxemia ("struck") in sheep in Britain; necrotic enteritis in humans and fowl
D	Epsilon	Classical enterotoxemia in sheep, goats, and possibly cattle
E	Iota	Pathogenicity in sheep and cattle unclear

Next in economic and pathologic importance are the hemorrhagic, or necrotic, enterotoxemias caused by *C. perfringens* type C. Different varieties of this type have been found in different geographic locations throughout the world (Brooks et al. 1957; Niilo 1980). It affects, with high mortality, 1- to 10-day-old lambs, calves, foals, piglets, and chickens. The greatest incidence occurs in piglets. Another acute disease, geographically limited to Britain and locally known as struck, affects young ewes. A serious form of human necrotic enteritis caused by a different variety of type C occurs in Papua New Guinea, and in the past has been reported in different parts of the world (McDonel 1980).

C. perfringens type B causes lamb dysentery, a hemorrhagic and rapidly fatal enterotoxemia of 1- to 2-week-old lambs, and may occasionally produce a similar syndrome in neonatal calves and foals. It has been isolated in Europe, South Africa, and the Middle East but not yet in North America, Australia, or New Zealand. Another toxigenic variety of type B has been reported to cause hemorrhagic enterotoxemia in adult sheep and goats in the Middle East. Although case reports of type B enterotoxemia still appear in the literature from the United Kingdom, there are no scientific studies available, probably because it is difficult to study multiple simultaneous activities of two major toxins produced by this type. The only study that has been reported was concerned with preparation of a vaccine against lamb dysentery (Jansen 1961).

Despite a few reports about the association of *C. perfringens* type A with enterotoxemia in animals, the pathogenic role of classical type A strains in enterotoxemic conditions has not yet been proven. Alpha-toxin is sometimes cited as the pathogenic factor in cases of presumed enterotoxemia; it was also erroneously considered to be responsible for human food poisoning symptoms before enterotoxin was discovered in 1968-1970. The enterotoxigenic variety of type A causes one of the most common bacterial human food poisonings, characterized by diarrhea, vomiting, and general malaise. Although the enterotoxigenic type A is important in meat hygiene and food preparation (Genigeorgis 1975), it has been found in some diarrheal conditions in several animal species (Niilo 1980; Estrada and Taylor 1989). Enterotoxin, produced only by strains of type A carrying the enterotoxin gene, is the sole factor responsible for the nonfatal, self-limiting enteritis in humans and, apparently, in animals.

In Britain, *C. perfringens* type E was reported to cause enteritis in a calf in 1943 and in lambs in 1949. No further reports of this type causing disease in farm animals have appeared.

PATHOGENESIS

In all clostridial enterotoxemias, the mode of entry of toxigenic *C. perfringens* is by ingestion. In some cases the causative strain may be present in the intestinal tract long before predisposing conditions trigger the onset of the disease. It is the microbe-toxin-host interaction that determines the initiation and outcome of the various enterotoxemias. Table 11.2 compares the general pathways of pathogenesis of enterotoxemias involving *C. perfringens* types A, C, and D; these three organisms and their toxins have been studied to a considerable extent; they will be discussed in this chapter, with conclusions based on experimental data.

Enterotoxigenic Type A

The source of infection with this organism is food that is heavily contaminated with enterotoxigenic *C. perfringens*. Under favorable conditions, the organism further multiplies and sporulates in the small intestine. These conditions have to provide sufficient nutrients for growth, but not promote a luxuriant growth, because the bacterium does not sporulate in highly nutritious environments. Sporulation is necessary for synthesis of enterotoxin, which is produced intracellularly and not secreted. The enterotoxin is released by cell lysis.

TABLE 11.2. Comparative pathways of pathogenesis of enterotoxemias caused by *C. perfringens* types A, C, and D

	Pathways of Pathogenesis		
Aspects of Infection	Type A	Type C	Type D
Intestine, diet factors	Growth, sporulation	Growth	Growth
Toxinogenesis	Enterotoxin	Beta-toxin	Epsilon prototoxin
Effects by digestive enzymes on toxin	None	Inactivation	Activation to epsilon-toxin
Effect by immunity	None	Protection (toxin neutralized)	Protection (toxin neutralized)
Action on intestinal mucosa	Binding, fluid outpouring; absorption?	Necrosis, hemorrhage; absorption?	Capillary permeability; absorption
Target organ	Cell membranes of intestinal mucosa	Intestine?	Vascular endothelium of brain and other organs
Clinical signs, lesions	Diarrhea	Necrotic enteritis, dysentery sudden death	Brain edema; CNS disturbance; hyperglycemia; glycosuria;
Species affected	Humans	Newborn lambs, calves, foals, piglets; fowl; sheep; humans	Sheep, goats, cattle
Commonly known disease	Human food poisoning	Hemorrhagic or necrotic enteritis	Enterotoxemia
Prognosis	Self-limiting	Fatal	Fatal

The enterotoxin is a single polypeptide with a unique amino acid sequence, having a molecular weight of approximately 35 kDa. It acts rapidly (in 5-30 minutes) on the intestinal mucosa, mainly in the jejunum and ileum, causing profuse fluid outpouring and diarrhea, along with associated clinical signs (Fig. 11.1). There is no invasion of the intestinal mucosa by the bacterial cells.

The molecular mode of action of the enterotoxin is a multi-step process, involving the brush border membrane of the villus epithelial cells. The initial step in this process is specific binding of the enterotoxin to a proteinaceous receptor on the cell membranes, followed by insertion of the molecule into the plasma membrane of the host cells. The next step is formation of a complex between the enterotoxin and membrane proteins; this complex is composed of one toxin molecule and two membrane proteins of 70 and 50 kDa, with a final size of 160 kDa (Wnek and McClane 1989). Evidence so far indicates that enterotoxin associates only with the plasma membrane and not with the cytoplasm. Recent studies on mapping of functional regions of the enterotoxin molecule (Hanna et al. 1992) have shown that the initial receptor-binding activity is located within a 30-amino-acid sequence at the C-terminal region of the toxin, which itself is a 319-residue polypeptide, with at least five distinct epitopes. The region between amino acid sequences 26 and 171 is involved with insertion and cytotoxicity of the toxin; removal of the N-terminal region, composed of the first 25 sequences, increases cytotoxicity.

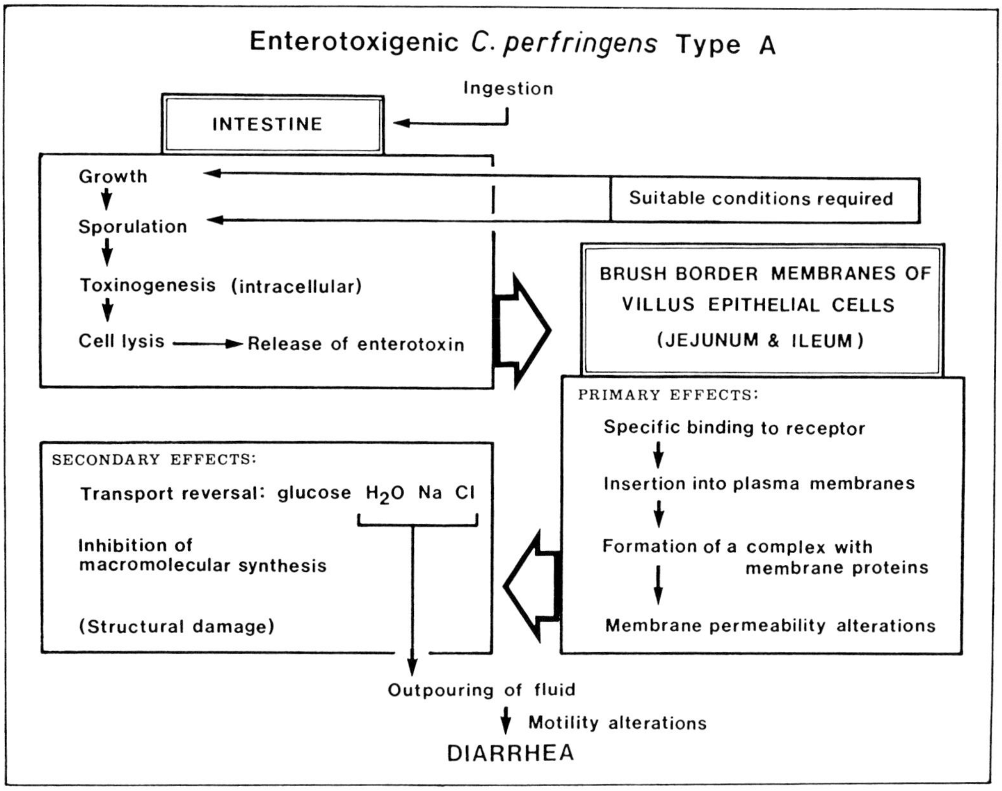

FIG. 11.1. Flow chart of pathogenesis in enterotoxigenic *Clostridium perfringens* type A enteritis.

Due to the toxin-membrane interaction, the cells develop extensive membrane permeability alterations that lead to rapid changes in intracellular levels of ions and other small molecules. The secondary effects are inhibition of macromolecular synthesis and energy metabolism, along with transport reversal, which is manifested in greatly increased secretory flux: outpouring of water, sodium, and chloride and inhibition of glucose uptake (McDonel 1979). In addition, there may be alterations in intestinal motility (Justus et al. 1981). The clinical effect of this toxic action is diarrhea (Fig. 11.1).

Indirect evidence suggests that little enterotoxin is absorbed from the intestine into the bloodstream. When experimentally introduced into the systemic circulation, the enterotoxin causes vasodilation and profound decrease in blood pressure, with resultant secondary effects in many organs (Niilo 1980). The enterotoxin is only feebly immunogenic by the intestinal route but highly antigenic, though nonprotective, by parenteral inoculation. Clinical disease, either natural or experimental, nevertheless, results in some detectable serum antibody, as demonstrated in humans, pigs, sheep, cattle, and horses (Niilo and Bainborough 1980; Niilo and Cho 1985; Estrada and Taylor 1989). Recent research indicates that biological control of this disease may be directed toward development of vaccines aimed at blocking the initial receptor-binding stage of the enterotoxin action (Hanna et al. 1992).

Type C

Studies on pathogenesis of *C. perfringens* type C enterotoxemia have been concerned mainly with susceptibility, clinical disease, histopathology, and pharmacology. No detailed studies on its mode of action at the molecular level have been done. Although beta-toxin has been known for a long time, it was not until 1978 that it was first prepared in a highly purified form. It appears to be a single polypeptide chain with a molecular weight of approximately 30 kDa and a 50% mouse-lethal dose of 1.87 μg.

The initial activity of beta-toxin in the intestine differs from the activity of enterotoxin of type A and epsilon-toxin of type D. Beta-toxin is sensitive to proteolytic enzymes; thus its full effect on the intestinal mucosa, mainly the jejunum, can take place only under conditions of diminished proteolytic enzyme activity, such as a lack of pancreatic secretion or presence of protease inhibitors in the ingesta (Niilo 1988). The presence of a trypsin inhibitor has been demonstrated in sow colostrum; this may promote the pathogenesis of *C. perfringens* type C enteritis in newborn piglets. In the newborn, there may be a short period after birth when pancreatic secretion has not reached its full potential, placing the host at increased risk. Other factors pertaining to susceptibility to this disease may be lack of colonization of the intestinal tract by a competitive normal bacterial flora, and alteration of an existing flora by sudden changes of diet.

Both the bacterial cells and their toxin are needed to initiate disease (Bergeland 1972; Niilo 1988). In piglets, Arbuckle (1972) showed that *C. perfringens* type C cells first adhere to and proliferate on the jejunal villi and thus provide their toxin an intimate contact with the villus epithelial cells. However, ultrastructural studies show damage to microvilli, degeneration of mitochondria, and damage to terminal capillaries as a primary lesion prior to any bacterial adhesion (Johannsen et al. 1986). Following the initial effect at the tips of the villi, where the fully functional, mature epithelial cells are more susceptible to damage than are the young undifferentiated crypt cells, the pathogenesis of progressive necrosis of the mucosa assumes a self-propagating pattern: destruction and desquamation of epithelium, further invasion by bacteria, proliferation and more local toxin production for structural tissue breakdown, and hemorrhage. Beta-toxin, as a potent necrotizing agent, promotes rapid extension of the pathogenic process through the mucosa, involving the crypt epithelium, mesenchymal elements of lamina propria, and muscularis mucosa. The necrotic tissue is infiltrated and densely covered with layers of *C. perfringens* cells. The intestinal lesions may vary from necrotic to hemorrhagic, the latter being associated with peracute clinical disease as seen in neonatal calves, foals, piglets, and lambs, while necrotic lesions occur in subacute and chronic cases, more often seen in older piglets and weaners.

Beta-toxin causes an increase in capillary permeability which allows the passage of toxin into the blood circulation with subsequent systemic effects. Experimental administration of purified beta-toxin causes a rise in blood pressure and a simultaneous fall in heart rate, accompanied by electrocardiographic disturbances suggestive of atrioventricular block (Sakurai et al. 1981)

The disease process in human necrotic enteritis in Papua New Guinea differs somewhat from that in animals. It affects mainly children rather than infants (Lawrence and Cook 1980). It has been noted that low-protein diets of these patients contain sweet potato, which acts as a trypsin inhibitor. Reduced pancreatic activity and sudden changes to meals containing large quantities of meat are considered essentially contributory to this disease.

Although the major gross lesions are in the mucosa of the small intestine, circulating beta-antitoxin is still protective against development of disease; the mechanism is not clearly understood. It appears reasonable to assume that the critical site of toxin-antitoxin reaction is in the initial penetration site on villi immediately beneath the epithelium, where the toxin would first come in contact with circulating antibody. For control, vaccination of dams is practical. Figure 11.2 shows the steps in the pathogenesis of peracute *C. perfringens* type C enterotoxemia and the immune factors that may function protectively at various levels of this disease.

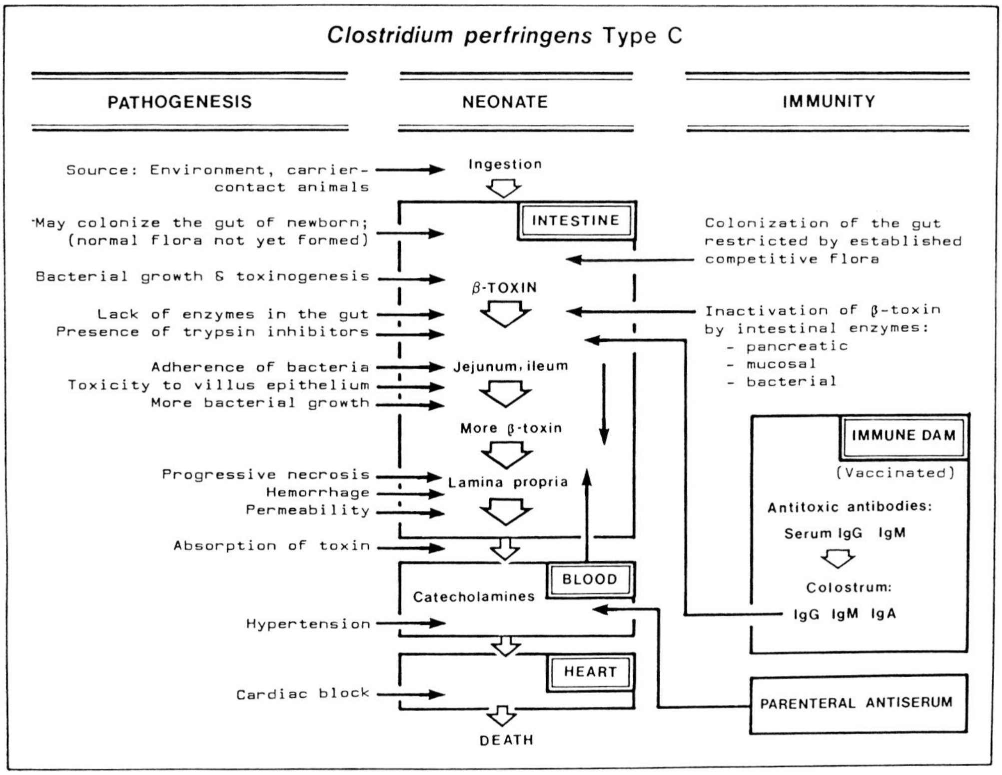

FIG. 11.2 Flow chart of pathogenesis and immunity in *Clostridium perfringens* type C enterotoxemia.

Type D

In enzootic areas, many normal sheep harbor *C. perfringens* type D in various parts of the alimentary tract. These strains can provide the source for a fatal intoxication if conditions are favorable for rapid and profuse growth of organisms in the intestine. Such conditions are associated with a sudden change from relatively poor diets to rich diets or gorging on palatable feed; hence the term overeating disease. When sudden change occurs, the rumen flora, which normally consists mostly of cellulose-digesting microbes, becomes unbalanced and much undigested food passes into the small intestine. The presence of undigested starch in the intestine favors profuse growth of the organism and the consequent production of toxin (Bullen 1970). It is not known why *C. perfringens* type D strains can outgrow type A strains, for they are initially outnumbered by the latter. Epsilon-prototoxin, of relatively low toxicity, is released by the multiplying bacteria in the lumen of the small intestine and then rapidly activated by digestive enzymes, such as trypsin. Maximum activation into epsilon-toxin is achieved by a combination of trypsin and chymotrypsin, increasing the toxicity by at least 1000-fold. Purified active epsilon-toxin has a molecular weight of approximately 32 kDa, and a toxicity of about 3.2×10^6 minimum lethal dose (MLD) / mg when tested intravenously in mice.

Production of epsilon-toxin in the bacterial cells is encoded by a gene that is carried on a large plasmid present only in *C. perfringens* types D and B. The toxin's molecular functional structure is not known, but some amino acid residues important for its function have been identified. Conversion of prototoxin to fully active toxin occurs by the removal of the N-terminal peptide. Current studies are aimed at structure-function relationship of the toxin by way of nucleotide sequencing (Hunter et al. 1992).

Epsilon-toxin does not exert direct pathogenic action on the intestinal mucosa, and small amounts of the toxin in the lumen appear to be harmless. When high concentrations of epsilon-toxin persist in the intestine for several hours, an increase in the permeability of the intestine occurs, leading to the absorption of large amounts of toxin into the systemic circulation (Bullen 1970) (Fig. 11.3). The active toxin selectively affects the brain and, to a certain degree, kidney and other tissues. The effect in the brain is manifested in the alteration of brain capillaries, making them permeable to plasma proteins and water; this results in rapid extravasation of fluid into the brain and increased intracerebral pressure. The increased pressure accounts for clinical signs of central nervous system derangement, such as incoordination and convulsions (Griner 1961). Edema and effusions also occur to varying degrees in the lungs, pericardium, and other serous cavities.

At the cellular level, epsilon-toxin has been shown to bind to receptors in the luminal surface of the vascular endothelium, particularly in the brain, and to the cells lining the loops of Henle, distal convoluted tubules of the kidney, and to the hepatic sinusoids. A high-affinity binding site for epsilon-toxin may be a sialoglycoprotein located in synaptosomal membranes, as shown in a recent laboratory study (Nagahama and Sakurai 1992). The resultant degeneration of the vascular endothelial cells allows alterations of fluid dynamics and leakage of plasma proteins. Ultrastructural studies of the brain have shown that tight junctions in the vascular endothelium break down, causing perivascular astrocyte processes to swell and rupture; this leads to the development of perivascular and intercellular edema, followed by focal to diffuse areas of degeneration and necrosis (Buxton and Morgan 1976). Macroscopic foci of bilaterally symmetrical encephalomalacia have been observed in the corpus striatum, thalamic area, midbrain, hippocampus, and cerebellar peduncles (Griner 1961). The extent of the lesions varies with acuteness of the disease.

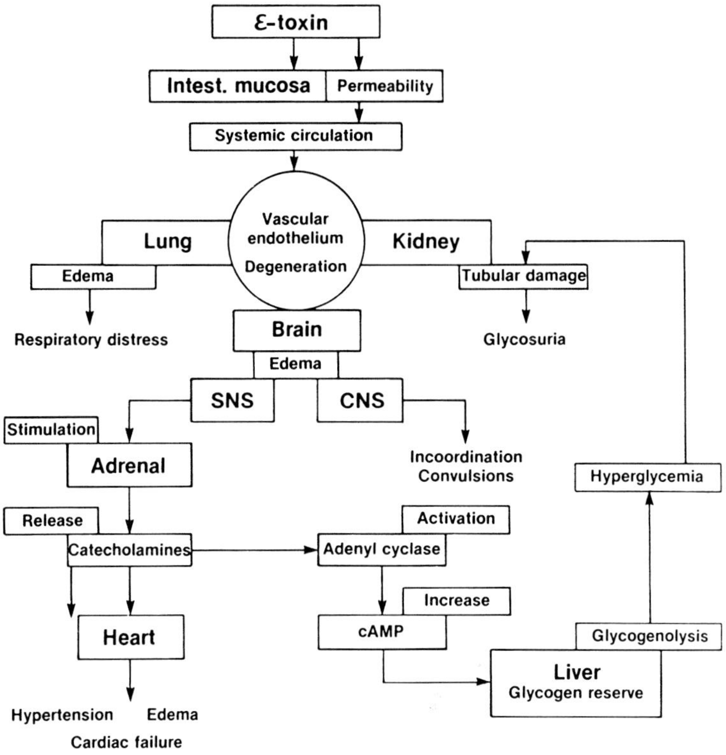

FIG. 11.3. Diagrammatic flow chart of mechanism of action of epsilon-toxin in *Clostridium perfringens* type D enterotoxemia, based on experimental data.

In type D enterotoxemia there are pathognomonic signs of hyperglycemia and glycosuria. Experimental work on sheep and laboratory animals has demonstrated that hyperglycemia is the result of glycogenolysis of liver glycogen reserves (Gardner 1973). Somewhat different schemes have been proposed for the mechanism leading to glycogenolysis (Buxton 1978; Worthington et al. 1979). One proposal is that hepatocyte-bound epsilon-toxin stimulates the production of cyclic adenosine 3′,5′-monophosphate (cAMP), which in turn causes breakdown of glycogen to glucose. A more plausible explanation appears to be that brain edema, as a primary lesion acting on the sympathetic division of the autonomic nervous system, stimulates a release of catecholamines (epinephrine, norepinephrine, dopamine) via the adrenal medulla. These activate adenyl cyclase, boosting the production of cAMP, which in turn causes glycogenolysis and hyperglycemia. Hyperglycemia, coupled with tubular damage in the kidneys, leads to glycosuria. This proposed sequence of events takes place in rapid succession, probably in less than 1 hour after epsilon-toxin enters the systemic circulation (Fig. 11.3). The pulpy kidney of sheep is a postmortem autolytic change, occurring rapidly in damaged tissue.

Epsilon-toxin is antigenic and stimulates the production of specific antibody, even when infection with type D organisms is subclinical. Experimental work with a monoclonal antibody has shown that immunity to a single epitope on epsilon-toxin may be sufficient to protect against the toxin and clinical disease (Percival et al. 1990).

The foregoing discussion on pathogenesis centered on type D enterotoxemia as seen in sheep. In goats, the clinical disease varies from peracute to chronic forms, which exhibit bloody diarrhea, associated with hemorrhagic or necrotic enterocolitis (Blackwell et al. 1991).

Also, immunity to this disease is poor in goats. A caprine-specific sensitivity of the intestinal tract to *C. perfringens* type D enterotoxemia is assumed to be responsible for this difference in the pathology between sheep and goats.

REFERENCES

Arbuckle, J. B. R. 1972. The attachment of *Clostridium welchii (Cl. perfringens)* type C to intestinal villi of pigs. J Pathol 106:65-72.

Bergeland, M. E. 1972. Pathogenesis and immunity of *Clostridium perfringens* type C enteritis in swine. J Am Vet Med Assoc 160:568-73.

Blackwell, T.E.; Butler, D.G.; Prescott, J.F.; and Wilcock, B.P. 1991. Differences in signs and lesions in sheep and goats with enterotoxemia induced by intraduodenal infusion of *Clostridium perfringens* type D. Am J Vet Res 52:1147-52.

Brooks, M. E.; Sterne, M.; and Warrack, G. H. 1957. A re-assessment of the criteria used for type differentiation of *Clostridium perfringens*. J Pathol Bacteriol 74:185-95.

Bullen, J. J. 1970. Role of toxins in host-parasite relationships. In Microbial Toxins, Vol. 1. Ed. S. J. Ajl, S. Kadis, and T. C. Montie, pp. 233-76. New York: Academic Press.

Buxton, D. 1978. Further studies on the mode of action of *Clostridium welchii* type D epsilon toxin. J Med Microbiol 11:293-98.

Buxton, D., and Morgan, K. T. 1976. Studies of lesions produced in the brains of colostrum deprived lambs by *Clostridium welchii (Cl. perfringens)* type D toxin. J Comp Pathol 86:435-47.

Estrada, A.E.; and Taylor, D.J. 1989. Porcine *Clostridium perfringens* type A spores, enterotoxin, and antibody to enterotoxin. Vet Rec 124:606-10.

Gardner, D. E. 1973. Pathology of *Clostridium welchii* type D enterotoxaemia. 3. Basis of the hyperglycaemic response. J Comp Pathol 83:525-29.

Genigeorgis, C. 1975. Public health importance of *Clostridium perfringens*. J Am Vet Med Assoc 167:821-27.

Griner, L. A. 1961. Enterotoxemia of sheep. 1. Effect of *Clostridium perfringens* type D toxin on the brains of sheep and mice. Am J Vet Res 22:429-42.

Hanna, P.C.; Wieckowski, E.U.; Mietzner, T.A.; and McClane, B.A. 1992. Mapping of functional regions of *Clostridium perfringens* type A enterotoxin. Infect Immun 60:2110-14.

Hunter, S.E.C.; Clarke, I.N.; Kelly, D.C.; and Titball, R.W. 1992. Cloning and nucleotide sequencing of the *Clostridium perfringens* epsilon-toxin gene and its expression in *Escherichia coli*. Infect Immun 60:102-10.

Jansen, B. C. 1961. The beta toxin of *Clostridium welchii* type B, Wilsdon, in relation to the production of a vaccine against lamb dysentery. Onderstepoort J Vet Res 28:495-549.

Johannsen, U.; Menger, S.; Erwerth, W.; and Koehler, B. 1986. Untersuchungen zur *Clostridium perfringens* Type C Enterotoxaemie (nekrotisierenden Enteritis) der Saugferkel. 3. Arch Exp Veterinaermed 40:895-909.

Justus, P. G.; Mathias, J. R.; Martin, J. L.; Carlson, G. M.; Shields, R. P.; and Formal, S. B. 1981. Myoelectric activity in the small intestine in response to *Clostridium perfringens* A enterotoxin: Correlation with histologic findings in an in vivo rabbit model. Gastroenterology 80:902-114.

Lawrence, G., and Cooke, R. 1980. Experimental pigbel: The production and pathology of necrotizing enteritis due to *Clostridium welchii type* C in the guinea-pig. Br J Exp Pathol 61:261-71.

McDonel, J. L. 1979. The molecular mode of action of *Clostridium perfringens* enterotoxin. Am J Clin Nutr 32:210-18.

__________. 1980. *Clostridium perfringens* toxins (type A, B, C, D, E). Pharmacol Ther 10:617-55.

Nagahama, M., and Sakurai, J. 1992. High-affinity binding of *Clostridium perfringens* epsilon-toxin to rat brain. Infect Immun 60:1237-40.

Niilo, L. 1980. *Clostridium perfringens* in animal disease: A review of current knowledge. Can Vet J 21:141-48.

__________. 1988. *Clostridium perfringens* type C enterotoxemia. Can Vet J 29:658-64.

Niilo, L.; and Bainborough, A. R. 1980. A survey of *Clostridium perfringens* enterotoxin antibody in human and animal sera in western Canada. Can J Microbiol 26:1162-64.

Niilo, L.; and Cho, H. J. 1985. Clinical and antibody responses to *Clostridium perfringens* type A enterotoxin in experimental sheep and calves. Can J Comp Med 49:145-48.

Percival, D. A.; Shuttleworth, A. D.; Williamson, E. D.; and Kelly, D. C. 1990. Anti-idiotypic antibody-induced protection against *Clostridium perfringens* type D. Infect Immun 58:2487-92.

Sakurai, J.; Fujii, Y.; Matsura, M.; and Endo, K. 1981. Pharmacological effect of beta toxin of *Clostridium perfringens type* C on rats. Microbiol Immunol 25:423-32.

Smith, L. DS. 1975. *Clostridium perfringens*. In The Pathogenic Anaerobic Bacteria, 2d ed. pp. 115-76. Springfield, Ill.: Charles C. Thomas.

Wnek, A. P., and McClane, B. A. 1989. Preliminary evidence that *Clostridium perfringens* type A enterotoxin is present in a 160,000-M complex in mammalian membranes. Infect Immun 57:574-81.

Worthington, R. W.; Bertschinger, H. J.; and Mulders, S. G. 1979. Catecholamine and cyclic nucleotide response of sheep to the injection of *Clostridium welchii* type D epsilon toxin. J Med Microbiol 12:497-502.

12 / *Nocardia; Actinomyces; Dermatophilus*

BY C. L. GYLES

***N**OCARDIA, Actinomyces*, and *Dermatophilus* are genera of gram-positive, filamentous bacteria, often considered with pathogenic fungi. These genera include species that are opportunistic pathogens in animals and humans. Infection is either from soil or of endogenous origin and leads to a localized lesion, which may spread.

NOCARDIA

The genus *Nocardia* consists of aerobic, gram-positive bacteria that may appear as long, slender, branching filaments; short rods; cocci; or combinations of these forms. In tissue, *Nocardia* are seen as weakly gram-positive, beaded filaments. These organisms are saprophytes, which occur commonly in soil and decaying organic matter, but certain species are opportunistic pathogens that produce a noncontagious infection. Unlike *Actinomyces, Nocardia* are not fastidious and grow readily on simple media, such as Saboraud's dextrose agar. *N. asteroides* is the species most frequently associated with disease and has been recovered from lesions in humans, dogs, cats, cattle, goats, horses, pigs, marine mammals, and fish (Walton and Libke 1974; Orchard 1979). Cattle and dogs are the animal species most frequently affected, but even in these species the prevalence of disease is low.

There is much uncertainty about the classification of several species of *Nocardia* and closely related organisms; *N. farcinica,* originally described in association with bovine farcy, is probably *N. asteroides* (Riddell 1975; Minniken et al. 1977). Certain isolates identified as *N. farcinica* undoubtedly belong to the genus *Mycobacterium* (Riddell 1975). Furthermore, failure to distinguish properly between *Nocardia* and *Actinomyces* in earlier studies has confused the literature on nocardiosis in animals.

N. asteroides is partially acid fast and retains fuchsin in a modified Ziehl-Neelsen staining procedure, in which 1% sulphuric acid is used instead of acid alcohol for decolorization. This feature of the organism is best demonstrated in filaments in tissue or on initial isolation. *N. brasiliensis* is occasionally recovered from disease in dogs, and *N. caviae* has been reported as a cause of disease in goats, but precise speciation of isolates from disease is often not pursued.

Typically, *Nocardia* infection originates from organisms aspirated into the respiratory tract or introduced into skin wounds, possibly leading to development of necrotizing pneumonia or superficial skin lesions. Less frequently, the organisms are ingested and establish in the abdomen. It is common to have multiple sites of infection arise from direct extension or hematogenous dissemination. *N. asteroides* is also a cause of bovine mastitis and abortion in cattle and pigs (Walton and Libke 1974; Orchard 1979; Koehne and Giles 1981); abortions occur sporadically but mastitis sometimes occurs in large outbreaks.

N. asteroides infections in dogs occur most frequently in dogs less than 1 year of age (Ackerman et al. 1982), but whether this represents greater exposure or greater susceptibility to infection in relation to age is not known. Canine distemper and other viral infections, as well as immunosuppression, may predispose to nocardiosis (Walton and Libke 1974; Ackerman 1982).

The characteristic *Nocardia* lesion is an abscess or cluster of abscesses poorly localized by fibrous tissue. Formation of only minimal amounts of fibrous tissue may account for the ease of spread of infection to adjacent tissues and by the hematogenous route. Whereas *N. brasiliensis* and *N. caviae* will produce granules in tissue, *N. asteroides* will not (Walton and Libke 1974), and the lesion produced by *N. asteroides* is not a granuloma.

Little is known about structures and products of *Nocardia* that may be important in pathogenicity. However, the amount of peptidoglycan in the cell wall has been proposed to be related to virulence (Beadles et al. 1980). In mouse tissues, a highly virulent strain of *N. asteroides* was shown to produce a 15-nm peptidoglycan layer, whereas a weakly virulent strain produced a 5-nm layer of peptidoglycan. Strain-to-strain variation in the thickness of the peptidoglycan produced in vitro has also been demonstrated. It was suggested that increased amounts of peptidoglycan could cause resistance to degradation by phagosomal hydrolases, especially lysozyme. Beadles et al. (1980) noted that certain strains of *N. asteroides* exhibited delayed growth in vitro and postulated that an extended lag period in in vivo growth might contribute to greater susceptibility of some strains to normal host defense mechanisms.

The stage of growth of the bacterium has been shown to greatly affect the virulence of *N. asteroides* (Beaman and Maslan 1978; Beaman and Moring 1988). More than 1000 times as many stationary phase cells were required to achieve the same mortality in mice as was obtained with log-phase cells. Differences in fatty acid and mycolic acid composition have been observed in stationary-phase and log-phase organisms. In contrast, the virulence of *N. brasiliensis* did not vary with the growth cycle (Conde et al. 1982).

N. asteroides has been reported to both inhibit phagosome-lysosome fusion and resist the microbicidal effects of polymorphonuclear leukocytes (PMNs). Production of superoxide dismutase and catalase appear to be virulence factors of *N. asteroides*; virulent strains of the organism are highly resistant to hydrogen peroxide, myeloperoxidase, and halide (Beaman and Beaman 1990).

Normal host defenses such as skin, mucociliary apparatus of the tracheobronchial tree, and phagocytosis (Beaman and Smathers 1976) undoubtedly prevent most contacts with *Nocardia* from developing into infection. The nocardiae contain the toxic and immunostimulating glycolipid trehalose dimycolate (cord factor) in their cell walls, and this component has been proposed to play a major role in activation of macrophages and release of tumor necrosis factor, which is important in host resistance to infection (Silva and Faccioli 1992). Established infections may be localized at the site of introduction of the organisms, but there is a tendency for *Nocardia* to invade blood vessels and spread by the hematogenous route. Osteomyelitis or encephalitis may result from secondary establishment of the organisms.

Cell-mediated immunity (CMI) appears to be of critical importance in the normal elimination of these bacteria from the body (Deem et al. 1982), and impairment of CMI is likely to predispose to disease. Although B-cell response and humoral immunity may play a role in protection against disease (Beaman and Smathers 1976; Beaman 1977; Beaman and Maslan 1977), Deem et al. (1982) noted that antibody is produced in low titers and irregularly in human and experimental murine infections, and that immunity is not transferred by the serum of immunized mice. In contrast, adoptive immunity to infection with *N. asteroides* can be transferred with either specifically primed spleen cells or splenic T lymphocytes.

It is evident that *N. asteroides* can function as a facultative intracellular parasite, and in lung infections its interaction with alveolar macrophages must be important. Beaman and Smathers (1976) studied the interaction between *N. asteroides* and cultured rabbit alveolar macrophages and showed that a weakly virulent strain was readily phagocytized and destroyed, but a highly virulent strain grew rapidly within the macrophages. They concluded that strain virulence was a major factor in the interaction of *N. asteroides* and nonimmune, unstimulated alveolar macrophages. Beaman (1977) reported the formation of multinucleate giant cells by in vitro macrophage fusion in response to virulent *N. asteroides*. He suggested that membrane fusion occurred in response to specific components of the bacterium.

ACTINOMYCES

Actinomyces are gram-positive, branching, filamentous bacteria, including anaerobic species *(A. bovis)* as well as facultatively anaerobic species *(A. viscosus)*. Members of *Actinomyces* require enriched media for growth, and their development is stimulated by CO_2. *A. israelii,* a pathogen in humans, is only rarely a cause of disease in animals. The species previously known as *Corynebacterium pyogenes* but recently reclassified as *A. pyogenes* has been discussed in Chapter 5. *Actinomyces* species discussed in this section are oral commensals that only occasionally invade damaged tissue.

The granulomatous lesions that develop as a result of actinomycotic infection are characterized by the presence of yellowish tissue microcolonies in hard "sulphur granules." When the granules are crushed and Gram-stained, gram-positive branching filaments, short rods, and cocci are evident. Red-staining club-shaped structures are also seen in rosettes at the periphery of the granule. The sulphur granule is composed of bacterial filaments and mineralized calcium phosphate of host origin. The clubs represent encapsulation of the organism by layers of a calcium phosphate-protein complex (Pine and Overman 1966), a result of phosphatase activity in response to inflammation. The presence of the clubs has been related to tissue resistance, and it has been suggested that they may represent antigen-antibody complexes. Clubs are seen most commonly in cattle infected with *A. bovis*. Sulphur granules have also been reported in lesions from the tonsils of pigs, with the only difference being that the organisms were morphologically like *Dermatophilus* rather then *Actinomyces* (Bak and Azuma 1991).

The infection usually remains localized in the tissues at the site of inoculation of organisms or in the affected body cavity. In the typical lesion, there is a central zone of polymorphonuclear leukocytes (PMNs) in association with bacteria and an outer zone of mononuclear cells.

The disease "lumpy jaw" of cattle, caused by *A. bovis,* has been known for over a century. *A. bovis*, present as part of the normal flora of the mouth, invades a variety of tissues and often produces lesions in bone. Growth of the organism may involve maxillary bone, tongue, pharynx, lungs, lymph nodes, and subcutaneous tissues of the head and neck. Trauma to the tissues is the initiating event in disease and may occur as a result of shedding of teeth or as a result of coarse feed. The lesion tends to remain localized and produces ill health by interference with eating but occasionally spreads to the lungs, liver, or vertebrae. Lumpy jaw is sometimes confused with "woody tongue," a condition in which hard lesions develop in the tongue of cattle as a result of infection with *Actinobacillus lignieresii*. Since *A. lignieresii* is an aerobic, gram-negative, rod-shaped bacterium, a bacteriologic differentiation is simple. Ehler (1977) has shown that direct needle aspiration is valuable in obtaining samples for examination and culture.

Actinomyces bovis-like organisms, called *A. suis*, cause porcine mastitis, which is characterized by the formation of much granulation tissue and enlargement of the gland. *A. suis* has also been recovered from purulent vaginal discharges and aborted fetuses from sows. *A. bovis* is recovered with *Brucella abortus* or *B. suis* from "poll evil" and "fistulous withers," infections of the supraspinous bursa in horses.

A. viscosus is the most frequent cause of actinomycosis of dogs (Swerczek et al. 1968; Davenport et al. 1975; Ehler 1977), a disease characterized by the formation of chronic pyogranulomatous lesions. Abscesses may develop under the skin, in the submandibular region, in the lungs, or in the abdomen, but thoracic infections are most common. Dog bites frequently lead to subcutaneous lesions but osteomyelitis may develop in some cases. *A. viscosus* is less frequently implicated in disease in cats, pigs, goats, cattle, and horses. The lesion formed by *A. viscosus* is similar to that produced by *Nocardia* species and lacks the hard actinomycotic sulphur granules (Pine and Overman 1966). The lesion is soft and does not contain acidophilic club-shaped structures. Older literature often uses the term actinomycosis to describe a particular type of lesion without reference to the etiologic agents; this leads to confusion, and it is clear that bacteriologic diagnosis can not adequately be replaced by histologic examinations (Swerezek et al. 1968). Isolates from actinomycosis in dogs and cats were sometimes called *A. baudetti,* and it appears that the organism was *A. viscosus* (Georg et al. 1972).

Actinomyces are not known to elaborate protein toxins nor to have toxic cell wall components, and only a limited number of studies have been conducted to investigate important factors in host-parasite interactions. A weanling-mouse model has been used to study experimental infections with various *Actinomyces* (Behbehani and Jordan 1982; Behbehani et al. 1983). Trauma and possibly some other abrogation of the host defenses are required to initiate disease by these endogenous organisms. For the microaerophilic species, the lowering of the redox potential in traumatized tissue undoubtedly facilitates growth of the bacteria. Frequently, other bacteria are recovered from lesions due to *Actinomyces*. Although their contribution to the disease process is unknown, there is evidence that accompanying gram-negative bacteria may facilitate development of disease (Jordan et al. 1984). Jordan et al. (1984) demonstrated that inclusion of *Eikenella corrodens* in the inoculum for experimental infection of mice with *A. israelii* resulted in a 10,000-fold reduction in the minimal infecting dose of *A. israelii* and in a longer duration of the lesions.

Interest in pathogenic mechanisms of *A. viscosus* has been aroused, because this organism synthesizes extracellular polymers that facilitate adherence to tooth surfaces and formation of dental plaques. Furthermore, fimbriae, which may play a role in disease, have been characterized. Two antigenically and functionally distinct types of fimbriae have been described: type 1 mediate adherence to salivary proline-rich protein and stratherin on tooth surfaces, whereas type 2 promote lactose-sensitive binding to certain oral streptococci (coaggregation) and to buccal epithelial cells (Cisar et al. 1979). Isolates of *A. viscosus* from animals have been shown to be serologically related to human isolates (Georg et al. 1972), but the role, if any, of these fimbriae in diseases in animals is not known. Recently, Yeung (1992) reported that a DNA probe specific for the structural gene of the type 1 fimbriae genes of a human isolate of *A. viscosus* hybridized with neither of two isolates of *A. viscosus* from dogs but did hybridize with three of three rodent isolates of *A. viscosus* and with two of three bovine isolates of *A. bovis*.

Engel et al. (1976, 1977) have shown that homogenates of *A. viscosus* have a direct chemotactic effect on human polymorphonuclear cells, generate chemotactic activity when incubated with fresh human serum, and stimulate host immune cells to produce and release mediators of inflammation. Heat inactivation of serum abolishes the effect of the *A. viscosus* product, and it has been suggested that complement is involved. It has been demonstrated that the monocytic infiltrate in the experimental lesion in mice appeared at 48 hours only in immunized mice (Engel et al. 1976), and it was proposed that this may be due to a delayed hypersensitivity reaction. Homogenates of *A. viscosus* were shown to be B-cell mitogens, and this activity may play a role in eliciting the plasma cell component of the lesion (Engel et al. 1977).

Behbehani and Jordan (1982) studied the comparative pathogenicity of *Actinomyces* species in weanling mice injected intraperitoneally with *A. israelii*, *A. viscosus*, and *A. naeslundii*. Older mice are resistant. Researchers observed that there was an initial acute phase of about 6 weeks during which the infections with *A. viscosus* and *A. naeslundii* were slightly less severe than those with *A. israelii,* but there were striking differences in the subsequent chronic phase. *A. israelii* persisted and resulted in a slowly progressive chronic infection, whereas *A. viscosus* and *A. naeslundii* were eliminated. Size of the bacterial aggregates was significant, and strains of *A. israelii* that grew as rough aggregates in liquid media regularly produced lesions, whereas smooth strains or homogenized inocula of rough strains did not.

Behbehani et al. (1983) studied the comparative histopathology of lesions in mice infected with *A. israelii*, *A. viscosus*, and *A. naeslundii* and related their observations to pathogenesis. They consider that the early lesion, which is dominated by a massive accumulation of PMNs, is due to the strong chemotaxis of *Actinomyces* for PMNs. Subsequent developments differed because the relationship of the mobilized PMNs to the organisms varied with the different species of bacteria. PMNs were unable to penetrate and invade the central granule in the acute *A. israelii* lesions. Growth of this organism as densely packed filaments could present physical difficulties for PMNs and macrophages. A physical barrier to engulfment by PMNs might also account for the importance of size of aggregates in the establishment of infection. Surface fibrils and microcapsules produced by *A. israelii* may also play a role. Workers noted that the eosinophilic fringe at the edge of the *A. israelii* lesion may be due to arginine-rich proteins and may result from degranulation of PMNs and subsequent release of their lysosomal enzymes in response to *A. israelii*. In contrast, *A. viscosus* and *A. naeslundii* in the core of the lesion are destroyed, possibly by antigen-antibody reactions triggering the classic complement pathway.

DERMATOPHILUS CONGOLENSIS

Dermatophilus congolensis is a branching, filamentous, gram-positive bacterium that undergoes a life cycle involving the release of spores from the filaments and the growth of these spores to produce germ tubes and new filaments (Roberts 1961) (Fig. 12.1). The narrow filaments that arise from the germinated spores (Fig. 12.1 A, C) undergo transverse and longitudinal septation to form wide filaments, which may contain up to eight coccoid structures (Fig. 12.1 A, B, C). Released spores form flagella and are motile. Chemotaxis of the motile zoospores for CO_2 may be used in separation of the spores from contaminating organisms in lesions. Impression smears of moist scabs from lesions may be stained with Giemsa or methylene blue to demonstrate the characteristic structures. Rich culture media are necessary for growth of the organism, which has been recovered only from lesions of animals and humans and not from the environment. Colonies of the organism show a clear hemolysis on sheep blood agar.

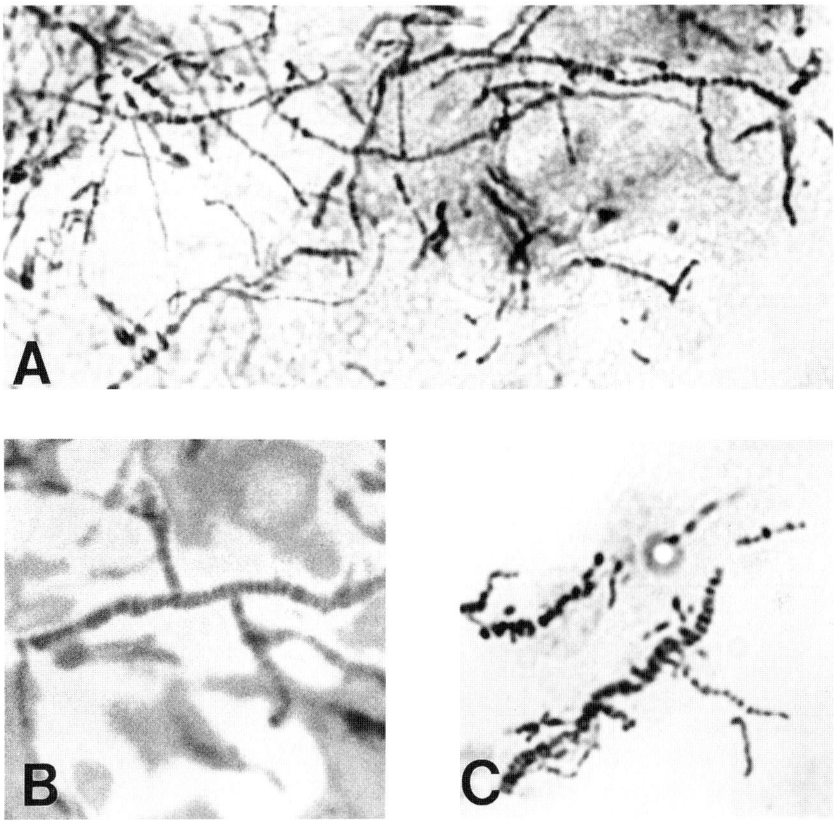

FIG. 12.1 Appearance of *Dermatophilus congolensis* in a Giemsa-stained smear of exudate from a skin lesion. **A.** Long filaments of varying widths contain coccoid elements. **B.** Broad filaments show striations, which represent rows of cocci. **C.** Coccoid structures are evident as parts of old, wide filaments or as parts of slender, newly formed germ tubes.

Dermatophilosis is an exudative dermatitis characterized by the development of hard crusts with exudation. The disease occurs most commonly in cattle, sheep, and horses but has been reported in 27 animal species including seals, lizards, and polar bears (Richards et al. 1973; Merchant 1990). Infections in humans have been reported after handling animals with dermatophilosis. The lesions commence as papules and pustules, which soon develop into exudative scab formations. Typically, both experimental and natural disease show a confinement to the superficial layers of the skin, but there have been reports of isolation of this organism from chronic inflammatory lesions in tissues other than skin in the cat (Gordon and Perrin 1971; Jones 1976; Stewart 1977). The disease in cattle and sheep occurs most commonly in tropical areas and in association with heavy rainfall. Maceration of the epidermis during heavy rainfall is considered to reduce resistance to invasion, and the moisture facilitates release of infective zoospores and their penetration into the skin.

D. congolensis survives long periods in dry crusts but does not remain viable for long in moist soil. It has been suggested that *D. congolensis* is a commensal on the skin and disease occurs only when the balance between bacterium and host is disturbed. Using an enzyme-linked immunosorbent assay, Lloyd (1981) demonstrated that antibodies to *D. congolensis* were detectable in the sera of cattle from herds with no clinical dermatophilus infection and suggested that subclinical infection was common.

Very little is known about bacterial factors that contribute to ability of the organism to maintain itself in the host, but there is interest in the production of keratinase. It has been suggested that the ability to produce and secrete keratinase is directly related to the ability to survive in keratinized tissue, and quantitation of keratinolytic activity has shown that *D. congolensis* produces substantial quantities of the enzyme (Hanel et al. 1991).

Injuries, insect bites, and prolonged wetting of the skin predispose to disease (Stewart 1977). Ticks have been suggested as vectors, but experimental studies indicate that the organism fails to survive in or on ticks for any length of time (Thoen et al. 1980). However, trauma associated with tick bites may predispose to infection. In a recent study in the West Indies, Morrow et al. (1989) reported that the prevalence of dermatophilosis was higher on animals infected with *Amblyomma variegatum* ticks than on animals that were not infested. Also, greater severity of disease was related to heavy infestation with adult ticks, but the distribution of skin lesions was not related to feeding sites of the ticks.

The host response to infection with *D. congolensis* is marked by a massive leukocytic infiltration into the epidermis and upper dermis (Oduye 1976; Amakiri and Nwufoh 1981). There is a marked increase in the number of cutaneous vessels per unit area in affected skin and a predominantly neutrophilic and lymphocytic response is observed. Histopathologic features are similar to those seen in other chronic skin infections. Lloyd and Noble (1982) studied factors influencing infection of the skin of mice by *D. congolensis* and concluded that intact mouse skin was highly resistant to infection but susceptibility could be increased by abrasion or by treatment with certain organic solvents. Treatment of the skin with ether greatly enhanced susceptibility of the skin to infection, possibly as a result of removal of cornified cells and sebaceous lipids (Lloyd et al. 1979). Lloyd and Noble (1982) demonstrated differences in virulence of strains of the organism and in susceptibility of various strains of inbred mice. High natural resistance has been reported in native West African cattle (Stewart 1977).

Jenkinson et al. (1989) investigated whether specific antibodies on the skin had a localized protective effect against *D. congolensis*. They determined that the antibodies did not immobilize zoospores and they could not attribute a specific function to these antibodies. They suggested that serum immunoglobulins, which accompany granulocytes on their migration into the epidermis during the early inflammatory response to infection, may serve to limit invasion by binding, immobilizing, clumping, and opsonizing the bacteria.

Several studies have evaluated vaccines against dermatophilosis (Ellis et al. 1991; Sutherland et al. 1991). To date there has been no effective vaccine. Two difficulties seem to account for this failure: antigenic heterogeneity among strains of the organism and protection of the site of infection from antibodies. Sutherland et al. (1991) vaccinated sheep with either live crude filaments of *D. congolensis* or a vaccine containing a 45-kDa zoospore protein plus mucoid material coating the filaments. Neither vaccine was effective in challenge experiments of vaccinated sheep placed in contact with sheep with active dermatophilosis. Ellis et al. (1991) evaluated two types of vaccines prepared from two strains of *D. congolensis*. One vaccine was heat-inactivated formalinized filaments inoculated intramuscularly, and the other consisted of live filaments inoculated intradermally. Both types of vaccine from both strains showed protection against challenge with one strain but not with the other. They attributed these findings to differences in antigenicity and virulence between the two strains.

Dermatophilus-like organisms have been reported to cause actinomycotic lesions in the tonsils of pigs (Bak and Azunia 1991). The organisms are morphologically like *Dermatophilus*, but they differ significantly in biochemical behavior. A new species, *Tonsillophilus suis*, has been proposed for these bacteria. They induce formation of typical sulphur granules, which contain filamentous bacteria, club-shaped amorphous structures, and macrophages.

REFERENCES

Nocardia

Ackerman, N.; Grain, E.; and Casteman, W. 1982. Canine nocardiosis. J Am Anim Hosp Assoc 18:147-153.

Beadles, T. A.; Land, G. A.; and Knezek, D. J. 1980. An ultrastructural comparison of cell envelopes of selected strains of *Nocardia asteroides* and *Nocardia brasiliensis*. Mycopathologia 70:25-32.

Beaman, B. L. 1977. In vitro response of rabbit alveolar macrophages to infection with *Nocardia asteroides*. Infect Immun 15:925-37.

Beaman, L., and Beaman, B. L. 1990. Monoclonal antibodies demonstrate that superoxide dismutase contributes to protection of *Nocardia asteroides* within the intact host. Infect Immun 58:3122-28.

Beaman, B. L., and Maslan, S. 1977. Effect of cyclophosphamide on experimental *Nocardia asteroides* infection in mice. Infect Immun 16:995-1004.

_________. 1978. Virulence of *Nocardia asteroides* during its growth cycle. Infect Immun 20:290-95.

Beaman. B. L., and Moring, S. E. 1988. The relationship among cell wall composition, stage of growth, and virulence of *Nocardia asteroides* GUH-2. Infect Immun 56:557-63.

Beaman, B. L., and Smathers, M. 1976. Interaction of *Nocardia asteroides* with cultured alveolar macrophages. Infect Immun 13:1126-35.

Conde, C.; Melendro, E. I.; Fresan, M.; and Ortiz-Ortiz, L. 1982. *Nocardia brasiliensis:* Mycetoma induction and growth cycle. Infect Immun 38:1291-95.

Deem, R. L.; Beaman, B. L.; and Gershwin, M. E. 1982. Adoptive transfer of immunity to *Nocardia asteroides* in nude mice. Infect Immun 38:914-20.

Koehne, G. and Giles, R.C. 1981. *Nocardia asteroides* abortion in swine. J. Am Vet Med Assoc. 179:478-79.

Minniken, D. E.; Patel, P. V.; Alshamaony, L.; and Goodfellow M. 1977. Polar lipid composition in the classification of *Nocardia* and related species. Int J Syst Bacteriol 27:104-17.

Orchard, V. A. 1979. Nocardial infections of animals in New Zealand, 1976-78. N Z Vet J 27:159-65.

Riddell, M. 1975. Taxonomic study of *Nocardia farcinica* using serological and physiological characters. Int J Syst Bacteriol 25:124-32.

Silva, C. L., and Faccioli, L. H. 1992. Tumor necrosis factor and macrophage activation are important in clearance of *Nocardia brasiliense* from the livers and spleens of mice. Infect Immun 60:3566-70.

Walton, A. M., and Libke, K. G. 1974. Nocardiosis in animals. Vet Med Small Anim Clin 69:1105-7.

Actinomyces

Bak, U. B., and Azuma, R. 1991. Pathology of infection caused by *Dermatophilus*-like organisms in porcine tonsils. J Comp Pathol 105:255-62.

Behbehani, M. J., and Jordan, H. V. 1982. Comparative pathogenicity of *Actinonmyces* species in mice. J Med Microbiol 15:465-73.

Behbehani, M. J.; Heeley, J. D.; and Jordan, H. V. 1983. Comparative histopathology of lesions produced by *Actinomyces israelii*, *Actinomyces naeslundii*, and *Actinomyces viscosus* in mice. Am J Pathol 110:267-74.

Cisar, J. O.; Kolenbrander, P. E.; and McIntire, F. C. 1979. Specificity of coaggregation reactions between human oral streptococci and strains of *Actinomyces viscosus or Actinomyces naeslundii.* Infect Immun 24:742-52.

Davenport, A. A.; Carter, G. R.; and Beneke, E. 1975. *Actinomyces viscosus* in relation to the other *Actinomyces* and actinomycosis. Vet Bull 45:313-16.

Ehler, W. J. 1977. Actinomycosis in the dog. Canine Pract 4:15-17.

Engel, D. G.; Epps, E.; and Clagett, J. 1976. In vivo and in vitro studies on possible pathogenic mechanisms of *Actinomyces viscosus*. Infect Immun 14:548-54.

Engel, D. G.; Clagett, J.; Page R.; and Williams, B. 1977. Mitogenic activity of *Actinomyces viscosus*. 1. Effects on murine B and T lymphocytes and partial characterization. J Immunol 118:1466-71.

Georg, L. K.; Brown, J. M.; Baker, H. J.; and Cassell, G. H. 1972. *Actinomyces viscosus* as an agent of actinomycosis in the, dog. Am J Vet Res 33:1457-70.

Jordan, H. V.; Kelly, D. M.; and Heeley, J. D. 1984. Enhancement of experimental actinomycosis in mice by *Eikenella corrodens*. Infect Immun 46:367-71.

Pine, L., and Overman, J. R. 1966. Differentiation of capsules and hyphae clubs of bovine sulphur granules. Sabouraudia 5:141-43.

Swerczek, T. W.; Schiefer, B.; and Nielsen, S. W. 1968. Canine actinomycosis. Zentralbl Veterinaermed [B]15:955-70.

Yeung, M.K. 1992. Conservation of an *Actinomyces viscosus* T14V type 1 fimbrial subunit homolog among divergent groups of *Actinomyces* spp. Infect Immun 60: 1047-52.

Dermatophilus

Amakiri, S. F., and Nwufoh, K. J. 1981. Changes in cutaneous blood vessels in bovine dermatophilosis. J Comp Pathol 91:439-42.

Bak, U. B., and Azuma, R. 1991. Pathology of infection caused by *Dermatophilus*-like organisms in porcine tonsils. J Comp Pathol 105:255-62.

Ellis, T. M.; Sutherland, S. S.; and Davies, G. 1991. Strain variation in *Dermatophilus congolensis* demonstrated by cross-protection studies. Vet Microbiol 28:377-83.

Gordon, M. A., and Perrin, U. 1971. Pathogenicity of *Dermatophilus* and *Geodermatophilus*. Infect Immun 4:29-33.

Hanel, H.; Kalisch, J.; Keil, M.; Marsch, W. C.; and Buslau, M. 1991. Quantification of keratinolytic activity from *Dermatophilus congolensis*. Med Microbiol Immunol Berl 180: 45-51.

Jenkinson, D. M.; Menzies, J. D.; Pow, I. A.; Inglis, L.; Lloyd, D. H.; and Mackie, A. 1989. Actions of bovine skin washings and sera on the motile zoospores of *Dermatophilus congolensis*. Res Vet Sci 47:241-46.

Jones, R. T. 1976. Subcutaneous infection with *Dermatophilus congolensis* in the cat. J Comp Pathol 86:415-20.

Lloyd, D. H. 1981. Measurement of antibody to *Dermatophilus congolensis* in sera from cattle in the west of Scotland by enzyme-linked immunosorbent assay. Vet Rec 109:426-27.

Lloyd, D. H., and Noble, W. C. 1982. *Dermatophilus congolensis* as a model pathogen in mice for the investigation of factors influencing skin infection. Br Vet J 138:51-60.

Lloyd, D. H.; Dick, W. B. D.; and Jenkinson, D. M. 1979. The effects of some surface sampling procedures on the stratum corneum of bovine skin. Res Vet Sci 26:250-52.

Merchant, S. R. 1990. Zoonotic diseases with cutaneous manifestations in food animals. Comp Cont Ed Pract Vet 12: 1675-82.

Morrow, A. N.; Heron, I. D.; Walker,A. R.; and Robinson, J. L. 1989. *Amblyomma variegatum* ticks and the occurrence of bovine streptothricosis in Antigua. J Vet Med B 36:241-49.

Oduye, O. O. 1976. Histopathological changes in natural and experimental *Dermatophilus congolensis* infection of the bovine skin. In Dermatophilus Infection in Animals and Man. Ed. G. H. Lloyd and K. C. Sellers, pp. 172-81. London: Academic Press.

Richard, J. L.; Pier, A. C.; and Cysewski, S. J. 1973. Experimentally induced canine dermatophilosis. Am J Vet Res 34:797-99.

Roberts, D. S. 1961. The life-cycle of *Dermatophilus dermatonomus,* the causal agent of ovine mycotic dermatitis. Aust J Exp Biol Med Sci 39:463-76.

Stewart, G. H. 1977. Dermatophilosis: A skin disease of animals and man. Vet Rec 91:537-44.

Sutherland, S. S.; Ellis, T. M.; and Edwards, J. R. 1991. Evaluation of vaccines against ovine dermatophilosis. Vet Microbiol 27:91-99

Thoen, C. O.; Jarnagin, J. L; Saari, D. A.; Ortiz, B.; and Harrington, R. 1980. Pathogenicity and transmission of *Dermatophilus congolensis* isolated from cattle in Puerto Rico. Proc 83rd Annu Meeting US Anim Health Assoc pp. 232-37. Louisville, Ky.

13 / *Salmonella*

BY R. C. CLARKE AND C. L. GYLES

SALMONELLA has long been recognized as an important zoonotic pathogen of worldwide economic significance in humans and animals. Interest in this pathogen has heightened in recent years due to the increased susceptibility of AIDS patients to salmonellosis, the devastating effects of *S. enteritidis* in the poultry industry, and the globalization of agricultural trade.

Infection of animals with various species of *Salmonella* sometimes results in serious disease and always constitutes a vast reservoir for the disease in humans. The interplay of *Salmonella* with its host takes a variety of forms, including remarkable host specificity, inapparent infections, recovered carriers, enteritis, septicemia, abortion, and combinations of disease syndromes. *Salmonella* are readily transferred from animal to animal, animal to humans, and human to human by direct or indirect pathways.

The genus *Salmonella* is a member of the family *Enterobacteriaceae* and shares characteristics of the family. In recent years there have been several modifications to the scheme for classification of these organisms. Currently, the genus *Salmonella* is divided into two species, *S. enterica* and *S. bongori*, with *S. enterica* subdivided into six subspecies, *enterica, salamae, arizonae, diarizonae, indica*, and *houtenae* (LeMinor and Popoff 1987; Reeves et al. 1989). Most *Salmonella* belong to *S. enterica* subsp. *enterica*; members of this subspecies are given a name, which is usually based on the name of the place where the serovar was first isolated. Other isolates are identified only by the subspecies, followed by the antigenic formula. Although the new nomenclature indicates that a serovar of *S. enterica* subsp. *enterica* should be called by the names of the species, subspecies and serovar (e.g. *S. enterica* subsp. *enterica* serovar Typhimurium), some researchers use the form *Salmonella* serovar Typhimurium, and it is common practice to identify the organisms by the genus followed by the serovar (e.g., *S. typhimurium*). The simple, common practice, will be used in this chapter.

Typically, *Salmonella* are nonlactose fermenters or slow lactose fermenters; most schemes for the detection of the organism are based on this property. However, an occasional strain ferments lactose rapidly. Identification of an isolate as a member of the genus *Salmonella* or as a member of one of the six subspecies is based on biochemical tests. Serotyping is used to identify the organism beyond the level of subspecies. Approximately 2300 serovars have been described on the basis of somatic (O), flagellar (H), and capsular (Vi) antigens (LeMinor and Popoff 1987), but less than 50 of these are encountered at significant frequency in disease.

On the basis of the O antigens, serogroups have been identified, each having a group-specific O antigenic factor. Within each O group, serovars are determined by means of the combination of O and H antigens. A polysaccharide microcapsule responsible for the Vi antigen is frequently present on *S. typhi, S. hirschfeldii,* and certain strains of *S. dublin.* Phage typing, biotyping, drug resistance, and plasmid profile analysis may be used to identify isolates beyond the level of serovar and are useful for epidemiologic studies. Phage typing is limited to a few serovars, notably *S. typhi, S. typhimurium, S. dublin, S. enteritidis, S. heidelberg,* and *S. schottmuelleri,* for which schemes have been established.

HOST SPECIFICITY

Salmonella may be considered in three groups based on their association with human and animal hosts (Wray and Sojka 1977; Turnbull 1979). One group is characterized by specificity for the human host (Table 13.1). Members of this group are *S. typhi, S. paratyphi, S. schottmuelleri, S. hirschfeldii* (agents of typhoid and paratyphoid fever), and *S. sendai. S. typhi, S. paratyphi,* and *S. hirschfeldii* show strict adaptation to humans, but *S. schottmuelleri* is occasionally recovered from animals. A second group consists of organisms that are more or less adapted to specific animal hosts (Table 13.1). *S. arizonae* is now frequently isolated from diarrheal disease in lambs, and it has been suggested that this serotype is now becoming adapted to its ovine host. The third group consists of unadapted *Salmonella* that cause disease in humans and a variety of animals. Most *Salmonella* fall into this group, but a small number of serovars, including *S. typhimurium* and *S. enteritidis,* tend to be involved in most cases of disease.

TABLE 13.1. Examples of host-adapted *Salmonella*

Host	Serovar
Human	*S. typhi, S. paratyphi*[a], *S. schottmuelleri*[a], *S. hirschfeldii*[a], *S. sendai*
Cattle	*S. dublin*
Swine	*S. choleraesuis, S. typhisuis*
Poultry	*S. pullorum, S. gallinarum*
Sheep	*S. abortusovis*
Horse	*S. abortusequi*

[a] Formerly identified as *S. paratyphi* A, *S. paratyphi* B, and *S. paratyphi* C, respectively.

SOURCES OF INFECTION

Salmonella inhabit the intestinal tract of vertebrates; excretion results in contamination of water, food, and the environment (Wray and Sojka 1977; Turnbull 1979). The organism has been recovered from the intestine of a wide range of animals including fish, reptiles, birds, and mammals. Fertilizers and feeds containing animal products are sometimes a source of infection for animals. Fish meal, bone meal, and meat meal have all been shown to be frequently contaminated with *Salmonella,* but pelleting these products results in a marked reduction in the frequency of detection of *Salmonella.* Contaminated milk and milk products are other sources of organisms, particularly for calves (Wray and Sojka 1977).

Salmonellosis is an important zoonosis and, although human-to-human transmission does occur, animals and their products constitute the most important source of the organism for humans (D'Aoust 1989). Meats, particularly poultry, are frequently incriminated as sources of infection. In recent years, eggs contaminated with *S. enteritidis* have been a major source of infection for humans. Extensive studies of *S. enteritidis* in England and the United States have shown that the predominant phage types recovered from humans were those commonly recovered from eggs and poultry. Raw milk is a frequent source of infection for farm families, and contamination of pasteurized milk has been the cause of a number of extensive outbreaks. Pet turtles are an important source of the organism, particularly for young children.

DISEASE

Salmonellosis is manifested in animals in three major forms: enteritis, septicemia, and abortion; however, in an outbreak, or even in a single animal, any combination of the three may be observed (Wray and Sojka 1977). Fever, inappetence, and depression are commonly observed in acutely ill animals. Enteritis caused by *Salmonella* results in the passage of foul-smelling, watery feces, which may contain fibrin, mucus, and sometimes blood. When the enteric disease is severe, death may result from dehydration, electrolyte loss, and acid-base imbalance.

Common clinical signs of the septicemic aspect of the disease include fever, inappetence, and depression. In dairy cattle a precipitous decline in milk production is usually seen, and pneumonia may develop in calves suffering from septicemic salmonellosis. Localizations of *Salmonella* may result in meningitis or polyarthritis. The septicemic disease may be severe and run an acute course, which is highly fatal in untreated animals, or it may be mild and run a subacute course with a slow resolution.

Abortion often occurs in pregnant animals that develop septicemia, but certain serovars are more prone to cause abortion than are others. *S. dublin* has been associated with outbreaks of abortion in cattle, and several other serovars adapted to animal hosts have a particular association with abortion (Wray and Sojka 1977).

Although poultry are often infected with *Salmonella*, disease does not usually develop. Both the rate of carriage and the occurrence of disease appear to be low in dogs; clinical disease is frequently seen in cattle, sheep, and pigs. In cattle the disease is important in both calves and adults. *S. typhimurium* is the commonest serotype recovered from diseased cattle, but in some areas, notably Europe, *S. dublin* is a major cause of disease (Wray and Sojka 1977). Interestingly, *S. dublin* has been rare in North America except for California and regions west of the Rocky Mountains, but in recent years there has been an eastward movement of this organism across the United States. In both adults and calves the disease usually takes the form of enteritis and septicemia. In calves the disease is rarely seen before 2 weeks of age and often occurs in outbreaks with an acute course and high mortality. Management systems involving purchase of young calves from several sources and the mixing and transporting of these animals are especially susceptible to outbreaks of the disease. In adult cattle the disease is frequently sporadic, less acute, and may be associated with abortions.

Salmonellosis in swine usually occurs after weaning and is seen most commonly in pigs 8-16 weeks old. In North America *S. choleraesuis* var *kunzendorf* is associated with the septicemic form and *S. typhimurium* with the enteric form of the disease. *S. choleraesuis* is a highly invasive organism, and interstitial pneumonia and multifocal hepatic necrosis are the most frequent systemic lesions; necrotic and ulcerative lesions are observed in the mucosa of the colon (Reed et al. 1986). Pigs are frequently healthy carriers.

Pullorum disease, caused by *S. pullorum,* is a predominantly enteric disease of poultry; fowl typhoid, caused by *S. gallinarum,* is a septicemic disease of poultry. Both have been eradicated from flocks in the major poultry-producing countries, and clinical salmonellosis in poultry is not a common problem in most countries. Nonetheless, the high frequency of carriage of other *Salmonella* by poultry creates an important source of infection for humans. Transmission of infection through eggs and contamination in the rearing environment and at shipping and slaughter are significant contributors to the problem. Rapid increases in the frequency of egg-associated outbreaks of human disease due to *S. enteritidis* have been noted in many countries; contamination of the egg may be derived from the ovaries or oviducts, or by penetration from the environment through the shell.

Salmonellosis in horses is often associated with stress such as surgery; parasitism; and hot, humid, weather. Outbreaks of equine salmonellosis sometimes occur in veterinary hospitals and may require closing of the hospital. Furthermore, persistent shedding by recovered animals represents a problem that continues after discharge of the animals. Dogs and cats are only infrequent carriers of the organism and even less frequent victims of disease. Tortoises and terrapins are commonly healthy carriers that shed large numbers of *Salmonella* and may be a source of infection, particularly for children.

Humans who are taking oral antibiotics are at increased risk of developing salmonellosis following exposure to *Salmonella*; interference with the normal flora is believed to be the basis for this increased susceptibility to disease. A similar situation exists for animals.

CARRIERS

Although *Salmonella* may survive for long periods in the environment, it is the carrier state that provides the major source of infection for animals and humans (Wray and Sojka 1977). The carrier state is characterized by the absence of evidence of disease in animals that are able to transmit infection to susceptible individuals. Carrier animals can shed large numbers of *Salmonella* (up to 10^5/g) in feces. Carriers develop as a result of the interaction of several factors, including the serovar, the age of the animal, and the number of bacteria ingested. Certain serovars appear more likely to induce the carrier state than are others. Young animals often shed *Salmonella* only during convalescence, whereas adults are more likely to become chronic shedders. A low dose of *Salmonella,* insufficient to cause disease, may result in the carrier state.

Various types of carriers have been identified (Wray and Sojka 1977). The active carrier excretes *Salmonella* for months or years. The active-carrier state may follow recovery from clinical disease; this is common for adult cattle that recover from salmonellosis due to *S. dublin*. These animals are sometimes referred to as persistent excretors and usually excrete organisms at a rate up to 10^5/g of feces. Persistent excretion may occur in animals with high serum-antibody levels against the O and H antigens of the excreted organism. Prolonged excretion following disease has also been demonstrated for calves and adults infected with *S. saint-paul* (Aitken et al. 1983). Treatment of adult animals with antibiotics during the course of the disease is reported to be ineffective in eliminating the carrier state in cattle infected with *S. dublin* (Wray and Sojka 1977). However, treatment of calves with disease due to *S. saint-paul* reduced the duration of excretion of the organism (Aitken et al. 1983). Following recovery from disease due to *S. typhimurium,* adult cattle usually do not shed organisms for a prolonged time (Wray and Sojka 1977).

Passive carriers are animals that ingest *Salmonella* and pass the organisms through the intestine into the feces with little or no invasion of the mesenteric lymph nodes. These animals cease shedding *Salmonella* shortly after they are removed from a contaminated environment.

Latent carriers are animals that have *Salmonella* organisms in their tissues but generally do not excrete them in their feces; these animals may become intermittent excretors.

Certain stress factors have been shown to promote excretion of *Salmonella* by carriers and to lead to activation or reactivation of disease in carrier animals (Wray and Sojka 1977). Transportation of animals, overcrowding, administration of corticosteroids, parturition, and concurrent viral and protozoan infections have all been shown to increase susceptibility of animals to disease.

PATHOGENESIS

Intestinal Infection

Although *Salmonella* may enter the body through the pharynx, respiratory tract, or the conjunctiva, the organisms usually gain entrance to the host by the oral route and are deposited in the intestine, where they invade the enterocytes (Wray and Sojka 1977). Nonspecific host defense factors that affect the ability of the organism to establish include gastric acidity, peristalsis, intestinal mucus, lysozyme in secretions, lactoferrin in the gastrointestinal tract, and the normal intestinal flora. It appears that large numbers of organisms in the intestine are required to initiate disease; thus experimental and natural disease are facilitated by elevation of gastric pH, impairment of peristalsis, and interference with the normal intestinal flora.

Greater susceptibility of the very young may be due in part to the absence of a well-established bacterial intestinal flora, and there is much experimental data showing the importance of the normal flora in limiting the establishment and growth of *Salmonella* in the intestine. The LD_{50} for *S. enteritidis* administered orally to germ free mice is 3-5 organisms, whereas the comparable value in conventional mice is 10^6 (Collins and Carter 1978). Interestingly, there is no significant difference in LD_{50}s in germ-free and conventional mice when the organisms are administered intravenously or intraperitoneally. Further indications of the significance of the normal intestinal flora are provided by enhanced susceptibility of animals following oral administration of antibiotics and by the resistance to experimental challenge that is conferred on day-old chicks following the oral inoculation of normal adult chicken intestinal flora.

Current concepts of penetration of the intestine by *Salmonella* are still influenced by the work of Takeuchi (1967), who conducted electron microscopic studies of experimental *Salmonella* infection in starved, opium-treated guinea pigs. Takeuchi found that when *Salmonella* come within a critical distance of the brush border of enterocytes, the microvilli and tight junctions undergo degeneration. It is possible that a translucent area around the bacterium, which was interpreted as the critical distance, represented exopolysaccharides or pili and that it was contact with the brush border that triggered the successive events. Bacteria enter the cells either through the microvilli or via the junction complexes between enterocytes and are usually enclosed in a membrane-bound vacuole that migrates to the basal region of the cell. The microvilli and apical cytoplasm undergo degeneration in the internalization process but are subsequently repaired. The bacteria pass through the enterocytes to the lamina propria, where they stimulate an inflammatory response and are engulfed by macrophages and neutrophils (Takeuchi and Sprinz 1967). The inflammatory reaction is characterized by the presence in the intestine and lumen of large numbers of polymorphonuclear leukocytes, many of which are laden with engulfed *Salmonella*. During the next 24 hours, bacteria in the lumen and in enterocytes are cleared, but those in the lamina propria persist and multiply within phagocytic cells.

Recent evidence, based on tissue culture models, suggests that the invasion process has two parts: first the bacterium adheres to an unidentified receptor on the epithelial cell; then it activates cellular functions through the receptor for epidermal growth factor (EGF) (Galan et al. 1992b; Portnoy and Smith 1992). Invasion of tissue culture cells by *Salmonella* induces tyrosine phosphorylation of the EGF receptor, which is thought to be a key initiating factor in a process that leads to increased calcium levels, depolarization of microvilli, formation of blebs, and internalization of the organism (Galan et al. 1992; Portnoy and Smith 1992). Invasion is affected by numerous regulatory systems triggered by environmental signals such as low oxygen levels, temperature, and osmolarity (Galan and Curtiss 1990; Lee and Falkow

1990; Jones et al. 1992). Studies of mutations that enhance invasion have shown that chemotaxis, motility, and the orientation of flagella can influence the invasion of *Salmonella* in tissue culture and in mice (Jones et al. 1992). Plasmids associated with virulence of *Salmonella* have not been shown to affect invasion (Gulig and Curtiss 1987).

The ability of *Salmonella* organisms to survive and multiply inside phagocytes is critical to the outcome of the infection. In the intracellular location, the bacterium is protected from injurious substances such as antibiotics, antibody, and complement. To survive, the bacterium must be able to resist the destructive systems present in macrophages, namely, reactive oxygen intermediates, reactive nitrogen intermediates, low pH, limitation of intracellular iron, and defensins. There are reports that salmonellae may inhibit phagosome-lysosome fusion, but there is also evidence that salmonellae are well adapted to evade the killing effects of the phagolysosome. It is evident that a high percentage of *Salmonella* organisms are eliminated by host defenses and only a proportion of the bacterial population survives for a prolonged time, leading to chronic infection and the carrier state.

Enteritis results in a shortening of villi, degeneration and abnormal extrusion of enterocytes, increased emptying of goblet cells, and a neutrophilic reaction in the lamina propria accompanied by transepithelial migration of neutrophils into the lumen (Takeuchi and Sprinz 1967) (Figs. 13.1-13.3). Early in the course of the enteritis, thrombi are visible in the vessels in the lamina propria, and later, damage to the vessel walls is associated with the thrombi. Bacteria are found in the Peyer's patches and are accompanied by an infiltration of neutrophils. Similar observations have been made in studies of the invasion of the intestine of chickens and calves infected with *Salmonella*.

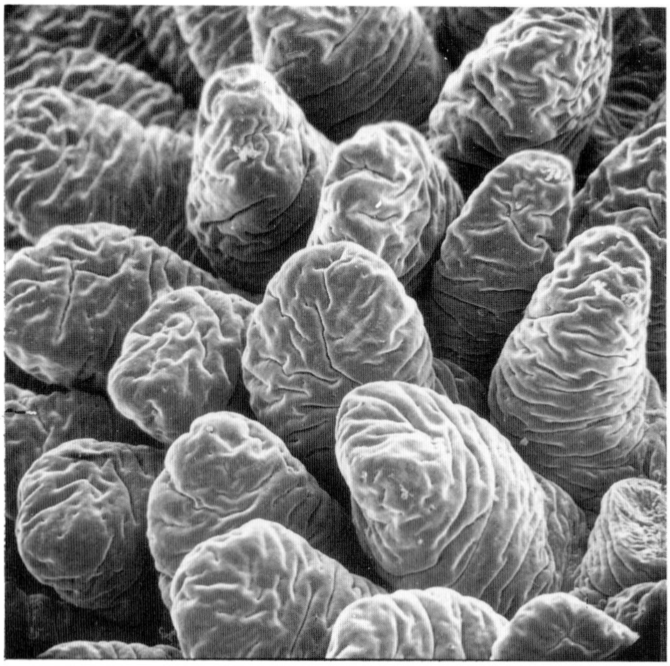

FIG. 13.1. Scanning electron micrograph of villi in the duodenum of a calf with severe *Salmonella typhimurium* infection. The duodenum appears normal although there is massive mucosal damage in the jejunum, ileum, and colon. (Clarke 1985.)

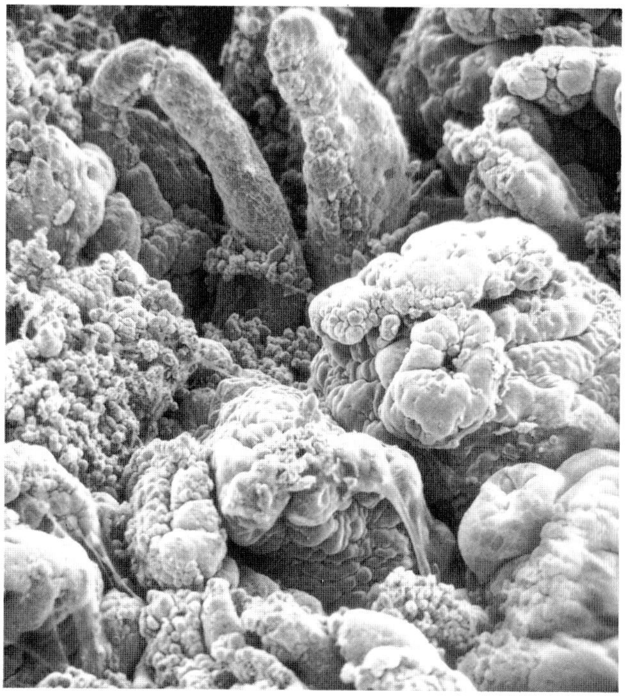

FIG. 13.2. Scanning electron micrograph of villi from the ileum of a calf with severe *Salmonella typhimurium* infection. Lesions were evident at 12 hours postinfection. Massive numbers of extruded enterocytes are visible as small, rounded cells on the mucosal surface. (Clarke 1985.)

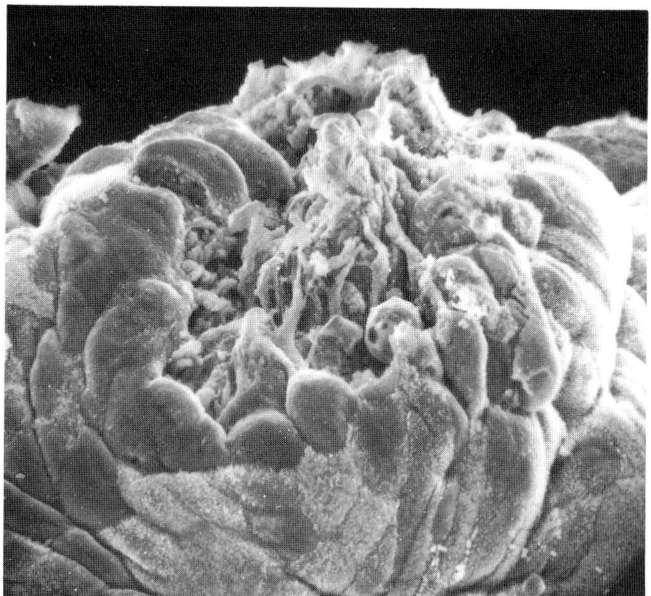

FIG. 13.3. Scanning electron micrograph of the tip of a villus in the ileum of a calf infected with *Salmonella typhimurium*. Necrosis of the villus tip is evident. (Clarke 1985.)

Subsequent to Takeuchi's work, researchers have investigated the attachment of *Salmonella* to epithelial cells and have shown that a nonfimbrial, mannose-resistant (MR) adhesin can mediate attachment of *Salmonella* to mammalian tissue culture cells in vitro; the in vivo mechanisms have not been identified. Many serovars of *Salmonella* lack the MR adhesin; the serovars that are less invasive in humans, *S. enteritidis* and *S. typhimurium,* possess the MR adhesin, whereas the more invasive ones, such as *S. dublin, S. choleraesuis*, and *S. typhi*, lack this adhesin. Most serovars of *Salmonella* produce mannose-sensitive hemagglutinating pili (type 1) that can bind to mannose derivatives on eukaryotic cells, but these do not appear to play a significant role in adherence of the bacterium to the ileal mucosa (Finlay and Falkow 1988).

Studies of experimental infection of calves with *Salmonella* show that the ileum is damaged more rapidly and more severely than are other parts of the intestine (Figs. 13.1-13.3). The villi are distorted and sometimes denuded of enterocytes. It appears that dying enterocytes lose their microvillous covering, detach, and are extruded. As the disease progresses, the jejunum and colon become invaded and damaged. Clearly, invasion of the mucosa is a critical part of the disease process (Fig. 13.4), and damage to infected intestine is not observed in the absence of invasion.

Intestinal Inflammation, Fluid Production

Gianella (1979) found that the intestinal fluid response seen in the disease is related to the inflammatory reaction to invasion of the intestine by the organism. This conclusion was based, in part, on the observation that experimental *Salmonella* infection in rabbits resulted in activation of the adenylate cyclase system in the intestinal mucosa and that pretreatment of the rabbits with indomethacin prevented the fluid response of the intestine. It was postulated that stimulation of adenylate cyclase was a response to prostaglandins released during the inflammatory process. However, enterotoxins that are capable of inducing fluid accumulation in the intestine of rabbits have been described (see Enterotoxins); the extent to which inflammation and enterotoxins contribute to diarrhea is not known. It is likely that loss of epithelial cells results in impaired absorption from the lumen of the intestine, and that enterotoxins could induce secretion of fluid from intact epithelial cells.

In animals in which an acute enteritis develops, death can occur in the absence of marked systemic involvement. In these cases death appears to be due to disturbances in fluid and electrolyte metabolism, which result in dehydration and acidosis. Animals that survive for long periods may develop severe gut lesions as well as septicemia.

Extraintestinal Infection

The struggle between bacteria and host usually involves more than the intestine. Free and engulfed bacteria in the mucosa and submucosa are transported by the lymphatics to the regional lymph nodes, which become involved in mounting an inflammatory response. From the lymph nodes, *Salmonella* travel via the efferent lymph vessels, drain into the blood circulation, and then are filtered out of circulation by the reticuloendothelial system, particularly in the spleen and liver. Release of endotoxin into the circulation can account for many of the systemic effects of disease, including fever and vascular damage. Thrombosis of small blood vessels may lead to ischemic necrosis of the extremities and tips of the ears. Failure to contain the infection leads to septicemia and sequelae, notably pneumonia, meningitis, and septic arthritis (Wray and Sojka 1977).

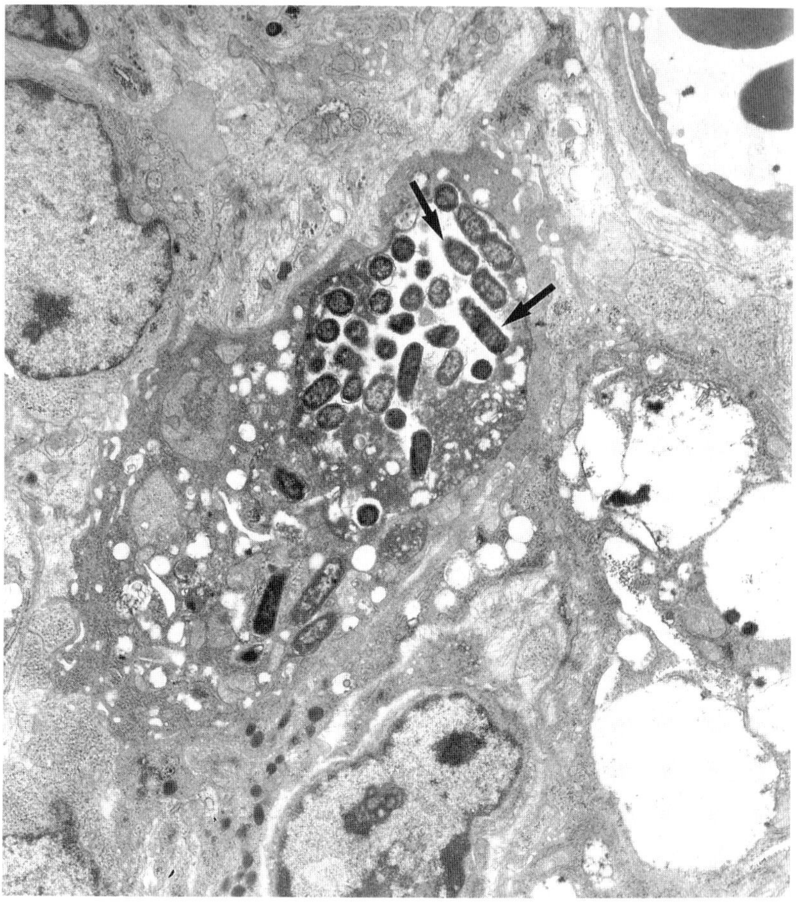

FIG. 13.4. Transmission electron micrograph of the lamina propria of the ileum of a calf infected with *Salmonella typhimurium*. Many intact *Salmonella* organisms (→) are visible inside a degenerating phagocytic cell. x7500. (Clarke 1985.)

Certain serotypes of *Salmonella* tend to produce disease characterized by septicemia and its sequelae rather than by significant intestinal disorder (Wray and Sojka 1977; Turnbull 1979). Host-adapted *Salmonella* are particularly prone to this form of disease; thus, *S. typhi* in humans, *S. choleraesuis* in pigs, *S. dublin* in cattle, *S. abortusovis* in sheep, and *S. abortus-equi* in horses produce disease in which septicemia is the dominant feature. Abortion is often a prominent feature of the disease but no specific predilection for the gravid uterus has been demonstrated.

BACTERIAL VIRULENCE FACTORS

Several potential virulence factors have been suggested for *Salmonella* but very few have been tested critically for their contribution to virulence. Tissue culture systems have contributed substantially to our understanding of many of these factors, but the use of a variety of cell types and strains of various serovars of *Salmonella* make it difficult to relate the data to disease in animals and humans. Furthermore, most studies of experimental salmonellosis have involved *S. typhimurium* in mice as a model for typhoid fever in humans, and only a few studies have investigated the diarrheal disease, although diarrhea and damage to intestinal epithelial cells are consistent changes in the intestine of animals with salmonellosis.

Enterotoxins

Poorly characterized enterotoxic activity has been reported to be present in several serovars of *Salmonella*. A cholera toxin (CT)-like enterotoxin has been identified in *S. typhimurium*; it is related in structure, function, and antigenicity to CT and *Escherichia coli* heat-labile enterotoxin (LT) (Finkelstein et al. 1983; Prasad et al. 1990, 1992). The enterotoxin is heat labile and cell associated, induces fluid secretion in ligated rabbit intestine, causes elongation of Chinese hamster ovary (CHO) cells, elevates intracellular cyclic AMP (cAMP) and prostaglandin E_2 (PGE_2) in rabbit intestine, binds to ganglioside G_{M1}, and its biological activity is neutralized by specific anti-CT antibodies. When the CT-like enterotoxin was examined by sodium dodecylsulfate polyacrylamide gel electrophoresis, 25 kDa and 12 kDa bands were observed; these likely correspond to A and B subunits, respectively. There may be at least one other enterotoxin that is produced by *Salmonella*. Clarke et al. (1988) found a *Salmonella* enterotoxin that did not bind G_{M1} ganglioside, and Rahman et al. (1992) recently reported the characterization of a *Salmonella* enterotoxin that caused elongation of CHO cells and accumulation of fluid in ligated rabbit intestine but was not neutralized by anti-CT.

Very little is known about the role of the enterotoxin(s) in disease. However, if the parallels to CT and LTI are maintained, it is expected that the B subunit of CT-related enterotoxin produced by *Salmonella* organisms external to the intestinal epithelium binds to G_{M1} ganglioside on the membrane of intestinal epithelial cells; the A subunit is internalized leading to activation of intracellular cAMP and PGE_2. These changes then cause increased secretion of electrolytes and movement of water into the lumen by osmotic effects.

Cytotoxins

Damage to intestinal epithelial cells is a striking feature of natural and experimental *Salmonella*-induced enteritis (Clarke 1985), and cytotoxins are likely agents of this damage. At least three cytotoxins have been described. A heat-labile, trypsin-sensitive cytotoxin has been detected in a wide variety of serovars of *Salmonella* by Ashkenazi et al. (1988), who demonstrated production of a HeLa-cell cytotoxin by all 131 isolates they examined. *S. typhi* produced lower amounts of the cytotoxin in vitro than did serotypes associated primarily with enteric disease. The cytotoxin inhibited protein synthesis in Vero cells but its activity was not neutralized by antibodies, which neutralized the cytotoxicity of Shiga toxin or Shiga-like toxins. Cytotoxicity was associated with a molecule in the range of 56-78 kDa. An earlier study (Koo et al. 1984) demonstrated that there was inhibition of protein synthesis in epithelial cells from *Salmonella*-infected rabbit intestine.

One report identified a cytotoxin that is a low molecular weight component of the bacterial outer membrane (Reitmeyer et al. 1986). It appears to be a second cytotoxin. More recently Libby et al. (1990) cloned a cytotoxin gene from *S. typhimurium* and showed by Southern hybridization that highly related sequences were present in all serovars of *Salmonella* tested, in all species of *Shigella*, and in enteroinvasive *E. coli* but was absent from over 40 other bacterial species. This third cytotoxin was purified to near homogeneity and shown to be a 26-kDa cell-associated contact hemolysin that was lethal for Vero cells and several other cell lines in which it caused a rapid lysis. The relationship of this hemolytic cytotoxin to the previously described cytotoxin was not addressed, but this hemolysin is clearly different from the others in molecular size and in mode of action. Contact hemolysin of *Shigella flexneri* is a 62-kDa protein, which causes lysis of the endocytic vacuole. Rahman et al. (1992) also identified a *Salmonella* factor that was cytolytic for CHO cells.

Lipopolysaccharide

Lipopolysaccharide (LPS) is an essential component of the outer membrane of *Salmonella* and is a major determinant of virulence. LPS is composed of an internal lipid A embedded in the outer membrane, a core region, and an O antigenic region, which consists of repeated oligosaccharide units. Mutants with defects in LPS (rough mutants) lack the O antigen and are less virulent than their parents, which possess complete LPS. Much of this difference in virulence can be attributed to greater susceptibility of rough strains to phagocytosis and killing by macrophages; compared with their smooth parents, rough strains are less protected against lysosomal attack and more sensitive to granule contents. Smooth LPS may also serve to provide distance between the cell membrane and the destructive effects of antigen-mediated reactions at the bacterial surface. Thus, length of the LPS molecule contributes to resistance to complement-mediated killing and rough *Salmonella* are highly serum sensitive compared with their smooth ancestors. The chemical characteristics of the O antigen also affect susceptibility to killing because different O antigens activate complement by the alternate pathway at different rates and thereby promote engulfment by macrophages at different rates (Liang-Takasaki et al. 1983; Saxen et al. 1987). Indeed, a low rate of complement activation and long LPS, which shields the organism from the damaging effects of the complement cascade, are considered to account for the greater virulence of *S. typhimurium* in mice compared with other *Salmonella* (Robbins et al. 1992).

An interesting study by Nnalue and Lindberg (1990) found that rough derivatives of *S. choleraesuis* were similar in virulence to their parents when they were tested in mice by the intraperitoneal or the intravenous route, but that they were avirulent by the oral route. These researchers concluded that a major role of smooth LPS of this serovar in mice was to promote survival in the gastrointestinal tract. These findings are different from those for other serovars of *Salmonella* and indicate that there may be diversity among serovars in the ways in which they cause disease. Other researchers have demonstrated that rough strains of *Salmonella* are less effective than their smooth counterparts in penetration of intestinal mucus and that smooth LPS contributes to invasion of intestine or polarized monolayers of epithelial cells. Robbins et al. (1992) note that *Salmonella* bacteremias in humans are almost all due to strains of groups A, B, C, or D, and they suggest that the structure of the LPS is a factor in this aspect of disease.

LPS also contributes to disease by virtue of its endotoxic properties. Vascular damage and thrombosis, which are observed in the intestine in salmonellosis (Takeuchi and Sprinz 1967; Clarke 1985), may be caused by LPS, which induces a vast array of inflammatory mediators and immunoregulatory cytokines. LPS-induced alterations in blood vessels could contribute to degeneration and death of intestinal epithelial cells. Systemic changes observed in disease may also be attributed to LPS. Thus, fever, disseminated intravascular coagulation, circulatory collapse, and shock are considered to be the result of LPS produced by the organism.

Flagella

Reports in the literature suggest that the role of flagella may vary with the species of *Salmonella*. Motility appears to be necessary for *S. typhi* but not *S. typhimurium* to invade epithelial cells. The function of motility appears to be to increase the chance of the organism coming into contact with the epithelial cell (Finlay and Falkow 1989). Jones et al. (1992) have recently shown increased invasiveness of chemotaxis mutants of *S. typhimurium* that were "smooth" swimmers and decreased invasiveness of "tumbly" mutants. Smooth swimmers have their flagella organized into a bundle at one pole, so that most of the bacterial surface is free of projecting flagella; tumbly swimmers have their flagella projecting from most of the bacterial surface. Furthermore, strains that did not produce flagella were as invasive as wild-type organisms, provided they were centrifuged on to the epithelial cell monolayer. These workers have advanced the fascinating hypothesis that salmonella organisms in the swimming mode have increased opportunity for interaction of the bacterial cell surface with the epithelial cell, compared with tumbling bacteria, whose flagella place the bacterial cell wall at some distance from the epithelial cell.

Siderophores

Humans and animals have mechanisms for binding iron absorbed in the diet so that the iron is not available to infecting microorganisms. Host iron-binding proteins that have been identified include transferrin, lactoferrin, ferritin, and hemoglobin. To compete with the host for the limited supply of iron available in vivo, bacteria produce iron chelators (siderophores) that have a high affinity for iron. Siderophores synthesized by the bacteria are exported into host fluids, where they capture iron and return to the bacteria through specific outer-membrane receptor proteins produced in response to low concentrations of iron (Finkelstein et al. 1983).

S. typhimurium produces enterochelin (enterobactin), a phenolate siderophore, which was suggested to be an important virulence factor. However, Benjamin et al. (1985) demonstrated that the production of an iron-binding siderophore was not essential for full virulence of mouse-virulent *S. typhimurium* given parenterally, which cast doubt on the significance of enterobactin in the virulence of *S. typhimurium* in experimental mouse typhoid. Their findings were in agreement with those of other workers who have shown that intracellular pathogens do not need high-affinity iron-gathering mechanisms for virulence. It is likely that, in natural disease, the extent of extracellular growth that occurs may determine the importance of iron-chelating systems for virulence of *Salmonella*. In pigs with experimental *S. choleraesuis* infection, there is a reduction of serum iron, total iron-binding capacity, and transferrin. The environment inside the phagocytic cell is one of low iron, and it is possible that the organism has nonsiderophore mechanisms for scavenging iron.

Heat-Shock Proteins

Many bacteria, including *Salmonella*, respond to a variety of stresses by producing increased levels of certain proteins called heat-shock proteins (HSPs). The environment inside murine macrophages induces *S. typhimurium* to produce increased levels of HSPs; mutants of the bacterium that are defective in their ability to make this response do not survive well in macrophages and are less virulent in mice.

VIRULENCE GENES

Virulence genes refer to genes that are not required for "house-keeping" functions and encode products that assist the organism in establishing in the host and/or causing damage to the host. Both plasmid and chromosomal virulence genes have been identified in *Salmonella*. Salmonellae need to be very versatile and adaptable, since they encounter a wide variety of environments, including a range of conditions outside the host, the intestine, the epithelial cell, and the phagocytic cell of a wide variety of animal hosts. It is becoming evident that they often depend on environmental cues to trigger appropriate responses, mediated by coordinate regulation of several widely dispersed genes in a regulon. Environmental cues that have been identified include osmolarity, starvation, stress, pH, growth phase, low oxygen levels and temperature (Mekalanos 1922).

S. abortusovis, S. choleraesuis, S. dublin, S. enteritidis, S. gallinarum, S. pullorum, and *S. typhimurium* all possess large virulence-associated plasmids that range in size from 50 kb to 96 kb (Jones et al. 1982; Terakado et al. 1983; Popoff et al. 1984; Helmuth et al. 1985). Strains possessing the virulence plasmid have LD_{50}s for mice that are much less than their counterparts that lack the plasmid. For example, a strain of *S. dublin*, cured of its 50-MDa plasmid, was shown to be 100- to 1000-fold less virulent than its plasmid-containing parent (Terakado 1983). Virulence plasmids are not essential for pathogenicity of *Salmonella*, and many strains that are free of plasmids are recovered from outbreaks of salmonellosis (Popoff et al. 1984).

A region of about 10 kb was reported to contain the virulence genes and to be highly conserved among virulence plasmids of different serovars. Recently, an 8.2-kb region has been shown to be sufficient for expression of the virulence phenotype associated with the plasmid (Krause et al. 1992). The nomenclature for genes in this locus is confusing, as different laboratories have applied different names to the same genes, but the use of the salmonella plasmid virulence (*spv*) gene nomenclature could be a means of standardizing the terminology (Caldwell and Gulig 1991; Krause et al. 1992). In a recent study, Krause et al. have identified the genes in this region and have shown how the *spv* nomenclature applies to genes previously assigned different names.

Functions attributed to genes in the virulence region include resistance to the complement-mediated bactericidal activity of serum, immune suppression, intracellular survival and growth, and stimulation of splenomegaly. These are all functions that are important in invasiveness and strains that possess the virulence plasmid are more invasive than their counterparts that lack the plasmid. Thus, the primary virulence phenotype associated with the plasmid is ability to disseminate from the intestine following oral infection in mice (Gulig 1990). The plasmid is not required for its interaction with epithelial cells (adherence, invasion, intracellular multiplication) and is presumed to be important in its interactions with phagocytic cells of the reticuloendothelial system (Finlay and Falkow 1989).

There are conflicting reports on association of serum resistance with the virulence plasmid, some laboratories claim that serum resistance is conferred by the plasmid and others claim that this is not the case (Jones et al. 1982; Gulig and Curtiss 1987). Sukupolvi et al. (1992) have suggested explanations for these disagreements. They found that strains that lacked the plasmid were serum resistant, but that the plasmid increased the resistance to killing by serum. They also found differences in serum resistance among strains of the same serovar and differences among serovars. For example, the virulence plasmid of *S. dublin* did not mediate resistance to killing by serum; the *tra*T gene, located on the virulence plasmid of other serovars, including *S. typhimurium*, but lacking from the *S. dublin* plasmid, was responsible for serum resistance. However, Heffernan et al. (1992) have identified a gene for resistance to killing by complement (*rck*) in the virulence plasmid of *S. typhimurium*. The gene product, Rck, had homology with *S. typhimurium* PagC and *Y. enterocolitica* Ail.

An 8.2-kb DNA fragment of the virulence region has been shown to encode five genes (*spv*R, *spv*A, *spv*B, *spv*C, *spv*D) in an operon, regulated by the product of *spv*R, which acts as a transcriptional activator (Krause et al. 1992). SpvB is produced only in the stationary growth phase and it has been suggested that growth limitation, induced by starvation in the phagolysosome, may be the trigger for expression of the plasmid virulence genes.

There was one report of a virulence plasmid carrying genes that regulated length of the O-side chain of the LPS in a strain of *S. dublin*. Loss of the plasmid resulted in a reduction in O-side chain length and in loss of virulence. Interestingly, change of *S. enteritidis* from virulent phage type 4 to avirulent phage type 7 is also associated with a loss in ability to produce LPS with a long O-side chain. This change is unrelated to the virulence plasmid.

Chromosomal genes are necessary for expression of enhanced virulence associated with the virulence plasmid. Thus, transfer of the plasmid to other salmonellae usually does not confer increased virulence. The serovars of *Salmonella* that typically possess a virulence plasmid are host-adapted; they undoubtedly possess the chromosomal genes that are necessary for the effects of the virulence plasmid genes to be expressed.

Genetic control of invasion is complex and involves at least nine loci and numerous genes located on the chromosome (Galan et al. 1992a; Lee et al. 1992; Stone et al. 1992). The complexity of invasion systems of *Salmonella* may be due to the ability to utilize more than one entry pathway or to some redundancy in function. Chromosomal genes associated with virulence and recognized in recent years include the genes for the invasin proteins, the hyperinvasion locus, and genes controlled by the two-component regulatory system *pho*P/*pho*Q (Fields et al. 1989; Groisman et al. 1989; Miller 1991). One locus that has been studied extensively in Galan's laboratory is the invasin (*inv*) locus. The *inv*E gene appears to be necessary for *Salmonella* to trigger the intracellular changes in epithelial cells that are needed for internalization of the bacterium. Thus, mutations in this gene prevent changes in distribution of polymerized actin and in intracellular free calcium, features of the cell invaded by *Salmonella*. Interestingly, there is high homology between the amino acid sequence of InvE and the YopN *Yersinia enterocolitica* outer membrane protein. Another invasin gene, *inv*A, encodes a protein which is related to *Yersinia* and *Shigella* proteins that are involved in virulence and translocate proteins across the bacterial cell membrane (Galan et al. 1992). Mutants in *inv*E or *inv*A adhere to epithelial cells but fail to invade.

Another locus that has been identified as affecting invasion is the hyperinvasion (*hil*) locus (Lee et al. 1992), which is located close to the *inv* locus and appears to encode either an invasion protein or an activator for expression of an invasion protein.

PhoP-regulated genes include genes that are activated and genes that are repressed by PhoP. PhoQ is a periplasmic sensor that transmits an intracellular signal by phosphorylation of PhoP, which binds to specific regions of DNA and activates a number of genes. One such PhoP-activated gene is *pag*C, which encodes a surface protein with high amino acid homology to the Ail protein of *Yersinia enterocolitica*. Mutations in the *pho*P locus result in high sensitivity to host defensins, and inability to survive in macrophages. Interestingly, these mutants are able to invade epithelial cells and multiply in them. It is likely that *pho*Q responds to the environment in the phagocytic cell and that PagC confers protection against destruction of the organism in the phagolysosome. There are also PhoP-repressed genes; they may encode surface products to which toxic compounds may bind.

Genes for other virulence factors include genes for enterotoxins, cytotoxins, LPS, flagella, and siderophores. Chromosomal genes for the CT-like enterotoxin have been cloned and used for studies on the enterotoxin (Prasad et al. 1990, 1992). Genes for a 26-kDa cytotoxin, which appeared to be a contact hemolysin, were cloned by Libby et al. (1990). There have been extensive studies on the numerous genes required for production of LPS, flagella, and chemotaxis; the genes for synthesis of enterobactin and iron uptake by this siderophore have been characterized. Genes for heat shock proteins (HSPs) have been identified in *S. typhimurium* and have been mutagenized for investigation of their role in virulence.

HOST RESISTANCE

Several aspects of host resistance to *Salmonella* have been observed, but most studies of the mechanisms that are involved have been conducted in mice. The outcome of the interaction of *Salmonella* with the host varies considerably with the serotype of *Salmonella* and the age of the host; generally, resistance to salmonellosis increases as the animal host ages (Wray and Sojka 1977; Turnbull 1979). For example, 1-day-old chicks are readily infected with 10^2

Salmonella, whereas 10^6 *Salmonella* are insufficient to induce infection in 50% of 8-wk-old chickens. In calves, however, natural disease is not usually observed before 2 weeks of age, but the disease tends to be more severe than in adults. In one carefully studied outbreak of disease in young calves, *Salmonella* were usually detected in the first 72 hours of life of the calves and were recovered from feces as early as 1 hour after birth, but disease was not observed before 20 days of age (Jones et al. 1983).

Studies in mice have shown profound effects of certain genes on resistance to *Salmonella* (O'Brien 1983; Hsu 1989); the alleles *Ity* and *Lps* regulate early growth of *S. typhimurium* in mice. Macrophages from Ity^r mice are more effective in limiting the growth of *Salmonella* than are macrophages from Ity^s mice; there is no significant difference in their ability to ingest the organisms. Strains of mice such as C57BL/6 and BALB/c that are homozygous for Ity^s allele are highly susceptible to *S. typhimurium* septicemia. Mice that have the lps^d (defective for LPS response) gene are extremely resistant to endotoxin, whereas those with the lps^n (normal response to LPS) gene exhibit the usual response to endotoxin. Surprisingly, mice with the lps^d genes are highly susceptible to *Salmonella,* thus, mice that have either or both the ity^s and lps^d genes are unusually susceptible to *Salmonella.* A third allele, *xid* (X-linked immune deficiency), controls the antibody response to polysaccharide antigens of *S. typhimurium.* Mice with this gene produce only low levels of IgG anti-*Salmonella* antibodies and succumb in the late phase of the disease. These and other studies suggest that early in the infection, the critical defense involves containment of the organism by the reticuloendothelial system by means of nonspecific immune mechanisms, and in the later phase, resistance is mediated by specific immune reactions involving both antibody and cell-mediated immune systems.

IMMUNITY

Both humoral antibody and cell-mediated immunity are important in the resistance to *Salmonella.* Antibodies against *Salmonella* are common in the sera of animals exposed to these organisms in their environment, and adults passively transfer antibodies to their offspring (Royal et al. 1968). Calves 2 weeks of age are capable of mounting a modest immune response to parenterally administered *Salmonella* antigen; at 3 weeks of age the response is markedly higher. When cattle are exposed to *Salmonella* in the environment or are vaccinated with the organisms, the humoral response is largely the IgM type of antibody. Specific IgG, IgM, and IgA antibodies against *Salmonella* LPS have been demonstrated in variable quantities in the sera of young calves, which were susceptible to experimental challenge with high doses of *Salmonella.* Evidence indicates that specific antibodies in colostrum and milk of cows vaccinated with *Salmonella* may interact with organisms in the lumen of the gut of calves and thus influence the outcome of infection (Royal et al. 1968). Secretory IgA may also be found in the intestine of animals that have recovered from disease or that have been vaccinated orally with live attenuated organisms. These antibodies in the intestine constitute the first line of specific immune defense; by binding to surface components of the organism, they may be able to interfere with attachment to and penetration of intestinal epithelial cells by the bacteria.

Natural and antibody-dependent antibacterial mechanisms may be important in defense against *Salmonella,* particularly in the gastrointestinal tract. Any moderating effects of specific antibodies are short-lived, since calves that receive immune colostrum but are removed from their dams and reared on milk-replacer are susceptible to experimental challenge at 3 weeks of age. Furthermore, antibody levels in serum do not correlate with protection against experimental challenge of calves vaccinated with either a live avirulent mutant strain of *Salmonella* or a heat-killed preparation of the organism (Habasha 1981; Lindberg and Robertsson et al.

1983). A number of calves immunized with live vaccine lacked serum titers of specific antibody but survived challenge exposure; calves immunized with killed vaccine failed to be protected, although they often had high titers of specific antibodies.

Recently, however, serum IgG antibodies against the O-specific polysaccharide of LPS have been suggested to be capable of conferring protection against salmonellosis. It is noted that such antibodies are protective in animals experimentally infected with *Salmonella* and it is argued that extravasation of serum IgG could result in protection at the level of the intestinal submucosa (Robbins et al. 1992).

Cell-mediated immunity is of considerable importance in protection against salmonellosis (Habasha 1981; Lindberg and Robertsson 1983). Transfer of sensitized T cells confers protection, whereas transfer of macrophages, B cells, and hyperimmune sera are not protective in the absence of immune T cells. Delayed-type hypersensitivity (DTH), measured by increase in skin-fold thickness associated with massive infiltration of mononuclear cells in the dermis at the site of antigen deposition, develops in calves with natural and experimental salmonellosis (Robertsson et al. 1982a,b; Lindberg and Robertsson 1983). DTH responses may be elicited by complex antigenic mixtures such as whole bacteria, ribosome preparations, and supernatants of homogenized *Salmonella;* by LPS with O-antigen polysaccharide chains homologous to those of the infecting organism; or by a porin preparation (Robertsson et al. 1982a,b). Lipid A and a macromolecular complex of the O polysaccharide are essential for the response to LPS. The porin preparation caused skin reactions regardless of the serotype of the *Salmonella* involved in the infection.

Macrophage migration-inhibition factor (MIF) has also been measured as an indicator of cell-mediated immune response in vaccinated and unvaccinated calves. The MIF test was reported to be a good in vitro test, which correlated well with protection of calves against challenge exposure (Habasha 1981).

Generally, cell-mediated immune reactivity correlates well with protection afforded calves immunized with live vaccines and challenged with large numbers of *Salmonella* (Habasha 1981; Lindberg and Robertsson 1983). Cross-protection develops for *S. typhimurium* and *S. dublin* and may be attributed to a shared O-antigenic determinant and shared porin antigens. It appears that a certain minimal systemic antigenic stimulation is necessary for skin reactivity to be detected; in calves vaccinated orally with a live vaccine, protection may be achieved in the absence of detectable skin reactions.

In calves and poultry, infection with *Salmonella* is accompanied by marked depletion of lymphocytes in the Peyer's patches and bursa of Fabricius, respectively, in the early stages of infection; this is followed by rapid repopulation of these sites. Neither the basis nor the significance of these effects on lymphocytes is understood. It is interesting that *Escherichia coli* Vero cell cytotoxin (which acts by inhibition of protein synthesis) is cytotoxic for certain lymphocytes and that *Salmonella* also produces a cytotoxin that inhibits protein synthesis.

Variation has been noted in the response of individuals to exposure to *Salmonella*. One component of this variation is genetic; in addition to the specific genes mentioned earlier, other aspects of immune function may affect the outcome of infection. The overall ability of animals to produce antibody, the avidity of the antibody that is produced, and the extent of the cell-mediated immune response that is mounted are all factors that may influence the course of infection. Some researchers are exploring the relationship of the major histocompatability complex to markers for host response with a view to being able to identify animals that respond in a superior fashion to a wide range of infections.

The basis for immunity induced by an *aro*A mutant of *S. typhimurium* was investigated in a mouse model of typhoid fever. The vaccine induced natural killer (NK) cell activity that was important in defence against the early stages of infection. It was suggested that these cells may also contribute to defense at later stages because they cause release of alpha and gamma interferon (Schafer and Eisenstein 1992).

VACCINES

Live attenuated, orally-administered *Salmonella* vaccines provide the best protection against *Salmonella* infection, especially when large challenge doses are used. This superior protection is due to the ability of live vaccines to stimulate a more effective cell-mediated immune response when compared with killed preparations. The oral route allows the attenuated vaccine strain to utilize natural routes of invasion, thereby presenting antigen to lymphocytes in the gut-associated lymphoid tissue, and inducing production of secretory IgA (sIgA) on mucosal surfaces throughout the body.

Advances in molecular techniques have facilitated creation of specific, nonreverting mutations that can be used to construct vaccine vehicles with multiple defects (Chatfield et al. 1992). The ability to create identical defects in a variety of host strains has allowed comparative studies to demonstrate differences in immunogenicity and virulence and thereby optimize selection of host strains for construction of a vaccine vehicle. The invasive ability of these attenuated strains has prompted numerous investigations of their use as a vehicle to carry heterologous antigens for stimulation of protective immunity against a number of bacterial, viral, and eukaryotic pathogens (Cardenas and Clements 1992) or for delivery of recombinant therapeutic proteins such as interleukin 1B (Carrier et al. 1992).

A number of strategies have been used to design mutations for vaccine constructs. One approach involves auxotrophic mutants that require metabolites not available in animal tissues. Double-deletion mutations in genes that are in separate locations on the chromosome and that independently attenuate virulence of the organism are usually used to provide an added measure of safety. Aromatic mutants, which have a complete block in the aromatic biosynthetic pathway and have a requirement for aromatic metabolites such as para-aminobenzoate and 2,3-dihydroxybenzoate (not present in mammalian tissues), are good examples of this kind of mutant. Oral vaccination of calves with a double (*aro*A, *aro*D) mutant of *S. typhimurium* was safe and elicited good protection against lethal doses of *Salmonella* (Jones et al. 1991). Other animal species that have been vaccinated orally with aromatic mutants of *S. typhimurium* and challenged with the homologous organism include mice, pigs, sheep, and chickens. An aromatic mutant of *S. dublin* has also been evaluated in mice and calves. In all cases, the vaccines were shown to have minimal undesirable effects (typically transient, mild diarrhea and fever in a small percentage of vaccinated animals) and to induce protection against challenge. Variations in degree of protection, shedding, and adverse reactions have been observed when different vaccine host strains or animal species were used.

Another approach for creation of live attenuated *Salmonella* involves the use of mutants that are attenuated by deletions in the genes for adenylate cyclase (*cya*) and for cAMP-receptor protein (*crp*). Cyclic AMP and cAMP-receptor protein regulate at least 200 genes, many of which are required for transport and breakdown of catabolites. Mutants with deletions in *cya* or *crp* are therefore unable to survive in animal tissues. *Salmonella* with deletion mutations in *cya* and *crp* have been shown to be safe and effective in eliciting protective immunity in mice, chickens, and pigs.

Given by the oral route, the aromatic mutants and the *cya*, *crp* mutants induce cell-mediated, systemic humoral, and mucosal immunity. Both types of mutants have been shown to reduce fecal shedding of challenge *Salmonella* organisms and subsequent development of the carrier state. Protection induced by these vaccines has been evaluated most extensively with the homologous organism as the challenge organism, but a few studies have demonstrated no significant cross-serovar protection.

DRUG RESISTANCE

Multiple-drug resistance is common in *Salmonella* and there is evidence that the use of antimicrobial drugs in animal feeds has contributed to selection for drug-resistant *Salmonella*. The patterns of drug resistance of *Salmonella* isolates from animals vary with the patterns of drug use in animal production. Consequently, there are variations in the predominant patterns of drug resistance of *Salmonella* from different countries, from different animal species, and from different farms. Many *Salmonella* isolates are resistant to streptomycin, tetracycline, and sulfonamides; a moderately high percentage are resistant to ampicillin, kanamycin, neomycin, and chloramphenicol; and only a low percentage are resistant to the fluoroquinolones, third-generation cephalosporins, gentamicin, or ampicillin or ticarcillin combined with a beta-lactamase inhibitor. Some serovars have remained relatively drug susceptible, but *S. typhimurium* is commonly resistant to several drugs and the resistance is often plasmid-encoded.

TREATMENT AND CONTROL

Treatment with an antibacterial drug to which the organism is susceptible is valuable in dealing with individual cases and outbreaks of salmonellosis. Such treatment is effective in reducing mortality in outbreaks and in reducing the contamination of the environment by organisms excreted during and after disease. In humans, salmonellosis frequently takes the form of mild food poisoning, which resolves without treatment; septicemic forms of the disease are treated vigorously with appropriate antibacterial drugs. Administration of antibacterial drugs to affected animals must be designed to destroy organisms in the intestine as well as in tissues. The intracellular location of the bacteria is a formidable barrier to antibacterial drugs, and there is dependence on host defense to destroy or contain such organisms. Fluids and electrolytes constitute an important part of treatment where there is dehydration. However, treatment failure is common and is associated with a rapid onset and progression of disease, massive intestinal tissue damage, and toxemia and septicemia.

Control is based on reducing the contamination of the environment by sick and carrier animals, finding and eliminating the source of infection, and minimizing the stresses placed on the animals. These objectives may require quarantine, test and slaughter programs, vaccines, biosecurity (making animal facilities rodent and bird proof), or irradiation of feeds.

REFERENCES

Aitken, M. M.; Brown, G. T. H.; Jones, P. W.; and Collins, P. 1983. *Salmonella saint-paul* infection in calves. J Hyg 91:259-65.

Ashkenazi, S.; Cleary, T. G.; Murray, B. E.; Wanger, A.; and Pickering, L. K. 1988. Quantitative analysis and partial characterization of cytotoxin production by *Salmonella* strains. Infect Immun 56:3089-94.

Benjamin, W. H.; Turnbough, C. L.; Posey, B. S.; and Briles, D. E. 1985. The ability of *Salmonella typhimurium* to produce the siderophore enterobactin is not a virulence factor in mouse typhoid. Infect Immun 50:392-97.

Caldwell, A. L., and Gulig, P. A. 1991. The *Salmonella typhimurium* virulence plasmid encodes a positive regulator of a plasmid-encoded virulence gene. J Bacteriol 173:7176-85.

Cardenas, L., and Clements, J. D. 1992. Oral immunization using live attenuated *Salmonella* spp. as carriers of foreign antigens. Clin Microbiol Rev 5:328-42.

Carrier, M. J.; Chatfield, S. N.; Dougan, G.; Nowicka, U. T.; O'Callaghan, D.; Beesley, J. E.; Milano, S.; Cillari, E.; and Liew, F. Y. 1992. Expression of human IL-1 beta in *Salmonella typhimurium*. A model system for the delivery of recombinant proteins in vivo. J Immunol 148:1176-81.

Chatfield, S.; Li, J. L.; Sydenham, M.; Douce, G.; and Dougan, G. 1992. *Salmonella* genetics and vaccine development. In Molecular Biology of Bacterial Infection: Current Status and Future Perspectives. Ed. C. E. Hoermache, C. W. Penn and C. J. Smyth, pp. 299-312. Cambridge: Cambridge University Press.

Clarke, G. J.; Qi, G. -M.; Wallis, T. S.; Starkey, W. G.; Collins, J.; Spencer, A. J.; Haddon, S. J.; Osborne, M. P.; Worton, K. J.; Candy, D. C. A.; and Stephen, J. 1988. Expression of an antigen in strains of *Salmonella typhimurium* which reacts with antibodies to cholera toxin. J Med Microbiol 25:139-46.

Clarke, R. C. 1985. Virulence of wild and mutant strains of *Salmonella typhimurium* in calves. PhD diss., University of Guelph, Ontario, Can.

Collins, F. M., and Carter, P. B. 1978. Growth of Salmonellae in orally infected germ-free mice. Infect Immun 21:41-47.

D'Aoust, J. -Y. 1989. *Salmonella*. In Foodborne Bacterial Pathogens. Ed. M. P. Doyle, pp. 327-445. New York: Marcel Dekker.

Fields, P. I.; Groisman, E. A.; Heffron, F. 1989. A *Salmonella* locus that controls resistance to microbicidal proteins from phagocytic cells. Science 243:1059-62.

Finkelstein, R. A.; Sciortino, C. V.; and McIntosh, M. A. 1983. Role of iron in microbe host interactions. Rev Infect Dis 5S:759-77.

Finlay, B. B., and Falkow, S. 1988. Virulence factors associated with *Salmonella* species. Microbiol Sci 11:324-28.

___________.1989. *Salmonella* as an intracellular parasite. Mol Microbiol 3:33-41.

Galan, J. E., and Curtiss III, R. 1990. Expression of *Salmonella typhimurium* genes is regulated by changes in DNA supercoiling. Infect Immun 58:1879-85.

Galan, J. E.; Ginocchio C.; and Costeas, P. 1992a. Molecular and functional characterization of the *Salmonella* invasion gene *inv*A: Homology of InvA to members of a new protein family. J Bacteriol 174:4338-49.

Galan, J. E.; Pace, J.; and Hayman, J. 1992b. Involvement of the epidermal growth factor receptor in the invasion of cultured mammalian cells by *Salmonella typhimurium*. Nature 357:588-89

Gianella, R. A. 1979. Importance of the intestinal inflammatory reaction in *Salmonella*-mediated intestinal secretion. Infect Immun 23:140-45.

Groisman, E. A.; Chiao, E.; Lipps, C. J.; and Heffron, F. 1989. *Salmonella typhimurium* phoP virulence gene is a transcriptional regulator. Proc Natl Acad Sci USA. 86:7077-81.

Gulig, P. A. 1990. Virulence plasmids of *Salmonella typhimurium* and other salmonellae. Microb Pathog 8:3-11

Gulig, P.A., and Curtiss III, R. 1987. Plasmid-associated virulence of *Salmonella typhimurium*. Infect Immun 55:2891-2901.

Habasha, F. 1981. Cell-mediated immunity in calves vaccinated with *Salmonella typhimurium*. PhD diss., University of California, Davis.

Heffernan, E. J.; Harwood, J.; Fierer, J.; and Guiney, D. 1992. The *Salmonella typhimurium* virulence plasmid complement resistance gene *rck* is homologous to a family of virulence-related outer membrane protein genes, including *pag*C and *ail*. J Bacteriol 174:84-91.

Helmuth, R.; Stephan, C.; Bunge, B.; Steinbeck, A.; and Bulling, E. 1985. Epidemiology of virulence-associated plasmids and outer membrane protein patterns within seven common *Salmonella* serotypes. Infect Immun 48:175-82.

Hsu, H. S. 1989. Pathogenesis and immunity in murine salmonellosis. Microbiol Rev 53:390-409.

Jones, G. W.; Robert, D. K.; Svinarich, D. M.; and Whitfield, H. J. 1982. Association of adhesive, invasive, and virulent phenotypes of *Salmonella typhimurium* with autonomous 60-megadalton plasmid. Infect Immun 38:476-86.

Jones, P. W.; Collins, P.; Brown, T. H.; and Aitken, M. M. 1983. *Salmonella saint-paul* infection in two dairy herds. J Hyg 91:243-57.

Jones, P. W.; Dougan, G.; Hayward, C.; Mackensie, N.; Collins, P.; and Chatfield, S. 1991. Oral vaccination of calves against experimental salmonellosis using a double *aro* mutant of *Salmonella typhimurium*. Vaccine 9:29-34.

Jones, B. D.; Lee, C. A.; and Falkow, S. 1992. Invasion of *Salmonella typhimurium* is affected by the direction of flagellar rotation. Infect Immun 60: 2475-80.

Koo, F. C.; Peterson, J. W.; Houston, C. W.; and Molina, N. C. 1984. Pathogenesis of experimental salmonellosis: Inhibition of protein synthesis by cytotoxin. Infect Immun 43:93-100.

Krause, M.; Fang, F. C.; and Guiney, D. G. 1992. Regulation of plasmid virulence gene expression in *Salmonella dublin* involves an unusual operon structure. J Bacteriol 174:4482-89.

Lee, C. A., and Falkow, S. 1990. The ability of *Salmonella* to enter mammalian cells is affected by bacterial growth state. Proc Natl Acad Sci USA 87:4304-8.

Lee, C. A.; Jones, B. D.; and Falkow, S. 1992. Identification of a *Salmonella typhimurium* locus by selection for hyperinvasive mutants. Proc Natl Acad Sci USA 89:1847-51.

LeMinor, L., and Popoff, M. Y. 1987. Designation of *Salmonella enterica* sp. nov. as the type and only species of the genus *Salmonella*. Int J Syst Bacteriol 37:465-68.

Liang-Takasaki, C.; Grossman, N.; and Lieve, L. 1983. Salmonellae activate complement via the alternate pathway depending on the structure of their lipopolysaccharide O-antigen. J Immunol 130:1867-70.

Libby, S. J.; Goebel, W.; Muir, S.; Songer, G.; and Heffron, F. 1990. Cloning and characterization of a cytotoxin gene from *Salmonella typhimurium*. Res Microbiol 141:775-83.

Lindberg, A. A., and Robertsson, J. A. 1983. *Salmonella typhimurium* infection in calves: Cell-mediated and humoral immune reactions before and after challenge with live virulent bacteria in calves given live or inactivated vaccines. Infect Immun 41:751-57.

Mekalanos, J. J. 1992. Environmental signals controlling expression of virulence determinants in bacteria. J Bacteriol 174:1-7.

Miller, S. I. 1991. PhoP/PhoQ: Macrophage-specific modulators of *Salmonella* virulence? Mol Microbiol 5:2073-78.

Nnalue, N. A., and Lindberg, A. A. 1990. *Salmonella choleraesuis* strains deficient in O antigen remain fully virulent for mice by parenteral inoculation but are avirulent by oral administration. Infect Immun 58:2493-501.

O'Brien, A. 1983. Genetic regulation and mechanisms of natural resistance to infectious diseases. In The Pathogenesis of Microbial Infection. Washington, D.C.: American Society for Microbiology.

Popoff, N. Y.; Miras, I.; Coynault, C.; Lasselin, C.; and Pardon, P. 1984. Molecular relationships between virulence plasmids of *Salmonella* serotypes *typhimurium* and *dublin* and large plasmids of other *Salmonella* serotypes. Ann Microbiol (Inst Pasteur) 135(A):389-98.

Portnoy, D. A., and Smith, G. A. 1992. Devious devices of *Salmonella*. Nature 357:536-37.

Prasad, R.; Chopra, A. K.; Peterson, J. W.; and Pericas, R. 1990. Biological and immunological characterization of a cloned cholera toxin-like enterotoxin from *Salmonella typhimurium*. Microb Pathog 9:315-29.

Prasad, R.; Chopra, A. K.; Charry, P.; and Peterson, J. W. 1992. Expression and characterization of the cloned *Salmonella typhimurium* enterotoxin. Microb Pathog 13:109-21.

Rahman, H.; Singh, V. B.; Sharma, V. D.; and Harne, S. D. 1992. *Salmonella* cytotonic and cytolytic factors, their detection in Chinese hamster ovary cells, and antigenic relatedness. Vet Microbiol 31:379-87.

Reed, W. M.; Olande, H. J.; and Thacker, L. H. 1986. Studies on the pathogenesis of *Salmonella typhimurium* and *Salmonella choleraesuis* var *kunzendorf* infection. Am J Vet Res 47:75-83.

Reeves, M. W.; Evins, G. M.; Heiba, A. A.; Plikaytis, B. D.; and Farmer, J. J. 1989. Clonal nature of *Salmonella typhi* and its genetic relatedness to other salmonellae as shown by multilocus enzyme electrophoresis, and proposal of *Salmonella bongori* comb. nov. J Clin Microbiol 27:313-20.

Reitmeyer, J. C.; Peterson, J. W.; and Wilson, K. J. 1986. *Salmonella* cytotoxin: A component of the bacterial outer membrane. Microb Pathog 1:503-10.

Robbins, J. B.; Chu, C.; and Sneerson, R. 1992. Hypothesis for vaccine development: Protective immunity to enteric diseases caused by nontyphoidal salmonellae and shigellae may be conferred by serum IgG antibodies to the O-specific polysaccharide of their lipopolysaccharides. Clin Infect Dis 15: 346-61.

Robertsson, J. A.; Svenson, S. B.; and Lindberg, A. A. 1982a. *Salmonella typhimurium* infection in calves: Delayed specific skin reactions directed against the O-antigenic polysaccharide chains. Infect Immun 37:737-48.

Robertsson, J. A.; Fossum, C.; Svenson, S. B.; and Lindberg, A. A. 1982b. *Salmonella typhimurium* infection in calves: Specific immune reactivity against O-antigenic polysaccharide detectable in vitro assays. Infect Immun 37:728-36.

Robertsson, J. A.; Lindberg, A. A.; Hoiseth, S.; and Stocker, B. A. D. 1983. *Salmonella typhimurium* infection in calves: Protection and survival of virulent challenge bacteria after immunization with live or inactivated vaccines. Infect Immun 41:742-50.

Royal, W. A.; Robinson, R. A.; and Duganzich, D. M. 1968. Colostral immunity against *Salmonella* infection in calves. N Z Vet J 16:141-45.

Saxen, H.; Reima, I.; and Makela, P. H. 1987. Alternative complement pathway activation by *Salmonella* O polysaccharide as a virulence determinant in the mouse. Microb Pathog 2:15-28.

Schafer, R., and and Eisenstein, T. K. 1992. Natural killer cells mediate protection induced by a *Salmonella aro*A mutant. Infect Immun 60:791-97.

Stone, B. J.; Garcia, C. M.; Badger, J. L.; Hasset, T.; Smith, R. I. F.; and Miller, V. L. 1992. Identification of novel loci affecting entry of *Salmonella enteritidis* into eucaryotic cells. J Bacteriol 174:3945-52.

Sukupolvi, S.; Riikonen, P.; Taira, S.; Saarilahti, H.; and Rhen, M. 1992. Plasmid-mediated serum resistance in *Salmonella enterica*. Microb Pathog 12:219-25.

Takeuchi, A. 1967. Electron microscope studies of experimental *Salmonella* infection. 1. Penetration into the intestinal epithelium by *Salmonella typhimurium*. Am J Pathol 50:109-19.

Takeuchi, A., and Sprinz, H. 1967. Electron microscope studies of experimental *Salmonella* infections in the preconditioned guinea pig. 2. Response of the intestinal mucosa to the invasion by *Salmonella typhimurium*. Am J Pathol 51:137-46.

Terakado, N.; Sekizaki, T.; Hashimoto, K.; and Naitoh, S. 1983. Correlation between the presence of a fifty megadalton plasmid in *Salmonella dublin* and virulence for mice. Infect Immun 41:443-44.

Turnbull, P. C. B. 1979. Food poisoning with special reference to *Salmonella*: Its epidemiology, pathogenesis, and control. In Clinics in Gastroenterology. Infections of the GI Tract. Ed. H. P. Lambert, pp. 663-714. Toronto: W. B. Saunders.

Wray, C. W., and Sojka, W. J. 1977. Reviews of the progress of dairy science: Bovine salmonellosis. J Dairy Sci 44:383-425.

14 / *Shigella*

BY C. L. GYLES

SHIGELLA are nonmotile, noncapsulated, lactose-negative members of the family *Enterobacteriaceae* and are an important cause of dysentery in primates (Rowe and Gross 1984). The organism is relatively fragile, and careful laboratory procedures are required to isolate it from fecal specimens, in which it is usually outnumbered by *Escherichia coli*. These procedures involve the use of selective plating media such as MacConkey agar and *Shigella-Salmonella* agar. On the basis of serologic and biochemical tests, the genus *Shigella* is divided into four species: *S. dysenteriae, S. flexneri, S. boydii,* and *S. sonnei*, which correspond to four subgroups, A, B, C, and D, respectively. There are 10, 6, and 15 serovars of *S. dysenteriae, S. flexneri, and S. boydii,* respectively, identified by arabic numerals and two serovars of *S. flexneri* that are identified as X and Y. Four of the serovars (1-4) of *S. flexneri* are divided into subserovars (la, lb; 2a, 2b; 3a, 3b, 3c; 4a, 4b). *S. sonnei* is not subdivided.

DISEASE

Shigella is a common and serious cause of dysentery in children and in monkeys. It is estimated that there are more than 100 million cases of shigellosis and over 600,000 deaths due to shigellosis in humans each year. *Shigella* dysentery may account for as much as 80% of all deaths among newly imported monkeys. Dogs may excrete the organism after being infected by humans, but they are resistant to disease and only occasionally are *Shigella* species associated with dysentery in puppies. Disease occurs because *Shigella* organisms invade the colonic epithelium and induce an inflammatory response that leads to formation of abscesses in the colon. Shigellosis in all species is characterized by frequent passage of low volumes of stools that contain blood, mucus, and inflammatory cells. Disease is often more severe with *S. dysenteriae* compared with other *Shigella* species.

One complication that may develop in children as a sequel to enteric infection with *S. dysenteriae* is the hemolytic uremic syndrome (HUS), characterized by renal failure, thrombocytopenia, and microangiopathic hemolytic anemia. Neurologic symptoms of lethargy, headache, and confusion are frequently observed. Shiga toxin appears to be responsible for HUS, but other toxins could also induce the syndrome. Recently, Ashkenazi et al. (1990) noted that Shiga toxin production was not necessary for development of neurologic symptoms and proposed that a 120-kDa cytotoxin, distinct from the Shiga toxin family of toxins, is sometimes responsible. A second complication of shigellosis is arthritis, seen particularly in individuals of the HLA-B27 histocompatibility group.

PATHOGENESIS

Studies of oral infection in humans and monkeys indicate the following sequence of events in shigellosis (Formal et al. 1984; Sansonetti 1991, 1992; Yoshikawa and Sasakawa 1992) (Fig. 14.1). Disease may be initiated by a dose as low as 10 bacterial cells; this low infective dose is related to the ability of the organisms to resist gastric acidity. The organisms multiply in the small intestine and reach concentrations of up to 10^9 organisms per ml of bowel contents; fever, abdominal discomfort, and watery diarrhea often are associated with this early phase of the disease. Within 2-3 days, the organisms leave the small intestine and establish in the colon. This latter phase is associated with a reduction in temperature, continued abdominal pain, and the frequent passage of small volumes of stool containing blood, mucus, and numerous leukocytes.

Studies of oral infection in monkeys given *S. flexneri* 2a show that some monkeys develop diarrhea, some develop dysentery, and others both diarrhea and dysentery. All affected animals show reduced colonic absorption or net colonic secretion, and this is the only transport defect in those animals with dysentery alone. Monkeys with diarrhea alone or diarrhea and dysentery show net jejunal secretion. There is severe colitis in all affected monkeys. When monkeys are infected with *S. flexneri* 2a by the intracecal route, a much higher percentage of inoculated monkeys develop disease and almost all develop dysentery without diarrhea. This is consistent with the notion that dysenteric disease is the essential feature of shigellosis and that the early diarrheal phase seen in some cases is due to fluid disturbances in the jejunum, associated with the presence of the bacteria in that part of the intestine.

The major factor in virulence of *Shigella* is their ability to penetrate and multiply within epithelial cells of the colon. Mutants that have lost the ability to invade the intestinal epithelium are nonpathogenic. Development of keratoconjunctivitis in guinea pigs (Sereny test) (Sereny 1957) and invasion of cultured epithelial cells are two tests that are useful in studying invasiveness of *Shigella*.

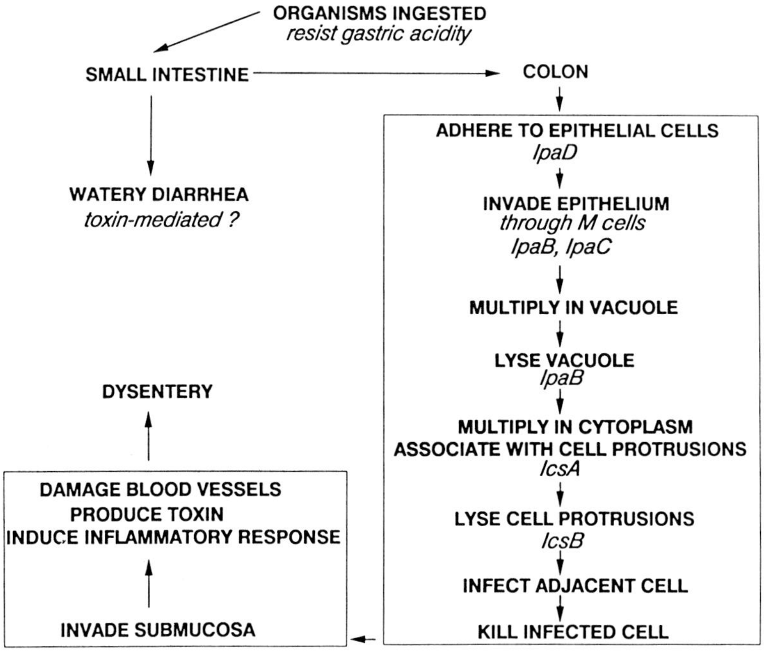

FIG. 14.1. Overview of pathogenesis of enteric disease due to *Shigella*.

Some stimulus from virulent *Shigella* induces intestinal epithelial cells to ingest the bacteria. Internalization occurs by a phagocytosislike process in the nonphagocytic epithelial cells (Fig. 14.2); it involves polymerization of actin and accumulation of myosin in a process that can be blocked by cytochalasin B. *Shigella* organisms are taken up in membranous vacuoles in epithelial cells (stages 1 and 2) but quickly lyse the vacuoles and escape (stage 3). The bacteria multiply in the cytoplasm of epithelial cells and spread laterally to adjacent cells in the epithelial lining (stage 4); abnormal and accelerated extrusion of damaged epithelial cells (stage 5) leads to formation of microabscesses in the colon. The cellular inflammatory reaction is confined to the mucosa and the submucosa is involved by edema only.

In cultured intestinal epithelial cells, the bacterium enters by endocytosis (Fig. 14.2), which requires energy production by the bacterium and the epithelial cell. Following association of the bacterium with cells, actin becomes polymerized into filaments in the region of the cell below the bacterium. It has long been assumed that the bacteria enter through the apical surface of colonic epithelial cells, but Mounier et al. (1992) provided evidence that *S. flexneri* enter cultured colonic Caco-2 cells by their basolateral poles. In vivo, *Shigella* appear to invade through M cells overlying lymphoid aggregates in the colon and spread to villus epithelial cells from these sites of invasion (Hale and Keren 1992).

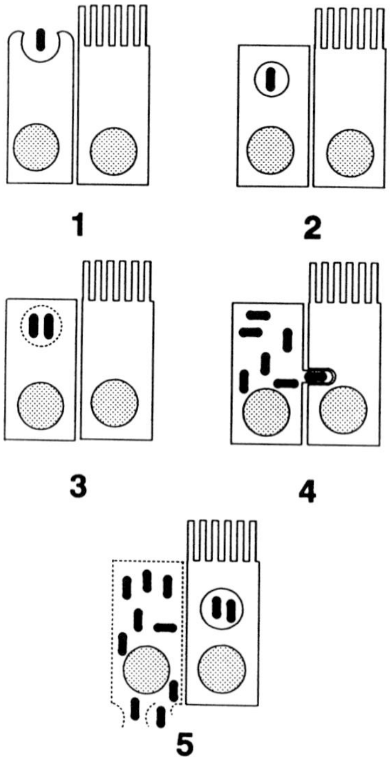

FIG. 14.2. Invasion of epithelial cells by *Shigella*. **1.** Bacterium is engulfed by an endocytic process. **2.** Internalized organism is enclosed in an endocytic vacuole. **3.** Bacterium multiplies and lyses the vacuole. **4.** Free organisms continue to multiply and move within the cytoplasm; organisms enclosed in protrusions of the cell membrane are passed on to adjacent cells. **5.** Infected cells are destroyed and *Shigella* organisms are released into the lamina propria where they are met by phagocytic cells; multiplication of bacteria occurs inside a vacuole in the newly infected cell and a new round of infection starts again.

Intracellular *Shigella* multiply in the vacuole, then escape from the vacuole into the intracellular compartment in which they continue to multiply at a rapid rate (Fig. 14.2). Initially, the bacteria express organellelike movement (Olm) along actin stress cables of the cell cytoskeleton (Vasselon et al. 1991) and subsequently become covered with filamentous actin, whose movement propels the bacteria so that they are transported in a rapid, random manner inside the cell (intracellular spread, Ics). Formation of long membrane-bound actin protrusions into adjacent cells allows bacteria to move from one cell to another without being exposed to the immune system (Vasselon et al. 1992) (Fig. 14.2).

Intracellular multiplication of the bacteria is accompanied by epithelial cell death attributed to blockage of respiration. The mechanism responsible for shutdown of host cell respiration and for cell death is not known, but mitochondria have been suggested to be a likely target.

Shiga toxin appears to increase the severity of dysentery by damaging small blood vessels of the large intestine, resulting in edema, hemorrhage, and an increased inflammatory response (Fontaine et al. 1988). Because lipopolysaccharide (LPS) is often absorbed from the damaged intestine and can also cause vascular damage it has been suggested that LPS may contribute to the systemic disturbances in shigellosis (Tesh and O'Brien 1991). Toxin may also contribute to fluid loss into the intestine (Keusch et al. 1991); it inhibits Na^+, Cl^- absorption and promotes premature expulsion of mature, absorptive villus epithelial cells in rabbit small intestine (Keenan et al. 1986; Kandel et al. 1989). Selectivity for villus rather than crypt cells is associated with a greater concentration of receptors in the villi. An alternative explanation for the diarrhea is that prostaglandins, synthesized as a part of the inflammatory response to invasion, may be responsible.

Shiga toxin is likely responsible for HUS, but the lack of suitable animal models has hampered attempts to resolve this issue. It is believed that glomerular endothelial cell injury or death causes decreased glomerular-filtration rate and impairment of kidney function. Hemolysis was earlier considered to be due to microangiopathy, but it may be due to release of von Willebrand multimers as a result of cell injury. Arthritis is sometimes seen as a sequel to shigellosis. It is likely due to cross-reaction between a protein produced by *Shigella* and a portion of the HLA-B27 protein expressed on lymphocytes.

VIRULENCE FACTORS

Invasion Factors

The genetic approach has yielded much of the recent information on invasion factors of *Shigella*. The ability of *Shigella* species to enter epithelial cells and cause disease depends on several unlinked chromosomal genes, as well as virulence genes on a large virulence plasmid. These genes encode functions such as attachment to host cells, induction of endocytosis, intracellular multiplication, and spread to adjacent cells. The process of invasion and intracellular multiplication leads to epithelial cell death, inflammation, and ulcerative lesions of the colon, resulting in the characteristic disease symptoms of severe dysentery and/or diarrhea.

Smooth LPS is required for virulence. Rough strains are unable to spread to neighboring cells and are probably more readily eliminated (Hale 1991). The chemical nature of the O antigen also affects virulence; thus, changes in the O antigen are sometimes associated with changes in virulence of smooth organisms. Changes in the LPS core can alter virulence, possibly because of the influence of the core on outer-membrane organization.

In *S. sonnei*, a surface antigen called form I antigen is readily lost on subculture of isolates to irreversibly generate form II cells, which are avirulent. The form I to form II change represents a smooth to rough transition, which is associated with the loss of a 120-MDa plasmid and of 2-amino-deoxy-L-altruonic acid from the LPS. The plasmid carries the genes for O antigen biosynthesis. A 140-MDa plasmid related to the form I plasmid of *S. sonnei* is also present in *S. flexneri* (and in enteroinvasive *E. coli*). Loss of the plasmid results in a loss of virulence and loss of ability to penetrate HeLa cells and to produce a positive Sereny test. In *S. flexneri* 2a, chromosomal DNA associated with *his* and *pro* markers encodes somatic antigens that are required for virulence. In *S. dysenteriae*, a 140-MDa plasmid, a 6-MDa plasmid, and chromosomal genes associated with the *his* locus are necessary for formation of LPS O antigen and for virulence (Haider et al. 1990).

The large *Shigella* virulence plasmid (120 or 140 MDa) is essential for penetration of mammalian epithelial cells, and for intracellular replication and spread after penetration. Genes for Congo red binding (*crb*) (= *pcr*, pigmentation of congo red) are carried on the virulence plasmid. Spontaneous mutants which are Crb⁻ are readily isolated from wild-type strains (mutants seem to overgrow the wild type on artificial media). Mutations are associated with loss or deletions of the plasmid, or, occasionally, with more subtle undefined changes. Congo red binding appears to be effected by a 101-kDa heme-binding protein on the bacterial surface, and hemin can substitute for Congo red in binding studies. The role of Crb in virulence is not clear, but there is evidence that binding to Congo red can induce expression of *Shigella* virulence genes (Maurelli et al. 1992). It has been suggested that Crb may bind to a host cell factor inside epithelial cells and that Congo red mimics this factor.

Five contiguous loci in a 30-kb region of the virulence plasmid and at least two other plasmid genes outside the cluster are involved in virulence (Fig. 14.3). Functions of the gene products are listed in Table 14.1.

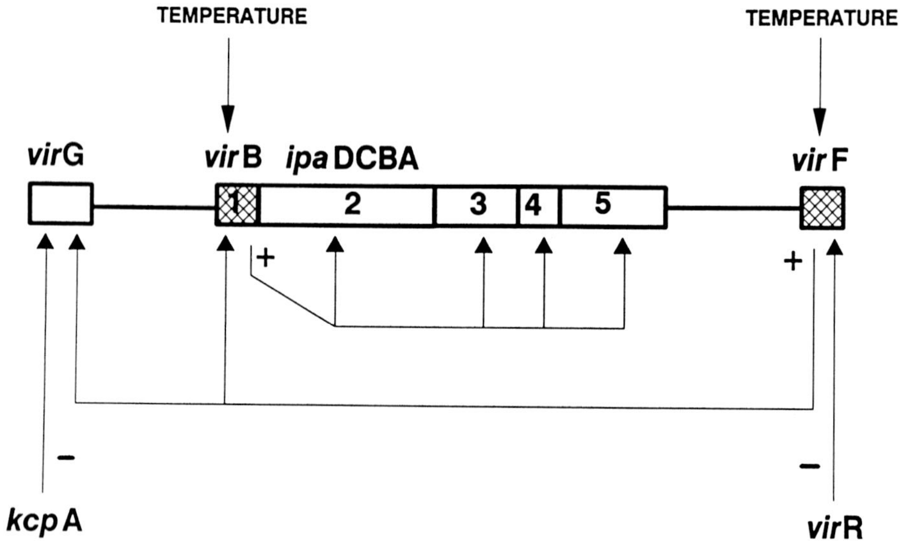

FIG. 14.3. Virulence genes on the large *Shigella* virulence plasmid. Five contiguous regions (1-5) and two additional loci (*vir*G and *vir*F) on the virulence plasmid contain virulence genes. *vir*F and *vir*B are regulatory genes: *vir*F encodes a positive activator (+) of transcription of *vir*G and of *vir*B, which in turn encodes a positive activator for the genes in regions 2-5. Both *vir*F and *vir*B are temperature regulated. Chromosomal loci *kcp*A and *vir*R encode negative regulators (-) of *vir*G and the genes of regions 1-5, respectively.

TABLE 14.1. Bacterial products implicated in virulence of *Shigella*

Bacterial Products	Activities
Products of plasmid genes	
Congo red binding (101-kDa protein)	Binds hemin; induces expression of virulence genes
Invasion protein antigens	Uptake of *Shigella* and lysis of vacuole
IpaA (78 kDa)	Nonessential
IpaB (62 kDa)	Invasin, cytolysin
IpaC (42 kDa)	Invasin
IpaD (38 kDa)	Adherence?
IcsA (VirG) (120 kDa)	Positive Sereny test; intracellular survival and multiplication
IcsB	Lysis of cellular protrusions
MxiA, MxiB	Export of invasin proteins
VirB	Positive regulator of invasion genes
VirF (30 kDa)	Positive regulator of plasmid virulence genes
Products of chromosomal genes	
Hydroxamate siderophore system	Promotes multiplication in tissues
KcpA	Positive Sereny test
LPS synthesis[a]	Required for invasion
OmpB	Modulation of expression of plasmid virulence genes?
Superoxide dismutase	Resistance to killing by polymorphonuclear leukocytes and macrophages
Shiga toxin (*S. dysenteriae* 1)	Vascular damage in the intestine; systemic effects?
VirR	Repressor of plasmid virulence genes

[a] Some or all of the genes may be on a plasmid.

Locus 1 is a positive regulatory gene, called *vir*B, in *S. flexneri* serotype 2a (*ipa*R in *S. flexneri* 5 and *inv*E in *S. sonnei*). A second positive regulatory gene, called *vir*F, regulates *vir*B and is responsible for induction of invasion genes in regions 2-5 (Hale 1991; Sasakawa et al. 1992). VirF is a 30-kDa protein that acts as a positive regulator of not only *vir*B but also of *vir*G, which is more suitably named *ics*A (intercellular spread). The expression or function of VirF is likely regulated by environmental signals, including temperature. A 120-kDa immunogenic surface protein is encoded by *ics*A and this protein is essential for intra- and intercellular spread. Mutants defective in *ics*A are unable to accumulate polymerized actin on their surfaces and to spread from one cell to another. IcsA appears to be inactivated by phosphorylation in host cells; alteration of the protein so that it is resistant to phosphorylation leads to increased ability of the bacterium to spread (d'Hauteville and Sansonetti 1992). Phosphorylation of this bacterial protein is proposed to be a host defense against invasion. Recently, *ics*B has been identified, upstream of *ipa*B. It appears that IcsB is responsible for lysis of the protrusions, which permit spread from cell to cell (Allaoui et al. 1992).

Locus 2 contains six genes organized as an operon. Four of these genes encode highly immunogenic Ipas (invasion protein antigens B, C, D, A) (Fig. 14.3). IpaB and IpaC are invasions and insertion mutations, which inactivate *ipa*B or *ipa*C, result in a loss of invasion but not of adherence. Ipa B is also a cytolysin responsible for lysis of the endocytic vacuole and release of bacteria into the cytoplasm (High et al. 1992). IpaB and IpaC are expressed at cell surface, but IpaB, IpaC, and IpaD do not have signal peptide sequences at their amino termini and they therefore require some special transport mechanism. IpaA is not essential for invasion and its role in pathogenesis has yet to be determined. IpaD has been suggested to play a role in adherence.

Multiple copies of genes for an antigen called IpaH have been identified on the virulence plasmid, and antibodies to this antigen have been detected in the serum of convalescent patients. Similarity of IpaH to YopM of *Yersinia pestis* suggests that it interferes with platelet-mediated events in the host (Sasakawa et al. 1992).

Genes of loci 3-5 are less well characterized than are those of loci 1 and 2, but they are known to include genes for adhesion and for transport and expression of the invasion protein antigens. Among these genes are the *mxi*A and *mxi*B (**m**embrane e**x**pression of **I**pa antigens) genes, which appear to be necessary for export of the invasin proteins (Andrews et al. 1991). Mutations in either locus results in failure of IpaB and IpaC to be expressed on the surface and in culture supernatant and in accumulation of IpaB and IpaC in the cytoplasm and inner membrane. Interestingly, MxiA has a high homology with the *Y. pestis* LcrD, which is involved in facilitating export of outer-membrane proteins of that organism. It has been suggested that MxiA either directly affects excretion of virulence factors or regulates expression of genes required for export of these factors (Andrews and Maurelli 1992).

Genes located in the chromosome are also required for penetration of epithelial cells by *Shigella*, and it is possible for strains to possess the plasmid and still be noninvasive. Recently, pulsed-field gel electrophoresis and Tn5 mutagenesis were used to identify nine virulence-associated loci on the chromosome of *S. flexneri* (Okada et al. 1991). Chromosomal virulence genes that have been identified in *S. flexneri* include *rfa* and *rfb*, which are required for O antigen production (near *his*); genes in the arginine-mannitol (*arg-mtl*) region that facilitate bacterial survival in the intestinal mucosa; genes necessary for a positive Sereny test and for fluid production in rabbit ileal loops; and the *kcp* locus (**k**erato**c**onjunctivitis **p**rovoking locus), near *pur*E. The *arg-mtl* region is associated with genes that code for production of the hydroxamate siderophore aerobactin and a 76-kDa, iron-regulated, outer-membrane protein. Aerobactin production is not required for intracellular growth, nor for the early killing of infected cells. However, it does provide a selective advantage and may operate at the stage of multiplication within tissues, when bacteria lie within the extracellular compartment of the intestinal villus. Chromosomal genes for iron-containing superoxide dismutase and the positive regulator of *omp*C and *omp*F promote survival of the organisms in the cytoplasm of infected cells.

Hale (1991) has categorized these chromosomal genes into three groups as follows. One group consists of virulence factors that have a direct effect on ability of the organism to survive in the intestine: they include genes for expression of the somatic antigen, aerobactin, and superoxide dismutase. A second group consists of genes that encode cytotoxins that are not essential for disease but increase the severity of disease. The genes for Shiga toxin (*stx*) fall into this group. The third group consists of regulatory genes, which determine expression of plasmid virulence genes: *omp*B, *vir*R, and *kcp*A.

Osmolarity affects expression of virulence genes. The transmembrane sensor of osmolarity, *omp*B, modulates expression of porin protein genes *omp*C and *omp*F through phosphorylation of OmpR. Loss of EnvZ or OmpR function results in reduction of invasiveness of *Shigella*; Hale (1991) has speculated that the *env*Z-*omp*R locus may regulate expression of virulence genes on the plasmid.

Expression of *Shigella* virulence is regulated by the temperature at which the organisms are grown. Strains that are phenotypically virulent when grown at 37°C are phenotypically avirulent when grown at 30°C. The chromosomal gene *vir*R, which is required for the temperature-dependent regulation of *Shigella* species, has been identified and cloned. However, temperature regulation of expression of virulence factors also occurs directly through temperature regulation of transcription of *vir*B. Tobe et al. (1991) have reported that, whereas transcription of *vir*B depended on both temperature and *vir*F, transcription of *ipa* and the other invasion operons depended on VirB but not on temperature. Based on these findings, they concluded that transcription of *vir*B is the main target for thermoregulation of the invasive phenotype in *Shigella*.

As with other pathogens, different proteins are expressed when the organism is in an intracellular, compared with an extracellular, environment. Headley and Payne (1990) reported a high expression of a 140-kDa protein (believed to be VirG) and an 80-kDa protein (believed to be IpaA) during intracellular growth of *S. flexneri* compared with earlier stages of infection of Hela cells.

Shiga Toxin

High levels of a protein cytotoxin (Shiga toxin) are produced by *S. dysenteriae* type 1 (Shiga bacillus) and released into culture supernatants. The toxin is the prototype of the Shiga and Shiga-like family of toxins (Chapter 15). Much lower levels of cell-bound cytotoxin are produced by other *Shigella*. The toxin is cytotoxic, enterotoxic, and lethal and inhibits protein synthesis at the level of EF-2-dependent binding of amino-acyl-tRNA to ribosomes.

Inhibition of eukaryotic protein synthesis can lead to cell death. Cytotoxicity has been demonstrated at the subfemtogram level per mammalian cell. Toxin binds to the glycolipid receptor, globotriosyl ceramide, and is internalized by receptor-mediated endocytosis. The toxin-containing vesicle probably fuses with lysosomes and is transported to the Golgi apparatus before release into the cytosol. Shiga toxin is a virulence factor, and inactivation of the toxin gene results in a reduction in severity of experimental disease in monkeys (Fontaine et al. 1988).

IMMUNITY

Local immunity in the large intestine is critical to resistance to shigellosis. Antibodies to LPS appear to be ineffective in inhibition of uptake of *Shigella* by intestinal epithelial cells, but a combination of antibodies against the O antigen on the surface plus cell-mediated immunity against infected host cells appears to be effective (Hale and Keren 1992). Antibodies to the surface proteins IpaA, B, C, D, H, and VirG are found in the serum of patients who have recovered from shigellosis, but the role of these antibodies in protection has not been investigated.

Parenterally administered vaccines have not been effective, but orally administered live attenuated vaccines have conferred type-specific protection. The presence of type-specific somatic antigens and the ability to penetrate the intestinal epithelium are associated with ability of organisms to induce protection against shigellosis. Transfer of the 140-MDa plasmid from *S. flexneri* 2a to *E. coli* K-12 allows the *E. coli* organism to invade HeLa cells. The addition of chromosomal genes that encode the O antigen of *S. flexneri* results in a transconjugant, which penetrates rabbit intestinal epithelium and produces mild inflammation in the lamina propria. This *E. coli-Shigella* hybrid confers protection on rhesus monkeys orally challenged with virulent *S. flexneri 2a* (Formal et al. 1984), but there are problems of diarrhea in a percentage of humans who receive this vaccine.

Certain live attenuated cultures have been reported to induce immunity that extended protection beyond the homologous serotype. *S. flexneri* serotype Y, attenuated by inactivation of *aro*D, induced protection against challenge with *S. flexneri* of serotypes Y, 2a, and 1b in monkeys that had been vaccinated orally (Lindberg et al. 1990).

REFERENCES

Allaoui, A.; Mounier, J.; Prevost, M. C.; Sansonetti, P. J.; and Parsot, C. 1992. *ics*B: A *Shigella flexneri* virulence gene necessary for the lysis of protrusions during intercellular spread. Mol Microbiol 6:1605-16.

Andrews, G. P., and Maurelli, A. T. 1992. *mxi*A of *Shigella flexneri* 2a, which facilitates export of invasion plasmid antigens, encodes a homolog of the low-calcium-response protein, LcrD, of *Yersinia pestis*. Infect Immun 60:3287-95.

Andrews, G. P.; Hromockyj, A. E.; Coker, C.; and Maurelli, A. T. 1991. Two novel viru- lence loci, *mxi*A and *mxi*B, in *Shigella flexneri* 2a facilitate excretion of invasion plasmid antigens. Infect Immun 59:1997-2005.

Ashkenazi, S.; Cleary, K. R.; Pickering, L. K.; Murray, B. E.; and Cleary, T. G. 1990. The association of Shiga toxin and other cytotoxins with the neurologic manifestations of shigellosis. J Infect Dis 161:961-65.

d'Hauteville, H., and Sansonetti, P. J. 1992. Phosphorylation of IcsA by cAMP-dependent protein kinase and its effect on intracellular spread of *Shigella flexneri*. Mol Microbiol 6:833-41.

Fontaine, A.; Arondel, J.; and Sansonetti, P. J. 1988. Role of Shiga toxin in the pathogenesis of shigellosis as studied using a Tox$^-$ mutant of *Shigella dysenteriae* 1. Infect Immun 56:3099-109.

Formal, S.; Hale, T. L.; Kapfer, C.; Cogan, J. P.; Snoy, P. J.; Chung, R.; Wingfield, M. E.; Elisberg, B. L.; and Baron, L. S. 1984. Oral vaccination of monkeys with an invasive *Escherichia coli* K-12 hybrid expressing *Shigella flexneri* 2a somatic antigen. Infect Immun 46:465-69.

Haider, K.; Azad, A. K.; Qadri, F.; Nahar, S.; and Ciznar, I. 1990. Role of plasmids in virulence-associated attributes and in O-antigen expression in *Shigella dysenteriae* type 1 strains. J Med Microbiol 33:1-9.

Hale, T. L. 1991. Genetic basis of virulence in *Shigella* species. Microbiol Rev 55: 206-24.

Hale, T. L., and Keren, D. F. 1992. Pathogenesis and immunology in shigellosis: Applications for vaccine development. Curr Top Microbiol Immunol 180:117-37.

Headley, V. L., and Payne, S. M. 1990. Differential protein expression by *Shigella flexneri* in intracellular and extracellular environments. Proc Natl Acad Sci USA 87:4179-83.

High, N.; Mounier, J.; Prevost, M. C.; and Sansonetti, P. J. 1992. IpaB of *Shigella flexneri* causes entry into epithelial cells and escape from the phagocytic vacuole. Eur Mol Biol Organ J 11:1991-99.

Kandel, G.; Donohue-Rolfe, A.; Donowitz, M.; and Keusch, G. T. 1989. Pathogenesis of *Shigella* diarrhea. 16. Selective targeting of Shiga toxin to villus cells of rabbit jejunum explains the effect of the toxin on intestinal electrolyte transport. J Clin Invest 84:1509-17.

Keenan, K. P.; Sharpnack, D. D.; Collins, H.; Formal, S. B.; and O'Brien, A. D. 1986. Morphologic evaluation of the effects of Shiga toxin, and *E. coli* shiga-like toxin on the rabbit intestine. Am J Pathol 125:69-80.

Keusch, G.T.; Jacewicz, M.; Mobassaleh, M.; and Donohue-Rolfe, A. 1991. Shiga toxin: Intestinal cell receptors and pathophysiology of enterotoxic effects. Rev Infect Dis 13 (Suppl 4):S304-10.

Lindberg, A. A.; Karnell, A.; Pal, T.; Sweiha, H.; Hultenby, K.; and Stocker, B. A. D. 1990. Construction of an auxotrophic *Shigella flexneri* strain for use as a live vaccine. Microb Pathog 8:433-40.

Maurelli, A. T.; Hromockyj, A. E.; and Bernardini, M. L. 1992. Environmental regulation of *Shigella* virulence. Curr Top Microbiol Immunol. 180:95-116.

Mounier, J.; Vasselon, T.; Hellio, R.; Lesourd, M.; and Sansonetti, P. J. 1992. *Shigella flexneri* enters human colonic Caco-2 epithelial cells through the basolateral pole. Infect Immun 60:237-48.

Okada, N.; Sasakawa, C.; Tobe, T.; Talukder, K. A.; Komatsu, K.; and Yoshikawa, M. 1991. Construction of a physical map of the chromosome of *Shigella flexneri* 2a and the direct assignment of nine virulence-associated loci identified by Tn5 insertions. Mol Microbiol 5:2171-80.

Rowe, B., and Gross, R. J. 1984. Genus *Shigella*. In Bergey's Manual of Systematic Bacteriology, Vol. 1. Ed. N. R. Krieg and J. G. Holt. Baltimore: Williams and Wilkins.

Sansonetti, P. J. 1991. Genetic and molecular basis of epithelial cell invasion by *Shigella* species. Rev Infect Dis 13 (Suppl 4):S285-92.

__________. 1992. Molecular and cellular biology of *Shigella flexneri* invasiveness: From cell assay systems to shigellosis. Curr Top Microbiol Immunol 180:1-19.

Sasakawa, C.; Buysse, J. M.; and Watanabe, H. 1992. The large virulence plasmid of *Shigella*. Curr Top Microbiol Immunol 180:21-44.

Sereny, B. 1957. Experimental keratoconjunctivitis shigellosa. Acta Microbiol Acad Sci Hung 4:367-76.

Tesh, V. L., and O'Brien, A. D. 1991. The pathogenic mechanisms of Shiga toxin and the Shiga-like toxins. Mol Microbiol 5:1817-22.

Tobe, T.; Nagai, S.; Okada, N.; Adler, B.; Yoshikawa, M.; and Sasakawa, C. 1991. Temperature-regulated expression of invasion genes in *Shigella flexneri* is controlled through the transcriptional activation of the *vir*B gene on the large plasmid. Mol Microbiol 5:887-93.

Vasselon, T.; Mounier, J.; Prevost, M. C.; Hellio, R.; and Sansonetti, P. J. 1991. Stress fiber-based movement of *Shigella flexneri* within cells. Infect Immun 59:1723-32.

Vasselon, T.; Mounier, J.; Hellio, R.; and Sansonetti, P. J. 1992. Movement along actin filaments of the perijunctional area and de novo polymerization of cellular actin are required for *Shigella flexneri* colonization of epithelial Caco-2 cell monolayers. Infect Immun 60:1031-40.

Yoshikawa, M., and Sasakawa, C. 1992. Molecular pathogenesis of shigellosis: A review. Microbiol Immunol 35:809-24.

15 / *Escherichia coli*

BY C. L. GYLES

ESCHERICHIA COLI derives its name from its discoverer (Theobald Escherich) and its principal habitat, the colon. The organism becomes established in the intestine shortly after birth, when the sterile intestine of the fetus is seeded with bacteria derived from the mother and the environment. *E. coli* continues throughout adult life as the major facultatively anaerobic species of bacteria in the intestine and is usually the dominant isolate on aerobic culture of feces or intestinal contents. Most strains of *E. coli* are harmless saprophytes, but others are virulent pathogens that affect the intestine or extraintestinal sites. The major diseases caused by *E. coli* are enteric infections, septicemia, urinary tract infection, and mastitis (Table 15.1).

E. coli is a gram-negative, rod-shaped bacterium, which ferments lactose and produces characteristic colonies on differential media such as MacConkey's agar. Several schemes have been developed to characterize isolates of *E. coli* and to aid in the identification of pathogenic strains: serotyping, biotyping, phage-typing, electrophoretic-typing, colicin-typing, and testing for virulence factors.

Serotyping is the most widely used of these methods and an international scheme has been established on the basis of O (cell wall), K (capsular), and H (flagellar) antigens (Orskov 1984). Approximately 173 O, 80 K, and 56 H antigens have been identified and numbered in this scheme. Most capsulated *E. coli* produce a microcapsule, but some strains of O groups 8, 9, 20, 64, and 101 produce an abundant capsule, similar to that produced by *Klebsiella*. For most strains of *E. coli*, only the O and H antigens are determined. The H antigens are sometimes important markers of pathogenicity among strains of the same O group. For example, this distinction is seen with O group 157; strains of serotype O157:H7 produce a cytotoxin and are associated with hemorrhagic colitis in humans, whereas strains of serotype O157:NM (nonmotile) or O157 (with an H antigen other than 7) do not produce the cytotoxin but produce enterotoxins and are implicated in diarrheal diseases in pigs.

Biotype, based on the patterns of reaction in selected biochemical tests, can be related to serotype, but the absence of a universal biotyping scheme severely limits this approach; the method is not widely used for characterizing isolates of pathogenic *E. coli*. Phage-typing is carried out by determining the susceptibility of isolates to a standard set of bacteriophages.

TABLE 15.1. *Escherichia coli* in diseases of animals

Disease	Designation of *E. coli*	Virulence factors
Diarrhea (most species)	ETEC	Enterotoxins Pili
Diarrhea (pigs)	EPEC	Attachment/effacement Pili?
Edema disease (pigs)	VTEC	Verotoxin VTe Pili
Hemorrhagic colitis (calves)	VTEC	VT1, VT2 Attachment/effacement
Septicemia (most species)	Septicemic	Serum resistance Iron scavenging LPS ?
UTI (dogs and cats)	Uropathogenic	Pili Iron scavenging Alpha hemolysin ? LPS ?
Pyometra (cats)	None	?
Mastitis	Mastitic	Opportunist

Note: ETEC = enterotoxigenic *E. coli;* EPEC = enteropathogenic *E. coli;* VTEC = verotoxigenic *E. coli*; VT = verotoxin; LPS = lipopolysaccharide; UTI = urinary tract infection.

Electrophoretic-typing is based on genetic variations in loci that encode a selection of enzymes. Proteins in extracts from bacterial lysates are separated by horizontal starch gel electrophoresis and stained for specific enzymes. Based on different mobilities of variants of an enzyme, distinctive patterns can be identified and used to group and to relate isolates. This system has been very effective in determining diversity within a serotype and close relationships among isolates of different serotypes. Colicin-typing involves detecting the production of colicin(s) by an isolate, then determining the effect of the colicin(s) on a standard set of indicator strains of *E. coli*. Neither of these schemes is in common use for identifying pathogenic isolates of *E. coli*.

VIRULENCE FACTORS

There have been rapid advances in identification and characterization of specific properties responsible for virulence of *E. coli* in both enteric and extraintestinal diseases. These advances have contributed to our understanding of the disease processes and our ability to differentiate *E. coli* that are normal gut flora from *E. coli* that are involved in enteric disease. Two major classes of virulence factors have been identified for enterotoxigenic *E. coli* (ETEC): pili, which are involved in colonization of the intestine, and enterotoxins, which are responsible for derangements in fluid and electrolyte movement across the gut epithelium. Advances in molecular genetics now permit very fine distinctions to be made among types of *E. coli* that cause enteric diseases. Verotoxins or Shiga-like toxins have been incriminated in a number of *E. coli* diseases that have both enteric and systemic manifestations; several virulence-associated factors have been demonstrated for septicemic and urinary tract pathogens.

Several structures and products of *E. coli* have either a demonstrated or a potential role in virulence in the gut and other tissues. The structures include capsule, cell wall, and pili (fimbriae); the products include enterotoxins, cytotoxins, hemolysins, and aerobactin.

Capsule

Certain strains of *E. coli* that cause diarrhea in calves and a subset of strains that cause diarrhea in pigs produce abundant capsular polysaccharide that may aid in colonization of the intestine (Hadad and Gyles 1982). The polysaccharide capsule is produced in vivo and appears to be a virulence factor in these strains. Spontaneous acapsular mutants of these strains failed to colonize the intestine and to produce diarrhea in experimentally infected calves (Hadad and Gyles 1982). It is uncertain, however, whether the acapsular mutants were deficient in structures and/or products other than capsular polysaccharide. Studies on the ultrastructure of the capsulated *E. coli* in association with the intestine of calves suggest that the capsular material contributes to the formation of microcolonies attached to the intestinal epithelium (Hadad and Gyles 1982; Acres 1985).

Capsule may contribute to virulence of invasive strains by being antiphagocytic and by being poor antigens. Among *E. coli* strains that cause septicemia in humans, there is a high prevalence of strains with the Kl capsular polysaccharide. Furthermore, $K1^+$ strains are more virulent than Kl^- strains in animal models of infection. K1 is a polysialic acid capsule that inhibits the progression of the alternative pathway of complement activation by promoting binding of factor H to C3b deposited on the bacterial surface. Thus, the K1 capsular polysaccharide enhances bacterial resistance to complement-mediated killing in serum and to phagocytosis by polymorphonuclear leukocytes. The end result is increased bacterial invasiveness. Interestingly, serum resistance of $K1^+$ *E. coli* varies, depending on the O group of the organism; some O:K1 combinations are serum resistant, while others are serum sensitive. Because K1, K9, and K92 are related in structure to the carbohydrate component of the neural cell adhesion molecule found in neural cells and in kidney, the immune response is muted and there is little or no specific antibody produced in response to these capsules. Similarly, K4, K5, and K54 are structurally related to chondroitin, a precursor of heparin, and are poorly antigenic.

Cell Wall

Cell wall lipopolysaccharide (LPS) is implicated in virulence, particularly in invasive strains of *E. coli*. LPS is the major constituent of the outer leaflet of the outer membrane of gram-negative bacteria, including *E. coli*. LPS consists of a lipid A moiety, buried in the membrane; a core; and a side chain, which occupies the free end of the molecule. Most of the biological activities of the molecule are due to the lipid A component, but the side chain, which constitutes the O antigen, is also an important factor in interactions with the host. Both the length and the chemical composition of the O side chain influence reaction with host cells and defense systems. Thus, long O side chains contribute to virulence, because the membrane attack complex formed by activation of complement is deposited at too great a distance from the cytoplasmic membrane to be lytic. Side chains of certain chemical composition activate the alternative complement pathway at a slower rate than do others, resulting in a slow rate of opsonization and killing.

LPS stimulates macrophages and other cells to produce a wide range of cytokines and to cause profound changes in the host. Induction of tumor necrosis factor (TNF) and interleukin-1 (IL-1) by LPS can account for some of the alterations that are observed. Thus, fever and disseminated intravascular coagulation are often prominent among the symptomatology of septicemic disease due to *E. coli*.

Proteins of the outer membrane may also contribute to virulence. TraT is an outer-membrane lipoprotein specified by certain F-like plasmids; it confers resistance to the bactericidal action of serum. TraT has been purified and shown to inhibit the formation of the membrane-attack complex primarily at the C6 step.

Pili

Type 1 Pili. Pili (fimbriae) are slender, proteinaceous filaments that project from the surface of the bacteria and confer adhesive properties on the organisms (Gaastra and De Graaf 1982). Most *E. coli*, whether they are pathogenic or not, produce type 1 pili, which are relatively long (2 μm) and mediate adherence to many surfaces, including human type A and guinea pig red blood cells. The attachment mediated by type 1 pili is inhibited by D-mannose (mannose-sensitive). Type 1 pili do not appear to be a virulence factor in enteric disease, but they seem to contribute to virulence of urinary tract pathogens.

The adhesin molecule is found at the tip of the type 1 pilus and is distinct from the major pilin subunit, which makes up the shaft of the pilus. In contrast, the major pilin subunit appears to be the adhesin for pili that mediate colonization of the intestine. Pilus expression is a complex matter and typically 10-11 genes are required to encode pilus subunits, tip components, pilus anchor, adhesin, periplasmic chaperone, outer membrane-usher, and regulators of pilus synthesis and tip length (De Graaf 1990; Jones et al. 1992).

K88 Pili (F4). Several specific adhesins, whose attachment to red blood cells and/or epithelial cells is not impaired by D-mannose (mannose-resistant), have been shown to be critical virulence factors in enteric disease caused by *E. coli* (Gaastra and De Graaf 1982). The K88 pilus was the first of these to be discovered. This adhesin was assigned a K antigen number because it was first demonstrated as a prominent surface antigen that was neither O nor H, and was detected by methods used to identify K antigens. A new system of naming pili has been proposed; in this scheme K88, K99, and 987P pili are called F4, F5, and F6, respectively. Subsequent studies showed that K88 exists as fine pili, which confer mannose-resistant adhesive properties on the bacteria. K88 pili are produced at body temperature (37°C), but not at room temperature (18°C), by certain serotypes of *E. coli* that cause diarrhea in pigs. The O groups most frequently associated with K88 pili are 8, 45, 138, 141, 147, 149, and 157. Three antigenic types of K88 protein have been recognized: K88a,b; K88a,c; and K88a,d. The K88 pili are determined by genes on a plasmid that usually carries genes for raffinose fermentation.

Strains of porcine ETEC that possess the K88 pili are able to adhere efficiently to the intestinal mucosa of neonatal pigs, thereby attaining high population densities in the gut. The K88 pili promote colonization particularly in the anterior small intestine in young pigs. Host specificity and resistance with increasing age of pigs are characteristic features of attachment mediated by K88 pili. Specific receptors in the intestinal epithelial cells are required for binding K88 pili; pigs lacking these receptors are resistant to infection by $K88^+$ ETEC but are susceptible to infection by ETEC with other adhesins. The gene for the K88 receptor is dominant, and homozygous recessive animals are resistant (Gibbons et al. 1977). Glycoproteins with terminal N-acetyl-lactosamine or N-acetyl-galactosamine inhibit binding of $K88^+$ ETEC to brush borders of enterocytes.

K99 Pili (F5). K99 pili are similar in structure and function to K88 pili but are found on bovine, ovine, and porcine ETEC (Gaastra and De Graaf 1982). Typically, K99 pili occur on ETEC that are heavily capsulated strains of O serogroups 8, 9, 20, 64, and 101. As with K88, the production of K99 antigen is plasmid mediated and temperature dependent, and adherence occurs with intestinal epithelial cells of younger but not of older animals. Synthesis of K99 antigen is repressed by certain components (notably alanine and glucose) in complex media, and a minimal medium supplemented with cassamino acids (Minca medium) is usually used to detect K99. The quantity of K99 produced by strains of O group 101 is considerably greater than that produced by strains of other O serogroups. The main component of K99 is also the adhesin and binds to its receptor, the ganglioside glycolipid Neu5Gc-α(2-3)-Gal -ß(1-4)Glc -ß(1-1) ceramide.

987P Pili (F6). The 987P pili are plasmid-determined pili found on porcine and bovine ETEC of O groups 9, 20, 101, and 141, which are associated with diarrhea in neonatal but not older pigs. A cluster of eight genes adjacent to the gene for heat-stable enterotoxin (STa) has been reported to encode the proteins necessary for synthesis and expression of 987P pili (Schifferli et al. 1991). It is difficult to demonstrate 987P pili on strains grown in vitro, and it is sometimes necessary to use bacteria grown in the intestine of pigs for this purpose. Some strains of porcine ETEC express both 987P and K88 pili.

F41 Pili. In addition to K99 pili, F41 pili frequently occur on bovine and porcine ETEC strains of O serogroups 9 and 101. F41 pili also occur on bovine and porcine ETEC that lack K99, K88, and 987P pili; they appear to be responsible for colonization of the intestine of pigs, lambs, and calves. Synthesis of F41 antigen is subject to repression by alanine and by low temperature (18°C); Minca medium is used to promote antigen production. The F41 operon is highly related to the K88 operon in those genes that encode accessory proteins but distinct in genes for fimbrial structural protein (Korth et al. 1992). These pili, and certain others, appear to have similar systems for assembly of pili but differ with respect to the specificity of their adhesins.

Pili Associated with Postweaning Diarrhea and Edema Disease in Pigs. Until recently, most strains of *E. coli* of O serogroups 138, 139, and 141, associated with postweaning diarrhea (PWD) in pigs, lacked characterized pili that were implicated in colonization. Recently, Kennan and Monckton (1990) and Salajka et al. (1992) described novel pili that were produced by these *E. coli*. The pili reported by Kennan and Monckton were associated with ETEC of O serogroup 141. The pili described by Salajka and associates were called "8813" and were associated with ETEC of O serogroups 25, 108, 138, 141, and 147; these pili mediated colonization of the intestine of weaned pigs. Nagy et al. (1992) also reported that fine pili, distinct from previously characterized types of fimbriae, mediate adherence of strains of O141 and O157 *E. coli* recovered from PWD. These pili adhere better to the intestine of weaned pigs compared with newborn pigs (Nagy et al. 1992).

Bertschinger et al. (1990) described adhesive fimbriae "107" that are produced in vivo by an O139:K12:H1 edema disease *E. coli* isolate. Subsequently, the gene (*fed*A), which encodes the major subunit of F107, was cloned and the nucleotide sequence determined (Imberechts et al. 1992). The gene was present in 20 of 24 edema disease strains of *E. coli* and was found in association with the genes for the edema disease toxin. The fimbriae were shown to mediate adherence to isolated porcine intestinal villi. The pili associated with *E. coli* from PWD and from edema disease were produced at 37°C but not at 18°C. The relationships among the pili reported by these four laboratories are not yet clear; however, pili on ETEC implicated in PWD are morphologically similar but antigenically and functionally different from the F107 pili on edema disease strains.

Colonization Factor Antigens I and II. Colonization factors I and II (CFA/I and CFA/II) are the best characterized of a number of specific pilus adhesins that promote colonization by strains of human ETEC. These plasmid-mediated pilus structures, which are mannose-resistant hemagglutinins, are antigenically distinct from each other and from the pili that promote attachment to the intestine of various animal species.

Vir Pili. Vir pili have been identified on a small percentage of strains of *E. coli* that cause septicemia in calves and lambs. These strains possess the vir plasmid that codes for a toxin and a fimbrial surface antigen. The toxin is lethal for experimental animals, but there is no information on the significance of the fimbriae.

Curli. Some strains of *E. coli* recovered from mastitic milk of dairy cows produce fine curly fimbriae (curli), which bind to fibronectin. It has been suggested that these pili may play a role in adherence of the bacteria to epithelial cells in the mammary gland, but the available evidence does not support this. There were no significant differences in response to challenge with *E. coli* in quarters of cows immunized with a curli-producing strain and quarters of cows immunized with a non-curli-producing strain of *E. coli*: all quarters developed mastitis (Todhunter et al. 1990). Curli have also been associated with certain avian pathogenic isolates.

Enterotoxins

Two classes of *E. coli* enterotoxin have been described (Smith and Gyles 1970): heat-labile *E. coli* enterotoxin (LT) and heat-stable enterotoxin (ST).

Heat-labile Enterotoxins (LTs). Enterotoxigenic strains of *E. coli* have been characterized for human, bovine, and porcine isolates. Only human and porcine strains produce LT (hLT and pLT, respectively). There appears to be a single major type of pLT produced by the wide variety of strains of porcine ETEC that produce the toxin. Among human isolates of ETEC, most produce the same type of hLT, but a second LT has been discovered that shares the mechanism of action of LT, although it differs in antigenic and biologic characteristics. This has resulted in the renaming of the original LT as LTI and the second LT as LTII.

Several biologic activities demonstrated for LT are used as tests for the toxin. The activity most directly related to the critical effect of the toxin in diarrheal disease is the ability to induce hypersecretion of fluid into the lumen of the small intestine. This may be detected directly by observing accumulation of fluid in ligated segments of small intestine, or by the development of diarrhea in response to oral or intraintestinal dosing of experimental animals with LT. It may be detected indirectly by measurement of changes in flux across the intestinal mucosa. The toxin causes morphologic changes in certain cell cultures, notably Y1 mouse adrenal cells, Chinese hamster ovary (CHO) cells and Vero (African green monkey kidney) cells, and these cell lines have been used in assays for LT. The ability of the toxin to increase vascular permeability has lead to the development of a rabbit skin test. Central to the activities of LT is its ability to elevate levels of cyclic adenosine monophosphate (cAMP) in tissues; changes in tissue levels of cAMP after exposure to LT may be used to assay for the toxin.

The LT molecule is a large protein (approximately 88 kDa), made up of one A and five B subunits (Robertson 1988). The A subunit (30 kDa) consists of an A_1 fragment (21 kDa), which contains the active site, and an A_2 fragment, which links the A_1 fragment to the B subunits. The B subunits contain the site for binding to intestinal epithelium. A and B subunits are synthesized intracellularly and holotoxin is assembled after the subunits have been transported through the cell membrane. LT is remarkably similar to cholera toxin (CT) in both structure and function; data obtained with CT are transposed to LT.

The basis for action of LT is its stimulation of adenylate cyclase in intestinal epithelial cells (Field et al. 1989; Fishman 1990). The elevated levels of adenylate cyclase result in hypersecretion of electrolytes and water.

The mode of action of LT in the gut is the same as that of CT (Field et al. 1989; Fishman 1990). LT binds to the enterocyte by its nontoxic B subunit, which recognizes specific cell membrane receptors containing the oligosaccharide of the GM1 ganglioside. LT also binds to glycoprotein receptors on intestinal brush borders. The A_1 fragment is transported across the mucosal surface of the cell, permitting interaction with the adenylate cyclase system, located intracellularly in the basal membrane region of the cell (Fig. 15.1). The adenylate cyclase system, involved in the conversion of adenosine triphosphate (ATP) to cAMP by means of the enzyme adenylate cyclase, consists of at least three components: one responsible for the enzymic conversion of ATP to cAMP, another for the guanosine triphosphate (GTP)-dependent regulation of enzyme function, and a third acts as hormone receptor. Normally, the system is activated by binding a hormone to its receptor, which induces binding of GTP to an active site on the regulatory component. Adenylate cyclase is active when it is complexed with GTP and the regulatory protein, but is inactive when the GTP is converted to guanosine diphosphate (GDP) by GTPase. The A_1 fragment of LT is an adenosine diphosphate (ADP) ribosyltransferase that transfers ADP-ribose from nicotinamide adenine dinucleotide (NAD) to the regulatory protein. ADP-ribosylation of the regulatory protein causes inhibition of the GTPase-mediated "turn-off" of adenylate cyclase. This fixes the complex in the active or."turned-on" state and results in an elevated level of cAMP. The abnormally high level of adenylate cyclase causes an efflux of Na^+ and Cl^- ions, which osmotically remove water from the cell.

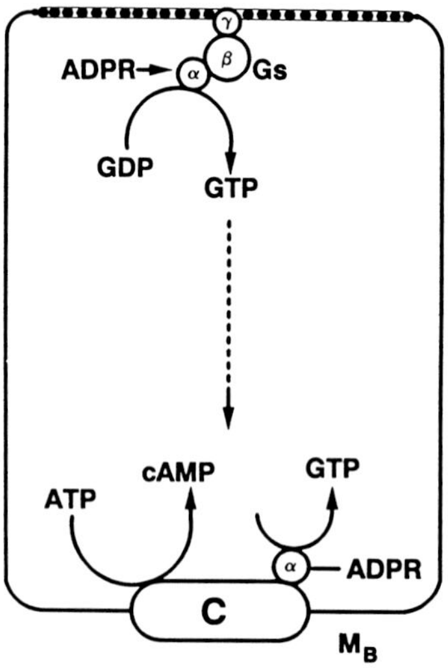

FIG. 15.1. Activation of adenylate cyclase by *Escherichia coli* heat-labile enterotoxin. **1.** Normally, when the GTP-binding protein Gs binds GTP, the α subunit separates and associates with adenylate cyclase (C), making it active. Bound GTP is subsequently hydrolyzed by GTPase activity carried by the α subunit, making the adenylate cyclase inactive. **2.** Linkage of ADP-ribose to the α subunit of Gs by the A_1 subunit of LT inhibits its GTPase activity, thereby preventing inactivation of the enzyme. Adenylate cyclase therefore remains in the turned-on mode.

Heat-stable Enterotoxins (STs). STs are heat-stable, low molecular weight toxins capable of inducing diarrhea (Gyles 1992). There are two major subsets of the ST group of toxins: STa or STI, and STb or STII.

STa is characterized by methanol solubility and an ability to induce accumulation of fluid in the intestine of suckling mice that have received the toxin orally or intragastrically. An infant-mouse assay is based on this property and is commonly used for demonstration of STa.

This form of ST may be produced by ETEC of human, bovine, or porcine origin; it is the only type of enterotoxin usually produced by bovine ETEC. STa causes a rapid elevation of cyclic guanosine monophosphate (cGMP) in intestinal epithelial cells, but this action is readily reversed when STa is removed. Measurement of the activation of guanylate cyclase in membranes from the intestinal mucosa of rats has recently been used as a rapid and sensitive assay for STa.

The gene for STa is part of a transposon and encodes a 72-amino acid product. Cleavage of an 18- or 19-amino acid signal peptide, followed by a 35-amino acid pro sequence leads to an 18- or 19-amino acid product. The terms STIa or STp are used to refer to the 18-amino acid STa peptide from *E. coli* of porcine, other animal, or human origin. STIb or STh refers to the 19-amino acid product from some strains of human origin.

STa from *E. coli* of human, bovine, and porcine origins has been purified and shown to exhibit slight molecular heterogeneity but remarkable similarity in chemical, biologic, and antigenic properties. The structure is characterized by the presence of six cysteine residues, which are involved in disulfide linkages; the active site is located in the C terminal 14-amino acid portion. STa has been produced synthetically as an 18-, 19-, or 14-amino acid peptide and the biologic and antigenic properties have been modified by rearrangement of the disulfide bridges. STa is resistant to low pH and to proteases.

The receptor for STa is reported to be protein or glycoprotein, and is responsible for specific binding of the toxin to enterocytes. There is evidence that guanylate cyclase is itself a receptor for STa (Schulz et al. 1991). The human receptor has been shown to be similar to receptors for natriuretic peptides, having an extracellular ligand-binding site and a cytoplasmic guanylate cyclase catalytic domain (De Sauvage et al. 1991). In pigs, the STa-binding capacity of intestinal brush border membranes varies with age and coincides with peaks in susceptibility of pigs to disease. Thus, binding capacity in pigs less than 1 week of age is approximately twice that in adults and in weaned pigs is almost three times that of unweaned pigs. Scatchard analysis has suggested that there is a single class of receptor; cross-linking with ^{125}I-STa has shown that STa binds to molecules of different sizes (Mezoff et al. 1991). There are reports of two classes of STa receptors: a low-affinity site, which has a dissociation constant (K_d) of approximately 10^{-9} M and a high-affinity site with a K_d of approximately 10^{-11} M. Binding to the high-affinity sites does not appear to be coupled to guanylate cyclase activation.

Villus cells of rat intestine contain approximately twice as many receptors as do crypt cells. In the rat, the concentration of toxin receptors is 3.5 times greater in colonocytes compared with ileocytes, and it has been suggested that impaired absorption in the colon combines with net secretion in the ileum to produce diarrhea in response to STa.

STa exhibits exquisite specificity with respect to its stimulation of guanylate cyclase; this activity is demonstrated only with particulate guanylate cyclase and only in intestinal cells. The effect is rapid, with no lag phase. The toxin is primarily antiabsorptive, inhibiting neutral Na^+ and Cl^- absorption, but it also stimulates secretion of Cl^-. The mechanism by which STa causes changes in absorption and secretion of ions is not known. There is evidence that thiol-disulfide exchange reactions are involved in the interaction of STa with its receptor (De Jonge et al. 1986; Robertson 1988). De Jonge et al. (1986) have suggested that cGMP activates an 86-kDa protein kinase in the microvillus border and this leads to phosphorylation of phosphatidyl-inositol, and subsequent formation of diacyl glycerol and inositol 1,4,5-triphosphate, and to activation of C-kinase. These three products result in mobilization of Ca^{++} from intracellular compartments, and Ca^{++} is responsible for impairment of absorption of Na^+ and Cl^- from the villi and for stimulation of secretion of Cl^- from the crypts.

STb is a heat-stable enterotoxin that is produced by certain strains of porcine ETEC and is sometimes found as the only type of enterotoxin produced by an isolate. It is often found in association with STa, LT, or STa and LT. This enterotoxin is methanol insoluble and is inactive in the infant-mouse assay. The ability to produce STb has recently been reported for two isolates of *E. coli* of human origin. STb is active in the intestine of mice, rats, calves, and rabbits, provided it is protected from proteases that are present in the intestine (Whipp 1991).

The gene for STb resides on plasmids and encodes a sequence of 71 amino acids, including a 23 amino acid leader peptide. STb has recently been purified and shown to be a 48 amino acid polypeptide with two disulfide bridges (Dubreuil et al. 1991; Hitotsubashi et al. 1992). Neutralizing antibodies were produced by conjugating STb to keyhole limpet hemocyanin and immunizing guinea pigs. STb caused an increase in the levels of prostaglandin E_2 (PGE_2) in intestinal mucosal cells and this has been proposed to account for the diarrheal effect. As supporting evidence, the authors noted that secretion was inhibited by aspirin and indomethacin, inhibitors of cyclooxygenase, and therefore of prostaglandin synthesis. The effect of STb is to stimulate secretion of bicarbonate into the lumen of the intestine.

The notion that stimulation of cyclic nucleotides accounts for the diarrhea associated with LT and with STa has not gone unchallenged. Some researchers have provided evidence that other mechanisms contribute to diarrhea (Peterson and Ochoa 1988; Stephen and Osborne 1988). It has been suggested that LT and CT have effects on enterochromaffin cells, involving hormones such as serotonin and vasoactive intestinal peptide, which can increase secretion. Also, there is an increased rate of loss of villus cells exposed to CT, and a mechanism of diarrhea similar to that for rotavirus infection, namely, loss of function and viability of absorptive cells, has been suggested to be in operation.

Verotoxins or Shiga-like Toxins

A heat-labile, toxic activity has long been associated with *E. coli* recovered from edema disease of pigs. In recent years, this edema disease toxin has been shown to be a member of the family of protein toxins called verotoxins (VT) or Shiga-like toxins (SLT). The name verotoxin derives from the fact that the toxins are lethal for Vero cells in culture; Shiga-like toxin indicates that the toxins are related in structure and function to Shiga toxin produced by *Shigella dysenteriae*. The toxin implicated in edema disease is called SLT-IIe (formerly SLT-IIv) or VTe. Verotoxigenic *E. coli* (VTEC) have also been implicated in disease in calves and in humans, and VTEC of bovine origin are an important source of VTEC for humans (Karmali 1989; Gyles 1992).

Several VTs have been purified and the genes that encode them have been cloned and sequenced. These toxins are all A-B subunit protein toxins, with an arrangement of one A and five B subunits and A_1 and A_2 fragments similar to that described for LT. VTs bind to globotriosyl- or globotetraosyl-ceramide in cell membranes and are toxic by virtue of an N-glycosidase activity on 28S ribosomal RNA. This action prevents the binding of amino-acyl-tRNA to ribosomes and thereby inhibits protein synthesis.

Verotoxins are implicated in edema disease of pigs, diarrhea and hemorrhagic colitis (HC) of calves, and HC and hemolytic uremic syndrome in humans. One feature common to all these diseases is that there is damage to vascular endothelium. Pathogenesis appears to involve colonization of the intestine by VTEC, production of VT in the intestine, absorption of VT from the intestine, and binding of VT to vascular endothelium in target organs. VT then causes damage to endothelium, leading to edema, hemorrhage, and thrombosis.

Verotoxins (and LTs) are toxins that are largely cell associated when the organisms are grown in vitro. A number of studies have shown that toxin synthesis and/or release may be enhanced by factors such as bile, trypsin, and antimicrobial agents that may be present in the intestine.

Hemolysins

Alpha-hemolysin is produced by certain serogroups of porcine *E. coli* that cause diarrhea or edema disease. The genes for alpha-hemolysin in these strains are carried on plasmids whose elimination has no effect on the ability of strains to produce diarrhea. Hemolysin production is also a common feature of human isolates of *E. coli* involved in extraintestinal infections. The genes are on the chromosome, but there is a high degree of sequence homology between the chromosomal and plasmid genes for alpha-hemolysin. It has been suggested that the role of alpha-hemolysin in extraintestinal infections is to increase the level of available iron in the host, but an alternative suggestion is that its toxicity for a variety of cell types contributes to the disease process.

Enterohemolysin is a weak hemolysin produced by some *E. coli* and demonstrable only on washed red blood cells. The genes for this hemolysin are carried on bacteriophages. A high percentage of VTEC of bovine or human origin produce enterohemolysin. A third hemolysin, called contact hemolysin, is produced by enteroinvasive *E. coli* that produce a *Shigella*-like dysentery in humans. The genes for the contact hemolysin are carried on the large virulence plasmid found in these strains (Haider et al. 1991); this hemolysin may play an important role in lysis of the endocytic vacuole to permit escape of the bacteria into the cytoplasm of colonic epithelial cells.

Cytotoxic Necrotizing Factors (CNF)

A protein toxin produced by a small percentage of highly virulent strains of *E. coli* O78:K80 that are septicemic for calves and lambs was originally called vir toxin, and shown to be lethal for mice, chickens, and calves (Smith 1974). In recent years, the vir toxin has been shown to be a member of a family of toxins called cytotoxic necrotizing factor (CNF) and has been renamed CNF2. CNF1 is the toxin originally associated with strains of *E. coli* recovered from diarrhea in infants and subsequently shown to be present in high frequency in isolates from bacteremia and urinary tract infections in humans. CNF (presumably CNF1) has also been associated with *E. coli* from diarrheal disease in pigs, calves, and humans (De Rycke et al. 1990). CNF2 is a 110-kDa protein that is immunologically related to the 115-kDa CNF1 protein. Both types of CNF induce multinucleation in HeLa cells and cause necrosis in rabbit skin. CNF1 has been reported to be associated with chromosomally encoded alpha-hemolysin in strains of *E. coli* recovered from dogs with diarrhea.

Iron-uptake System

Plasmids with genes for colicin V (Col V plasmids) have been associated with strains that cause bacteremia in calves, lambs, and humans. The critical contributor to virulence of these strains are plasmid genes that encode a specific high-affinity iron-uptake system. The mechanism allows the bacteria to multiply in an environment of limited concentration of free iron, which exists in tissues and fluids of the host. The system includes aerobactin (a hydroxamate siderophore) and an inducible outer-membrane protein, which acts as a receptor for ferric-aerobactin complex. In response to low levels of iron, the siderophore is synthesized and released. When it complexes with ferric iron, the complex binds to receptor and is internalized; the iron is subsequently removed so as to meet the needs of the bacterium for iron. Other iron- uptake systems are found in *E. coli*, but the aerobactin system is the one that is associated with invasive types of the organism.

PATHOGENESIS

Diseases Caused by Enterotoxigenic *E. coli*

The pathogenesis of diarrheal disease due to ETEC may be considered in two parts: colonization of the small intestine and production and action of enterotoxins. Colonization occurs when sufficient numbers of ETEC reach the small intestine, attach to the epithelial cells, and multiply to attain massive numbers. Whereas the number of *E. coli* per gram of intestine in the mid-jejunum in normal animals is of the order of 10^4, the number in animals with *E. coli* diarrhea is of the order of 10^9. Colonization pili play an essential role in attachment of ETEC to the surface of intestinal epithelial cells, and capsular polysaccharide may contribute to the formation of microcolonies by certain strains. Pili are involved in specific attachment of the bacteria to sites on the epithelial cells; capsular polysaccharide probably strengthens the bonds between organisms and between organisms and epithelium, thereby permitting formation of microcolonies. Attachment is only one aspect of colonization and little is known about factors that are involved in multiplication of ETEC in situ. The appearance of normal and colonized calf small intestine is shown in Figure 15.2.

FIG. 15.2. Scanning electron micrograph of colonization of the jejunum of the calf by enterotoxigenic *Escherichia coli*. **A.** Villi in the jejunum of a normal calf. **B.** Villi in the jejunum of a calf infected with enterotoxigenic *E. coli* (ETEC). **C.** Higher magnification, showing individual ETEC attached to enterocytes on the villi. (Hadad and Gyles 1982; reprinted with permission, American Journal of Veterinary Research.) (Fig. 15.2 B and C on facing page).

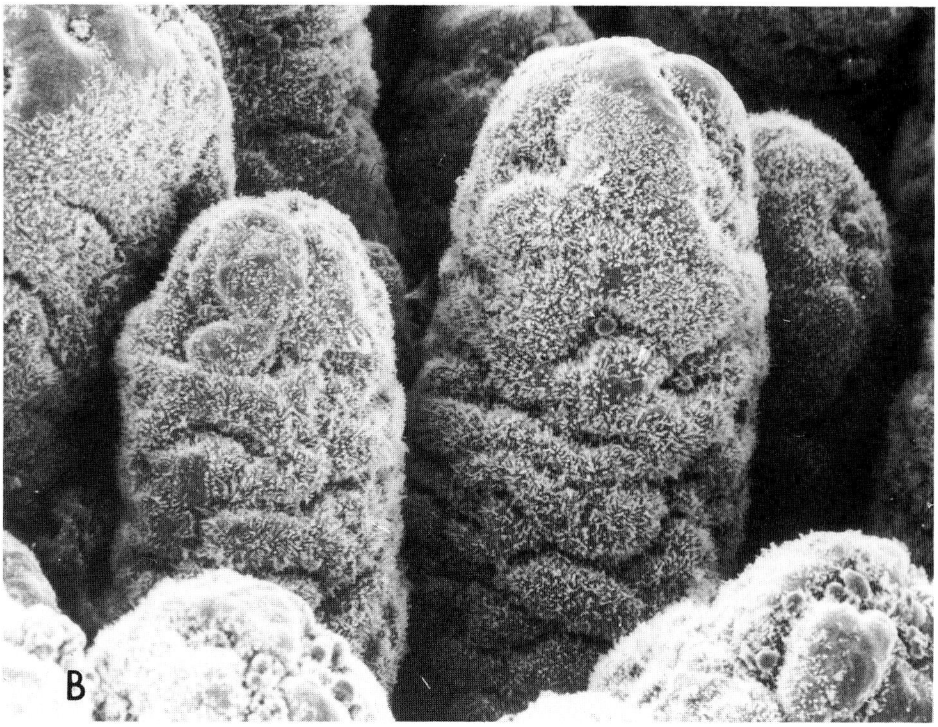
B

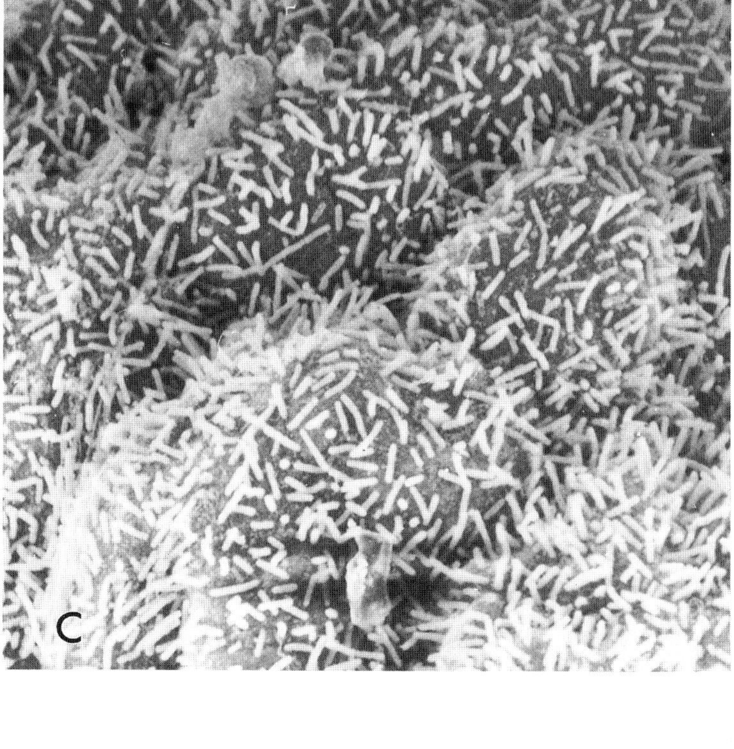
C

Host factors important in colonization include age, gastric pH, and the presence in the intestine of specific antibodies against surface antigens of the ETEC. High gastric pH, associated with young animals and with the ingestion of large volumes of milk or milk replacer, results in a loss of the normal bactericidal effect of the gastric environment. When antibodies against pilus adhesins or against the capsular polysaccharide are present in the intestine, the antibodies are effective in preventing disease or moderating the course of disease.

Diarrhea results from the action of enterotoxin(s) produced by ETEC in close apposition to the enterocytes. The principal effect of STa is antiabsorptive because it inhibits the coupled transport of Na^+ and Cl^- from the lumen into the intestinal epithelial cells; the net result is excessive secretion of an isotonic fluid. STa also impairs gut motility, a factor that could facilitate retention of ETEC in the gut lumen. There is little information on the physiologic disturbance caused by STb, but there is evidence that it stimulates electrogenic secretion of HCO_3^-. LT causes fluid loss by inhibiting the coupled absorption of Na^+ and Cl^- by villus epithelial cells and by stimulating active secretion of Cl^- by crypt cells. The mechanisms by which these enterotoxins act involve alterations in regulatory systems and do not result in structural lesions.

Although the essential features of pathogenesis are similar in most animal hosts, there are some species differences worth emphasizing.

Calves. Diarrhea due to ETEC occurs most commonly in calves less than 1 week of age but can be a problem in calves as old as 2-3 weeks (Acres 1985; Holland 1990). The remarkable association of susceptibility with age is such that experimental disease is readily produced in calves 1 day old or younger but not in older calves. Bovine ETEC belong to a limited number of OK serogroups and most strains are K99$^+$ isolates of O8:K25, O8:K85, O9:K30, O9:K35, O20:K?, O101:K28, or O101:K30. The lower small intestine is reported to be the first region to be colonized in unsuckled calves given ETEC orally; colonization spreads anteriorly.

ETEC are sometimes recovered from the feces of diarrheic calves, along with other agents that may cause diarrhea: rotavirus, coronavirus, BVD-virus, *Cryptosporidium*, *Salmonella*. Combinations of agents, such as ETEC and rotavirus, may cause diarrhea in older calves in which each agent by itself would not produce disease.

Mild lesions have been described in association with experimental infection of newborn calves with bovine ETEC. The lesions include stunting and fusion of villi and degeneration of microvilli (Hadad and Gyles 1982; Pearson and Logan 1983), but it is likely that other agents were responsible for these lesions.

Pigs. Although pigs may suffer from *E. coli* diarrhea from birth to 12 weeks of age, the first week of life and the first 2 weeks after weaning constitute two distinct peaks in occurrence of the disease. Several serologic types of ETEC are capable of causing disease; the dominant types vary over time, possibly in response to changes in natural and artificial protective immunity.

The first types of porcine ETEC that were recognized produced hemolytic nonmucoid colonies; had K88 or no recognized colonization pili; were ST$^+$, or ST$^+$LT$^+$; belonged to O serogroup 8, 45, 138, 141, 147, 149, or 157; and caused fluid accumulation in the ligated intestine of 1-week-old, as well as weaned, pigs. Later, porcine ETEC that bear remarkable resemblance to bovine ETEC were identified. These form nonhemolytic mucoid colonies; have K99, 987P, or F41 colonization pili; produce only STa; belong to O serogroup 8, 9, 20, 64, or 101; and cause fluid accumulation in ligated intestine in 1-week-old, but not in weaned, pigs (Soderlind et al. 1988; Harel et al. 1991). Many porcine strains of the first group produce both STa and STb and induce fluid accumulation in the intestine of both weaned and neonatal pigs.

However, the *E. coli* of the second group are associated almost exclusively with neonatal diarrhea.

Neonatal diarrhea caused by ETEC appears to be due to ingestion of ETEC from the surroundings of the farrowing pen, the absence of an effective gastric acid barrier and an established intestinal flora, the presence of receptors for colonization pili, and the high susceptibility to enterotoxins that are all features of the newborn pig. Sufficient levels of anti-pilus antibody will prevent disease and widespread use of vaccines seems to have reduced the prevalence of neonatal *E. coli* diarrhea in pigs.

Postweaning diarrhea due to *E. coli* may take the form of profuse diarrhea as seen in neonatal diarrhea. The pathogenesis is the same as for neonatal diarrhea except that the organisms that are involved are adapted to colonization of the intestine of older pigs. Some researchers have suggested that immune-mediated intestinal damage occurs in weaned pigs deprived of an adequate intake of solid food prior to weaning, and that a hypersensitivity to dietary antigen is a part of the clinical picture of postweaning diarrhea. Hemolytic *E. coli*, particularly of O serogroups 8, 138, 141, and 149, have been implicated. Villous atrophy develops in association with the dietary changes at weaning; the resulting malabsorption creates an intestinal environment favorable to overgrowth by hemolytic ETEC. Although hemolysin is not necessary for ETEC to induce PWD experimentally, it appears that hemolysin contributes to colonization of the intestine in weaned pigs. Another contributing factor to PWD is the loss of sow's milk with its antibacterial properties (transferrin, lactoferrin, antibodies, lectins that bind fimbrial adhesins).

A hemorrhagic gastroenteritis is sometimes seen in postweaning diarrhea due to *E. coli*. The basis for the structural lesions seen in this condition is not understood. There are reports that the disease sometimes develops in conjunction with rotavirus infection and some of the strains that cause this syndrome are known to produce a verotoxin; rotavirus may cause intestinal damage, and VT could lead to hemorrhage.

Danish workers noted that postweaning diarrhea usually occurred in herds in which there was no *E. coli* diarrhea in nursing pigs (Svendsen et al. 1977). They suggested that antibodies in the milk protected against *E. coli* diarrhea before weaning but prevented the pigs from mounting a sufficient immune response to provide protection following weaning. The fact that strains of a single OK group (O149:K91) constitute the overwhelming majority of porcine strains in Denmark undoubtedly contributed to the observation. Experiments were conducted to demonstrate that postweaning diarrhea could be readily produced following challenge of pigs from vaccinated sows. Svendsen et al. (1977) also demonstrated that the strains of O149:K91 recovered from weaned pigs differed from those recovered from suckling pigs; the former were usually $K88^-$ and colicinogenic, whereas the latter were typically $K88^+$ and less-frequently colicinogenic.

Other Animal Species. *E. coli* is responsible for diarrhea in the young of a wide variety of animal species, including cats, dogs, horses, camels, rabbits, and chickens. However, there is very little information on characteristics of ETEC from these species. Hemolytic *E. coli*, especially members of O group 4, have been noted in association with diarrhea in puppies. Recently, an LT^+STa^+ isolate of *E. coli* was identified as a likely cause of diarrhea in a 2-month-old puppy and LT^+STa^+ *E. coli* were recovered from the feces of 18-20% of dogs with diarrhea. Strains of hemolytic *E. coli* that produce only STa have also been reported from dogs with diarrhea, and *E. coli* recovered from the feces of two dogs with acute diarrhea produced an LT-like toxin.

E. coli, especially of O groups 2, 15, 26, 49, 92, 103, and 128, have been implicated as a major cause of diarrhea in rabbits. Interestingly, *E. coli* are either not recovered from fecal samples of most healthy young rabbits, or are present in very low numbers. The *E. coli* that cause diarrhea in rabbits are similar to human enteropathogenic *E. coli* (EPEC), in that they do not produce enterotoxins and are not enteroinvasive, but they attach to ileal and cecal epithelium and cause effacement of the microvilli (Okerman 1987). *E. coli* that were LT^+ or LT^+ and STa^+ have been implicated in outbreaks of diarrhea in broiler chicks in the Philippines.

Humans. ETEC cause diarrhea in children in developing countries and in travellers to these countries. The pathogenesis of disease in humans is as described for animals, the major difference being that the colonization pili of human ETEC are different from those of animal ETEC.

Diseases Caused by Verotoxigenic *E. coli*

Pigs. Edema disease (*E. coli* enterotoxemia) is a disease of weaned pigs due to VTEC and is characterized by sudden death; neurologic disturbances such as ataxia, convulsions, and paralysis; development of a peculiar squeal; and the occurrence of edema in several locations, notably the subcutis of the forehead and eyelids and the greater curvature of the stomach. VTEC that induce edema disease are usually hemolytic and belong to O serogroups 138, 139, and 141. The strains of O group 138 or 141 usually produce enterotoxin, but those of 139 may or may not be enterotoxigenic. VTEC are found in moderate numbers in normal, weaned pigs (Gannon et al. 1988) and colonize the intestine under special, poorly characterized conditions involving genetic susceptibility of the host, a high plane of nutrition, and stresses. Disease results from the effects of toxin (VTe), which is absorbed from the intestine and damages the vascular system (Bertschinger and Pohlenz 1983). The signs of disease are attributable to damage to the vasculature of the brain, particularly the cerebellum.

Edema disease may be produced experimentally by parenteral inoculation of weaned pigs with preparations of purified VTe (MacLeod et al. 1991). Genetic resistance to disease is associated with a single locus that controls susceptibility to colonization of the intestine. In experimental infection, *E. coli* are observed to adhere preferentially to the mid-jejunum and ileum; there are no attaching/effacing lesions (Bertschinger and Pohlenz 1983; Methiyapun et al. 1984). There is a lag of at least 2 days following oral inoculation before signs of disease develop. No changes are observed in enterocytes in association with colonization (Bertschinger and Pohlenz 1983), but vascular lesions consisting of swelling and vacuolation of endothelial cells, subendothelial deposition of fibrin, formation of microthrombi, and necrosis of the tunica media have been reported (Methiyapun et al. 1984)

Calves. Calves, like pigs, frequently carry VTEC in their intestine without showing any adverse effects; only a few O serogroups (5, 26, 111, 103) have been clearly associated with hemorrhagic colitis in calves (Schoonderwoerd et al. 1988; Janke et al. 1990) (Fig. 15.3). Several studies have shown an association between diarrhea in calves over 3 weeks of age and presence of VTEC in the feces.

Humans. In humans, VTEC are implicated in hemorrhagic colitis and hemolytic uremic syndrome (HUS). Those VTEC capable of causing a hemorrhagic colitis are called enterohemorrhagic *E. coli* (EHEC). Only certain VTEC, notably those of serotype O157:H7, are implicated in hemorrhagic colitis and HUS. Exposure of the bacteria, in vitro, to some antibiotics may result in marked increase in toxin that is released; it has been suggested that treatment with these drugs might increase the chances of development of HUS.

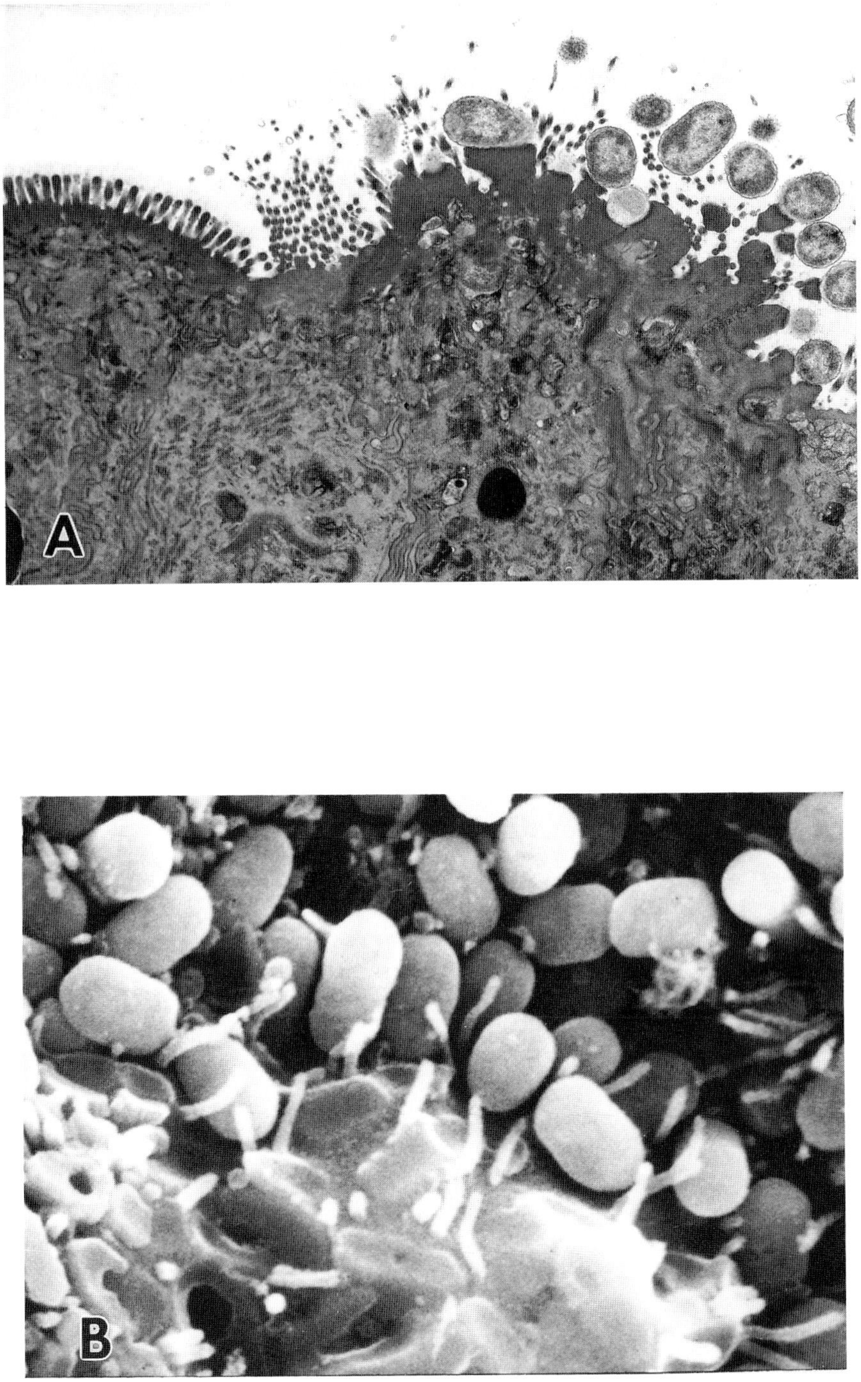

FIG. 15.3. Attaching/effacing lesions in the colon of a calf caused by a verotoxigenic *Escherichia coli*. **A.** Transmission electron micrograph showing close attachment of the bacteria to the epithelial cells and effacement of microvilli. **B.** Scanning electron micrograph showing cup-shaped depressions and formation of "pedestals." (Courtesy R. C. Clarke, Agriculture Canada Health of Animals Laboratory, Guelph, Ontario.)

Other Types of Enteric Disease

In humans, EPEC do not produce enterotoxins and are noninvasive but induce diarrhea in children. They adhere very closely to the microvilli of enterocytes and cause damage to the microvilli and are, therefore, called attaching and effacing (AE) *E. coli*, but the mechanism by which they cause diarrhea is not understood (Donnenberg and Kaper 1992). Initial attachment is mediated by bundles of pili, called bundle-forming pili, that also mediate attachment of organisms to one another (Giron et al. 1991). Genes for the pilus are carried on a 92-kb plasmid. Following attachment by the pili, the bacteria subsequently become closely adherent to the epithelial cell and cause signal transduction, involving stimulation of protein kinase activity in the epithelial cell. Other intracellular changes include accumulation of actin below the point of adherence of the bacteria and dramatic increases in Ca^{++} concentration. Changes in Ca^{++} concentration may be responsible for the fluid disturbances manifested as diarrhea. Recently, strains of *E. coli* of 045:K"E65" have been shown to behave like EPEC in pigs (Helie et al. 1991).

Enteroinvasive *E. coli* (EIEC) are another type of *E. coli* that cause enteric disease in humans. They invade the epithelium of the large bowel and cause a *Shigella*-like dysentery; they synthesize a Shiga-like cytotoxin, which has enterotoxic properties. No counterpart to EIEC has been reported from animals.

Septicemia

The *E. coli* that cause septicemia possess virulence attributes that are different from those of *E. coli* that cause enteric disease. Extensive studies have been conducted on the role of *E. coli* in septicemia in calves and in humans. Studies investigating the permeability to *E. coli* of the intestinal epithelium of young pigs, foals, and puppies have shown that the intestine of the young readily permits passage of *E. coli* from the lumen across the epithelium, and it is the ability or inability of the systemic immune system to eliminate these organisms that is critical.

Calves. Septicemic disease, called colisepticemia, occurs only in very young calves, usually in the first week of life. The bacteria gain entrance to the bloodstream through the nasopharyngeal area, the small intestine, or the umbilicus. Agammaglobulinemia or hypogammaglobulinemia is a prerequisite for disease, but a high percentage of calves may be susceptible because of a failure to receive sufficient colostrum early enough in life or because of a defect in absorption of gammaglobulin. Severity of the disease can be related to the degree of deficiency in serum gammaglobulins. The specific component of gammaglobulin that confers resistance to disease has not been identified, but it appears that specific antibodies against the *E. coli* are not required.

Most strains of septicemic *E. coli* belong to O group 15, 35, 78, 115, 117, or 137; strains of serogroup 078:K80 are frequently implicated. Septicemic strains have special properties enabling them to resist the defense mechanisms that destroy other *E. coli*, and to multiply in tissues. These bacterial factors include resistance to the bactericidal activity of normal serum and the ability to extract iron from serum and transport it into the cytoplasm. Many of the signs and symptoms observed in affected animals are attributable to endotoxin produced by the *E. coli*. Some calves die quickly of endotoxic shock with little evidence of illness; some cease drinking, develop diarrhea, show respiratory distress, and die; and others develop meningitis or localization of the organisms in the joints and/or kidneys.

Pigs. A septicemic disease due to *E. coli* occurs occasionally but has not been the subject of extensive studies. Strains of O groups 20 and 165 have been incriminated in a number of cases. Polyserositis is one clinical syndrome that may result from *E. coli* septicemia in pigs;

the syndrome is similar to that produced by *Haemophilus* species or *Mycoplasma* species.

Dogs. Septicemic colibacillosis in dogs usually takes the form of an acute or peracute septicemia, which is highly fatal to young puppies. Specific serotypes associated with the disease have not been identified.

Poultry. Colisepticemia of poultry is an important cause of morbidity and may result in 5-50% mortality. Strains of serogroups O1, O2 and O78:K80 are most frequently involved. Disease has been reported in chickens, turkeys, and ducks. Colisepticemia is believed to involve inhalation of pathogenic bacteria in feces-contaminated dust, followed by binding of the bacteria to the tracheal epithelium. Pili appear to be a virulence factor of strains of *E. coli* pathogenic for poultry; strains of O78 poultry pathogens have been reported to produce pili that have a subunit of molecular weight of 18 kDa and that bind to avian epithelial cells in vitro. Virulence-associated properties that have been reported include serum resistance, invasiveness for HeLa cells, production of colicin V, and possession of an aerobactin iron-uptake system. Production of hemolysin is not a characteristic of these organisms. Ability to bind the dye Congo red has also been associated with virulence, but this is disputed.

Urinary Tract Infections

E. coli is the most frequent cause of urinary tract infections (UTI) in dogs and cats (Oxenford et al. 1984). The organism is one of several bacterial species that constitute the resident flora of the lower genitourinary tract; development of significant bacteriuria (excretion of 10^5 or greater organisms per ml of voided urine) requires that the bacteria possess special virulence attributes and/or that there be impairment of the normal defense mechanisms. A major virulence factor of uropathogenic *E. coli* is the ability to adhere to epithelial cells of the urinary tract. Senior et al. (1992) reported that type 1 fimbriae were most frequently associated with canine uropathogenic *E. coli* and that these pili mediated adherence in vitro to canine uroepithelial cells. They suggested that type 1 pili were the most important pili found on canine uropathogenic *E. coli*. Garcia et al. (1988) observed that some canine uropathogenic *E. coli* possessed pili that are antigenically related to F12 and F13 fimbriae of human uropathogenic *E. coli*. The adhesin carried by the canine isolates was different from the P adhesin associated with human isolates; canine, but not human, isolates adhered in vitro to canine kidney epithelial cells (Gaastra et al. 1988). It is likely that the type 1 pili are important in attachment to bladder epithelium and in cystitis, while the F12- and F13-related fimbriae are important in attachment to kidney epithelial cells and in pyelonephritis.

As with *E. coli* from UTI in humans, isolates from UTI in dogs belong to a small cluster of serologic types. Organisms of O groups 2, 4, 6, and 83 are common in UTI in dogs; strains of O groups 4 and 6 are also common in human UTI. It appears, however, that canine uropathogenic *E. coli* are different from human strains. Whereas a high percentage of human UTI strains express the K1 antigen, strains of canine origin rarely express any defined K antigen (Senior et al. 1992). Alpha-hemolysin is a virulence-associated factor found on about 50% of canine UTI isolates (Wilson et al. 1988; Senior et al. 1992), but its role in disease has not been established.

Host defenses include normal micturition, anatomic structures (there is a higher prevalence of infections in females), mucosal defense barriers, and the antimicrobial properties of urine and prostatic secretions. Most infections develop as a result of bacteria ascending the urethra to the bladder.

The urine of dogs and cats is often bactericidal for *E. coli*, and there is some uncertainty about the appropriate numbers of *E. coli* in urine that should be considered to constitute significant bacteriuria. An *E. coli* count of 10^5 or greater in human urine is accepted as significant bacteriuria, and this number is generally used for dogs and cats. However, human urine supports the growth of *E. coli* and it is likely that significant bacteriuria may exist when a lower number of *E. coli* is present in the urine of dogs and cats. Despite this uncertainty, quantitative culture of urine samples is essential for the diagnosis of urinary tract infection. Osmolality may be a major factor in the bactericidal action of the urine of cats and dogs and it is noteworthy that cats, whose urine is usually of high osmolality, have a much lower frequency of bacterial UTI than do dogs.

Cystitis is the most common form of UTI, but urethritis, ureteritis, prostatitis, and pyelonephritis also occur in response to *E. coli* infection.

E. coli that lack specific virulence attributes may cause opportunistic infections of the urinary tract. There is always a predisposing factor, most commonly instrumentation, or some condition that prevents normal emptying of the bladder at regular intervals.

E. coli have also been implicated as an important agent of UTI in sows with bacteriuria, cystitis/pyelonephritis, or vaginal discharge. These *E. coli* are not well characterized but most seem to possess type 1 pili.

Pyometra

E. coli are the major cause of pyometra in dogs and cats; the infection is associated with progesterone-stimulated endometrium and myometrium. Strains of O groups 2, 4, 6, 42, 44, and 141 have been recovered from infected dogs, and strains of O groups 2 and 22 have been reported from cats.

Mastitis

Bovine mastitis due to *E. coli*, *Klebsiella*, or *Enterobacter* is referred to as coliform mastitis and may occur as a peracute, acute, chronic, or subclinical infection. Bacteria in the environment invade via the teat canal and establish a local infection that is confined to the udder. The peracute form of the disease occurs, commonly, shortly after parturition and is characterized by rapid onset, inflammation of the udder, fever, anorexia, and depression. This form of the disease is highly fatal. Occasionally, a gangrenous mastitis develops. Signs in the acute form of coliform mastitis are similar to those in the peracute form, but they are less severe and the animal is more likely to recover. Chronic and subclinical infections represent clinically milder forms of infection in which the organisms may be recovered from the gland but there is little or no response to their presence.

No markers have been identified that distinguish strains of *E. coli* that cause mastitis from strains in normal feces and the environment. *E. coli* recovered from the udders of cattle with mastitis belong to a wide range of O serogroups and exhibit biochemical properties indistinguishable from those of *E. coli* from the feces and environment (Nemeth 1992). Although capsule and serum resistance have been suggested to be virulence factors for mastitic *E. coli*, those *E. coli* that cause bovine mastitis appear to be simply opportunistic environmental organisms that have no special virulence factors. In a recent study, Nemeth (1992) demonstrated that six strains of *E. coli* from mastitic milk and six strains from bovine feces were all able to induce mastitis in cows in which a quarter was infused with approximately 200 colony-forming units of bacteria. There was no difference in severity of mastitis, but the quarters had been milked out at hourly intervals after the cow's temperature reached 41.5°C. Both sets of isolates reached similar concentrations in the milk but the mastitic milk isolates grew more rapidly and reached peak concentration earlier. It is possible that previous growth in milk conditioned and selected *E. coli* for adaptation to the mammary gland.

The antibacterial properties of milk, the unavailability of free iron in milk, and the levels of specific opsonins in milk undoubtedly function in defense against infection in the mammary gland. Polymorphonuclear leukocytes constitute the major antibacterial defense against bacteria invading the mammary gland and effectively phagocytose and destroy bacteria opsonized by IgM. Encapsulated *E. coli* are not as readily opsonized for phagocytosis as are noncapsulated ones unless the animal has antibodies against the polysaccharide capsule. Serum resistance is a common feature of mastitic *E. coli*, but it is also a common property of bovine fecal isolates (Nemeth 1992). Lactoferrin inhibits multiplication of *E. coli* by binding iron and rendering it unavailable to the bacteria; its effect is bacteriostatic. However, lactoferrin is probably ineffective in the lactating gland because of the presence of a high concentration of citrate. Natural antibody against the basal core of the LPS fraction of *E. coli* may contribute to effectiveness in opsonization of mastitic *E. coli*.

Bacterial endotoxin plays a major role in inducing the inflammatory response in the mammary gland as well as the profound toxemia, which characterizes the acute form of the disease. A second toxin, which is heat labile and cytotoxic, has been implicated in necrosis of the superficial layer of the teat and lactiferous sinuses (Frost and Hill 1982).

Inoculation of the mammary gland with *E. coli* endotoxin results in a disease that is similar to the natural disease produced in response to *E. coli* organisms. However, the epithelial cell necrosis seen in the natural disease does not develop. LPS is detoxified by acyloxyacyl hydrolase present in neutrophils, but little is known about variations in levels of this enzyme in milk neutrophils (McDermott et al. 1991). Both natural and experimental disease demonstrate that the mammary gland in the dry state or late in lactation responds differently to both endotoxin and cytotoxin, compared with the gland early in lactation. It appears that there is a gradual decline in susceptibility of the duct epithelial cells to these toxins throughout lactation so that they are least susceptible in the dry period.

Mastitis control programs that are effective in reducing mastitis due to staphylococci and streptococci are generally ineffective against *E. coli* mastitis. This may be due to the organisms reaching the teats at times other than at milking; the spread of infection is primarily from environment to teat rather than from one teat to another.

E. coli also causes mastitis in animal species other than cattle. It is of particular importance in swine, in which it is the most frequent cause of mastitis and agalactia; the effects of the disease are more severe on the newborn piglets than on the sow. The essential features of pathogenesis of the disease in pigs are similar to those in cattle.

LABORATORY DIAGNOSIS

Routine culture and identification of *E. coli* from relevant samples are all that are required for the laboratory diagnosis of septicemia and infection of the reproductive tract due to *E. coli*; quantitative culture is necessary in the case of UTI. For enteric diseases, it is necessary to use methods that distinguish the pathogenic from the nonpathogenic *E. coli*. Three approaches have been taken. The first involves initial culture, followed by methods to determine surface antigens or the production of enterotoxins. Slide agglutination tests with appropriate antisera are often used to determine O, K, and pilus antigens; production of enterotoxins may be detected by biologic tests or by immunologic methods, such as enzyme-linked immunosorbent assay (ELISA), radioimmunoassay (RIA), coagglutination, and the Biken test (a modified Elek test for LT). A second approach, applicable only when well-preserved intestine is available, is to demonstrate adherent *E. coli* attached to the intestinal epithelium. A third method involves demonstration that the isolate contains DNA sequences found in *E. coli* genes that code for enterotoxins. The development of polymerase chain-reaction amplification procedures involving oligonucleotide primers that can readily be purchased or synthesized has made it easier to detect *E. coli* with virulence factors (Woodward et al. 1992).

DRUG RESISTANCE

Drug resistance in *E. coli* from the intestinal tract of animals reflects their exposure to antibacterial drugs that select resistant clones. Thus there is marked variation in resistance, which can be related to the antibacterial drugs used in animals in an area. Initial choice of an antibacterial drug must therefore be based on knowledge of the usual patterns of resistance of *E. coli* isolates from a particular animal host in the area, and it is important to conduct antimicrobial susceptibility tests on each isolate. A high percentage of strains will likely be susceptible to fluoroquinolones, third generation cephalosporins, carbadox, gentamicin, trimethoprim-sulfonamide, and nitrofurans; isolates are likely to be resistant to streptomycin, sulfonamides, and tetracyclines; and a moderate percentage of strains are likely to be susceptible to chloramphenicol, ampicillin, neomycin and kanamycin. Ampicillin/clavulinic acid, ampicillin/sulbactam, and ticarcillin/clavulanic acid are also effective against a high percentage of isolates. The addition of a suicide inhibitor of beta-lactamase protects the beta-lactam antibiotic from some but not all beta-lactamases.

VACCINES

Several types of vaccines have been used with success against *E. coli* diarrhea in neonatal pigs and calves. In pigs, vaccines are designed for administration to the sow, which provides passive protection to her nursing offspring. The vaccines are based on surface antigens of ETEC and/or toxoids. The major types of vaccines administered parenterally to pregnant sows are bacterins that consist of one or more serogroups of ETEC known to be prevalent in an area; pilus vaccines that typically consist of four pilus types (K88, K99, 987P, and F41); combinations of bacterins and pili; and toxoids of LT or cholera toxin, with or without pili. It is common to use an adjuvant with vaccines that are administered parenterally; but one recent report has suggested that incorporation of pilus vaccines into an immunostimulating complex (ISCOM) may be more effective. Two types of vaccine have been administered orally to pigs. Live ETEC given orally, late in gestation, have proven effective in inducing specific protection of nursing pigs. In one protocol, heat-killed, selected serotypes of ETEC are administered to pregnant sows orally and by parenteral injection 18-25 days before farrowing and are provided to nursing pigs by means of creep feed. This protocol has been reported to be effective.

Limitations of the preceding vaccines, except the last one, are that they depend solely on the transfer of passive immunity from the dam and the piglets are susceptible to postweaning *E. coli* diarrhea. A number of vaccines designed to protect weaned pigs have been tested in pigs of 2-4 weeks of age. Formalin-killed, whole cell bacterins, administered parenterally or orally, and killed *E. coli* with K88, K99, 987P, and F41 pili, administered in the feed, have been reported to induce protection in weaned pigs. A toxoid of VTe was effective in stimulating protective antibodies against the effects of VTe of edema disease; a genetically altered VTe was also reported to induce protection against lesions resulting from challenge of weaned pigs with an edema disease strain of *E. coli*.

Vaccines used in calves have been based largely on the K99 antigen and are generally effective. The vaccine may incorporate killed K99-positive cultures, or extracted, purified K99 antigen. Passive protection against mortality due to ETEC may be obtained by administration of a high-titered K99 monoclonal antibody to young calves, but the duration of protection is short-lived.

Vaccination with the Rc mutant *E. coli* J5 (O111:B4) has been tested for its ability to protect cows against *E. coli* mastitis. The vaccine is based on the existence of a common LPS core among *E. coli* and on the exposure of this core on the surface during synthesis of LPS by growing organisms. There have been extensive California trials of the Rc mutant *E. coli* J5 as a vaccine against coliform mastitis in dairy cows. These trials indicate that this vaccine is very effective in reducing the occurrence of coliform mastitis in dairy herds.

Iron-regulated outer-membrane proteins of *E. coli* have also been identified as logical antigens that may be used to stimulate antibodies that may protect the mammary gland from coliform mastitis. Initial studies suggest that this approach is feasible.

REFERENCES

Acres, S. D. 1985. Enterotoxigenic *Escherichia coli* infections in newborn calves: A review. J Dairy Sci 68:229-56.

Bertschinger, H. U.; Bachmann, M.; Mettler, C.; Popischil, A.; Schraner, E. M.; Stamm, M.; Sydler, T.; and Wild, P. 1990. Adhesive fimbriae produced in vivo by *Escherichia coli* O139:K12(B):H1 associated with enterotoxaemia in pigs. Vet Microbiol 25:267-81.

Bertschinger, H. U., and Pohlenz, J. 1983. Bacterial colonization and morphology of the intestine in porcine *Escherichia coli* enterotoxemia (edema disease). Vet Pathol 20:99-110.

De Graaf, F. K. 1990. Genetics of adhesive fimbriae of intestinal *Escherichia coli.* Curr Top Microbiol Immunol 151:29-51.

De Jonge, H. R.; Bot, A. G. M.; and Vaandrager, A. B. 1986. Mechanism of action of *E. coli* heat-stable enterotoxin. In Bacterial Protein Toxins, Second European Workshop. Ed. P. Falmagne, J. E. Alouf, F. J. Fehrenbach, J. Jelijaszewics, and M. Thelestram, pp. 335-40. Stuttgart: Gustav Fischer.

De Rycke, J.; Gonzalez, E. A.; Blanco, J.; Oswald, E.; Blanco, M.; and Boivin, R. 1990. Evidence for two types of cytotoxic necrotizing factors in human and animal clinical isolates of *Escherichia coli.* J Clin Microbiol 28:694-99.

De Sauvage, F. J.; Camerato, T. R.; and Goeddel, D. V. 1991. Primary structure and functional expression of receptor for *Escherichia coli* heat-stable enterotoxin. J Biol Chem 266:17912-18.

Donnenberg, M. S., and Kaper, J. 1992. Minireview: Enteropathogenic *Escherichia coli.* Infect Immun 60:3953-61.

DuBreuil, J. D.; Fairbrother, J. M.; Lallier, R.; and Lariviere, S. 1991. Production and purification of heat-stable enterotoxin b from a porcine *Escherichia coli* strain. Infect Immun 59:198-203.

Field, M.; Rao, M. C.; and Chang, E. B. 1989. Intestinal electrolyte transport and diarrheal disease. N Engl J Med 321:879-83.

Fishman, P. H. 1990. Mechanism of action of cholera toxin. In ADP-ribosylating Toxins and G Proteins. Ed. J. Moss and M. Vaughan, pp. 127-40. Washington D.C.: American Society for Microbiology.

Frost, A. J., and Hill, A. W. 1982. Pathogenesis of experimental bovine mastitis following a small inoculum of *Escherichia coli.* Res Vet Sci 33:105-12.

Gaastra, E.; Hamers, A. M.; Bergmans, E. N.; van der Zeijst, B. A. M.; and Gaastra, W. 1988. Adhesion of canine and human uropathogenic *Escherichia coli* and *Proteus mirabilis* strains to canine and human epithelial cells. Curr Microbiol 17:333-37.

Gaastra, W., and De Graaf, F. 1982. Host-specific fimbrial adhesins of noninvasive enterotoxigenic *Escherichia coli* strains. Microbiol Rev 46:129-61.

Gannon, V. P. J.; Gyles, C. L.; and Friendship, R. W. 1988. Characteristics of verotoxigenic *Escherichia coli* from pigs. Can J Vet Res 52:331-37.

Garcia, E.; Bergmans, H. E. N.; Van der Bosch, J. F.; Orskov, I.; Van der Zeijst, B. A. M.; and Gaastra, W. 1988. Isolation and characterization of dog uropathogenic *Escherichia coli* strains and their fimbriae. Antonie van Leeuwenhoek 54:149-63.

Gibbons, R. A.; Sellwood, R.; Burrows, M; and Hunter, P. A. 1977. Inheritance of resistance to neonatal *E. coli* diarrhea in the pig: Examination of the genetic system. Theor Appl Genet 51:65-70.

Giron, J. A.; Ho, A. S.; and Schoolnick, G. K. 1991. An inducible bundle-forming pilus of enteropathogenic *Escherichia coli*. Science 254:710-13.

Gyles, C. L. 1992. *Escherichia coli* cytotoxins and enterotoxins. Can J Microbiol 734-46.

Hadad, J. J., and Gyles, C. L. 1982. Scanning and transmission electron microscope study of the small intestine of colostrum-fed calves infected with selected strains of *Escherichia coli*. Am J Vet Res 43:41-49.

Haider, K.; Albert, M. J.; Hossain, A.; and Nahar, S. 1991. Contact haemolysin production by enteroinvasive *Escherichia coli* and shigellae. J Med Microbiol 35:330-37.

Harel, J.; Lapointe, H.; Fallara, A.; Lortie, L. A.; Bigras-Poulin, M.; Lariviere, S.; and Fairbrother, J. M. 1991. Detection of genes for fimbrial antigens and enterotoxins associated with *Escherichia coli* serogroups isolated from pigs with diarrhea. J Clin Microbiol 29:745-52.

Helie, P.; Morin, M.; Jacques, M.; and Fairbrother, J. M. 1991. Experimental infection of newborn pigs with an attaching and effacing *Escherichia coli* O45:K"E65" strain. Infect Immun 59:814-21.

Hitotsubashi, S.; Fujii, Y.; Yamanaka, H.; and Okamoto, K. 1992. Some properties of purified *Escherichia coli* heat-stable enterotoxin II. Infect Immun 60:4468-74.

Holland, R. E. 1990. Some infectious causes of diarrhea in young farm animals. Clin Microbiol Rev 3:345-75.

Imberechts, H.; de Greve, H.; Schlicker, C.; Bouchet, H.; Pohl, P., Charlier, G.; Bertschinger, H.; Wild, P.; Vandekekerckhove, J.; van Damme, J.; van Montagu, M.; and Lintermans, P. 1992. Characterization of F107 fimbriae of *Escherichia coli* 107/86, which causes edema disease in pigs, and nucleotide sequence of the F107 major fimbrial subunit gene, *fed*A. Infect Immun 60:1963-71.

Janke, B. H.; Francis, D. H.; Collins, J. E.; Libal, M. C.; Zeman, D. H.; Johnson, D. D.; and Neiger, R. D. 1990. Attaching and effacing *Escherichia coli* infection as a cause of diarrhea in young calves. J Am Vet Med Assoc 196:897-901.

Jones, C. H.; Jacob-Dubuisson, F.; Dodson, K.; Kuehn, M.; Slonim, L.; Striker, R.; and Hultgren, S. J. 1992. Adhesin presentation in bacteria requires molecular chaperones and ushers. Infect Immun 60:4445-51.

Karmali, M. A. 1989. Infection by verocytotoxin-producing *Escherichia coli*. Clin Microbiol Rev 2:15-38.

Kennan, R. M., and Monckton, R. P. 1990. Adhesive fimbriae associated with porcine enterotoxigenic *Escherichia coli* of the O141 serotype. J Clin Microbiol 28:651-53.

Korth, M. J.; Apostol, J. M.; and Moseley, S. L. 1992. Functional expression of heterologous fimbrial subunits mediated by the F41, K88, and CS31A determinants of *Escherichia coli*. Infect Immun 60:2500-2505.

McDermott, C. M.; Morrill, J. L.; and Fenwick, B. W. 1991. Deacylation of endotoxin during natural cases of bovine mastitis. J Dairy Sci 74:1227-34.

MacLeod, D. L.; Gyles, C. L.; and Wilcock, B. 1991. Experimental reproduction of edema disease in pigs with Shiga-like toxin II variant toxin (edema disease toxin). Vet Pathol 28:66-73.

Methyiapun, S., Pohlenz, I. F. L.; and Bertschinger, H. U. 1984. Ultrastructure of the intestinal mucosa in pigs experimentally inoculated with an edema disease-producing strain of *Escherichia coli* (O139:K12:H1). Vet Pathol 21:516-20.

Mezoff, A. G.; Jensen, N. J.; and Cohen, M. B. 1991. Mechanisms of increased susceptibility of immature and weaned pigs to *Escherichia coli* heat-stable enterotoxin. Pediatr Res 29:424-28.

Nagy, B.; Casey, T. A.; Whipp, S. C.; and Moon, H. W. 1992. Susceptibility of porcine intestine to pilus-mediated adhesion by some isolates of piliated enterotoxigenic *Escherichia coli* increases with age. Infect Immun 60:1285-94.

Nemeth, J. 1992. A Comparative Study of Virulence of Bovine Mastitis and Fecal *E. coli* Isolates. PhD diss. University of Guelph, Ontario.

Okerman, L. 1987. Enteric infections caused by non-enterotoxigenic *Escherichia coli* in animals: Occurrence and pathogenicity mechanisms. A review. Vet Microbiol 14:33-46.

Orskov, F. 1984. Genus *Escherichia*. In Bergey's Manual of Systematic Bacteriology, Vol. 1. Ed. N. R. Krieg and J. G. Holt. Baltimore: Williams and Wilkins.

Oxenford, C. J.; Lomas, G. R.; and Love, D. N. 1984. Bacteriuria in the dog. Small Anim Pract 25:83-91.

Pearson, G. R., and Logan, E. F. 1983. The pathogenesis of enteric colibacillosis in neonatal unsuckled calves. Vet Rec 105:159-64.

Peterson, J. W., and Ochoa, L. G. 1989. Role of prostaglandins and cAMP in the secretory effects of cholera toxin. Science 245:857-59.

Robertson, D. C. 1988. Pathogenesis and enterotoxins of diarrheagenic *Escherichia coli*. In Virulence Mechanisms of Bacterial Pathogens. Ed. J. A. Roth, pp. 241-63. Washington, D.C.: American Society for Microbiology.

Salajka, E.; Salajkova, Z.; Alexa, P.; and Hornich, M. 1992. Colonization factor different from K88, K99, F41 and 987P in enterotoxigenic *Escherichia coli* strains isolated from postweaning pigs. Vet Microbiol 32:163-75.

Schifferli, D. M.; Beachey, E. H.; and Taylor, R. K. 1991. Genetic analysis of 987P adhesion and fimbriation of *Escherichia coli*: The *fas* genes link both phenotypes. J Bacteriol 173:1230-40.

Schoonderwoerd, M., Clarke, R. C.; van Dreumel, A. A.; and Rawluk, S. 1988. Colitis in calves: Natural and experimental infection with a verotoxin-producing strain of *Escherichia coli* O111:NM. Can J Vet Res 562:484-87.

Schulz, S.; Yuen, P. S. T.; and Garbers, D. L. 1991. The expanding family of adenylyl cyclases. Trends Pharmacol Sci 12:116-20.

Senior, D. F.; deMan, P.; and Svanborg, C. 1992. Serotype, hemolysin production, and adherence characteristics of strains of *Escherichia coli* causing urinary tract infection in dogs. Am J Vet Res 53:494-98.

Smith, H. W. 1974. A search for transmissible pathogenic characters in invasive strains of *Escherichia coli*: The discovery of a plasmid-mediated toxin and a plasmid-controlled lethal character closely associated, or identical, with colicine V. J Gen Microbiol 83:95-111.

Smith, H. W., and Gyles, C. L. 1970. The relationship between two apparently different enterotoxins produced by enteropathogenic strains of *Escherichia coli* of porcine origin. J Med Microbiol 3:258-66.

Soderlind, O.; Thafvelin, B.; and Molby, R. 1988. Virulence factors in *Escherichia coli* strains isolated from Swedish pigs with diarrhea. J Clin Microbiol 26:879-84.

Stephen, J., and Osborne, M. P. 1988. Pathophysiological mechanisms of diarrhoeal disease. In Bacterial Infections of Respiratory and Gastrointestinal Mucosae. Ed. W. Donachie, E. Griffiths and J. Stephens, pp. 149-69. Washington, D.C.: IRL Press.

Svendsen, J.; Riising, H. J.; and Christensen, S. 1977. Studies on the pathogenesis of enteric *E. coli* infections in weaned pigs. Nord Vetmed 29:212-20.

Todhunter, D. A.; Smith, K. L.; and Hogan, J. S. 1990. Intramammary challenge during lactation with a wild strain of *Escherichia coli* following immunization with a curli producing strain of *E. coli*. J Dairy Sci 73 (Suppl 1):152.

Wilson, R. A.; Keefe, T. J.; Davis, M. A.; Browning, M. T.; and Ondrusek, K. 1988. Strains of *Escherichia coli* associated with urogenital disease in dogs and cats. Am J Vet Res 49:743-46.

Whipp, S. C. 1991. Intestinal responses to enterotoxigenic *Escherichia coli* heat-stable toxin b in non-porcine species. Am J Vet Res 52:734-37.

Woodward, M. J.; Carroll, P. J.; and Wray, C. 1992. Detection of entero- and verocyto-toxin genes in *Escherichia coli* from diarrhoeal disease in animals using the polymerase chain reaction. Vet Microbiol 31:251-61.

16 / *Actinobacillus* and *Haemophilus*

BY **J. I. MACINNES** AND **N. L. SMART**

THE FAMILY *Pasteurellaceae* comprises a diverse group of organisms organized into the genera *Haemophilus, Actinobacillus*, and *Pasteurella* (Mannheim et al. 1984). Members of the family *Pasteurellaceae* are almost always found in association with vertebrates and a number are important pathogens of food animals. These organisms are small, gram-negative, nonmotile, pleomorphic, coccobacillary rods. They are facultative anaerobes with both respiratory and fermentative forms of metabolism, a low G + C ratio (38-44 mol%), and a relatively small genome. Traditionally, *Haemophilus* and *Actinobacillus* species have been classified according to their requirements for growth factors such as nucleotide adenine dinucleotide (NAD) and hematin, host range, and disease profile (Table 16.1).

TABLE 16.1 NAD requirement and ecology of selected *Actinobacillus* and *Haemophilus* spp.

Organism	Requirement for NAD	Host	Disease	Normal microflora
A. equuli	-	Horse	Foals: sleepy foal disease or joint ill (purulent nephritis, arthritis) Adults: abortion, septicemia, nephritis, endocarditis	+
		Swine	Piglets: arthritis Adults: abortion, endocarditis	-
A. lignieresii	-	Cattle	Wooden tongue	+
		Sheep	Suppurative lesions of skin and lungs	+
A. pleuropneumoniae	+/-	Swine	Pleuropneumonia	-
A. suis	-	Swine	Septicemia, pneumonia, nephritis, arthritis	+
		Horses[a]	Septicemia, polyarthritis	+
		Cattle	Abortion	+
Haemophilus minor group	+	Swine	?	+
H. paragallinarum	+	Chickens	Coryza, conjunctivitis	-
H. parasuis	+	Swine	Glasser's disease (polyserositis)	+
H. somnus	-	Cattle	Meningoencephalomyelitis, septicemia, abortion	+

[a] These isolates are frequently described as *A. suis*-like, Taxon 11, hemolytic *A. lignieresii* variants.

Recently, DNA hybridization and rRNA sequencing have been used to assess genetic relationships among *Actinobacillus* and *Haemophilus* spp. (Borr et al. 1991, Dewhirst et al. 1992). There is now good evidence that *Actinobacillus* (formerly *Haemophilus*) *pleuropneumoniae*, *A. lignieresii*, *A. suis*, and *A. equuli* are close relatives; *Haemophilus* minor group strain 202 and *Pasteurella haemolytica* also share some homology with these *Actinobacillus* spp. *H. parasuis*, *H. paragallinarum*, and *Haemophilus* taxon C, together with *P. multocida,* have also been clustered together based on their rRNA sequences. *H. somnus*, *A. seminis*, and *P. aerogenes* form another rRNA group, while *H. influenzae*, the type species of the genus *Haemophilus*, is in a fourth group. Many other taxa thought to be associated with the family *Pasteurellaceae* have yet to be classified. In the future, there will doubtless be many revisions to the classification of organisms that are currently assigned to the genera *Haemophilus* and *Actinobacillus*.

ACTINOBACILLUS PLEUROPNEUMONIAE

EPIDEMIOLOGY

Isolation of *A. pleuropneumoniae* has been reported throughout the world in countries where intensive pig rearing is practised (Nicolet 1992). To date, 12 serotypes and 2 biotypes of *A. pleuropneumoniae* have been described (Nielsen 1990). Isolates of biotype 1 require nicotinamide-adenine dinucleotide (NAD), whereas biotype 2 (formerly *Pasteurella* haemolytica-like) isolates are NAD-independent. The different serotypes of *A. pleuropneumoniae* have unique capsular structures, but cross-reactivity can be detected because some serotypes (e.g., 3, 6, and 8) share common lipopolysaccharide molecules and all have common outer-membrane proteins. There is usually a limited number of serotypes present in a given geographic area. For example, strains of serotypes 1 and 5 are prevalent in North America, whereas serotypes 2 and 9 have been isolated in many European countries.

Some attempts have been made to further differentiate among isolates of the same serotype. Using electropherotyping and restriction endonuclease fingerprinting, researchers have shown that a limited number of clones are associated with disease (MacInnes et al. 1990a; Møller et al. 1992). In particular, the genetic heterogeneity of serotypes 1, 2, 5, and 6 seems to be limited, whereas serotypes 7, 8, and 12 are more diverse. Like other pathogens, particular clones can be detected in different geographic areas and can be shown to persist for many years. Although some serotypes are reported to be more virulent than others, much of the evidence for this is indirect (Frey and Nicolet 1990; Komal and Mittal 1990).

DISEASE

A. pleuropneumoniae can cause an acute and rapidly fatal pleuropneumonia in immunologically naive animals (Sebunya and Saunders 1983). The acute form of the disease is characterized by extensive hemorrhage and fibrin deposition in the lungs. Affected animals show signs of severe respiratory distress, cyanosis, fever, and vomiting. Pigs of all ages can succumb to infection, and animals that survive infection are likely to become carriers that can spread the organism. In chronically infected animals, the organism may be sequestered in the lungs in necrotic lesions covered with a thick capsule of connective tissue, and it can reside in the tonsils and upper respiratory tract. In herds where the organism is widespread, animals can be infected without overt clinical signs. It has been generally thought that once an animal has been naturally infected with *A. pleuropneumoniae* it is resistant to reinfection with other serotypes (i.e., heterologous challenge). There is now both direct and indirect evidence to suggest that this is not always the case (Falk and Lium 1991; Cruijsen et al. 1992).

HOST-PATHOGEN INTERACTION

In acute infections, the organism enters the lungs and multiplies rapidly. During growth, the organism releases a large quantity of outer-membrane blebs. Lipopolysaccharide and perhaps cytotoxins or other factors stimulate the recruitment of neutrophils and production of an inflammatory response. Once at the site of infection, the neutrophils are lysed by one or more cytolysins. Consistent with this notion, large numbers of neutrophils can be seen from 3 to 9 hours following experimental infection. However, after this time, few intact neutrophils can be found. It is this destruction of the host neutrophils that is likely responsible for the massive and rapid tissue damage. During acute infection, the presence of an intact capsule helps the organism evade phagocytosis, and the capsule may also have an immunosuppressive effect. In animals that survive acute infection, the organism may become sequestered in fibrinous lesions and may be able to colonize the tonsillar crypts. In both locations, *A. pleuropneumoniae* probably synthesizes fimbriae and/or other adhesion molecules (Utrera and Pijoa 1991).

The humoral immune response, in particular IgG, appears to play an important role in protection against *A. pleuropneumoniae*. Transfer of large quantities of convalescent serum or even anticytotoxin antibodies can confer a significant degree of protection (Bossé et al. 1992b). The role of cell-mediated immunity (CMI) has not been examined thoroughly, but to date there is no evidence that cell cytotoxicity is necessary (Bhatia et al. 1991). Potentially, a strong CMI response may mediate damage to the host.

PREVENTION AND TREATMENT

Bacterin-type vaccines are able to reduce mortality but they do little to prevent morbidity or development of the carrier state. Since bacterins prepared against one serotype have little effect on infection with other serotypes, most commercial preparations contain several different serotypes. The present bacterins are likely not very effective, because they contain little of the various cytolysins, which are known to be important in protection, and inactivation procedures alter some of the antigens. In addition, bacterins are produced using bacteria grown under conditions that are very different from those found in the animal and, as a result, important antigens (e.g., transferrin-binding proteins) may not be expressed. In order to circumvent some of these problems, researchers are now studying the possibility of using live attenuated, conjugate, and recombinant *A. pleuropneumoniae* vaccines (Byrd and Kadis 1992).

A. pleuropneumoniae is sensitive to many antibiotics, but due to the acute nature of the disease, therapy must be initiated soon after onset to prevent mortality. Increasingly, many strains are becoming resistant to commonly used antibiotics (MacInnes et al. 1990a). Many serotype 5 isolates in Ontario are tetracycline resistant; serotype 7 isolates are frequently resistant to beta-lactam antibiotics. Resistance to penicillins appears to be plasmid mediated, whereas tetracycline resistance is chromosomally encoded. In addition to treatment of acute disease, antibiotics have also been used in medicated early weaning programs designed to eliminate the organism from chronically infected herds. In most cases, this approach has met with limited success. Serodiagnosis has also been used to detect and eliminate infected animals and to prevent the introduction of *A. pleuropneumoniae* infected animals into herds. Tests based on capsular antigens appear to be the most sensitive and specific (Bossé et al. 1992a).

VIRULENCE FACTORS

Capsule

In recent years, the virulence factors of *A. pleuropneumoniae* have been studied extensively. These include several cytotoxins, hemolysins, outer-membrane proteins, capsule, and lipopolysaccharide (Inzana 1991). There is no evidence that any of these virulence factors is plasmid encoded. Several studies have shown that the capsule of *A. pleuropneumoniae* plays an important role in defense against the host immune system. Some, but not all, serotypes are resistant to killing by complement and antibody. Organisms with an intact capsule are resistant to phagocytosis, but they can be taken up in the presence of specific opsonizing antibodies. Mutants that are capsule deficient are not able to cause disease and can provide good protection against subsequent challenge with virulent organisms. These mutants have been considered for the development of live attenuated vaccines (Rosendal and MacInnes 1990). Capsular vaccines provide some protection against mortality but not morbidity.

Lipopolysaccharide

The different serotypes of *A. pleuropneumoniae* all contain unique lipopolysaccharide (LPS) molecules, with distinct core regions, and an O side chain that varies in length. The chemical composition of all LPS side chains has been reported (Perry et al. 1990). *A. pleuropneumoniae* LPS molecules are good mitogens, can activate the alternative complement system, clot *Limulus* amoebocyte lysates, and give a positive Shwartzman reaction (Fenwick 1990). When purified LPS is introduced into the lungs of a pig, there is a marked inflammatory cell infiltration. This infiltration is mediated via activation of complement and induction of cytokines such as IL-1, IL-6, and tumor necrosis factor (TNF) in alveolar macrophages. With LPS alone, the hemorrhagic and necrotic lesions characteristic of pleuropneumonia are not seen. These lesions are most likely the result of the release of lytic enzymes and reactive oxygen radicals, which are released from macrophages and neutrophils that are damaged by cytotoxin. In strains with little capsule, LPS is reported to mediate adhesion by binding to respiratory mucus (Bélanger et al. 1992). The protective efficacy of LPS has been tested, but neither purified LPS nor O side chains conjugated to tetanus toxoid provide complete protection (Inzana 1991).

Toxins

There is growing evidence that a number of toxins play a key role in virulence. The older literature is filled with conflicting reports of an "*A. pleuropneumoniae* toxin" with varying degrees of hemolytic and cytotoxic properties (Inzana 1991). "The toxin" was variously described as RNA dependent, calcium dependent, calcium independent, 105 kDa, 104 kDa, 109 kDa, and 120 kDa. We now know there are three different but related toxins produced by *A. pleuropneumoniae* serotypes 1 to 12 isolates. The distribution and activities of these cytotoxins are summarized in Table 16.2.

At high concentrations, the *A. pleuropneumoniae* toxins are cytolysins with the ability to lyse red blood cells and/or kill lymphocytes, epithelial cells, T lymphocytes, and macrophages (Devenish et al. 1992). They act by forming pores in the cell membranes. At sublytic concentrations, the toxins affect oxidative metabolism of phagocytic cells and probably have other biological effects (Udez and Kadis 1992).

TABLE 16.2. Cytotoxins of *A. pleuropneumoniae*

Cytotoxin[a] (synonyms)	Apparent M.W.[b] (kDa)	Activity	Present in Serotypes
ApxIA (ClyIA, HlyIA)	105	Strongly hemolytic, strongly cytotoxic	1, 5, 9, 10, 11
ApxIIA (ClyIIA, HlyIIA, AppA)	103	Moderately cytotoxic, weakly haemolytic	All except 10
ApxIIIA (PtxA, ClyIIIA, Mat, pleurotoxin)	120	Strongly cytotoxic	2, 3, 4, 8

[a] The proposed revised nomenclature of Frey et al. (1993). [b] M.W. = molecular weight.

The *A. pleuropneumoniae* cytotoxins belong to the repeats-in-toxin (RTX) family of toxins (Devenish et al. 1992; Dom et al. 1992; Jansen et al. 1992a,b; Frey et al. 1992). The RTX family of toxins is widely distributed throughout gram-negative bacteria. Their acquisition by *A. pleuropneumoniae* appears to involve some form of transposition, but the mechanism is not well understood. The *apx*III operon of *A. pleuropneumoniae* is similar to that encoding the *E. coli* alpha-hemolysin with the gene order *CABD*. The *A* gene encodes the cytotoxin that is acylated by the *C* gene product. The activated cytotoxin is exported from the cells by the membrane-associated *B* and *D* gene products. Like *E. coli*, the major species of mRNA includes the *C* and *A* genes and a second minor transcript encodes all four.

In *A. pleuropneumoniae*, all the *apx*III operons are complete, whereas the *apx*II operons contain only *C* and *A* genes. The *apx*I operons in serotypes 1, 5, 9, 10, and 11 are complete, but in serotypes 2, 4, 6, 7, 8, and 12, only the *B* and *D* genes are present (Jansen et al. 1992b). Transport is associated with the carboxyl terminal portion of ApxA and, despite substantial sequence differences, the B and D proteins encoded by one operon appear to be able to transport the A protein from another (McWhinney et al. 1992). Indeed, the *B* and *D* gene products of *E. coli hly* operon are able to transport *A. pleuropneumoniae* ApxA proteins. By contrast, *C* genes from different operons are not interchangeable.

Several other domains have been described in the *A. pleuropneumoniae* toxins. One domain in the amino-terminal half comprises four hydrophobic regions and is associated with pore formation in target membranes. A second domain, which consists of glycine-rich nonapeptide repeats, is thought to bind calcium, a requirement for target cell binding and biological activity. The cytolytic activity of ApxAII is contained within the C-terminal portion of the molecule and is separate from the region associated with hemolytic activity. In addition to the RTX toxins, there are almost certainly other cytolysins or bioactive molecules that have not yet been fully characterized (Bertram 1990; Inzana 1991). The toxin story is also confused by the presence of the *hly*X (*cfp*) gene, which was originally reported to encode a CAMP factor. Rather than encoding a cytolysin, this gene encodes a homologue of the global regulatory protein of *E. coli*, FNR, and it activates the expression of a latent haemolysin in *E. coli* (MacInnes et al. 1990b). Its role in regulating virulence factors in *A. pleuropneumoniae*, if any, remains to be determined. We now know that the CAMP factor activity is associated with the various cytotoxins (Devenish et al. 1992).

Outer-Membrane Proteins

The outer membranes of *A. pleuropneumoniae* contain three to five major proteins and approximately 15 minor ones (Inzana 1991). The precise pattern of proteins depends on growth and electrophoresis conditions and on the method of isolation. The outer-membrane proteins

(OMPs) from different serotypes are similar in size and contain cross-reactive epitopes. Some of these epitopes are also shared by other members of the *Pasteurellaceae* and, in the case of the 17-kDa antigen, by *E. coli* (MacInnes and Rosendal 1987). A number of iron-regulated OMPS (IROMPs) have also been reported. The gene for a 60-kDa IROMP has been cloned and shown to encode a protein that is able to bind hemin and porcine, but not bovine or human, transferrin (Gerlach et al. 1992). The fact that the 60-kDa transferrin-binding protein is specific for porcine transferrin is consistent with the notion that *A. pleuropneumoniae* is species specific, but it is possible that there are other mechanisms of iron acquisition. Immunization with OMPs provides partial protection against *A. pleuropneumoniae*. Interestingly, vaccination with proteinase K-treated cells is reported to provide better protection than vaccination with either untreated or periodate-treated OMPs (Chiang et al. 1991).

ACTINOBACILLUS SUIS

EPIDEMIOLOGY

There have been reports of sporadic *Actinobacillus suis* infection of pigs from a number of countries (Sanford 1992). Isolated cases have also been reported from wild birds and an elk. *A. suis*-like organisms have been isolated from horses, but the taxonomic status of these organisms is uncertain (Samitz and Biberstein 1991). In conventional swine, *A. suis* is a commensal organism in the tonsils and upper respiratory tract, and an opportunistic pathogen. Although there have been no systematic studies, it is unlikely that the organism survives in the environment for any appreciable length of time; *A. suis* rapidly loses viability in pathological samples and culture media.

DISEASE

A. suis can infect pigs of all ages, but infection is most serious in very young animals. High-health status herds are more likely to experience problems with this organism. Once herd immunity has been established, however, the number of disease outbreaks decreases (Sanford et al. 1990). In neonates and suckling pigs, *A. suis* can cause an acute and rapidly fatal septicemia; death can occur within 15 hours. Affected animals may show signs of cyanosis, petechial hemorrhage, fever, respiratory distress, neurological disturbances, and arthritis. In slightly older animals, the disease is less severe and may be characterized by fever, anorexia, and persistent cough. Although mortality is much lower, these animals tend to be poor doers. Similar symptoms have been reported in piglets infected with *A. equuli* or *A. equuli*-like organisms. In mature animals, *A. suis* infection can be confused with erysipelas. These animals may have erythematous skin lesions, fever, and inappetence; abortion, metritis, and meningitis have also been reported in sows.

HOST-PATHOGEN INTERACTION

A. suis infection occurs via the aerosol route or by close contact. The organism may also gain entry through breaks in the skin (e.g., upon castration). Once the organism has entered the bloodstream, it spreads rapidly throughout the body. This spread is presumably facilitated by cytotoxin and other, as yet, undefined factors. Capsule and LPS undoubtedly play a role in the survival of the organism in the bloodstream but, to date, there are no reports of characterization of these molecules. *A. suis* (and *A. equuli*) produce enterobacterial common antigen (ECA), an LPS-like molecule, but its role in virulence is not known (Bottger et al. 1987). *A. suis* can form microcolonies on vessel walls that lead to regions of hemorrhage and necrosis. Gross lesions are usually seen in the lungs, kidney, heart, spleen, intestines, and skin. The lungs may also be filled with a serous or serofibrinous exudate and superficially look like lungs of animals with pleuropneumonia. Occasionally, animals are seen with an acute necrotizing myocarditis that is reminiscent of mulberry heart disease.

PREVENTION AND TREATMENT

At present, there are no commercial vaccines for *A. suis*; however, autogenous bacterins are thought to be useful. *A. pleuropneumoniae* bacterins likely confer some cross-protection, but in the absence of significant quantities of cytolysin, this protection is probably limited. *A. suis* isolates are susceptible to most antibiotics and treatment with beta-lactams or other drugs is usually effective if initiated in time. Oxytetracycline and streptomycin in the feed are also reported to be useful to limit the spread of infection.

VIRULENCE FACTORS

With the exception of the *ash* operon, which encodes an RTX toxin, relatively little is known about virulence factors of *A. suis*. The nucleotide sequences of the *ash*C and *ash*A genes are almost identical to those reported for *apx*IIC (*cly*IIC) and *apx*IIA (*app*C) of *A. pleuropneumoniae* serotype 5 (Burrows and Lo 1992). Like the *apx*II operon, the *B* and *D* genes are not linked and it is likely that *A. suis* isolates also contain a second complete RTX operon. *apx*III-like operons have not been found in any of the strains tested. Given similarities that this organism has with *A. pleuropneumoniae*, other virulence factors may be the same. However, there are a number of differences between these two organisms. Although *A. suis* appears to be less pathogenic, it seems better able to invade and persist in the bloodstream than *A. pleuropneumoniae*.

HAEMOPHILUS PARASUIS

Glasser's disease, caused by the bacterium *Haemophilus parasuis*, was first described by Glasser in 1910. At that time, the infection was associated with sporadic cases of polyserositis in pigs. *H. parasuis* is commonly found as part of the normal flora of the upper respiratory tract of conventionally raised swine; however, pigs raised under specific-pathogen-free (SPF) conditions often do not harbor *H. parasuis*. SPF pigs are highly susceptible to Glasser's disease, presumably due to the lack of protective antibodies as a result of their *H. parasuis*-free status. Primary exposure of SPF pigs of any age to *H. parasuis* can lead to outbreaks of acute disease associated with high morbidity and mortality. Postmortem lesions seen in Glasser's disease may range from no grossly observable pathology to large depositions of fibrin in joints and on any or all of the serosal surfaces of the body. The course of clinical disease may range from acute death to a more chronic course where pigs are most often observed to be lame, pyrexic, depressed, and anorexic. Survivors often show evidence of chronic lesions such as arthritis or pleuritis.

Attachment of these bacteria to the mucosa of the upper respiratory tract is likely mediated by fimbriae. The expression of these structures in vitro seems to be highly dependent on growth conditions, for they are not observed after conventional cultivation. Under certain circumstances, *H. parasuis* bacteria may invade the mucosal barrier and enter the blood system. The invasive mechanism is not known. Stress factors such as moving and/or mixing of pigs have previously been identified as predisposing factors for Glasser's disease in conventional pigs. Most often, young, weaned, conventional pigs are affected because maternal antibody protection is low at the time of weaning. Outbreaks of Glasser's disease in SPF pigs are most often associated with primary exposure to *H. parasuis*, but additional factors, such as environmental stress and the presence of particular *H. parasuis* serovars, have been shown to influence morbidity and mortality rates.

Serotyping studies have shown that there is more antigenic heterogeneity among strains of *H. parasuis* than was first suspected. Four serovars (A-D) were described by Bakos in 1955, but many serovars have been added in recent years. In 1986, Morozumi and Nicolet defined 7 serovars to which 6 more serovars (Jena 6-Jena 12) were added by Kielstein in 1991. In a recent study, Kielstein and Rapp-Gabrielson (1992) named 5 new serovars (ND1-ND5). In this scheme, the serovars described by Morozumi and Nicolet (1986) remain unchanged, while there is some overlap between the newly defined ND serovars and the Jena serovars described previously by Kielstein et al. (1991). A total of 15 serovars were described by Kielstein and Rapp-Gabrielson (1992), however, 26% of the 243 isolates studied were nontypable. This problem has been encountered in all studies where *H. parasuis* isolates were investigated by serological methods. Smart et al. (1989) used restriction endonuclease analysis to group field strains of *H. parasuis*. Twenty-nine restriction endonuclease patterns were identified among 61 isolates.

There is some evidence that certain serovars are more pathogenic than others. Pathogenicity has been associated with encapsulation, outer-membrane protein profiles, and whole-cell protein profiles. Experimental inoculations of pigs and guinea pigs have shown that virulence differences can be demonstrated between serovars (Kielstein and Rapp-Gabrielson 1992). However, in a study of field isolates from the nasal cavity of pigs (asymptomatic carriers) and pigs which had died of polyserositis, no association was found between serovar, site of isolation, and pathogenic potential (Rapp-Gabrielson and Gabrielson 1992).

There are conflicting reports in the literature as to whether the capsule is a virulence factor. Expression of capsule may be affected by in vivo and in vitro procedures and this may explain the apparent confusion. There is also some evidence that endotoxin plays a role in pathogenesis. Pigs that die acutely of Glasser's disease show evidence of septicemia, cyanosis, and renal glomerular necrosis with microthrombi in other tissues (Peet et al. 1983). *H. parasuis* is reported to produce (Cu, Zn) superoxide dismutase activity, which might help to protect the organism from killing by phagocytic cells. Some strains of *H. parasuis* contain plasmids, but their role in disease is not known.

Protection against Glasser's disease may be achieved with the use of a formalin-killed bacterin (Riising 1981). Cross-protection between selected strains of *H. parasuis* has been demonstrated, and for this reason, several commercial Glasser's bacterins contain more than one strain. The antigen responsible for stimulating protective immunity in vaccinated pigs has not been identified.

HAEMOPHILUS PARAGALLINARUM

Haemophilus paragallinarum causes infectious coryza in chickens. The major reservoir of disease is chronically infected birds or apparently healthy carriers. This bacterium is highly host associated and does not survive for any length of time in the environment. *H. paragallinarum* colonizes the upper respiratory tract and sinuses and causes acute rhinitis and sinusitis, with edema, facial swelling, and conjunctivitis (Reid and Blackall 1984). The length and severity of outbreaks in flocks may show considerable variation.

Page (1962) described 3 serogroups (I, II, III). Blackall et al. (1990) renamed these serogroups A, C, B, respectively and added 4 serovars to each of A and C. This scheme is based on agglutination of hemagglutinating antigens that are present on the outer membrane.

These antigens are important factors for colonization and have been shown to stimulate protective immunity.

Several other virulence factors of *H. paragallinarum* have been described. Organisms cultured under iron-limiting conditions have been observed to express additional outer-membrane proteins (Ogunnariwo and Schryvers 1992). Siderophore production has not been observed. The hyaluronic acid capsule is considered to be an important virulence determinant. The LPS of *H. paragallinarum* is toxic to chickens and, when injected, causes hydropericardium (Iritani et al. 1981).

HAEMOPHILUS SOMNUS

Haemophilus somnus is part of the normal bacterial flora of the male and female bovine genital tract. It is less commonly isolated from the normal upper respiratory tract, although it does adhere to nasal turbinate epithelium, and thus may be considered a transient colonizer of this site. *H. somnus* spreads by contact and it appears that calves are infected early in life from their dams. The most common diseases caused by *H. somnus* are acute thromboembolic meningoencephalitis (TME), pneumonia, and myocarditis. In addition, *H. somnus* has been associated with the following syndromes: abortion, metritis/infertility, arthritis, otitis externa, mastitis, orchitis, conjunctivitis, and laryngitis/tracheitis (Harris and Janzen 1989). Usually, clinical disease caused by *H. somnus* occurs in immature cattle and is associated with environmental stressors such as moving, mixing of animals from different sources, and climatic conditions. Morbidity can vary from low to high and affected animals usually die.

Unlike the other organisms discussed in the chapter, *H. somnus* is an intracellular pathogen. In an extensive review by Corbeil (1990), the virulence factors associated with *H. somnus* were summarized as attachment, growth stimulation by normal flora, serum resistance, Fc receptors, interference with phagocytic function, and toxicity for several different bovine cell types. Attachment of organisms facilitates colonization, which may be asymptomatic in carrier animals. However, the attachment of virulent strains to vascular endothelial cells may precipitate the development of vasculitis and thrombosis, characteristic of *H. somnus* infection. The mechanism by which this pathology develops is unknown, except for the proposal that LPS may play a role in the development of this lesion. Recent studies have shown that *H. somnus* shows a particular specificity for cells of bovine origin. Presumably, this effect is mediated via the Fc receptors present on the surface of the bacteria. Reduced immune cell function has been demonstrated only in bovine cells, and *H. somnus* cultured under iron-limiting conditions are able to utilize only bovine transferrin (Ogunnariwo et al. 1990). The ability of these organisms to utilize transferrin as an iron source has been attributed to the appearance of outer-membrane proteins, which are expressed only when iron is limited in the environment. Siderophore production has not been observed.

The ability of *H. somnus* to cause systemic disease is dependent on serum resistance, which facilitates the survival and dissemination of bacteria throughout the animal body. There is a strong correlation between the presence of this property and strain virulence. Cole et al. (1992) showed that strains not associated with disease lacked two linked genes that encoded for outer membrane proteins which appear to confer serum resistance.

Recent studies have focused on the outer-membrane proteins of *H. somnus* because of their apparent role as virulence determinants, diagnostic antigens, and immunogens. Several OMPs have been investigated, but the most promising one is a 40-kDa OMP that has been consistently found in convalescent serum of experimentally infected animals (Corbeil et al. 1991). It is also cross-reactive with members of the family *Pasteurellaceae*. Antibodies directed against this antigen have been shown to prevent disease.

FUTURE PROSPECTS

Despite considerable efforts by researchers around the world, the role of various virulence factors in many *Actinobacillus* and *Haemophilus* species is still poorly understood. This lack of understanding is due, in large part, to the absence of good genetic tools to generate isogenic strains and other useful mutants. Efforts are currently underway in a number of laboratories to develop shuttle vectors and transposon mutagenesis systems, so it will be possible to have a real understanding of these interesting, but at times, frustrating organisms.

REFERENCES

Actinobacillus spp.

Bélanger, M.; Roiux, S.; Foiry, B.; and Jacques, M. 1992. Affinity for porcine respiratory tract mucus is found in some isolates of *Actinobacillus pleuropneumoniae*. Fed Eur Microbiol Soc Microbiol Lett 97:119-26.

Bertram, T. A. 1990. *Actinobacillus pleuropneumoniae*: Molecular aspects of virulence and pulmonary injury. Can J Vet Res 54:S53-56.

Bhatia, B.; Mittal, K. R.; and Frey, J. 1991. Factors involved in immunity against *Actinobacillus pleuropneumoniae* in mice. Vet Microbiol 29:147-58.

Borr, J. D.; Ryan, D. A. J.; and MacInnes, J. I. 1991. Analysis of *Actinobacillus pleuropneumoniae* and related organisms by DNA-DNA hybridization and restriction endonuclease fingerprinting. Int J Syst Bacteriol 41:121-29.

Bossé, J. T.; Friendship, R.; and Rosendal, S. 1992a. Evaluation of a pooled-antigen ELISA for serodiagnosis of *Actinobacillus pleuropneumoniae* serotype 1, 5, and 7 infections. Proc 12th Int Pig Vet Soc, August 17-20, The Hague, Neth, p. 220.

Bossé, J. T.; Johnson, R. P.; Nemec, M.; and Rosendal, S. 1992b. Protective local and systemic antibody response of swine exposed to an aerosol of *Actinobacillus pleuropneumoniae* serotype 1. Infect Immun 60:479-84.

Bottger, E. C.; Jurs, A; Barrett, T.; Wachsmuth, K.; Metzger, S; Bitter-Suermann, D. 1987. Qualitative and quantitiative determination of enterobacterial common antigen (ECA) with monclonal antibodies: Expression of ECA by two *Actinobacillus* species. J Clin Microbiol 25:377-82.

Burrows., L. L., and Lo, R. Y. 1992. Molecular characterization of an RTX toxin determinant from *A. suis*. Infect Immun 60:2166-73.

Byrd, W., and Kadis, S. 1992. Preparation, conjugation, and immunogenicity of conjugate vaccines directed against *Actinobacillus pleuropneumoniae* virulence determinants. Infect Immun 60:3042-51.

Chiang, Y. -W.; Young, T. F.; Rapp-Gabrielson, V. J.; and Ross, R. F. 1991. Improved protection of swine from pleuropneumonia by vaccination with proteinase K-treated outer membranes of *Actinobacillus (Haemophilus) pleuropneumoniae*. Vet Microbiol 27:49-62.

Cruijsen, T.; van Leengoed, L.; Kroon, P.; and Verheijden, J. 1992. Protective immunity against *Actinobacillus pleuropneumoniae* in convalescent pigs. Proc 12th Int Pig Vet Soc, August 17-20, The Hague, Neth, p. 217.

Devenish. J.; Brown, J. E.; and Rosendal, S. 1992. Association of the RTX proteins of *Actinobacillus pleuropneumoniae* with hemolytic, CAMP, and neutrophil cytotoxic activities. Infect Immun 60:2139-42.

Dewhirst, F. E.; Pasteur, B. J.; Olsen, I.; and Fraser, G. J. 1992. Phylogeny of 54 representative strains of species in the family *Pasteurellaceae* as determined by comparison of 16S rRNA sequences. J Bacteriol 174:2002-13.

Dom, P.; Haemsbrouck, F.; Kamp, E. M.; and Smits, M. A. 1992. Influence of *Actinobacillus pleuropneumoniae* serotype 2 and its cytolysins on porcine neutrophil chemiluminescence. Infect Immun 60:4328-34.

Falk, K., and Lium, B. M. 1991. An abattoir survey of pneumonia and pleuritis in slaughter weight swine from 9 selected herds. 3. Serological findings and their relationship to pathomorphological and microbiological findings. Acta Vet Scand 32:79-88.

Fenwick, B. W. 1990. Virulence attributes of the lipopolysaccharide of the HAP group organisms. Can J Vet Res 54:S28-32.

Frey, J., and Nicolet, J. 1990. Hemolysin patterns of *Actinobacillus pleuropneumoniae*. J Clin Microbiol 28:232-36.

Frey, J.; van den Bosch, H.; Segers, R.; and Nicolet, J. 1992. Identification of a second haemolysin (HlyII) in *Actinobacillus pleuropneumoniae* serotype 1 and expression of the gene in *Escherichia coli*. Infect Immun 60:1671-76.

Frey, J.; Bosse, J. T.; Chang, Y. F. et al. 1993. *Actinobacillus pleuropneumoniae* RTX-toxins: Uniform designation of hemolysins, cytolysins, pleurotoxin, and their genes. Submitted for publication.

Gerlach, G. F.; Anderson, C.; Potter, A. A.; Klashinsky, S; and Willson, P. J. 1992. Cloning and expression of a transferrin-binding protein from *Actinobacillus pleuropneumoniae*. Infect Immun 60:892-98.

Inzana, T. J. 1991. Virulence properties of *Actinobacillus pleuropneumoniae*. Microbiol Pathog 11:305-16.

Jansen, R.; Briaire, J.; Kamp, E. M.; and Smits, M. A. 1992a. Comparison of the cytolysin II genetic determinants of *Actinobacillus pleuropneumoniae* serotypes. Infect Immun 60:630-36.

Jansen, R.; Briaire, J.; Kamp, E. M.; and Smits, M. A. 1992b. The cytolysin genes of *Actinobacillus pleuropneumoniae*. Proc 12th Int Pig Vet Soc, August 17-20, The Hague, Neth, p. 197.

Komal, J. P. S., and Mittal, K. R. 1990. Grouping of *Actinobacillus pleuropneumoniae* strains of serotypes 1 through 12 on the basis of their virulence in mice. Vet Microbiol 25:229-40.

MacInnes, J. I., and Rosendal, S. 1987. Analysis of major antigens of *Haemophilus* (*Actinobacillus*) *pleuropneumoniae* and related organisms. Infect Immun 55:1626-34.

MacInnes, J. I.; Borr, J. D.; Massoudi, M.; and Rosendal, S. 1990a. Analysis of Southern Ontario *Actinobacillus* (*Haemophilus*) *pleuropneumoniae* isolates by restriction endonuclease fingerprinting. Can J Vet Res 54:244-50.

MacInnes, J. I.; Kim, J. E.; Lian, C.-J.; and Soltes, G. A. 1990b. *Actinobacillus pleuropneumoniae hlyX* gene homology with the *fnr* gene of *Escherichia coli*. J Bacteriol 172:4587-92.

Mannheim, W.; Carter, G. R.; Kilian, M.; Biberstein, E. L.; and Phillips, J. E. 1984. Family III *Pasteurellaceae*. In Bergey's Manual of Systematic Bacteriology, Vol I. Eds. N. R. Krieg and J. G. Holt, pp. 550-75. Baltimore: Williams and Wilkins.

McWhinney, D. R.; Chang, Y. F.; Young, R.; and Struck, D. K. 1992. Separable domains define target cell specificities of an RTX toxin from *Actinobacillus pleuropneumoniae*. J Bacteriol 174:291-97.

Møller, K.; Nielsen, R.; Andersen, L. V.; and Kilian, M. 1992. Clonal analysis of the *Actinobacillus pleuropneumoniae* population in a geographically restricted area by multilocus enzyme electrophoresis. J Clin Microbiol 30:623-27.

Nicolet, J. 1992. *Actinobacillus pleuropneumoniae*, p.401-8. In Diseases of swine. 7th Ed. Ed. A. D. Leman, B. E. Straw, W. L. Mengeling, S. D'Allaire, and D. J. Taylor, pp. 401-8. Ames: Iowa State University Press.

Nielsen, R. 1990. New diagnostic techniques: A review of the HAP group of bacteria. Can J Vet Res 54:S68-72.

Perry, M. B.; Altman, E.; Brisson, J.-R.; Beynon, L. M.; and Richards, J. C. 1990. Structural characteristics of the antigenic capsular polysaccharides and lipopolysaccharides involved in the serological classification of *Actinobacillus* (*Haemophilus*) *pleuropneumoniae*. Serodiagn Immunother Infect Dis 4:299-308.

Rosendal, S., and MacInnes, J. I. 1990. Characterization of an attenuated strain of *Actinobacillus pleuropneumoniae*, serotype 1. Am J Vet Res 1:711-17.

Samitz, E. M., and Biberstein, E. L. 1991. *Actinobacillus suis*-like organisms and evidence of hemolytic strains of *Actinobacillus lignieresii* in horses. Am J Vet Res 52:1245-51.

Sanford, S. E. 1992. *Actinobacillus suis*. In Diseases of swine, 7th Ed. Ed. A. D. Leman, B. E. Straw, W. L. Mengeling, S. D'Allaire, and D. J. Taylor, pp. 633-36. Ames: Iowa State University Press.

Sanford, S. E.; Josephson, G. K. A.; Rehmtulla, A. J.; and Tilker, A. M. E. 1990. *Actinobacillus suis* infection in pigs in southwestern Ontario. Can Vet J 31:443-47.

Sebunya, T. N. K., and Saunders, J. R. 1983. *Haemophilus pleuropneumoniae* infection in swine: A review. J Am Vet Med Assoc 182:1331-37.

Udez, F. A., and Kadis, S. 1992. Effects of *Actinobacillus pleuropneumoniae* hemolysin on porcine neutrophil function. Infect Immun 60:1558-67.

Utrera, V., and Pijoan, C. 1991. Fimbriae in *A. pleuropneumoniae* strains isolated from pig respiratory tracts. Vet Rec 128:357-58.

Haemophilus spp.

Bakos, K.; Nilsson, A.; and Thal, E. 1952. Untersuchungen ueber *Haemophilus suis*. Nord Vet Med 44:558

Blackall, P. J.; Eaves, L. E.; and Rogers, D. G. 1990. Proposal of a new serovar and altered nomenclature for *Haemophilus paragallinarum* in the Kume hemagglutination scheme. J Clin Microbiol 28:1185-87.

Cole, S. P.; Guiney, D. G.; and Corbeil, L. B. 1992. Two linked genes for outer membrane proteins are absent in four non-disease strains of *H. somnus*. Mol Microbiol 6:1895-1902.

Corbeil, L. B. 1990. Molecular aspects of some virulence factors of *Haemophilus somnus*. Can J Vet Res 54:S57-S62.

Corbeil, L. B.; Kania, S. A.; and Gogolewski, R. P. 1991. Characterization of immunodominant surface antigens of *Haemophilus somnus*. Infect Immun 59:4295-301.

Harris, F. W., and Janzen, E. D. 1989. The *Haemophilus somnus* disease complex (Hemophilosis): A review. Can Vet J 30:816-22.

Iritani, Y.; Iwaki, S.; and Yamaguchi, T. 1981. Biological activities of crude polysaccharide extracted from two different immunotype strains of *Haemophilus paragallinarum* in chickens. Avian Dis 25:29-37.

Kielstein, P.; Rosner, H.; and Muller, W. 1991. Typing of heat-stable *Haemophilus parasuis* antigen by means of agar gel precipitation and the dot-blot procedure. J Vet Med [B] 38:315-20.

Kielstein, P., and Rapp-Gabrielson, V. J. 1992. Designation of 15 serovars of *Haemophilus parasuis* on the basis of immuno-diffusion using heat-stable antigen extracts. J Clin Microbiol 30:862-65.

Morozumi, T., and Nicolet, J. 1986. Some antigenic properties of *Haemophilus parasuis* and a proposal for serological classification. J Clin Microbiol 23:1022-25.

Nicolet, J.; Morozumi, T.; and Bloch, I. 1986. Proposal for serological classification of *H. parasuis*. Proc 9th Int Pig Vet Soc, Barcelona, Spain, p. 260.

Ogunnariwo, J. A.; Cheng, C.; Ford, J.; and Schryvers, A. B. 1990. Response of *Haemophilus somnus* to iron limitation: Expression and identification of a bovine-specific transferrin receptor. Microb Pathog 9:397-406.

Ogunnariwo, J. A., and Schryvers, A. B. 1992. Correlation between the ability of *Haemophilus paragallinarum* to acquire ovotransferrin-bound iron and the expression of ovotransferrin-specific receptors. Avian Dis 36:655-63.

Page, L. A. 1962. *Haemophilus* infections in chickens. 1. Characteristics of 12 *Haemophilus* isolates recovered from diseased chickens. Am J Vet Res 23:85-95.

Peet R. L.; Fry, J.; Lloyd, J.; Henderson, J.; Curran, J.; and Moir, D. 1983. *Haemophilus parasuis* septicemia in pigs. Aust Vet J 60:187.

Reid, G. G., and Blackall, P. J. 1984. Pathogenicity of Australian isolates of *Haemophilus paragallinarum* and *Hemophilus avium* in chickens. Vet Microbiol 9:77-82.

Rapp-Gabrielson, V. J., and Gabrielson, D. A. 1992. Prevalence of *Haemophilus parasuis* serovars among isolates from swine. Am J Vet Res 53:659-64.

Riising, H. J. 1981. Prevention of Glasser's disease through immunity to *Hemophilus parasuis*. Zentralbl Veterinaermed [B] 28:630-38.

Smart, N. L.; Miniats, O. P.; and Friendship, R. M. 1989. Glasser's disease and prevalence of subclinical infection with *Haemophilus parasuis* in swine in Southern Ontario. Can Vet J 30:339-43.

17 / *Bordetella*

BY **D. A. BEMIS** AND **E. H. BURNS, JR.**

THE GENUS *Bordetella* contains four species of well-adapted respiratory tract parasites. The prototype species, *B. pertussis*, produces natural disease solely in humans while the most cosmopolitan species, *B. bronchiseptica*, produces disease in many mammalian species. *B. avium* is primarily associated with respiratory disease of poultry. *B. parapertussis*, a rare cause of disease in humans, has also been recently isolated from sheep (Cullinane et al. 1987).

Taxonomically, *Bordetella* represents a cluster of aerobic, gram-negative rods whose familial relationship to other bacterial genera is uncertain. Several *Alcaligenes* species, *Achromobacter xylosoxidans*, and unnamed Centers for Disease Control (CDC) group IV organisms appear to be most closely related to *Bordetella*. Species in the genus *Bordetella* are closely related phenotypically, possess common antigens, and share a high degree of DNA similarity. In fact, some have suggested that only *B. avium* is genetically divergent enough to be given a separate species status. The overall guanine-plus-cytosine content of *Bordetella* DNA is high (61.6-62.6% for *B. avium* and 68.3-69.0% for *B. pertussis*, *B. parapertussis*, and *B. bronchiseptica*) (Kersters et al. 1984).

Bordetella spp. have relatively simple growth requirements and can be readily isolated from infected animals. Only *B. pertussis* is incapable of growing on ordinary peptone-containing medium, due to the presence of growth inhibitory substances such as short-chain fatty acids, colloidal sulfur, or secondary metabolites. Various combinations of fresh red blood cells, starch, charcoal, glycerol, ion exchange resins, and cyclodextrin have been incorporated into *B. pertussis* isolation media to absorb or inactivate inhibitory substance. Unlike *Haemophilus*, *Pasteurella*, and a number of other bacterial respiratory tract pathogens, bordetellae do not require complex blood-borne nutrients for growth. Bordetellae grow somewhat more slowly than other commonly encountered bacteria. Mature colonies appear on most agar media in 2-3 days. *B. bronchiseptica* and *B. avium* strains grow more rapidly and are more resistant to various physicochemical conditions than *B. pertussis* or *B. parapertussis*. Both *B. bronchiseptica* and *B. avium* grow on enteric isolation media (e.g., MacConkey agar) and Cetrimide agar. Resistance of *B. bronchiseptica* strains to nitrofuran antimicrobials provides an additional basis for selective isolation of these organisms from infected animals. The bordetellae also are asaccharolytic; they use amino and other organic acids, instead of sugar, as primary carbon and energy sources. Substrate alkalinization resulting from such metabolism is often a feature applied in selective-differential media and tests for further identification of these *Bordetella* of veterinary importance.

HABITAT

The natural habitat of *Bordetella* (and the major source of infection) is the respiratory tract of susceptible animals. Here, the bordetellae appear to replicate freely before succumbing to the host's immune defenses. Preferred amino acids, proline and glutamic acid, are abundant in respiratory secretions. Bordetellae produce hydroxamate siderophores and binding proteins, which indirectly and directly mobilize iron from transferrin and lactoferrin (Agiato and Dyer 1992). None of the *Bordetella* spp. are thought to survive for long periods outside their animal hosts. Recent demonstration of *B. bronchiseptica* growth in natural fresh and salt water samples has led to speculation of possible environmental reservoirs of *B. bronchiseptica* (Porter et al. 1991), but isolation of bordetellae from natural water sources has not been reported. While water, air, and soil may serve as vehicles for transmission of *Bordetella* infections, their presence in these natural environments is apt to be short-lived and directly correlated with the number of infected animals in that particular environment. Studies measuring airborne infection pressure in animal-housing air seem to bear this out (Stehmann et al. 1991). Although bordetellae have simple growth requirements, they are readily killed by decreased oxidation/reduction potentials, UV irradiation, pH and temperature extremes, and many common chemicals. The comparatively slow growth rate of bordetellae also makes them poor competitors for nutrients in most natural environments.

DISEASES

Bordetelloses are highly contagious respiratory diseases that affect animals worldwide. In some populations, the prevalence of infection may approach 100%. Young animals rarely escape infection. *Bordetella* spp. each exhibit a tropism for ciliated respiratory epithelium (Fig. 17.1). Infection results in varying degrees of acute exudative inflammation along part of or the entire network of conducting airways. Symptoms may include oculonasal discharge, torticollis, sneezing, coughing, dyspnea, decreased weight gain, and tracheal collapse. For all but the paroxysmal stage of coughing in human pertussis, bacteria are usually present in large numbers when disease occurs. High morbidity and low mortality is the rule in bordetellosis. Most infections are probably asymptomatic or only mildly symptomatic. Under field conditions, severe disease with increased mortality may be associated with stress and coinfection with other pathogens.

B. bronchiseptica infects many domestic and wild animal species. Prevalence of infection is high in dogs, pigs, and guinea pigs; cats, rabbits, rats, and horses are also commonly infected. In dogs, *B. bronchiseptica* is thought to be an important component of the syndrome referred to as infectious canine tracheobronchitis, or kennel cough. In pigs, *B. bronchiseptica* is variably felt to play a preliminary role in development of the disease atrophic rhinitis. Infections with *B. bronchiseptica* are rare in humans (Woolfrey and Moody 1991) are noticeably uncommon in ruminants. *B. bronchiseptica* infections have occasionally been seen in large sheep species housed in zoos, and *B. parapertussis* has been isolated from cases of chronic nonprogressive ovine pneumonia in New Zealand. Atypical *B. bronchiseptica* infections (wound infections, abortion, bacteremia, peritonitis) are occasionally observed in compromised human and domestic animal hosts. *B. bronchiseptica* is commonly isolated from poultry but is avirulent for these species. *B. avium* is a primary agent of turkey coryza and is also pathogenic for quail and possibly other avian species. *B. avium* infections outside the respiratory tract or in humans have not been reported.

FIG. 17.1. Scanning electron micrograph of *Bordetella bronchiseptica* attached to canine tracheal cilia. x20,000. (Reprinted with permission from J. Infect. Dis. 144:352 © 1982 by The University of Chicago. All rights reserved.)

VIRULENCE FACTORS

Infection and colonization by *Bordetella* spp. is mediated by several proteins. Among these are filamentous hemagglutinin, pertactin, and fimbriae. These proteins are each involved in attachment of *Bordetella* spp. to host tissue.

Bordetellae, in addition to attachment factors, also produce several toxins. Among these are an extracellular adenylate cyclase toxin, dermonecrotic toxin, tracheal cytotoxin, and osteotoxin. *B. pertussis* also produces pertussis toxin, which exhibits ADP-ribosyltransferase activity and plays a role in attachment by *B. pertussis*. Intact structural genes for pertussis toxin have been detected in *B. parapertussis* and *B. bronchiseptica*, but pertussis toxin is not produced by species other than *B. pertussis* and is, therefore, not involved in animal disease (Arico and Rappuoli 1987). For that reason, it will not be discussed here. The major virulence factors of *Bordetella* spp. are listed in Table 17.1.

TABLE 17.1. Major virulence factors of *Bordetella* species

Virulence Factor	Species	Size Range (kDa)	Location	Function
Pertussis toxin	*B. pertussis*	105	Extracellular	ADP-ribosyltransferase
FHA	All except *B. avium*[a]	150-220	Extracellular	Attachment
Pertactin	All	68-70	Outer membrane	Attachment
Fimbriae	All	14.4-24	Outer membrane	Attachment?
Adenylate cyclase toxin	All except *B. avium*	~216	Extracellular	Inhibits phagocytosis
Dermonecrotic toxin	All	155-190	Intracellular	Impairs osteogenesis
Tracheal cytotoxin	All	0.921	Extracellular	Destruction of ciliated cells
Osteotoxin	All	80	Extracellular	Toxic to osteoblasts

[a] *B. avium* has hemagglutinating activity that correlates with virulence.

Bordetella virulence factors, except tracheal cytotoxin, are under control of the *Bordetella* virulence gene (*bvg*) regulatory system, formerly called *vir*. This system is similar to two-component regulatory systems present in other bacteria. Due to this regulation, virulence factors are not always expressed by *Bordetella* but are only present when cells are in phase I, also called virulent phase.

Filamentous Hemagglutinin

Filamentous hemagglutinin (FHA) is a high molecular weight protein (220 kD in *B. pertussis,* 150 kD in *B. bronchiseptica*), which has been demonstrated in all *Bordetella* spp. except *B. avium*. *B. avium* does, however, have marked hemagglutinating activity associated with virulence. FHA is secreted from the cell but remains associated with the outer membrane and is a major attachment factor of *Bordetella*. The protein contains three domains involved in adhesion of *Bordetella* to eukaryotic cells: the arg-gly-asp (RGD) sequence, a carbohydrate-binding site, and a heparin-binding site (Relman et al. 1989; Menozzi et al. 1991). The RGD sequence is found on the extracellular matrix proteins fibronectin, laminin, and collagen and is recognized by integrin receptors. In FHA, the RGD sequence is known to bind to the integrin CR3 on macrophages and promote phagocytosis by the macrophages (Relman et al. 1990). *B. pertussis* has been shown to survive intracellularly and can inhibit phagosome-lysosome fusion in neutrophils and the respiratory burst in macrophages (Pearson et al. 1987;

Steed et al. 1992). The RGD sequence, however, is not involved in binding of bordetellae to ciliated cells. Such attachment is mediated by the carbohydrate-binding domain. Binding to macrophages can occur through the carbohydrate-binding domain as well, but such binding does not induce internalization. Recently, a heparin-binding site has been discovered on the FHA molecule. The significance of this site in colonization is not known as yet.

FHA was previously known as fimbrial hemagglutinin. Although now known not to be fimbrial, FHA is closely related to fimbriae. FHA is encoded by a five-gene operon consisting of *fha*B, D, A, E, and C. Three of the five genes in this operon have significant similarity with fimbrial genes. *fha*A is involved in regulation of synthesis and export and has approximately 26.5% identity with the *Escherichia coli pap*C gene, which encodes an anchor protein for *E. coli* pyelonephritis-associated pili. *fha*D exhibits similarity with the *E. coli pap*D gene, which encodes a chaperonin protein, and *fha*E is homologous to the *Bordetella* fimbrial genes, *fim*2, *fim*3, and *fim*X. *fha*E is believed to encode a minor fimbrial subunit (Locht et al. 1992). Mutations in either *fha*A, D, or E eliminated or reduced expression of both FHA and fimbriae, indicating that production and secretion of FHA and fimbriae involve common accessory genes. These genes have been designated *fim*B, C, and D by other researchers (Willems et al. 1992). The other two genes in this operon do not show significant similarity with fimbrial genes; *fha*B is the structural gene for FHA, and *fha*C is involved in export.

FHA, in addition to its adherence properties, causes hemagglutination of erythrocytes from many species. The clinical significance of hemagglutination is not known and is most likely to be merely a laboratory phenomenon. However, agglutination of bovine red blood cells does correlate with virulence of *B. bronchiseptica* for pigs, and agglutination of guinea pig erythrocytes correlates with pathogenicity of *B. avium* for turkeys, although *B. avium* does not react with antibodies against FHA (Gentry-Weeks et al. 1988). FHA is highly immunogenic and the presence of antibodies against FHA is protective against respiratory challenge in mouse models.

Pertactin

Pertactin is a 68- to 70-kDa protein produced by all species of *Bordetella*. It is an outer-membrane protein, nonfimbrial in nature, which functions as an adhesin. This protein is produced from a single gene, designated *prn*, as an approximately 93-kDa precursor, which is modified to the mature form. The mature protein is actually around 60.5 kD but appears larger on polyacrylamide gels due to its high proline content. The precursor protein contains two RGD sequences but only one is maintained in the mature form. The RGD sequence is believed to be the major mechanism through which pertactin promotes adherence of *Bordetella* to host cells (Leininger et al. 1991). Pertactin also contains a nicotinamide-adenine dinucleotide (NAD)-binding site similar to that present on fragment A of diphtheria toxin. This site is not present in pertactin, which is produced by *B. parapertussis*, but is present on that from the other species (Li et al. 1991). Despite the presence of this binding site, binding of NAD by pertactin has not been demonstrated. Pertactin has been shown to be a protective antigen that can induce immunity to atrophic rhinitis in pigs and can stimulate protection against *B. bronchiseptica* infection in mice.

Fimbriae

Fimbriae are produced by all four species of *Bordetella* but their role in attachment is still uncertain. Figure 17.2 shows the presence of fimbriae on a *B. bronchiseptica* strain. Fimbriae are involved in attachment in many other species and there is evidence for their involvement in attachment by *B. bronchiseptica* and *B. avium* (Lee et al. 1986; Jackwood and Saif 1987). As with most *Bordetella* virulence factors, however, the majority of studies performed have used *B. pertussis*, in which fimbriae have been found to be unnecessary for attachment. These studies do not rule out the possibility of fimbriae functioning in addition to the other adhesins, or as primary adhesins in the other species, or as determinants of host specificity. Therefore, further studies on fimbriae from *Bordetella* spp. other than *B. pertussis* are necessary.

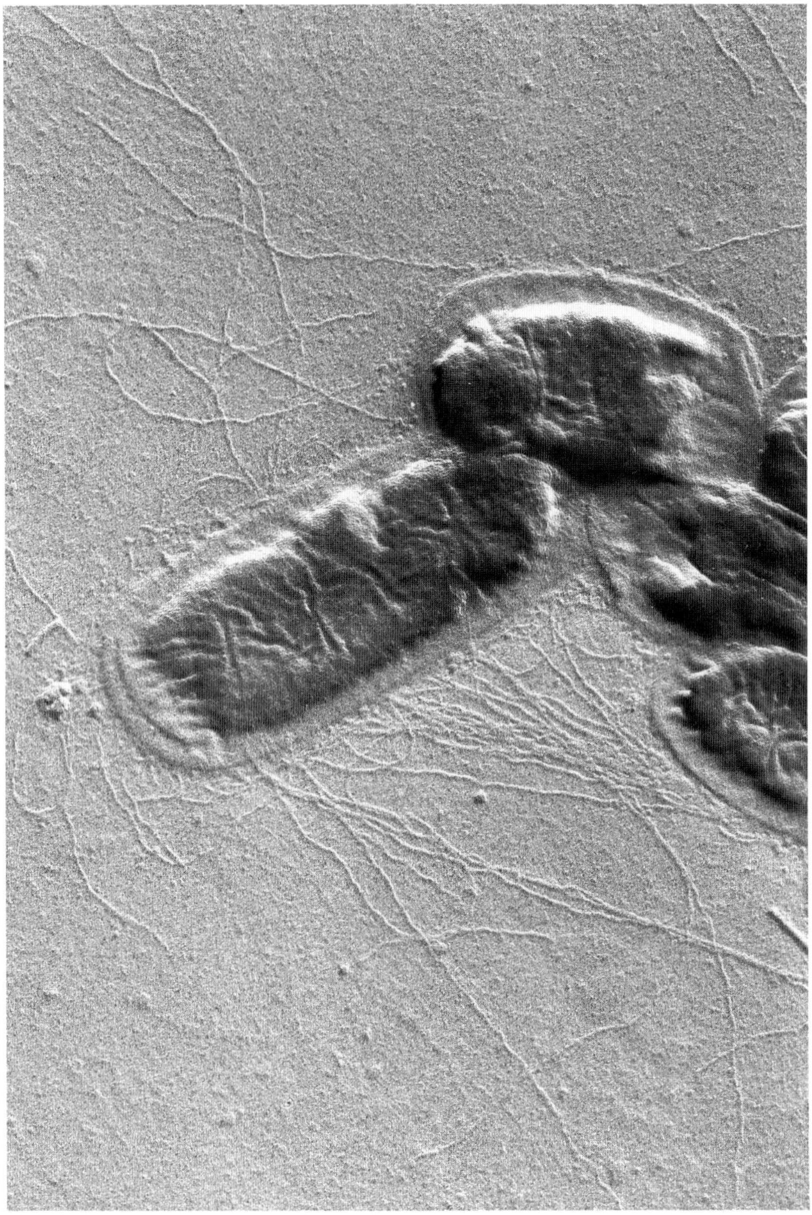

FIG. 17.2. Transmission electron micrograph of shadow-casted *Bordetella bronchiseptica* virulent-phase organisms showing numerous fimbriae. x34,000.

Fimbriae of *B. pertussis* have been classified into two serotypes based on agglutination with standard typing sera. In the Eldering serotyping scheme, agglutinogens 2 and 3 correspond to fimbriae. These sera cross-react with *B. bronchiseptica* and *B. avium* strains, indicating that at least two fimbrial types exist in these species (Mooi et al. 1987). In *B. pertussis*, fimbriae are composed of subunits of 22.5 kDa for serotype 2 and 22 kDa for serotype 3. In *B. bronchiseptica*, three proteins of molecular weights 21.5, 22, and 24 kDa have been identified as being fimbrial subunits, and in *B. avium,* three proteins of 14.4, 17, and 24 kDa have been identified as fimbrial proteins. It is not known if each of these proteins represents a distinct monomeric fimbrial type, or if each is a subunit of a multimeric type of fimbriae or some combination of the two.

Three fimbrial genes have been identified in *B. pertussis* and are designated *fim*2, *fim*3, and *fim*X. Each is present as a single-copy gene. *fim*X is not expressed in *B. pertussis* but activity has been observed from the *fim*X promoter when cloned into *B. bronchiseptica* (Riboli et al. 1991). In *B. bronchiseptica* and *B. parapertussis*, five to six *Sal*I fragments of genomic DNA hybridize with an oligonucleotide probe specific for one end of *fim*2, indicating that the fimbrial genes are present in multiple copies in these species. Interestingly, this probe did not hybridize with genomic DNA from *B. avium* despite the fact that antibodies against serotype 2 fimbriae reacted on Western blots. It is not known if each copy of the fimbrial genes present is expressed.

Genes coding for accessory proteins involved in regulation of expression and transport of fimbriae have been identified and designated *fim*B, C, and D. *fim*B and *fim*C exhibit similarity to *E. coli pap*D and *pap*C genes, respectively, and there is some speculation that *fim*D may encode a fimbrial adhesin protein (Willems et al. 1992). As already indicated, these genes are present immediately downstream of the FHA operon, and mutations in these genes affect production of both FHA and fimbriae.

Fimbriae are highly immunogenic and antibodies against fimbriae are common in convalescent serum. In mouse studies, fimbriae are protective antigens and have been used in acellular and component pertussis vaccines.

Adenylate Cyclase Toxin/Hemolysin

Adenylate cyclase toxin, which has also been called cyclolysin, is a bifunctional protein exhibiting both adenylate cyclase and hemolysin activity. It is produced by all species of *Bordetella* except *B. avium.* The native form of the protein is approximately 216 kDa and is secreted from the bacterial cell. Smaller, catalytically active fragments ranging from 43 to 70 kDa have been found, but these are believed to be degradation products formed during purification procedures. Extracellular adenylate cyclase toxin causes buildup of cAMP inside affected macrophages and neutrophils; this buildup inhibits the respiratory burst of macrophages and prevents phagocytic activity of neutrophils. Therefore, the action of adenylate cyclase toxin helps the bacterium evade host immune system defenses.

After release from the bacterial cell, adenylate cyclase toxin binds to and enters the host cell. Binding and facilitating entry into the cell is the main role for the hemolysin portion of the adenylate cyclase toxin molecule. This process is not mediated by endocytosis as it is for the *Bacillus anthracis* adenylate cyclase, but is believed to occur through binding to glycolipid moieties on the cell surface and subsequent formation of a pore (Gordon et al. 1988). The carboxy-terminal portion of the molecule is homologous to the *E. coli* alpha-hemolysin and the *Pasteurella haemolytica* leukotoxin. These molecules are both known to act through the formation of pores in the cell membrane. All three of these proteins belong to the repeat in toxin (RTX) family of toxins. RTX toxins exhibit repeated amino acid sequences in certain regions of the protein. *Bordetella* adenylate cyclase toxin contains 15 repeats of a 13-amino acid sequence and 5 repeats of a 9-amino acid sequence in the C-terminal hemolysin portion of the molecule (Glaser et al. 1988).

After entry into the cell, mediated by the hemolysin, the amino-terminal adenylate cyclase domain begins to catalyze the production of cAMP. This portion of the molecule is highly similar to mammalian adenylate cyclase and to *B. anthracis* adenylate cyclase. It has a calmodulin-binding domain and cAMP production is activated by calmodulin. Increases in intracellular cAMP levels due to the presence of this enzyme prevent the respiratory burst of macrophages, which may allow intracellular survival of bacteria (Pearson et al. 1987). In addition, adenylate cyclase toxin activity inhibits phagocytosis by neutrophils (Confer and Eaton 1982). Thus, through the action of this toxin, fewer bordetellae become internalized by cells of the immune system, and those that do become internalized have an increased chance for survival.

The adenylate cyclase toxin is encoded by a five-gene operon similar to that which encodes the *E. coli* alpha-hemolysin and *P. haemolytica* leukotoxin. *cya*A is the structural gene for the adenylate cyclase toxin (Glaser et al. 1988). *cya*B and *cya*D encode proteins involved in secretion. *cya*C is necessary for proper posttranslational modification of the protein to an active form (Barry et al. 1991). The nature of this modification is not as yet known for the *Bordetella* adenylate cyclase toxin. *cya*E is the only one of the adenylate cyclase toxin genes that does not show similarity with the *E. coli* alpha-hemolysin genes. The gene product of *cya*E is known to be involved in secretion.

Dermonecrotic Toxin

The dermonecrotic toxin is produced by all four species of *Bordetella*. It ranges in size from 155 kDa to 190 kDa, depending on the species; smaller, approximately 75-kDa forms have been reported but are believed to be proteolytic degradation products. The toxin is intracellular and is released upon lysis of the bacteria. Some investigators have reported that a portion of the dermonecrotic toxin molecule may be exposed on the cell surface. Dermonecrotic toxin has also been called heat-labile toxin since it can be inactivated by heating to 56°C.

Dermonecrotic toxin produces nonulcerating necrotic lesions when injected subcutaneously into animals and is lethal for mice at high doses. It is known to inhibit the sodium-potassium ATPase and cause contraction of smooth muscle surrounding peripheral blood vessels (Endoh et al. 1990). This vasoconstriction is responsible for the splenic atrophy observed in mice injected with dermonecrotic toxin. The toxin does not affect ciliary activity but does cause damage to cultured swine nasal tissue. The degree of turbinate atrophy induced by *B. bronchiseptica* in pigs correlates with the level of dermonecrotic toxin produced.

The dermonecrotic toxin of *Bordetella* spp. is not functionally the same as that of *Pasteurella multocida*, which is responsible for the most destructive and permanent form of atrophic rhinitis. The *P. multocida* dermonecrotic toxin increases activity of osteoclasts; this is not observed with the *Bordetella* dermonecrotic toxin. The *Bordetella* dermonecrotic toxin impairs osteogenesis by causing damage to osteoblasts and osteoprogenitor cells (Horiguchi et al. 1991). Cultured osteoprogenitor cells exposed to purified dermonecrotic toxin become irregular in shape with small blebs on their surfaces. The cells remain viable and continue to proliferate but at a reduced growth rate. These cells exhibit greatly reduced alkaline phosphatase activity and type I collagen levels, both of which would normally become elevated during differentiation to osteoblasts. Thus, turbinate atrophy caused by *Bordetella* is due to inhibition of osteogenisis not to osteoclastic resorption as in *P. multocida* infection. *B. bronchiseptica*-induced nasal turbinate atrophy is generally less severe than that induced by *P. multocida*, and regeneration to nearly normal turbinate structure occurs by the time *B. bronchiseptica*-infected pigs reach market weight. Experimental infection with toxigenic *B. bronchiseptica* enhances the ability of toxigenic *P. multocida* to cause atrophic rhinitis in pigs.

Osteotoxin

Recently, an 80-kDa protein from *B. avium* that is toxic to osteoblasts has been identified as beta-cystathionase, a homodimeric enzyme with one molecule of pyridoxal 5′ phosphate in each subunit (Gentry-Weeks et al. 1993). This osteotoxin does not exhibit dermonecrotic toxin activity or react with antibodies against dermonecrotic toxin. Osteotoxin has also been demonstrated in other *Bordetella* species, including *B. bronchiseptica*, but its role in disease has not yet been determined.

Tracheal Cytotoxin

The tracheal cytotoxin is a soluble monomer of peptidoglycan released from actively growing cells of all *Bordetella* spp. It has the chemical composition of N-acetylglucosaminyl-1, 6-anhydro-N-acetylmuramylalanyl-γ-glutamyldiaminopimelyl-alanine and is believed to be produced by cleavage of intact peptidoglycan, not by release of precursors (Goldman et al. 1990). The peptide portion of the molecule has a molecular weight of around 921 D and the tracheal cytotoxin of *Bordetella* is identical to a peptidoglycan monomer released by *Neisseria gonorrhoeae*, which is responsible for destruction of ciliated cells in human Fallopian tubes.

In the presence of the *Bordetella* tracheal cytotoxin, ciliary activity is decreased and ciliated cells lose their normal elongated shape, becoming rounded with a constriction of their apical end. In addition, mitochondria become swollen and exhibit decreased amounts of cristae. Intercellular junctions also deteriorate, leading to extrusion of the cell. This loss of ciliated cells causes accumulation of mucus, bacteria, and foreign particles, which leads to coughing and increased susceptibility to secondary infections. Besides causing loss of ciliated cells, the tracheal cytotoxin inhibits DNA synthesis in epithelial cells. Therefore, division and differentiation of the basal cell population is delayed preventing replacement of the ciliated cells extruded due to the action of the tracheal cytotoxin.

Tracheal cytotoxin has been shown to compete with serotonin for binding sites on hamster tracheal epithelial cells. Tracheal cytotoxin was found to bind to this site with low affinity, and it is not known if this is the only binding site present. The mechanism of toxicity has not been determined, but there has been speculation that tracheal cytotoxin may interfere with or mimic the effects of serotonin.

Tracheal cytotoxin is believed to be an excellent adjuvant like other muramyl peptides, but antibodies against tracheal cytotoxin itself are difficult to produce. There are two reasons for this rarity of antibodies. First, the tracheal cytotoxin is identical to the human sleep-promoting factor FSu, and similar molecules may exist in animals. Second, since tracheal cytotoxin is a component of bacterial cell walls, animals may become tolerant due to the presence of normal flora.

Other Factors

Bordetella produce several other compounds that could be considered virulence factors. One of these factors is lipopolysaccharide (LPS). *Bordetella* LPS has biologic activities similar to LPS from *Enterobacteriaceae*; it is lethal for galactosamine-sensitized mice, it is pyrogenic and mitogenic, and it causes macrophage activation and induction of tumor necrosis factor production (Amano et al. 1991). There have been reports that LPS composition in *Bordetella* may be different in avirulent and virulent phase. *Bordetella* also produce several outer-membrane proteins that could prove to play a role in virulence. One of these is a 40-kDa porin protein, which is similar to the GroEL proteins found in other species (Burns et al. 1991). There is also a suggestion that these porin proteins undergo qualitative changes between phases. The recently described hydroxamate siderophore, bordetellin, and the binding proteins for lactoferrin and transferrin may have a number of effects on virulence that have not been identified (Agiato and Dyer 1992).

Regulation

All the virulence factors mentioned above, with the exception of tracheal cytotoxin, are under control of the Bvg regulatory system. Due to the presence of this regulatory system, *Bordetella* exist in two phases. In the virulent phase, also called phase I, *Bordetella* produce the virulence factors FHA, pertactin, fimbriae, adenylate cyclase, and dermonecrotic toxin. In the avirulent phase, also called phase III, *Bordetella* do not produce these virulence factors but do produce other proteins that are normally repressed in the virulent phase, such as flagella in *B. bronchiseptica*. Phase change can occur by two mechanisms: phase variation and modulation. Phase variation involves an irreversible change from virulent phase to avirulent phase, which is caused by a spontaneous mutation in the *bvg*S gene whose function will be described later in this discussion (Monack et al. 1989). Modulation involves a reversible change between phases caused by environmental factors (Melton and Weiss 1989). Growth at low temperature or in the presence of high concentrations of magnesium sulfate or nicotinic acid causes modulation from virulent phase to avirulent phase. Removal of the modulating agent allows reversion to virulent phase.

The Bvg regulatory system, formerly called Vir, is similar to other bacterial two-component regulatory systems. It involves a 23-kDa transcriptional activator protein encoded by the *bvg*A gene and a 135-kDa transmembrane protein encoded by *bvg*S, which functions as an environmental sensor. The *bvg*S gene has been identified in all *Bordetella* spp. The *bvg*A gene has been found in all species except *B. avium* (Gentry-Weeks et al. 1991). BvgS is located in the inner membrane. Its periplasmic domain is the sensory domain and its cytoplasmic domain is a protein kinase that phosphorylates BvgA under nonmodulating conditions (Arico et al. 1991). The phosphorylated form of BvgA enhances transcription from the *bvg*AS genes, positively regulating its own production. Phosphorylated BvgA also binds to and increases transcription from the promoters for FHA and Act.

Act is a 23-kDa activator protein distinct from BvgA. The *act* gene was discovered in *B. pertussis* and its existence in the other species has not been confirmed. Act binds to promoters for adenylate cyclase toxin and pertussis toxin, increasing production of these virulence factors (Huh and Weiss 1991). The promoters for pertactin and fimbriae exhibit homology with the pertussis toxin promoter. These proteins are known to be regulated by the Bvg regulatory system and due to this homology, the mechanism of their regulation may be through the Act protein. Dermonecrotic toxin is also Bvg regulated. Cloning of the dermonecrotic toxin gene has been reported, but there has not been enough analysis of its promoter region for speculation as to whether its transcription is enhanced by BvgA directly or through the action of the Act protein. Tracheal cytotoxin production is not Bvg regulated and occurs in both virulent phase and avirulent phase.

Under modulating conditions, BvgA is not phosphorylated. Additional transcriptional enhancement of the *bvg*AS genes does not occur, the Act protein is not produced, and Bvg-regulated virulence factors are not expressed. The vir-repressed genes (*vrgs*), such as those that encode production of flagella in *B. bronchiseptica*, do become activated (Akerley et al. 1992). The mechanism of activation of *vrgs* is unknown. These genes could be repressed by phosphorylated BvgA or Act, or they could be activated by nonphosphorylated BvgA. Many of the *vrgs* encode proteins whose functions have not presently been determined. A *B. pertussis* *vrg*-6 mutant failed to induce lymphocytosis and was significantly less able to colonize lungs and trachea of mice than its parent strain, suggesting a possible role for *vrgs* in pathogenesis (Beattie et al. 1992). The regulation of *Bordetella* virulence factors under modulating and nonmodulating conditions is diagrammed in Figure 17.3.

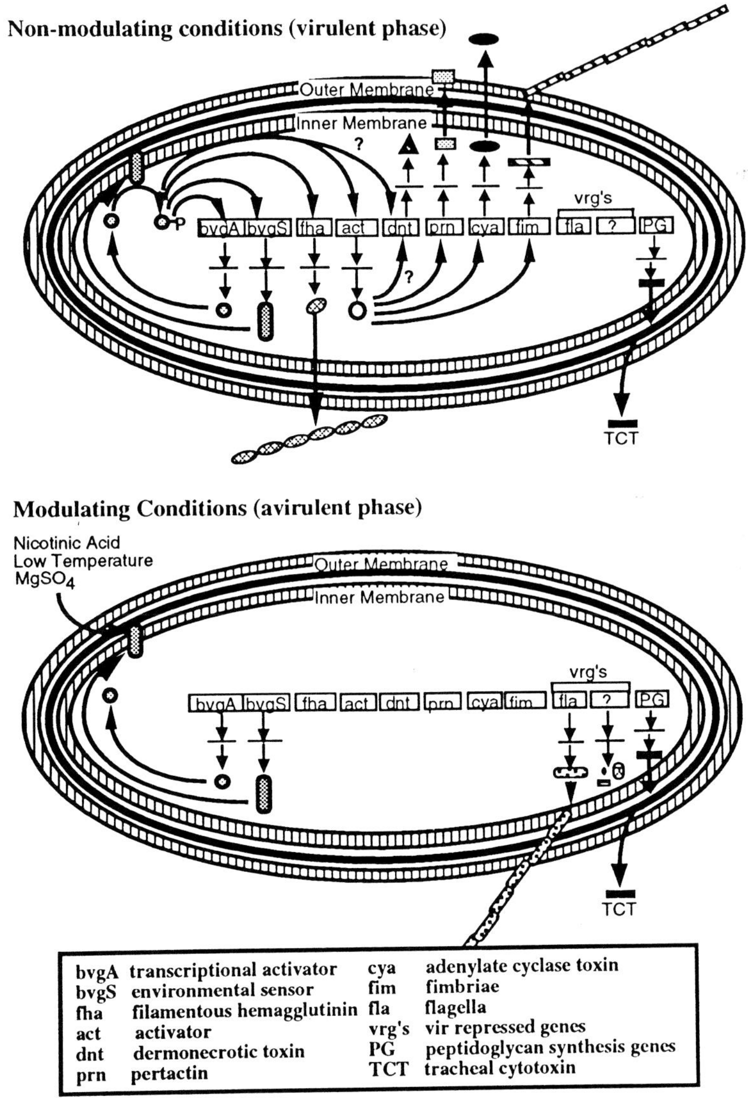

FIG. 17.3. Schematic of genetic regulation of virulence in *Bordetella*.

PATHOGENESIS

Exposure to *Bordetella* appears to be directly or indirectly dependent on contact with infected animals. Droplet aerosols, dust, and water can facilitate the spread of infection. Prevalence of infection is greatest in young animals, but infections may be seen in animals of all ages.

Bordetella spp. have marked predilection for ciliated tissue of the respiratory tract; rarely do they colonize other tissues of healthy animals. The mechanisms for this organotropism are not entirely known, but virulent-phase *Bordetella* tightly adhere to cilia in susceptible animals. FHA, pertactin, and pertussis toxin appear to mediate this function in *B. pertussis*. FHA also functions as an adhesin in *B. bronchiseptica*, and hemagglutinating activity and fimbriation are correlated with attachment in *B. bronchiseptica* and *B. avium*. It has been suggested that *B. bronchiseptica* and *B. avium* attach to sialoganglioside receptors in the respiratory tract, whereas *B. pertussis* attaches to asialoglycosphingolipids and sulfatides (Ishikawa and Isayama 1987; Arp et al. 1988; Brennan et al. 1991).

Once anchored to the epithelium, *Bordetella* find the pH, oxygenation level, and nutrient availability suitable for proliferation. As mentioned previously, proline and glutamic acid are found in high levels in respiratory secretions of certain mammals. *Bordetella* do not avoid recognition by professional phagocytes; in fact, to the contrary, they have several surface properties (FHA, pertactin) that provide for specific receptor-mediated uptake. However, the presence of phagocytes at the site and their efficiency in the moving environment of respiratory tract cilia are unknown. Although some *Bordetella* virulence factors produce destructive effects on phagocytes, infection is not easily established in body tissues that are readily accessible by phagocytes. Differences in host specificity are reflected in *Bordetella* attachment patterns. Full identity of these adhesins and their receptors and what effect other elements of colonization have on host specificity have yet to be determined.

Bacterial growth in the lumen of the airways results in increased attraction of polymorphonuclear leukocytes (PMNs) to the epithelium. Lipopolysaccharide alone can produce this effect in the airways and is likely to play a major role in PMN chemotaxis during infection. Inhibition of ciliary motility and tracheobronchial clearance occur early in infection. Presumably, tracheal cytotoxin and other factors such as ammonia released during bacterial metabolism contribute to the impairment of ciliary function. Dermonecrotic toxin and adenylate cyclase toxin do not appear to have pronounced effects on ciliary activity, but the effects of these toxins on ciliated epithelial cells require further study. Increased mucus production and diminished tracheobronchial clearance may lead to an accumulation of neutrophil-laden exudates, which can obstruct small airways and produce severe or chronic disease. Inhibition of alveolar macrophage function by adenylate cyclase toxin may facilitate secondary infections. Osteo- toxins and/or dermonecrotic toxin are responsible for nasal turbinate atrophy seen during *Bordetella* infections. Dermonecrotic toxin, adenylate cyclase toxin, and other *Bordetella* products may also diminish host immune responses to a wide variety of antigens (Horiguchi et al. 1992).

Extracellular virulence factors are thought to be of primary importance in *Bordetella* pathogenesis because these organisms normally remain on the lumenal surface of epithelial cells. *Bordetella* are rarely detected in the blood stream or deep body tissues except in compromised hosts. Virulent-phase *Bordetella* are more resistant to complement-mediated serum killing than avirulent *Bordetella*. Survival inside macrophages and PMNs, as well as uptake and survival (invasion) of virulent-phase *B. pertussis* in nonprofessional phagocytes, have recently been demonstrated (Confer and Eaton 1982; Ewanowich et al. 1989; Steed et al. 1992). Invasion efficiency was low and evidence for intracellular replication or intercellular spread is lacking. Whether intracellular survival occurs in vivo or plays a role in pathogenesis is not known.

IMMUNITY

Symptoms of uncomplicated bordetellosis in animals may resolve in a few days; however, complete clearance of *Bordetella* from the respiratory tract often requires several weeks or months. In human pertussis, paroxysmal coughing typically persists long after *B. pertussis* has been cleared from the respiratory tract. Presumably, the main difference in these symptoms is largely due to pertussis toxin.

Clearance and immunity to reinfection is mediated by locally produced antibodies, which begin to appear as early as 4 days after exposure (Bey et al. 1981; Arp and Hellwig 1988). Local antibodies block attachment of bacteria to cilia but do not remove bacteria that are already attached. Most animals, if they are kept free from sources of reexposure, will eventually become *Bordetella* free and be immune to reinfection. Local immunity is short-lived; as immunity wanes, animals may become infected again. Whether reinfected or chronically infected, a small percentage of all adult populations appears to constantly shed *Bordetella* and represents another important source of transmission. Frequent reexposure bolsters immunity and is the basis for live intranasal vaccines.

Serum antibodies are produced in variable but generally low quantities following infection. Systemic vaccination does not prevent infection but offers some protection against clinical disease. Pertactin, filamentous hemagglutinin, and perhaps other *Bordetella* virulence factors are major protective antigens in effective vaccines (Kobisch and Novotny 1990; Ohgitani et al. 1992). The role of *Bordetella*-specific cell-mediated immunity in protection against disease is unknown; a variety of cell-mediated immune responses have been demonstrated in infected animals. Whole cell bacterins and live attenuated intranasal vaccines have been widely employed to protect against animal bordetellosis. Acellular *B. bronchiseptica* vaccines have also become commercially available. As is already being investigated with *B. pertussis*, component and recombinant vaccines show promise for future control of animal bordetelloses.

REFERENCES

Agiato, L. A., and Dyer, D. W. 1992. Siderophore production and membrane alterations by *Bordetella pertussis* in response to iron starvation. Infect Immun 60:117-23.

Akerley, B.; Monack, D.; Falkow, S.; and Miller, J. 1992. The *bvg*AS locus negatively controls motility and synthesis of flagella in *Bordetella bronchiseptica*. J Bacteriol 174: 980-90.

Amano, K.; Wantanabe, M.; Takimoto, H.; and Kumazawa, Y. 1991. Structural and biological comparison of lipopolysaccharides (LPS) from *Bordetella* species. Dev Biol Stand 73:193-199.

Arico, B., and Rappuoli, R. 1987. *Bordetella parapertussis* and *Bordetella bronchiseptica* contain transcriptionally silent pertussis toxin genes. J Bacteriol 169:2847-53.

Arico, B.; Scarlato, V.; Monack, D.; Falkow, S.; and Rapuoli, R. 1991. Structural and genetic analysis of the *bvg* locus in *Bordetella* species. Mol Microbiol 5:2481-91.

Arp, L. H.; Hellwig, D. H.; and Huffman, E. L. 1988. Sialogangliosides as putative tracheal musocal receptors for *Bordetella avium* adhesins. Annu Meet Am Soc Microbiol (Abstr) B164, Miami Beach.

Arp, L. H., and Hellwig, D. H. 1988. Passive immunization versus adhesion of *Bordetella avium* to the tracheal mucosa of turkeys. Avian Dis 32:494-500.

Barry, E.; Weiss, A.; Ehrmann, I.; Gray, M.; Hewlett, E.; and Goodwin, M. 1991. *Bordetella pertussis* adenylate cyclase toxin and hemolytic activities require a second gene, *cyaC*, for activation. J Bacteriol 173:720-26.

Beattie, D. T.; Shahin, R.; and Mekalanos, J. J. 1992. A *vir*-repressed gene of *Bordetella pertussis* is required for virulence. Infect Immun 60:571-77.

Bey, R. F.; Shade, F. J.; Goodnow, R. A.; and Johnson, R. C. 1981. Intranasal vaccination of dogs with live avirulent *Bordetella bronchiseptica*: correlation of serum agglutination titer and the formation of secretory IgA with protection against experimentally induced infectious tracheobronchitis. Am J Vet Res 42:1130-32.

Brennan, M. J.; Hannah, J. H.; and Leininger, E. 1991. Adhesion of *Bordetella pertussis* to sulfatides and to the Gal NAc beta 4 Gal sequence found in glycosphingolipids. J Biol Chem 266:18827-31.

Burns, D.; Gould-Kostka, J.; Kessel, M.; and Aciniega, J. 1991. Purification and immunological characterization of a GroEL-like protein from *Bordetella pertussis*. Infect Immun 59:1417-22.

Confer, S., and Eaton, J. 1982. Phagocytic impotence caused by an invasive bacterial adenylate cyclase. Science 217:948-50.

Cullinane, L. C.; Alley, M. R.; Marshall, R. B.; and Manktelow, B. W. 1987. *Bordetella parapertussis* from lambs. NZ Vet J 35:175.

Endoh, M.; Nagai, M.; Burns, D. L.; Nakase, Y.; and Manclark, C. R. 1990. Effect of exogenous agents on the action of *Bordetella* heat-labile toxin. Proc 6th Int Symp Pertussis, pp. 30-36, Bethesda.

Ewanowich, C. A.; Melton, A. R.; Weiss, A. A.; Sherburne, R. K.; and Peppler, M. S. 1989. Invasion of HeLa 229 cells by virulent *Bordetella pertussis*. Infect Immun 57:2698-704.

Gentry-Weeks, C. R.; Cookson, B.; Goldman, W.; Rimler, R.; Porter, S.; and Curtis, R. 1988. Dermonecrotic toxin and tracheal cytotoxin, putative virulence factors of *Bordetella avium*. Infect Immun 56:1698-707.

Gentry-Weeks, C. R.; Provence, D.; Keith, J.; and Curtis, R. 1991. Isolation and characterization of *Bordetella avium* phase variants. Infect Immun 59: 4026-33.

Gentry-Weeks, C. R.; Keith, J. M.; and Thompson, J. 1993. Toxicity of *B. avium* beta-cystathionase toward MC3T3-E1 osteogenic cells. J Biol Chem 268:In press.

Glaser, P.; Danchin, A.; Ladant, D.; Barzu, O.; and Ullman, A. 1988. *Bordetella pertussis* adenylate cyclase: The gene and the protein. Tokai J Exp Clin Med 13(Suppl):239-52.

Goldman, W.; Collier, J.; Cookson, B.; Marshall, G.; and Erwin, K. 1990. Tracheal cytotoxin of *Bordetella pertussis*: Biosynthesis, structure, and specificity. Proc 6th Int Symp Pertussis, pp. 5-12. Bethesda.

Gordon, V.; Leppla, S.; and Hewlett, E. 1988. Inhibitors of receptor mediated endocytosis block the entry of *Bacillus anthracis* adenylate cyclase toxin but not that of *Bordetella pertussis* adenylate cyclase toxin. Infect Immun 56:1066-69.

Horiguchi, Y.; Matsuda, H.; Koyama, H.; Nakai, T.; and Kume, K. 1992. *Bordetella bronchiseptica* dermonecrotizing toxin suppresses in vivo antibody responses in mice. Fed Eur Microbiol Soc Microbiol Lett 69:229-34.

Horiguchi, Y.; Nakai, T.; and Kume, K. 1991. Effects of *Bordetella bronchiseptica* dermonecrotic toxin on the structure and function of osteoblastic clone MC3T3-E1 cells. Infect Immun 59:1112-16.

Huh, Y., and Weiss, A. 1991. A 23 kilodalton protein, distinct from BvgA, expressed by virulent *Bordetella pertussis* binds to the promoter region of *vir* regulated toxin genes. Infect Immun 59:2389-95.

Ishikawa, H., and Isayama, Y. 1987. Evidence for sialyl glycoconjugates as receptors for *Bordetella bronchiseptica* on swine nasal mucosa. Infect Immun 55:1607-9.

Jackwood, M., and Saif, Y. 1987. Pili of *Bordetella avium*: Expression, characterization, and role in in vitro adherence. Avian Dis 31:277-86.

Kersters, K.; Hinz, K. H.; Hertle, A.; Segers, P.; Lievens, A.; Siegmann, O.; and DeLey, J. 1984. *Bordetella avium* sp. nov., isolated from the respiratory tracts of turkeys and other birds. Int J Syst Bacteriol 34:56-70.

Kobisch, M., and Novotny, P. 1990. Identification of a 68 kilodalton outer membrane protein as the major protective antigen of *Bordetella bronchiseptica* by using specific pathogen free piglets. Infect Immun 58:352-57.

Lee, S.; Way, A.; and Osen, E. 1986. Purification and subunit heterogeneity of pili of *Bordetella bronchiseptica*. Infect Immun 51:586-93.

Leininger, E.; Roberts, M.; Kenimer, J.; Charles, I.; Fairweather, N.; Novotny, P.; and Brennan, M. 1991. Pertactin, an arg-gly-asp containing *Bordetella pertussis* surface protein that promotes adherence of mammalian cells. Proc Natl Acad Sci USA 88:345-49.

Li, L.; Dougan, G.; Novotny, P.; and Charles, I. 1991. P.70 pertactin, an outer membrane protein from *Bordetella parapertussis*: Cloning, nucleotide sequence, and surface expression in *Escherichia coli*. Mol Microbiol 5:409-17.

Locht, C.; Geoffroy, M.; and Renauld, G. 1992. Common accessory genes for the *Bordetella pertussis* filamentous hemagglutinin and fimbriae share sequence similarities with the *pap*D and *pap*C gene families. Eur Mol Biol Organ J 11:3175-83.

Melton, A., and Weiss, A. 1989. Environmental regulation of expression of virulence determinants in *Bordetella pertussis*. J Bacteriol 171:6206-12.

Menozzi, F. D.; Gantiez, C.; and Locht, C. 1991. Interactions of the *Bordetella pertussis* filamentous hemagglutinin with heparin. Fed Eur Microbiol Soc Microbiol Lett 62:59-64.

Monack, D.; Arico, B.; Rappuoli, R.; and Falkow, S. 1989. Phase variants of *Bordetella bronchiseptica* arise by spontaneous deletions in the *vir* locus. Mol Microbiol 3:1719-28.

Mooi, F.; van der Heide, H.; terAvest, A.; Welinder, K.; Livey, I.; van der Zeijst, B.; and Gaastra, W. 1987. Characterization of fimbrial subunits from *Bordetella* species. Microb Pathog 2:473-84.

Ohgitani, T.; Uchida, C.; Okabe, T.; and Sasaki, N. 1992. Protective effect by cell-free antigen obtained from culture supernatant of Phase I *Bordetella bronchiseptica*. J Vet Med Sci 54:37-42.

Pearson, R.; Symes, P.; Conboy, M.; Weiss, A.; and Hewlett, E. 1987. Inhibition of monocyte oxidative responses by *Bordetella pertussis* adenylate cyclase toxin. J Immunol 139:2749-54.

Porter, J. F.; Parton, R.; and Wardlaw, A. C. 1991. Growth and survival of *Bordetella bronchiseptica* in natural waters and in buffered saline without added nutrients. Appl Environ Microbiol 57:1202-6.

Relman, A.; Domenighini, M.; Tuomanen, E.; Rappouli, R.; and Falkow, S. 1989. Filamentous hemagglutinin of *Bordetella pertussis*: Nucleotide sequence and crucial role in adherence. Proc Natl Acad Sci USA 86:2637-41.

Relman, D.; Tuomanen, E.; Falkow, S.; Golenbock, D.; Saukkonen, K.; and Wright, S. 1990. Recognition of a bacterial adhesin by an integrin: Macrophage CR3 ($\alpha_M\beta_2$, CD11b/CD18) binds filamentous hemagglutinin of *Bordetella pertussis*. Cell 61:1375-82.

Riboli, B.; Pedroni, P.; Cuzzoni, A.; Grandi, G.; and de Ferra, F. 1991. Expression of*Bordetella pertussis* fimbrial (*fim*) genes in *Bordetella bronchiseptica*: *fimX* is expressed at a low level and *vir* regulated. Microb Pathog 10:393-403.

Steed, L. L.; Akporiaye, E. T.; and Friedman, R. L. 1992. *Bordetella pertussis* induces respiratory burst activity in human polymorphonuclear leukocytes. Infect Immun 60:2101-205.

Stehmann, R.; Mehlhorn, G.; and Neuparth, V. 1991. Characterization of strains of *Bordetella bronchiseptica* isolated from animal housing air and of the airborne infection pressure proceeding from them. Dtsch Tieraerztl Wochenschr 98:448-50.

Willems, R. J. L.; van der Heide, H. G. J.; and Mooi, F. R. 1992. Characterization of a *Bordetella pertussis* fimbrial gene cluster which is located directly downstream of the filamentous hemagglutinin gene. Mol Microbiol 6:2661-71.

Woolfrey, B. F., and Moody, J. A. 1991. Human infections associated with *Bordetella bronchiseptica*. Clin Microbiol Rev 4:243-55.

18 / *Pasteurella*

BY P. E. SHEWEN AND J. A. RICE CONLON

PASTEURELLAE are gram-negative, nonmotile, fermentative, facultatively anaerobic coccobacilli or rods that show bipolar staining, particularly as fresh isolates stained with Romanofsky stains like Wright's or Giemsa (Kilian and Frederiksen 1981). Primarily animal pathogens, these organisms have adapted to parasitic life on the oral and upper respiratory epithelia of apparently healthy animals and occasionally humans. Most strains have marked host species specificity, being associated almost exclusively with one or two specific hosts. Infection is more common than disease, which usually occurs as a consequence of stress such as overcrowding, chilling, transportation, or intercurrent infection. Nevertheless, the "pasteurelloses," as these diseases are called, are of major pathologic and economic significance in veterinary medicine.

The type species, *Pasteurella multocida*, was extensively investigated in 1880 by Pasteur as the cause of fowl cholera and was subsequently identified in association with rabbit septicemia, swine plague, bovine pneumonia, and hemorrhagic septicemia. It has been isolated as part of the normal oral and pharyngeal flora of many species of animals including dogs, cats, wild and domestic ruminants, horses, swine, rabbits, opossums, rodents, birds, and reptiles. Growth on artificial media is best achieved with the addition of serum or blood, and it is distinguished from the other main species, *Pasteurella haemolytica*, by lack of hemolysis on blood agar and failure to grow on MacConkey's agar. Although *P. multocida* and *P. haemolytica* are morphologically similar, modern taxonomy, based on DNA-DNA or DNA-ribosomal RNA hybridization, indicates that *P. haemolytica* should more appropriately be grouped with the *Actinobacillaceae* (Mutters et al. 1986), but this proposed reassignment of genera has not yet been adopted.

PASTEURELLA MULTOCIDA

Diseases attributed to *Pasteurella multocida* occur worldwide in virtually all species of animals (Biberstein 1981). They include septicemias of cattle, swine, birds and rabbits; bronchopneumonia in cattle, sheep, goats, pigs, and rabbits; rhinitis of piglets and rabbits; and wound infections in dogs, cats, and humans. Many infections result from the invasion of commensal organisms during periods of stress, but exogenous transmission may occur by aerosol or contact exposure. Ticks and fleas have also been incriminated as natural vectors.

Several serological classification systems have been devised in an attempt to correlate specific serotypes or strains with host specificity, virulence, or disease. Five capsular serogroups (Rimler and Rhoades 1987) and 16 somatic serotypes (Heddleston et al. 1972) are currently utilized to differentiate the organism. Carter's classification and its subsequent modifications group *P. multocida* into five serotypes (Table 18.1) based on capsular polysaccharide antigens, using a passive hemagglutination test. A sixth type (type C), shown in Table 18.1, is not a valid capsular serotype.

TABLE 18.1. Characteristics of strains of *Pasteurella multocida* Types A-F

Type	Characteristics
A	Part of the normal flora of the respiratory tract of many domestic animals; associated with disease, usually respiratory, in stressed animals; includes most isolates from fowl cholera
B	Noncommensal; associated with hemorrhagic septicemia of ruminants in Asia and Australia
C	Part of the normal flora of dogs and cats; not encapsulated and thus not a valid type
D	Like A, part of the normal flora; causes disease in stressed animals; includes strains exclusively associated with atrophic rhinitis of pigs
E	Noncommensal; causes hemorrhagic septicemia in ruminants in central Africa
F	Associated with disease in turkeys

Heddleston's system was devised primarily to type avian isolates of *P. multocida*, since many highly virulent strains were found to be nonencapsulated and, therefore, untypable using Carter's system. Sixteen serotypes have been identified based on heat-stable somatic antigens, using a gel diffusion precipitin test. Each of Heddleston's serotypes is capable of inducing antiinfection immunity against the homologous serotype but cross-protection is rare. Most *P. multocida* strains are designated by a number for the Heddleston somatic antigen type and a letter for the capsular antigen (e.g., types 5:A, 8:A, 9:A produce fowl cholera).

In birds, *P. multocida* induces fowl cholera, an acute to peracute septicemia with worldwide occurrence that results in severe losses in both domestic and wild fowl (Rhoades and Rimler 1984). Domestic turkeys and wild waterfowl are particularly susceptible. Outbreaks are sporadic but devastating, with high mortality in the acute form common in turkeys. A more chronic respiratory form with lower mortality but prolonged course is usually observed in chickens. The organism is transmitted from nasal, oral, or conjunctival secretions of infected birds, often unapparent or convalescent carriers, and by aerosol or contamination of drinking water and feed. Clinical disease is precipitated by stress such as overcrowding, laying or moulting, or severe climatic change.

Septicemia results from penetration of the pharyngeal mucous membranes by highly pathogenic organisms. The virulence of these strains, all type A, correlates with their ability to resist phagocytosis once they enter the tissues. In acute cases, the predominant lesions, widespread hemorrhages and necrotic foci in the liver, are presumed to be due to endotoxemia (Collins 1977). Free endotoxin and large numbers of organisms have been demonstrated in tissues and body fluids of moribund animals; disseminated intravascular coagulation, typical of endotoxemia, is frequently seen. Correlations have also been drawn between strain virulence and the presence of neuraminidase.

Birds that recover from natural infection are solidly immune to reinfection with any avian serotype. However, parenteral vaccination with inactivated bacteria results in immunity that is specific for homologous, but not heterologous, serotypes. Therefore, although vaccination reduces the incidence of disease, fowl cholera does occur in flocks that have been properly vaccinated with commercial, mixed serotype bacterins. More reliable cross-protective immunity is induced by a commercially available live attenuated vaccine (CU strain) given to turkeys via the drinking water or administered parenterally to chickens. A disadvantage of this vaccine is the relatively high mortality occasionally observed if administered to birds during periods of undefined or unapparent stress and also the ability of the vaccine strain to revert to virulence (Davis 1987). Interestingly, cross-protection, apparently lacking after immunization of turkeys with killed bacteria grown in vitro, has been induced using killed bacteria grown in vivo (Rimler and Rhoades 1981). The protection induced suggests the production of cross-protective immunogens during in vivo growth of the organism, the identification of which may result in safer, more effective vaccines in the future.

Pasteurella multocida induces a similar fatal septicemic disease of cattle known as hemorrhagic septicemia. Catastrophic epidemics with high morbidity and mortality have occurred in cattle and water buffalo in Asia, South America, Australia, Europe, and Africa (Bain 1963). Isolated cases in individual animals, including deer in North America and England, and horses and sheep in North Africa, have also been reported. High fever and depression are followed quickly by death. There may be edema of the head and neck and bleeding from body orifices. At necropsy, hemorrhages and edema are observed on serous and mucous surfaces and in lymph nodes, spleen, lungs, and other organs. The causative organisms are either capsular type B or E (Africa only); all possess the somatic O antigen designation 6. Although these types may be isolated from the nasopharynx of clinically healthy carriers, their presence is never considered normal. Transmission of infection is probably by aerosol, but as with other pasteurelloses, there is an important environmental component in the pathogenesis of the disease. Epizootics occur seasonally in association with significant changes in weather, for example, during the monsoon season in Southeast Asia.

Hemorrhagic septicemia is the only pasteurellosis for which a safe and effective bacterin is available. The efficacy of this vaccine probably relates to the restricted serotype specificity of causative organisms. Although the virulence factors responsible for the production of hemorrhagic septicemia are unknown, it has been shown that the highly virulent type B bacteria produce hyaluronidase (Carter and Chengappa 1980), an enzyme otherwise associated with gram-positive bacteria. Its possible significance in the pathogenesis of hemorrhagic septicemia is under investigation.

Pasteurella multocida is a common and serious pathogen of rabbits, producing a variety of syndromes. The principal sign of chronic enzootic pasteurellosis is recurrent purulent rhinitis ("snuffles"), which often results in sequelae such as conjunctivitis, otitis media, sinusitis, subcutaneous abscesses, and chronic bronchopneumonia. Chronic genital infection leading to pyometra or orchitis has also been reported. When sporadic fatal epizootics occur, rabbits die as a result of acute fibrinous bronchopneumonia, septicemia, or endotoxemia (Flatt 1974). *P. multocida* is the predominant organism recovered from the nasal passages of rabbits. The organisms type mainly as capsular type A or are untypable somatic antigen types 3, 4, and 5 (DiGiacomo et al. 1991).

Clinical disease is precipitated by stress such as overcrowding, transportation, experimental manipulation, or concurrent infection. *Bordetella bronchiseptica* is frequently isolated with *P. multocida* in respiratory disease. Several recent studies of pathogenesis have demonstrated the peculiar ability of virulent type A bacteria to adhere to rabbit mucosal epithelium, which is apparently mediated by fimbriae, and the ability of virulent organisms to resist phagocytosis and killing by rabbit neutrophils. Resistance to phagocytosis has been attributed to the hyaluronic acid capsule or other cell-associated proteins (outer-membrane proteins), which have been shown to interfere with phagocytosis (Truscott and Hirsch 1988). Pasteurellosis in rabbit colonies is prevented by routine antibiotic therapy to reduce carriers, by culling carriers recognized by regular nasal sampling, or by the use of caesarian-derived stock. In experimental trials, live, streptomycin-dependent mutants have been used as vaccines. These protect against clinical disease after challenge with homologous serotypes but do not prevent the development of the carrier state in animals.

Pasteurella multocida types A and D are carried in the nasal cavity of clinically normal swine. Certain type D strains produce a heat-labile exotoxin, which is dermonecrotic in guinea pigs, lethal for mice, and cytotoxic for Vero cells. It is now generally accepted that atrophic rhinitis of piglets results most often from combined infection of the nasal turbinates with *Bordetella bronchiseptica* and toxigenic strains of *P. multocida* type D, designated AR^+ (DeJong et al. 1980). Pathogenic strains of *B. bronchiseptica* cause nonprogressive turbinate hypoplasia, followed by turbinate regeneration; complication with AR^+ *P. multocida* results in severe, progressive atrophic rhinitis. Infection with large numbers of AR^+ *P. multocida* alone, particularly in conjunction with external stress, can induce atrophic rhinitis in the absence of *B. bronchiseptica*. In such cases, the toxin has been incriminated as the cause of the pathologic changes; purified toxin alone was sufficient to produce characteristic lesions (Foget et al. 1987).

Commercial vaccines are available for the prevention of atrophic rhinitis. The use of *B. bronchiseptica* bacterins may reduce the incidence of atrophic rhinitis but not completely control the disease. Formalized *P. multocida* bacterins provide incomplete or poor protection. It has been reported that vaccination of sows with crude, untreated toxin significantly protects their offspring against atrophic rhinitis, but vaccination with toxoid gives poor protection (Pedersen and Barford 1982). Therefore, the potential exists for production of effective atrophic rhinitis vaccines if stable toxin preparations can be developed.

Infection with respiratory viruses, mycoplasma, or haemophilus or other severe stress may lead to pasteurella pneumonia in pigs of all ages. The predominant serotype of *P. multocida* isolated from pneumonic lungs is type A. Pathogenicity appears to be related to the ability of this capsular type to adhere to respiratory mucous membranes and resist phagocytosis by alveolar macrophages. Most isolates are Heddleston type 3, although types 5 and 12 are also recovered. Use of formalin-inactivated vaccines for the prevention of swine pneumonia has produced disappointing results, possibly because of the use of inappropriate serotypes in commercial bacterins.

Pasteurella multocida type A, and occasionally D, are also commonly implicated in bacterial pneumonias in a number of other species including dogs, cats, goats, sheep, and cattle, with several somatic serotypes being isolated in each species. These serotypes are only weakly cross-reactive, if at all, and it is not surprising, for this reason alone, that commercial bacterins produced with a limited number of strains are generally not efficacious in the prevention of disease. *P. multocida* type A is a common isolate of the upper and lower respiratory tract of cattle and, in some instances, may be the most prevalent organism found (Allen et al. 1991). Its role in shipping fever pneumonia is still disputed, but together with *P. haemolytica*, pulmonary colonization is strongly associated with increased morbidity in feedlot cattle.

Pasteurella multocida is a common inhabitant of the oral bacterial flora of many animals and a frequent contaminant in wounds due to animal bites or scratches, along with other facultatively or strictly anaerobic gram-negative bacteria. Most *P. multocida* infections in humans result from animal bites or scratches. Localized suppuration is common, but bacteremia or endotoxemia may result in more serious disease.

PASTEURELLA HAEMOLYTICA

Pasteurella haemolytica is a commensal of the nasopharynx, and is confined almost exclusively to ruminants, with most adequately characterized strains originating from cattle, sheep, and goats (Biberstein 1981). There are 16 established serotypes, recognised on the basis of soluble or extractable surface antigens by a passive hemagglutination procedure or a rapid plate agglutination test; some strains are untypable. There are two distinct biotypes, A and T, based on fermentation of arabinose and trehalose, respectively. Serotypes 3, 4, 10, and 15 are biotype T; all others are A. The classification has biological significance, since biotype A strains are associated with ruminant pneumonias, septicemia in young lambs, and ovine mastitis, while biotype T strains are usually associated with septicemic infection in older lambs and sheep. Both biotypes and all serotypes are found on the nasal and oral epithelia of clinically normal ruminants.

Pneumonias of cattle and sheep are, economically, the most important *P. haemolytica* infections. Disease usually follows within 1-2 weeks of a stressing experience such as transportation, hence the name "shipping fever" in cattle. Although many believe pasteurellosis to be a secondary complication of virus infection, this relationship has not been demonstrated consistently in field studies, nor has vaccination of cattle against common respiratory viruses had a significant impact on the incidence of pasteurella pneumonia. Experimental infection of cattle with *P. haemolytica* alone has successfully reproduced the fibrinous bronchopneumonia and pleuritis typical of this disease. Under field conditions, the severity of the pneumonia varies and economic losses result both from acute fatalities and the poor productivity of chronically affected animals.

Serotype 1 (biotype A) predominates in bovine pneumonia, serotype 2 in ovine disease. Pathogenic serospecificity may relate to the presence of specific surface structures that permit adherence to and colonization of the lower respiratory epithelium. At least two types of fimbriae have been demonstrated on serotype 1, a rigid 12-nm-wide structure and a flexible 5-nm-wide fimbria (Morck et al. 1989). Each serotype also produces a characteristic polysaccharide capsule that is antiphagocytic (Czuprynski et al. 1989) and may mediate adhesion to respiratory epithelium (Brogden et al. 1989). While experimental studies in cattle show a correlation between production of antibody to capsule and protection against transthoracic challenge (Confer et al. 1989), a recent vaccination trial using purified capsular polysaccharide (CPS) revealed a relationship between vaccination with CPS and anaphylaxis. A serum IgE response to a component of the purified product was demonstrated, and it was suggested that this had the potential to exacerbate pneumonia through induction of pulmonary hypersensitivity on subsequent exposure to live organisms (Conlon and Shewen 1993).

All established serotypes produce a soluble, heat-labile leukotoxin with specificity for leukocytes of ruminants, exclusively (Shewen and Wilkie 1983a). At low concentrations, this toxin impairs phagocytosis and lymphocyte proliferation; higher concentrations result in cell death due to lysis (Clinkenbeard et al. 1989a; Majury and Shewen 1991). Lysis results from formation of 0.9- to 1.2-nm transmembrane pores in the target cell, which allow the movement of potassium, sodium, and calcium through transmembrane gradients (Clinkenbeard et al. 1989a,b). Both the enzymes released following cytolysis and the leukotoxin itself are chemotactic for other leukocytes and augment lung damage due to increased cell recruitment into the area (Slocombe et al. 1985). Recently, this toxin has also been shown to be lytic for ruminant platelets (Clinkenbeard and Upton 1991), lysis of which may induce the pulmonary vascular thrombosis and fibrin exudation typically associated with shipping fever pneumonia.

Cloning and sequencing of the genetic determinants responsible for leukotoxin production revealed close homology with those encoding the alpha-hemolysin of *Escherichia coli* (Lo et al. 1987; Strathdee and Lo 1987; Highlander et al. 1990), resulting in similar designation for genes *lkt*C, *lkt*A, *lkt*B, and *lkt*D. Both *lkt*A and *lkt*C are necessary for the production of active toxin, with *lkt*A encoding a protein of 101.9 kDa and *lkt*C encoding a protein of approximately 19.8 Kda. LktC probably activates LktA by acylation, as is the case for the alpha-hemolysin of *E. coli* (Issartel et al. 1991). The gene products of *lkt*B and *lkt*D facilitate secretion of the toxin (Strathdee and Lo 1989).

Pasteurella haemolytica leukotoxin is one of a family of functionally and genetically similar bacterial toxins, all of which contain repeating domains in the toxin molecule and are thus designated the RTX toxins (repeat domains in the structural toxin) (Lo 1990). Bacteria that produce RTX toxins include *Actinobacillus actinomycetemcomitans, A. equuli, A. lignieresii, A. pleuropneumoniae, A. suis, Bordetella pertussis, Escherichia coli, Morganella morganii, Pasteurella haemolytica, Proteus mirabilis, P. penneri, and P. vulgaris. P. haemolytica* leukotoxin (Strathdee and Lo 1987) and the hemolysin of *A. pleuropneumoniae* have 6 repeat domains (Chang et al. 1989), whereas the alpha-hemolysin of *E. coli* has 11 (Welch et al. 1983).

As in other gram-negative bacteria, the cell wall of *P. haemolytica* contains lipopolysaccharide (LPS) (endotoxin). Serotypes 2 and 8 lack O-polysaccharide side chains, while the other 14 serotypes have characteristic smooth LPS. Monoclonal antibodies, specific to the core region of serotype 1 LPS, reveal shared epitopes in the core oligosaccharides of several A biotypes (Lacroix et al. 1993). Early work with purified LPS injected either intravenously or intraarterially in sheep at sublethal concentrations showed increased pulmonary arterial pressure, decreased cardiac output; and decreased left arterial, pulmonary venous, and systemic blood pressure (Keiss et al. 1964). Later investigations of the effects of purified LPS administered intravenously in calves demonstrated release of thromboxane A_2, prostaglandins, serotonin, cyclic adenosine monophosphate (cAMP) and cyclic guanosine monophosphate (cGMP) (Emau et al. 1987). All or some of these mediators may be responsible for the clinical signs associated with endotoxic shock. Although *P. haemolytica* LPS has many and varied in vivo and in vitro effects, a high serum antibody response to this major surface antigen does not correlate with resistance to experimentally induced pneumonia in calves (Confer et al. 1986).

Serotype 1 *P. haemolytica* has also been shown to produce several antigenic proteins in iron-restricted cultures, including a transferrin-binding protein that is specific for bovine transferrin (Ogunnariwo and Schryvers 1990). In addition, organisms of some serotypes, including 1 and 2, produce neuraminidase and a unique neutral protease that specifically hydrolyses O-sialoglycoproteins (Abdullah et al. 1992). While the former is thought to facilitate mucosal colonization, the role of the latter in pathogenesis is undiscovered.

Immunity to *P. haemolytica* pneumonia resulting from natural infection in cattle can be correlated with antibody to the surface antigens of *P. haemolytica*, detected by bacterial agglutination or passive hemagglutination, and with antileukotoxic activity in serum (Shewen and Wilkie 1983b). Vaccination with formalized bacterins has been practised for almost 60 years with at best questionable efficacy. In fact, considerable evidence from both field trials and experimental studies shows an adverse effect of vaccination with these products (Mosier et al. 1989). Killed bacterins induce agglutinating antibody in vaccinated cattle but little or no antitoxic response. Better protection has been demonstrated after immunization with live organisms, presumably due to in vivo elaboration of toxin and other virulence factors. Several component vaccines have recently been introduced. Most are complex and composed mainly of soluble antigens, including capsular carbohydrate-protein antigens and leukotoxin.

Vaccination with one of these induced protection in experimental and field trials, apparently related to the stimulation of both antitoxic and serospecific agglutinating activities (Jim et al. 1988; Shewen et al. 1988).

Septicemia of sheep due to *P. haemolytica* takes two forms. One occurs in lambs less than 3 months old. Severe pleuritis and pericarditis are present and viable type A organisms can be isolated from all organs. The other occurs in lambs 5-12 months of age. It is acute or peracute and almost always caused by biotype T strains. Outbreaks of disease often coincide with a change from poor grazing to a comparatively rich diet. Other apparent stresses such as adverse climatic conditions or intercurrent infection may also precipitate systemic infection. It is postulated that organisms already present in the tonsils multiply and invade the adjacent tissues of the alimentary tract. Groups of organisms enter the bloodstream as emboli and lodge in the capillary beds of lung, liver, and spleen. Rapid multiplication of *P. haemolytica* in these tissues leads to death, presumably from the release of endotoxin (Dyson et al. 1981). Commercial bacterins do not induce satisfactory immunity; experimental component vaccines, similar to those developed for pneumonia, show promise (Gilmour et al. 1991).

P. haemolytica is the most common cause of ovine mastitis in North America. Several serotypes (1, 6-9, all biotype A) have been isolated from infected ewes in different outbreaks (Shoop and Myers 1984). These isolations probably reflect the predominant serotype carried on the oral and nasal mucous membranes of animals within the flock. The nursing lamb is suspected of both introducing the agent and providing the mechanical trauma needed for the development of clinical disease, which is characterized by severe, usually unilateral, necrotizing inflammation of the mammary gland ("blue bag"). Systemic reaction, likely due to endotoxin release, and death may follow. Vaccination with formalized bacterins has no apparent protective effect.

PASTEURELLA SPP.

Pasteurella pneumotropica has been recovered from the pharyngeal membranes of rodents, dogs, and cats. Originally isolated from pneumonia in laboratory mice, it has also been found contaminating bite wounds. Experimental and spontaneous *P. pneumotropica* pneumonia, developing secondary to experimental Sendai virus infection in mice, has been used as the model for studies of viral:bacterial synergism in the induction of pneumonias in all species (Jakab 1981).

P. aerogenes is a gas-producing pasteurella isolated from the respiratory and intestinal tracts of pigs. Its association with disease is unknown.

P. caballi is the name proposed for a newly-recognized species isolated from respiratory and other infections in horses. These strains differ from other pasteurellae in that they are aerogenic and catalase-negative (Schlater et al. 1989).

P. gallinarum has been identified in the respiratory secretions of poultry and is likewise of unknown pathogenicity. *P. avium*, also a poultry isolate, was previously designated *Haemophilus avium*.

P. anatipestifer is the cause of septicemia and serositis in turkeys, ducks, and other waterfowl. Signs are principally neurologic: tremors of the head and neck, ataxia, and in extreme cases, torticollis. Ducklings are most susceptible and mortality is high. Prophylactic or therapeutic antibiotic treatment is prohibitively expensive and most bacterins have proven ineffective, although an inactivated vaccine was reported to protect young turkeys against experimental challenge (Timms and Marshall 1989).

P. ureae is a urease-positive pasteurella, unique because of its exclusive carriage in the upper respiratory tract of humans. Its pathogenic significance is difficult to assess, although it has been associated occasionally with pneumonia and meningitis.

P. testudinis is associated with respiratory infections in turtles and other amphibians. Interestingly, it can be cultured at 25°C, but not at 37°C, reflecting the lower body temperature of its hosts.

Group EF-4 are *Pasteurella*-like organisms, similar to *P. multocida*, isolated from the mucous membranes of dogs and cats. They have been associated with infected bite wounds in humans and sporadically with acute granulomatous pneumonia in felidae.

REFERENCES

Abudullah, K. M.; Udoh, E. A.; Shewen, P. E.; and Mellors, A. 1992. A neutral glycoprotease of *Pasteurella haemolytica* A1 specifically cleaves O-sialoglycoproteins. Infect Immun 60:56-62.

Allen, J. W.; Viel, L.; Bateman, K. G.; Rosendal, S.; Shewen, P. E.; and Physick-Sheard, P. 1991. The microbial flora of the respiratory tract in feedlot calves: Associations between nasopharyngeal and bronchoalveolar lavage cultures. Can J Vet Res 55:341-46.

Bain, R. V. S. 1963. Haemorrhagic septicaemia. FAO Agricultural Studies 62. Rome: UN Food and Agricultural Organization.

Biberstein, E. L. 1981. *Haemophilus-Pasteurella-Actinobacillus*: Their significance in veterinary medicine. In *Haemophilus, Pasteurella* and *Actinobacillus*. Ed. M. Kilian, W. Frederiksen, and E. L. Biberstein, pp. 61-73. London: Academic Press.

Brogden, K. A.; Adlam, C.; Lehmkuhl, H. D.; Cutlip, R. C.; Knights, J. M.; and Engen, R. L. 1989. Effect of *Pasteurella haemolytica* (A1) capsular polysaccharide on sheep lung in vivo and on pulmonary surfactant in vitro. Am J Vet Res 50:555-59.

Carter, G. R., and Chengappa, M. M. 1980. Hyaluronidase production by Type B *Pasteurella multocida* from cases of haemorrhagic septicaemia. J Clin Microbiol 11:94-96.

Chang, Y.; Young, R.; Moulds, T.; and Struck, D. 1989. Cloning and characterization of a hemolysin gene from *Actinobacillus pleuropneumonia*. DNA 8:635-47.

Clinkenbeard, K., and Upton, M. L. 1991. Lysis of bovine platelets by *Pasteurella haemolytica* leukotoxin. Am J Vet Res 52:453-57.

Clinkenbeard, K.; Boon, A. W.; Mosier, D. A.; and Confer, A. W. 1989a. Effects of *Pasteurella haemolytica* leukotoxin on isolated bovine neutrophils. Toxicon 27:797-804.

Clinkenbeard, K.; Mosier, D. A.; Timko, A. L.; and Confer, A. W. 1989b. Effects of *Pasteurella haemolytica* leukotoxin on cultured bovine lymphoma cells. Am J Vet Res 50:271-75.

Collins, F. M. 1977. Mechanisms of acquired resistance to *Pasteurella multocida* infection. A review. Cornell Vet 67:103-36.

Confer, A. W., Panciera, R.; and Mosier, D. A. 1986. Serum antibodies to *Pasteurella haemolytica* lipopolysaccharide: Relationship to experimental bovine pneumonic pasteurellosis. Am J Vet Res 47:1134-38.

Confer, A. W.; Simons, K. R.; Panciera, R. J.; Mort, A. J.; and Mosier, D. A. 1989. Serum antibody response to carbohydrate antigens of *Pasteurella haemolytica* serotype 1: Relation to experimental bovine pneumonic pasteurellosis. Am J Vet Res 50:98-105.

Conlon, J. A., and Shewen, P. E. 1993. Clinical and serological evaluation of a *Pasteurella haemolytica* A1 capsular polysaccharide vaccine. Vaccine: In press.

Czuprynski, C. J.; Noel, E. F.; and Adlam, C. 1989. Modulation of bovine neutrophil antibacterial activities by *Pasteurella haemolytica* A1 purified capsular polysaccharide. Microbiol Pathol 6: 133-41.

Davis, R. B. 1987 Cholera and broiler breeders. Poult Dig 20:430-34.

DeJong, M. F.; Oei, H. L.; and Tentenburg, G. J. 1980. AR-pathogenicity-tests for *Pasteurella multocida* isolates. Proc Int Pig Vet Soc, p. 211, Copenhagen, Den.

DiaGiacomo, R. F.; Xu, Y.; Allen, V.; Hinton, M. H.; and Pearson, G. R. 1991. Naturally acquired *Pasteurella multocida* infection in rabbits: Clinicopathological aspects. Can J Vet Res 55: 234-38.

Dyson, D. A., Gilmour, N. J. L.; and Angus, K. W. 1981. Ovine systemic pasteurellosis caused by *Pasteurella multocida* biotype T. J Med Microbiol 14:89-95.

Emau, P.; Giri, S. N.; and Bruss, M. L. 1987. Effects of smooth and rough *Pasteurella haemolytica* lipopolysaccharide on plasma cyclic nucleotides and free fatty acids in calves. Vet Microbiol 15:279-92.

Flatt, R. E. 1974. Pasteurellosis. In Biology of the Laboratory Rabbit. Ed. S. H. Weesbroth, R. E. Flatt, and A. L. Kraus, pp. 194-205. New York: Academic Press.

Foget, N. T.; Pedersen, K. B.; and Elling, F. 1987. Characterization and biologic effects of the *Pasteurella multocida* toxin. Fed Eur Microbiol Soc Microbiol Lett 43:45-51.

Gilmour, N. J. L.; Donachie, W.; Sutherland, A. D.; Gilmour, G. S.; Jones, G. E.; and Quirie, M. 1991. Vaccine containing iron-regulated proteins of *Pasteurella haemolytica* A2 enhances protection against experimental pasteurellosis in lambs. Vaccine 9:137-40.

Heddleston, K. L.; Gallagher, J. E.; and Rebers, P. A. 1972. Fowl cholera: Gel diffusion precipitin test for serotyping *Pasteurella multocida* from avian species. Avian Dis 16:925-36.

Highlander, S. K.; Engler, M. J.; and Weinstock, G. M. 1990. Secretion and expression of the *Pasteurella haemolytica* leukotoxin. J Bacteriol 172:2343-50.

Issartel, J. P.; Koronakis, V.; and Hughes, C. 1991. Activation of *Escherichia coli* prohaemolysin to the mature toxin by acyl carrier protein-dependent fatty acylation. Nature 351:759-61.

Jakab, G. J. 1981. Interactions between Sendai virus and bacterial pathogens in the murine lung. A review. Lab Anim Sci 31:170-77.

Jim, K.; Guichon, T.; and Shaw, G. 1988. Protecting feedlot calves from pneumonic pasteurellosis. Vet Med 10:1084-87.

Keiss, R. E.; Will, D. H.; and Collier, J. R. 1964. Skin toxicity and haemodynamic properties of endotoxin derived from *Pasteurella haemolytica*. Am J Vet Res 25:935-41.

Kilian, M., and Frederiksen, W. 1981. Ecology of *Haemophilus*, *Pasteurella*, and *Actinobacillus*. In *Haemophilus, Pasteurella,* and *Actinobacillus*. Ed. M. Kilian, W. Frederiksen, and E. L. Biberstein, pp. 11-38. London: Academic Press.

Lacroix, R. P.; Duncan, J. R.; Jenkins, R. P.; Leitch, R. A.; Perry, J. A.; and Richards, J. C. 1993. Structural and serological specificites of *Pasteurella haemolytica* lipopolysaccharides. Infect Immun 61:170-81.

Lo, R. Y. C. 1990. Molecular characterization of the cytotoxins produced by *Haemophilus, Actinobacillus, Pasteurella*. Can J Vet Res 54:533-35.

Lo, R. Y. C.; Strathdee, C. A.; and Shewen, P. E. 1987. Nucleotide sequence of the leukotoxin genes of *Pasteurella haemolytica* A1. Infect Immun 55:1987-96.

Majury, A. L., and Shewen, P. E. 1991. The effect of *Pasteurella haemolytica* A1 leukotoxic culture supernate on the in vitro proliferative response of bovine lymphocytes. Vet Immunol Immunopathol 29:41-56.

Morck, D. W.; Olson, M. E.; Acres, S. D.; Daoust, P. Y.; and Costerton, J. W. 1989. Presence of bacterial glycocalyx and fimbriae on *Pasteurella haemolytica* in feedlot cattle with pneumonic pasteurellosis. Can J Vet Res 53:167-71.

Mosier, D. A.; Confer, A. W.; and Panciera, R. J. 1989. The evolution of vaccines for bovine pneumonic pasteurellosis. Res Vet Sci 47:1-10.

Mutters, R.; Bisgaard, M.; and Pohl, S. 1986. Taxonomic relationship of selected biogroups of *Pasteurella haemolytica* as revealed by DNA:DNA hybridizations. Acta Pathol Microbiol Immunol Scand 94:195-202.

Ogunnariwo, J. A., and Schryvers, A. B. 1990. Iron acquisition in *Pasteuerlla haemolytica*: Expression and identification of a bovine-specific transferrin receptor. Infect Immun 58:2091-97.

Pedersen, K. B., and Barford, K. 1982. Effect on the incidence of atrophic rhinitis of vaccination of sows with a vaccine containing *Pasteurella multocida* toxin. Nord Vet Med 34:293-302.

Rhoades, K. R., and Rimler, R. B. 1984. Avian pasteurellosis. In Diseases of Poultry, 8th Ed. Ed. M. S. Hofstad, H. J. Barnes, B. W. Calneck, W. M. Reid, and H. W. Yoder, Jr., pp. 141-56. Ames: Iowa State University Press.

Rimler, R. B., and Rhoades, K. R. 1981. Lysates of turkey-grown *Pasteurella multocida*: Protection against homologous and heterologous serotype challenge exposures. Am J Vet Res 42:2117-21.

____________ . 1987. Serogroup F, a new capsule serogroup of *Pasteurella multocida*. J Clin Microbiol 25:615-18.

Schlater, L. K.; Brenner, D. J.; Steigerwalt, A. G.; Moss, C. W.; Lambert, M. A.; and Packer, R. A. 1989. *Pasteurella caballi*, a new species from equine clinical specimens. J Clin Microbiol 27:2169-74.

Shewen, P. E., and Wilkie, B. N. 1983a. *Pasteurella haemolytica* cytotoxin: Production by recognized serotypes and neutralization by type-specific rabbit antisera. Am J Vet Res 44:715-19.

__________ . 1983b. *Pasteurella haemolytica* cytotoxin neutralizing activity in sera from Ontario beef cattle. Can J Comp Med 47:497-98.

Shewen, P. E.; Sharp, A.; and Wilkie, B. N. 1988. Efficacy testing a *Pasteurella haemolytica* extract vaccine. Vet Med 10:1078-83.

Shoop, D. S., and Myers, L. L. 1984. Serologic analysis of isolates of *Pasteurella haemolytica* and *Staphylococcus aureus* from mastitic ewes. Am J Vet Res 45:1944-46.

Slocombe, R.; Malark, J.; Ingersol, R.; Derksen, F.; and Robinson, N. 1985. Importance of neutrophils in the pathogenesis of acute pneumonic pasteurellosis in calves. Am J Vet Res 46:2253-58.

Strathdee, C. A., and Lo, R. Y. C. 1987. Extensive homology between the leukotoxin of *Pasteurella haemolytica* A1 and the alpha-hemolysin of *Escherichia coli*. Infect Immun 55:3233-36.

__________ . 1989. Cloning, nucleotide sequence and characterization of genes encoding the secretion function of the *Pasteurella haemolytica* leukotoxin determinant. J Bacteriol 171:916-28.

Timms, L. M., and Marshall, R. N. 1989. Laboratory assessment of protection given by experimental *Pasteurella anatipestifer* vaccine. Br Vet J 145:483-93.

Truscott, W. M., and Hirsh, D. C. 1988. Demonstration of an outer membrane protein with anti-phagocytic activity from *Pasteurella multocida* of avian origin. Infect Immun 56:1538-44.

Welch, R.; Hull, R.; and Falkow, S. 1983. Molecular cloning and physical characterization of a chromosomal haemolysin from *Escherichia coli*. Infect Immun 42:178-86.

19 / *Yersinia*

BY C. L. GYLES

THE YERSINAE share characteristics of the family *Enterobacteriaceae*. Thus, they are facultatively anaerobic, oxidase-negative, gram-negative rods that ferment glucose. The organisms typically show bipolar staining, particularly in smears from tissues, and show temperature-dependent expression of several phenotypic characteristics. The name Yersinia derives from Alexandre Yersin, who discovered the plague bacillus in Hong Kong in 1894. The species *Y. pestis* and *Y. pseudotuberculosis* have previously been included with the family *Pasteurellaceae* but were transferred to the *Enterobacteriaceae* because of their oxidase-negative reaction and their DNA relatedness to *E. coli*. The International Committee on Systematic Bacteriology has determined that *Y. pestis* should not be accorded separate species status but should be a subspecies of *Y. pseudotuberculosis*; however, the designation *Y. pestis* remains in popular usage and will be employed in this chapter. Both organisms are closely related and share several common antigens as well as susceptibility to phages, but there are differences in the kinds of disease that are induced and in the ability to cause disease in various host species.

Seven species in the genus *Yersinia* may be differentiated by biochemical tests: *Y. pestis*, *Y. pseudotuberculosis*, *Y. enterocolitica*, *Y. intermedia*, *Y. frederiksenii*, *Y. kristensenii*, and *Y. ruckeri* (Bercovier and Mollaret 1984). The name *Y. rohdei* has been proposed for an additional species, recovered from feces of dogs and humans. *Y. pestis*, the only member of the genus that has a requirement for animal hosts for its maintenance, is much less biochemically active than the others. *Y. enterocolitica* and *Y. pseudotuberculosis* have significant metabolic competence and can grow over a temperature range of 5-42°C and are therefore able to survive for long times outside an animal's body. Four species, *Y. pestis*, *Y. pseudotuberculosis*, *Y. enterocolitica*, and *Y. ruckeri*, are important in disease. *Y. pseudotuberculosis* is classified into 10 serovars and *Y. enterocolitica* into 57 serotypes and 5 biotypes.

DISEASE

The pathogenic *Yersinia* are all intracellular parasites. *Y. pestis* is responsible for an acute septicemic disease (bubonic plague) in rats and humans. In humans, severe subcutaneous hyperemia and hemorrhage give a dark appearance because of conversion of hemoglobin to hemosiderin, hence the term "black death." Urban or domestic plague accounts for epidemics in which disease is transmitted from rats to humans and from rats to rats in an urban setting. Normally, the cycle of transmission involves rat to flea to rat (or rat to rat by ingestion), but humans are also infected by fleas. Occasionally, humans will develop infection by handling contaminated animal tissues. Domestic animals, including cats and dogs, may become infected during an epidemic. Sylvatic plague involves epizootics of infection among wild animals such as squirrels, prairie dogs, rats, mice, chipmunks, bobcats, and rabbits; there is infrequent transmission from these animals to humans.

Bubonic plague is initiated by the bite of a flea that ingested bacteria when it drew blood from an infected rat. Bacteria deposited at the site of the bite are transported to the regional lymph nodes, in which they multiply and induce severe necrosis and hemorrhage. The bacteria subsequently disseminate via the bloodstream to all organs. If pneumonia develops, the bacteria may be spread by infected airborne droplets. Production of murine toxin, a capsule, fibrinolytic factor, and coagulase are unique to *Y. pestis* and undoubtedly contribute to the acute and highly invasive kind of disease produced by this organism.

Carnivores are relatively resistant to bubonic plague and coyotes have been used as sentinel animals in areas of sylvatic plague. Antibodies in the sera of coyotes can be used to monitor developments.

Y. enterocolitica and *Y. pseudotuberculosis* are less invasive and typically cause localized exudative infection. *Y. enterocolitica* was first recognized as a pathogen in 1939. The organism is carried in the tonsils of pigs, a major reservoir for humans (Tauxe et al. 1987); but cats and dogs are other animal sources of infection. Contaminated water and food of animal origin are other sources of the organism. Only a few serotypes are important in disease. The major O groups among human pathogens are O:3, O:8, and O:9; strains of O:1 and O:2 have been associated with disease in chinchillas, hares, and goats. Ingested organisms invade the intestine and the most frequent clinical sign is diarrhea. In humans, diarrhea is seen primarily in young children; ileitis or mesenteric adenitis in the absence of diarrhea is often seen in older children and adults. Pain in the lower right quadrant in humans sometime leads to a misdiagnosis of appendicitis. Extraintestinal forms of disease in humans include arthritis and erythema nodosum.

In animals, *Y. enterocolitica* has clearly been implicated as an agent of diarrhea in sheep and goats in which it has been reported to cause microabscessation in the intestine. Diarrheal disease has been reported in several additional species including dogs, chinchillas, hares, deer, cattle, sheep, and goats. *Y. enterocolitica* has also been reported as an agent of abortion in sheep.

Y. pseudotuberculosis causes disease in a wide variety of wild and domestic mammals and birds and rarely in humans. Affected species include deer, cattle, sheep, pigs, and wild birds. Following ingestion, the organisms are sometimes contained by the mesenteric lymph nodes, but they may progress beyond this level to cause a septicemia. A hallmark of septicemic disease is the formation of miliary nodules in organs.

Y. ruckeri is the cause of a disease of fish known as enteric redmouth disease, so-called because of the occurrence of subcutaneous hemorrhages around the mouth and throat of affected fish. Disease takes the form of a generalized septicemia and may cause substantial losses in the trout-farming industry in certain areas. Asymptomatic carrier fish maintain the organisms in the fish environment; stresses, particularly an increase in water temperature, lead to transmission to other fish. Reservoirs of the bacterium include aquatic invertebrates, such as crayfish, and terrestrial mammals, such as muskrats. A heat-sensitive lipid substance has been reported to be associated with virulence of strains of *Y. ruckeri* (Furones et al. 1990), but there is very little information on virulence genes and virulence factors of this organism.

The remainder of this chapter will be confined to a discussion of *Y. pestis, Y. pseudotuberculosis*, and *Y. enterocolitica*.

VIRULENCE FACTORS

Yersinia are facultative intracellular parasites and the hallmark of *Yersinia* infection is invasiveness. Less than 10 fully virulent *Y. pestis* will kill 50% of mice or guinea pigs inoculated intraperitoneally, and it is easy to observe increases in the LD_{50} that occur when virulence factors are eliminated from this species. A number of bacterial factors implicated in virulence of yersiniae are discussed below.

Fraction 1 antigen

Fraction 1 (F1) is a glycoprotein associated with the capsule of *Y. pestis* and produced during growth at 37°C but not at temperatures less than 26°C. There is molecular heterogeneity and a tendency to form aggregates, but there is a single immunogenic specificity. The gene *fra*, which encodes production of F1, is carried on a 60-MDa plasmid. F1 is antiphagocytic through an anticomplementary action, and antibodies against F1 are protective. Mutants that lack F1 are less virulent for guinea pigs but are unchanged in virulence for mice (Brubaker 1979).

V and W antigens

Antigens identified as V and W are produced by all three pathogenic species, but these antigens are not exported from the cytoplasm of *Y. pestis* (Butler 1983). V symbolizes virulence and W was the designation for another antigen that was associated with the V antigen. The designation *vwa*$^+$ is used for bacteria with genes for the V and W antigens, located on a 42- to 48-MDa virulence plasmid. The V antigen is a protein of 38-kDa molecular weight and the W antigen is a lipoprotein of 145 kDa. Production of V and W antigens is linked to temperature and Ca^{++} concentration. A shift in temperature from 26°C to 37°C in a Ca^{++}-deficient environment blocks synthesis of stable RNA, thereby restricting growth of the organisms. Simultaneously, the *vwa* genes are derepressed and synthesis of VW occurs. Temperature and Ca^{++} concentration probably act as signals that indicate to the bacterium that it is in an intracellular environment. VW antigens promote resistance to intracellular killing rather than enhance resistance to uptake by professional phagocytes, thereby assisting in dissemination of the organisms. Loss of either VW antigen or Ca^{++} dependence results in an increase in the LD_{50} by approximately 10^6 (Brubaker 1979). Antibody to V, but not to W, is protective in mice.

Pesticin

Pesticin (P1) is a 63-kDa peptidoglycan hydrolase that is produced by *Y. pestis* and inhibits growth of serovar 1 *Y. pseudotuberculosis* by interference with cell wall biosynthesis. P1 can be recovered from the supernatant of *Y. pestis* cultures grown at 30°C. Genes for P1 are found on a 10-kb plasmid. P1 production has been used as a marker for virulence; however, virulence is not due to P1 but to linked genes on the plasmid. Genes for coagulase and fibrinolysin have been proposed to be the relevant linked genes. Both activities are promoted by plasminogen activator, which enhances invasion of tissues. Coagulase is active at 25°C but not at 37°C and is probably important in promoting clotting of blood in the flea. Fibrinolysin, which dissolves blood clots, is more active at 37°C than at 25°C (Butler 1983) and is likely the important P1-associated virulence factor that contributes to invasiveness.

Absorption of Pigments

Absorption of pigments (P), or the ability to absorb exogenous pigments (Pgm), is commonly associated with virulent *Y. pestis* but not the other pathogenic *Yersinia*. Either Congo red dye or hemin may be used to test for ability to absorb pigments. Early reports (Wake and Morgan 1986) indicated that P was encoded by a plasmid gene *pgm* and that strains that lacked the ability to produce pigmented colonies on a hemin-containing medium were markedly reduced in virulence. Recent reports have localized the gene to the chromosome.

Virulence is restored to *pgm*⁻ mutants when iron is supplied exogenously. All *Yersinia* assimilate iron by siderophore-independent mechanisms and can use hemin as the sole source of iron. Pigmentation is believed to represent a unique ability to store iron in the form of an outer-membrane storage complex. P has a relationship with P1 in that P1 receptor is also involved in iron metabolism of the bacteria (Brubaker 1991).

Murine Toxin

Y. pestis produces a protein toxin, called murine toxin, because of its high toxicity for mice. Toxicity is associated with two proteins: toxin A (240 kDa) and toxin B (120 kDa). The toxin appears to cause a beta-adrenergic blockade and prevents the action of epinephrine. The murine toxin is part of the cytoplasmic membrane; there is doubt as to its release from the organism and its significance in disease.

Lipopolysaccharide

Cell wall lipopolysaccharide (LPS), or endotoxin, probably contributes to the symptomatology of disease caused by *Yersinia*. The massive numbers of bacteria in the bloodstream and organs of animals infected with *Y. pestis* would permit LPS to induce fever, vascular damage, and disseminated intravascular coagulation. LPS likely induces an inflammatory response in the localized lesions associated with other *Yersinia* species.

Enterotoxin of *Y. enterocolitica*

A heat-stable enterotoxin (YST), related to the heat-stable enterotoxin (STa) produced by enterotoxigenic *Escherichia coli*, is produced by some strains of *Y. enterocolitica*. Interestingly, the enterotoxin is produced at 25°C but not at 37°C in vitro. YST and STa are antigenically related and are similar in their mechanism of action; they cause diarrhea through activation of particulate guanylate cyclase in intestinal epithelial cells. Although YST is 30 amino acids and *E. coli* STa is 18 or 19 amino acids, the active regions of both toxins are highly related. The role of the enterotoxin in disease is not clear, since it is not produced at body temperature and some studies have shown no role for YST in experimental infections. However, in one study, all 89 pathogenic strains were positive in hybridization tests with a DNA probe for YST, whereas none of 51 nonpathogenic strains hybridized with the probe (Delor et al. 1990).

Invasins

Yersiniae differ from other invasive *Enterobacteriaceae* in that their invasive function can be transferred to *E. coli* by transfer of a single gene (*inv* or *ail*). In 1985, a novel cell-culture selection procedure was used to identify a clone for the *inv* gene derived from *Y. pseudotuberculosis* (Isberg and Falkow 1985). *E. coli* that carry chromosomal *Y. pseudotuberculosis* DNA cloned into a cosmid vector were spun down onto cell monolayers in microtitre plates. Following a brief incubation, the supernatant was discarded and the cells were washed, exposed to gentamicin, washed a second time, then lysed to release intracellular bacteria, which were tested for invasiveness. The *inv* gene discovered through this method was inactivated by transposon insertion then used to replace the gene in the wild-type *Y. pseudotuberculosis*. The mutant that lacked this gene function was markedly impaired in invasiveness.

The protein (invasin, Inv) encoded by the gene was determined to be a 957-amino acid outer-membrane protein, whose N-terminal two-thirds are required for secretion and orientation and whose C-terminal portion mediates invasion. The eukaryotic cell receptor for invasin is a class of integrins (β_1) that function as receptors for matrix proteins such as fibronectin and that transmit signals to the cell by their cytoplasmic domains. Interestingly, invasin has a much higher affinity for the receptor than does fibronectin and this is likely important in the interaction. Particles bound to integrin are usually maintained at the surface, but binding of *Yersinia* results in internalization, presumably because of the large number of receptors that are bound.

A second protein, Ail, that mediates attachment and invasion was discovered in *Y. enterocolitica* (Miller et al. 1988). Ail is a surface-exposed, 17-kDa outer-membrane protein; mutations that inactivate the *ail* function lead to a decrease in invasion of tissue culture and a decrease in virulence of the organism in mice. Ail mediates binding to a wide variety of cultured cells but promotes efficient internalization of bacteria in only a few cell types (Isberg 1990).

VIRULENCE GENES

Plasmid Genes

A class of virulence plasmids (pVY) of 40- to 48-MDa molecular weight is implicated in virulence of *Yersinia* (Straley 1988). Plasmid products appear to contribute to virulence through resistance to phagocytosis, and survival and growth in an intracellular environment. Phenotypes associated with the virulence plasmids include adherence to epithelial cells, resistance to phagocytosis, production of the V and W antigens, autoagglutinability (determined by *yad*A), temperature-inducible outer-membrane proteins (Yops), calcium dependence at 37°C, and a positive Sereny test.

The virulence plasmid of the yersiniae is called the **l**ow-**c**alcium **r**esponse (lcr) or Vwa$^+$ plasmid and codes for 16-20 proteins, including the V antigen, Yops, and regulators of expression of virulence determinants. The V antigen functions as a secreted virulence protein and as a regulator of the low-calcium response regulon. Genes on the plasmid are absolutely required for virulence and loss of the plasmid results in loss of virulence in mice. Temperature and Ca^{++} concentration are known regulators. In vitro, maximal expression of plasmid virulence genes occurs in low Ca^{++} at 37°C, when growth of the bacterium is restricted. The Yops are expressed at 37°C in vitro in *Y. enterocolitica* and *Y. pseudotuberculosis* but not in *Y. pestis*, in which they are expressed only in vivo.

As indicated above, virulent strains of pathogenic *Yersinia* require Ca^{++} in order to grow at 37°C. This requirement is met in the extracellular compartments in the body in which the Ca^{++} concentration is about 1 mM. However, in intracellular locations, the Ca^{++} concentration is less than 0.1 mM. High levels of Mg^{++} enhance the low Ca^{++} effect and nucleotides provide some relief from the low-Ca^{++} effect (Brubaker 1991; Straley 1991).

The *yop* genes are organized in several operons scattered around the plasmid. They too are coordinately regulated by a 20-kb region of the plasmid called the calcium region or low-calcium response (*lcr*) region. The Ca^{++} region of the plasmid is conserved in all three species, although the plasmid in *Y. enterocolitica* differs by about 50% in homology with the plasmids of *Y. pestis* and *Y. pseudotuberculosis*. The *lcr* region contains several loci, called *vir* in *Y. enterocolitica* or *lcr* in *Y. pestis* and *Y. pseudotuberculosis*. Induction of the *lcr* results in synthesis of several plasmid-encoded proteins. Regulation involves genes in or close to the *lcr* and is effected through negative regulation by Ca^{++} regulators and positive activation through temperature-sensitive regulators. *yop*N (which is part of the *lcr*E locus) encodes a surface-exposed outer-membrane protein of 33-kDa that functions as a sensor that interacts with Ca^{++} to send a signal to inhibit transcription of *yop* genes (Forsberg et al. 1990). *lcr*F imposes temperature regulation on the *yop* genes, through a 31-kDa transcriptional activator that belongs to the *ara*C family of transcriptional activators.

*lcr*D encodes a 70-kDa inner-membrane protein that regulates the low-calcium response; *lcr*Q acts in the control of the regulon by Ca^{++}. LcrK is required for transport of Yops to the surface of the bacterium and is negatively and positively regulated as are the *yop* structural genes. The *lcr*V gene, which encodes the V antigen (Yop 41) of *Y. enterocolitica* is part of a three-gene operon, similar to *lcr*GVH operon in *Y. pestis*. Transcription of this operon is activated by a temperature shift from 25°C to 37°C. The V antigen of *Y. pestis* is itself a regulatory protein that is required for Ca^{++} dependence and for optimal expression of *lcr* genes (Price et al. 1991).

Eleven Yops have been described for *Y. pestis* and 10 for *Y. enterocolitica*. Several have been shown to be required for virulence. YadA (YopA, Yop1, or P1) forms fibrillar structures that significantly increase the surface charge and hydrophobicity of the organism and may contribute to adherence of the bacteria to epithelial cells. YadA binds fibronectin (Tertti et al. 1992) and is known to enhance colonization of the mouse intestine and to confer resistance to complement-mediated killing of *Y. enterocolitica* by human serum. The *yad*A gene is not a part of the *yop* regulon and is not regulated by Ca^{++} concentration, but it is activated by LcrF, the temperature-regulated activator of the *yop* regulon. YopE is antiphagocytic and cytotoxic; inactivation of *yop*E results in loss of virulence by oral and intraperitoneal routes but not by the intravenous route (Rosqvist et al. 1990). YopH is also antiphagocytic and has protein tyrosine phosphatase activity. Bliska et al. (1991) have shown that YopH dephosphorylates 120- and 60-kDa proteins in macrophages. YopM inhibits platelet aggregation and probably functions to mute the inflammatory response (Leung et al. 1990).

In *Y. enterocolitica* and *Y. pseudotuberculosis*, Yops inhibit complement deposition on the bacteria, thereby providing resistance to phagocytosis. These two species of *Yersinia* release some Yops into the medium under certain culture conditions, which raises the possibility that these proteins may function as released proteins constituting immunological flak.

The virulence plasmid pYV of *Y. enterocolitica* encodes a lipoprotein, called YlpA, which is highly related to the TraT protein of *E. coli* (China et al. 1990). The gene for YlpA is regulated by *vir*F; however, YlpA does not appear to affect virulence for mice inoculated by the intravenous route.

Y. pestis has two additional plasmids whose genes encode proteins associated with resistance to phagocytosis and to dissemination of bacteria into the bloodstream and tissues (Brubaker 1991). A 10-kb plasmid carries genes for *pla* (**pl**asminogen **a**ctivator), which confers adhesive properties on this organism. When the gene was cloned into *E. coli*, the *E. coli* adhered to globo-series glycolipids (Kienle et al. 1992). Mutation in *pla* has been shown to cause a reduction in virulence. This plasmid also contains genes for the bacteriocin pesticin. A 60-MDa plasmid has the genes for the F1 capsular antigen and for murine toxin, but in some strains, this plasmid is integrated into the chromosome.

Chromosomal Genes

The virulence plasmid contributes to tissue invasion, but strains lacking the plasmid can also invade tissue culture cells. It is, therefore, not surprising that chromosomal loci have been shown to confer invasiveness; chromosomal loci *inv* and *ail* encode genes that individually confer the invasive phenotype on noninvasive *E. coli* (Miller et al. 1988).

Invasion genes for *Y. enterocolitica* have been identified by methods as described for the *inv* genes of *Y. pseudotuberculosis*. Two distinct genes were cloned from *Y. enterocolitica*: the counterpart of the *inv* gene of *Y. pseudotuberculosis* was called inv_{ent} and the other was called *ail*, for **a**ttachment **i**nvasion **l**ocus. Sequences that hybridize with a probe for *inv* are present in all yersiniae, but *inv*-related sequences are not always functional. On the other hand, sequences that hybridize with a probe for *ail* are highly correlated with invasiveness and with virulence.

There is considerable similarity between *inv* genes and invasin proteins of *Y. enterocolitica* and *Y. pseudotuberculosis*. The genes have 73% nucleotide sequence homology and the proteins have 77% amino acid sequence similarity. *Y. enterocolitica* mutants with a loss of function of inv_{ent} invade cells in culture only poorly but are not different from the wild type in virulence for mice. One explanation is that other factors, such as Ail and a low-level plasmid-encoded invasion system, are able to compensate in vivo. Rosenshine et al. (1992) have shown that invasin-mediated uptake of *Y. enterocolitica* and *Y. pseudotuberculosis* by HeLa cells was blocked by tyrosine protein kinase inhibitors, indicating that protein kinases are involved in internalization of the bacteria. Recently, Ail was also shown to impart resistance to complement-mediated killing, an activity reminiscent of the product of the *rck* gene of *Salmonella typhimurium*. Interestingly, *ail* and *rck* are highly homologous.

A high molecular weight iron-regulated protein (HWMP2), which is encoded by a chromosomal gene, has been found to be common to the pathogenic types and absent from nonpathogenic species of *Yersinia* (Carniel et al. 1992). It has been suggested that this protein may play a role in virulence.

*ymo*A (*Yersinia* modulator) is a chromosomal gene of *Y. enterocolitica* that modulates the expression of plasmid-encoded virulence functions (Cornelis et al. 1991). YmoA modulates thermoregulation of *vir*F and, therefore, of the *yop* regulon; it also thermoregulates expression of YST.

The pH6 antigen is the product of a chromosomal gene, *psa*A, in *Y. pestis;* it is expressed at 37°C at pH 6 but not at pH 8. It is not expressed at 26°C. Loss of function of the gene results in a 200-fold increase in the intravenous LD_{50} dose for mice (Lindler et al. 1990). It has been suggested that this antigen may be expressed in phagolysosomes of macrophages and in the extracellular environment of abscesses. The pH6 antigen is also found in *Y. pseudotuberculosis*.

PATHOGENESIS

The prevailing view is that the decisive aspect of virulence in *Yersinia* is the ability to resist killing by macrophages. In an excellent recent review, Brubaker (1991) concluded that the yersiniae are primarily extracellular parasites, because organisms are observed predominantly in extracellular sites following intravenous injection of mice. Nonetheless, the organisms are known to grow within macrophages but not inside polymorphonuclear leukocytes (Straley and Harmon 1984; Straley 1991), and it does seem that much of the virulence mechanism of these organisms is geared toward survival and multiplication in macrophages. In the case of *Y. enterocolitica* and *Y. pseudotuberculosis*, ability to invade across the intestinal epithelium is another critical factor.

The following picture can be pieced together from the substantial observations on the pathogenic *Yersinia* (Brubaker 1991; Straley 1991). Outside the mammalian host, the organisms experience a low-temperature environment and produce a spectrum of surface proteins suitable for that environment and for a preinfection state. Thus, expression of invasin is maximal at low temperature, indicating that invasin is important at an early stage in the disease process (Isberg 1990). Once in the mammalian host, adherence and invasin functions allow the bacteria to be taken up by nonphagocytic epithelial cells in the ileum and colon. After penetration of these cells, the bacteria encounter phagocytic cells in the mucosa and submucosa and are ingested by them. They are able to resist killing, likely because the intracellular environment of the phagocytes (37°C, low Ca^{++}) results in induction of *lcr* operons and in production of surface proteins that confer resistance to killing by phagocytes. Some bacteria pass on to the regional lymph nodes.

In both intestinal tissues and lymph nodes, bacteria and dead phagocytes lead to formation of abscesses; focal necrosis is observed in affected organs. The inflammatory response is less than would be expected, given the degree of tissue necrosis and it has been suggested that the organisms are able to suppress cell-mediated immunity. In *Y. enterocolitica* and *Y. pseudotuberculosis*, the infection is usually contained at this level, but the latter sometimes becomes septicemic and induces abscess formation in liver, spleen, and other organs. In the case of *Y. pestis*, the organisms are usually injected into the bloodstream and, therefore, by-pass the intestinal mucosal barrier. The anti-phagocytic capsule and other factors such as murine toxin enhance the ability of the organism to resist killing by phagocytes, and the bacteria rapidly spread throughout the body.

Straley (1991) has speculated that the role of the Lcr is to protect the organisms in extracellular niches and suggests that, under the influence of a 37°C temperature and high Ca^{++} levels in the extracellular compartment, yersiniae produce low quantities of V and Yops and are only weakly anti-phagocytic. When yersiniae are taken up by macrophages they respond to the low Ca^{++} environment in phagolysosomes with massive release of Yops and V antigen and resist killing. When yersiniae are released from the macrophages, they remain sufficiently anti-phagocytic that they resist ingestion by neutrophils and the recruitment of phagocytes is reduced by the action of Yops such as YopM and V antigen.

The mechanism by which diarrhea develops in response to infection with *Y. enterocolitica* is not known. Production of enterotoxin by organisms in the lumen of the intestine may contribute to diarrhea, but it is not evident that enterotoxin is produced in vivo. Loss of function of absorptive epithelial cells as a result of the invasion process could also account for diarrhea.

IMMUNITY

Formalin-killed *Y. pestis* organisms have been used as plague vaccines for almost a century and are still in use today. There is considerable evidence for effectiveness, based on challenge studies in animals and records of low occurrence of disease in large populations of vaccinated individuals. Live vaccines have also been used but suffer from more severe side effects and variability in immunogenicity and virulence. Antibacterial immunity, measured with the F1 antigen, is short-lived; revaccination is recommended after 6 months. F1 is a convenient antigen for such tests because it is specific for the plague bacillus. However, the nature of the infection suggests that cell-mediated immunity is more important than humoral immunity in protecting the host; tests of delayed-type hypersensitivity to pestin have also been used (Wake and Morgan 1986).

REFERENCES

Bercovier, H., and Mollaret, H. M. 1984. Genus XIV. *Yersinia* Van Loghem 1944. In Bergey's Manual of Systematic Bacteriology, Vol 1. Ed. N.R. Krieg and J. G. Holtz, pp. 498-506. Baltimore: Williams and Wilkins.

Bliska, J. B.; Guan, K. L.; Dixon, J. E.; and Falkow, S. 1991. Tyrosine phosphate hydrolysis of host proteins by an essential *Yersinia* virulence determinant. Proc Natl Acad Sci USA 88: 1187-91.

Brubaker, R. R. 1979. Expression of virulence in yersiniae. In Microbiology, 1979. Ed. D. Schlessinger, pp. 168-71. Washington, D.C.: American Society for Microbiology.

_________. 1991. Factors promoting acute and chronic diseases caused by yersiniae. Clin Microbiol Rev 4:309-24.

Butler, T. 1983. Plague and Other *Yersinia* Infections. New York: Plenum Press.

Carniel, E.; Guiyoule, A.; Guilvout, I., and Mercereau-Puijalon, O. 1992. Molecular cloning, iron-regulation and mutagenesis of the *irp*2 gene encoding HMWP2, a protein specific for the highly pathogenic *Yersinia*. Mol Microbiol 6:379-88.

China, B.; Michiels, T.; and Cornelis, G. R. 1990. The pYV plasmid of *Yersinia* encodes a lipoproten, YlpA, related to TraT. Mol Microbiol 4: 1585-93.

Cornelis, G. R.; Sluiters, C.; Delor, I.; Geib, D.; Kaniga, K.; Lambert, de Rouvroit, C.; Sory, M. P.; Vanooteghem, J. C.; and Michiels, T. 1991. *ymo*A, a *Yersinia enterocolitica* chromosomal gene modulating the expression of virulence functions. Mol Microbiol 5: 1023-34.

Delor, I.; Kaeckenbeeck, A.; Wauters, G.; and Cornelis, G. R. 1990. Nucleotide sequence of *yst*, the *Yersinia enterocolitica* gene encoding the heat-stable enterotoxin, and prevalence of the gene among pathogenic and nonpathogenic yersiniae. Infect Immun 58: 2983-88.

Forsberg, A.; Viitanen, A. M.; Skurnik, M.; Wolf-Watz, H and Straley, S. C. 1990. The surface-located YopN protein is involved in calcium signal transduction in *Yersinia pseudotuberculosis*. Mol Microbiol 5:977-86.

Furones, M. D.; Gilpin, M. J.; Alderman, D. J.; and Munn, C. B. 1990. Virulence of *Yersinia ruckeri* serotype I strains is associated with a heat-sensitive factor (HSF) in cell extracts. Fed Eur Microbiol Soc Microbiol Lett 66:339-44.

Isberg, R. R. 1990. Pathways for the penetration of enteroinvasive *Yersinia* into mammalian cells. Mol Biol Med 7:73-82.

Isberg, R. R., and Falkow, S. 1985. A single genetic locus encoded by *Yersinia pseudotuberculosis* permits invasion of culture animal cells by *Escherichia coli* K-12. Nature 317:262-64.

Kienle, Z.; Emody, L.; Svanborg, C.; and O'Toole, P. W. 1992. Adhesive properties conferred by the plasminogen activator of *Yersinia pestis*. J Gen Microbiol 138:1679-87.

Leung, K. Y.; Reisner, B. S.; and Straley, S. C. 1990. YopM inhibits platelet aggregation and is necessary for virulence of Yersinia pestis in mice. Infect Immun 58:3262-71.

Lindler, L. E., and Klempner, M. S. 1990. *Yersinia pestis* pH 6 antigen: Genetic, biochemical, and virulence characterization of a protein involved in the pathogenesis of bubonic plague. Infect Immun 58:2569-77.

Miller, V. L.; Finlay, B. B.; and Falkow, S. 1988. Factors essential for the penetration of mammalian cells by *Yersinia*. Curr Top Microbiol Immunol 138:15-39.

Price, S. B.; Cowan, C.; Perry, R. D.; and Straley, S. C. 1991. The *Yersinia pestis* V antigen is a regulatory protein necessary for Ca^{2+}-dependent growth and maximal expression of low-Ca^{2+} response virulence genes. J Bacteriol 173:2649-57 .

Rosenshine, I.; Duronio, V.; and Finlay, B. B. 1992. Tyrosine protein kinase inhibitors block invasin-promoted bacterial uptake by epithelial cells. Infect Immun 60:2211-17.

Rosqvist, R.; Forsberg, A.; Rimpilainen, M.; Bergman, T.; and Wolf-Watz, H. 1990. The cytotoxic protein YopE of Yersinia obstructs the primary host defence. Mol Microbiol 4:657-67.

Straley, S.C. 1988. The plasmid-encoded outer membrane proteins of *Yersinia pestis*. Rev Infect Dis 10:S323-326.

_________. 1991. The low-Ca^{2+} response virulence regulon of human-pathogenic yersiniae. Microb Pathog 10:87-91.

Straley, S. C., and Harmon, P. A. 1984. *Yersinia pestis* grows within phagolysosomes in mouse peritoneal macrophages. Infect Immun 45:655-59.

Tauxe, R. V.; Wauters, G.; Goosen, V.; VanNoyen, R.; Vandepitte, J.; Martin, S. M.; De Mol, P.; and Thiers, G. 1987. *Yersinia enterocolitica* infections and pork: The missing link. Lancet 1:1129-32.

Tertti, R.; Skurnik, M.; Vartio, T.; and Kuusela, P. 1992. Adhesion protein YadA of *Yersinia* species mediates binding of bacteria to fibronectin. Infect Immun 60:3021-24.

Wake, A., and Morgan, H. R. 1986. Host-Parasite Relationships and the *Yersinia* Model. New York: Springer-Verlag.

20 / *Brucella*

BY **C. O. THOEN**, **F. ENRIGHT**, AND **N. F. CHEVILLE**

B*RUCELLA* **SPP.** cause zoonoses of worldwide importance. Brucellae that cause disease in goats, cattle, swine, sheep, dogs, dolphins, and humans are small, nonmotile, gram-negative coccobacillary rods. The organisms are aerobes that require enriched media for primary isolation; in addition, *B. abortus* and *B. ovis* usually require an atmosphere of increased carbon dioxide (CO_2) tension. The six recognized species, *B. abortus, B. melitensis, B. suis, B. canis, B. ovis,* and *B. neotomae*, can be differentiated on the basis of CO_2 requirement, production of hydrogen sulfide (H_2S), phage susceptibility, cell-surface antigens, and sensitivity to azo dyes. DNA polymorphism due to highly repeated DNA has been found in *Brucella* spp. Differences in restriction-fragment patterns were found when Southern blots of DNA from both *B. ovis* and *B. abortus* isolates were hybridized to a 1.6-kilobase pair (kb) *Hin*d III fragment of *B. ovis* containing repeated DNA and to a 1.3-kb *Hin*d III fragment of *B. abortus* encoding BCSP31 (Halling and Zehrs 1990).

Identification of certain isolates may require tests to determine their ability to utilize selected amino acid and carbohydrate substrates as sources of carbon and energy as measured by oxygen uptake. The differential sensitivity of *Brucella* spp. to dyes is due to slight but detectable pore-diameter differences due to alteration in group 2 outer-membrane proteins. Despite considerable phenotypic differences among the *Brucella* spp., all share greater than 90% DNA homology. Very slight differences in DNA (single-base pairs) are expected to exist between strains of the same *Brucella* sp. Polymerase chain-reaction amplification techniques have been used to detect the presence of *Brucella* spp. (Bricker and Halling 1992).

Brucellae are facultative intracellular pathogens that are able to survive and multiply within phagocytic cells and lymphoid tissues. The persistence of brucellae within host cells is of prime importance in the evolution of the granulomatous reaction typical of brucellosis.

VIRULENCE

Virulence factors associated with brucellae are poorly defined. We have little understanding of the factors that determine the ability of brucellae to invade mucosal membranes, resist the lethal effects of normal blood plasma, promote entry into phagocytic cells, resist destruction within phagocytes, alter or prevent the induction of protective immune responses, and colonize and replicate within specialized cells of the placenta.

Historically, the level of virulence of *Brucella* spp. and strains has been determined by experimental infection of various hosts. Within a given *Brucella* sp., it has been repeatedly demonstrated that strains with smooth colonial morphology (bacteria with surfaces composed of lipopolysaccharide [LPS]) are more virulent than those with rough colonial morphology (bacteria lacking a major LPS component in their outer membrane) (Roop et al. 1991). The basis for this association of virulence with smooth LPS has been investigated in a variety of in vitro culture systems with both phagocytic and nonphagocytic cells from a number of hosts (Kreutzer et al. 1979; Riley and Robertson 1984; Canning et al. 1985; Bertram et al. 1986; Detilleux et al. 1990). In general, these studies clearly demonstrate that both smooth (virulent) and rough (nonvirulent) brucellae are able to enter host cells. In studies utilizing phagocytic cells (both neutrophils and macrophages), the smooth brucellae survive and even multiply while the rough bacteria are eventually eliminated. Survival of smooth bacteria has been related to their ability to prevent or limit lysosomal-phagosomal fusion, and their ability to resist the destructive effects of lysosomal enzymes after fusion has occurred (see above references).

An important property of *Brucella* spp. associated with virulence is the ability to survive and multiply intracellularly in host phagocytes (Cheville 1993). In vitro studies show that the organisms are ingested by macrophages but are partially protected in the phagosome and can multiply. Electron microscopy has demonstrated the presence of an external layer (envelope) on *B. abortus, B. melitensis,* and *B. suis*. The biologic significance of an external envelope interfering with attachment of brucellae to mucous membranes of the host is unclear.

The survivability of brucellae within phagocytic cells is associated with the composition of the cell wall of the bacteria. Initially, it was thought that the ability of smooth brucellae to resist intracellular killing by neutrophils was due to the bacteria's inability to stimulate oxidant production by the phagocytic cell. The lack of respiratory burst activity observed when neutrophils were exposed to smooth *Brucella* spp. was thought to be due to the fact that the bacteria were not opsonized with fresh serum (Kreutzer et al. 1979). Canning et al. (1985) have demonstrated that opsonized *B. abortus* would stimulate oxidant production in both bovine neutrophils and mammary gland macrophages. Irradiated *B. abortus* (strain 2308), opsonized with either pooled bovine reactor serum, normal bovine serum, or normal fetal calf serum, was shown to cause bovine neutrophils to produce moderate levels of hydrogen peroxide (H_2O_2); however, the opsonized bacteria resulted in very low levels of H_2O_2 being produced by bovine blood-derived macrophages (Bounous et al. 1992). The data on bovine macrophages are less clear and suggest that brucellae are not able to stimulate oxidant production necessary to initiate killing of the bacteria. This may be important; for all evidence to date suggests that brucellae persist within macrophages and not within neutrophils.

Elimination of bacterial agents by phagocytes typically requires fusion of lysosomes with phagosomes. Studies have demonstrated that smooth brucellae inhibit this fusion;. however, the bacterial factors responsible for the inhibition of lysosomal fusion are not known. In one study, a water extract of smooth *B. abortus* inhibited phagosomal-lysosomal fusion while water extracts from a nonvirulent *B. abortus* failed to inhibit fusion (Bertram et al. 1986). The only significant differences found within the aqueous extracts from both the virulent and nonvirulent strains were in the total amounts of sugars in each extract. Others have shown that crude supernatants from heat-killed *B. abortus* but not from washed, heat-killed *B. abortus* prevented lysosome-phagosome fusion in bovine neutrophils. Two low-molecular weight components of the crude supernatants were determined to be 5′-guanosine monophoshate and adenine. These two compounds inhibited the fusion of primary granules with phagosomes in bovine neutrophils (Canning et al. 1985; Bertram et al. 1986). Recently proteins produced by *B. abortus* in response to stress have been identified; however, their role in survival of bacteria in the host is unclear (Mayfield et al. 1988; Lin et al. 1992; Roop et al. 1992).

Virulent brucellae may also escape the killing mechanism of leukocytes by resisting the effects of both oxygen-dependent and -independent bactericidal systems of these cells. Smooth pathogenic strains of *B. abortus* have been shown to be more resistant to killing by the myeloperoxidase system than were rough strains. The basis for this differential susceptibility is unknown, but it is thought to be related to LPS components on the surface of the smooth bacteria (Riley and Robertson 1984). Studies conducted on the interactions of brucellae with African green monkey kidney (Vero) cells, have demonstrated that both smooth and rough brucellae are internalized. Rough brucellae, by virtue of the increased hydrophobicity of their cell wall, adhere more readily than smooth bacteria; however, the smooth bacteria demonstrate a greater rate of replication. Differences in the infectivity of smooth and rough brucellae for Vero cells are related to the ability of smooth cells to gain access to the rough endoplasmic reticulum (RER) of the host cells. Even though more rough bacteria than smooth bacteria entered these cells, the increased susceptibility of the rough bacteria to damage by lysosomal products may have prevented them from gaining access to the RER (Detilleux et al. 1990).

Gene-replacement and gene-deletion techniques have been widely used to construct vaccine strains that can be distinguished by marker deletions or additions. These vaccine strains have been used in work with *B. abortus*. Construction and testing of a Cu-Zn superoxide dismutase (SOD) deletion mutant of *B. abortus* strain 19 suggested that this enzyme plays a role in the survival and pathogenicity (Tatum et al. 1992). However, injection of SOD-deficient mutants into animals failed to demonstrate important differences in survivability of the mutants (Sriranganathan et al. 1991). Replacement of a gene encoding a 31-kDa protein with antibiotic resistance marker gene has also been reported (Halling et al. 1991); the loss of the 31-kDa gene does not significantly affect invasion, growth, or replication in vitro.

DISEASE

Establishment of infection by *Brucella* spp. depends on the number and virulence of bacteria and the relative resistance of a host, as determined by innate and acquired specific immune mechanisms. Approximately 18% of cross-bred cattle have been estimated to be innately resistant to infections with *B. abortus*. This resistance is inherited as a dominant trait and is influenced by one or several genes. Phenotypically, this resistance is manifested by the ability of macrophages from resistant cattle to limit the replication of *B. abortus*. The gene or genes that control host resistance to infection have not been identified (Price et al. 1990). A similar pattern of inheritance has been demonstrated in swine that are resistant to *B. suis* infections. Innate resistance to other intracellular pathogens in mice is controlled by one or only a few genes and is very similar to the pattern of resistance to *Brucella* spp. observed in both cattle and swine. In contrast, acquired resistance to brucellae infections in mice is influenced by a large complex of genes (Cheers 1984).

The probability of isolation of *B. abortus* at parturition increased from 0.22 to 0.9 as fetal age at the time of challenge of nonvaccinated heifers increased from 60 to 150 days of gestation (Crawford et al. 1987). The host mechanisms responsible for increased susceptibility to infection as pregnancy advances are not known, but they may be related to the differential susceptibility of placental trophoblasts during the middle and late stages of pregnancy (Samartino and Enright 1992). Organisms may penetrate the mucosa of nasal or oral cavities. Following penetration of mucosal epithelium, bacteria are transported, either free or within phagocytic cells, to regional lymph nodes, which become enlarged due to lymphatic and reticuloendothelial hyperplasia and inflammation. These changes may require several weeks to develop and may persist for months. If bacteria do not become localized and are not killed in regional lymph nodes, spread to other organs via lymph and blood may occur. Brucellae gain access to the uterus via a hematogenous route and the bacteria initially localize within erythrophagocytic trophoblasts of the placentome (Fig. 20.1). Adjacent chorioallantoic trophoblasts become infected and support massive growth of the bacteria. These cells eventually rupture and ulcerate the chorioallantoic membrane. Ulceration of the membrane is accompanied by an inflammatory reaction within the membrane. Bacteria and inflammatory cells both are present within the lumen of the uterus. Bacteria spread via a hematogenous route to the fetus and to the placentome. Other than a diffuse submucosal inflammatory reaction, it is important to note that the endometrium is not infected with brucellae and remains largely intact.

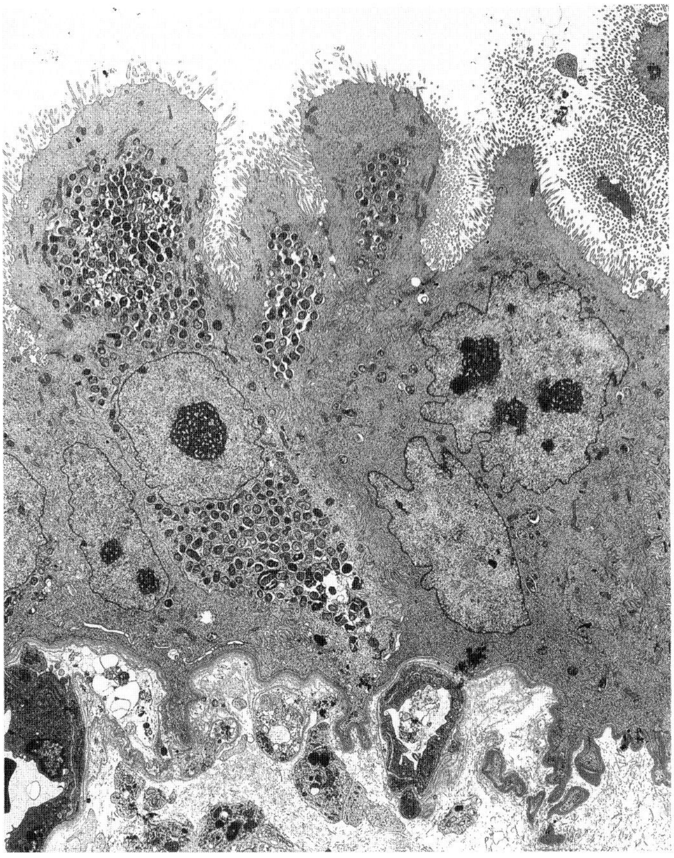

FIG. 20.1. Electron micrograph of trophoblast from the placenta of a cow with brucellosis. Large numbers of *Brucella abortus* are in the cytoplasm.

The presence of elevated amounts of erythritol in uterine tissues of cattle and other species, including sheep, goats, and swine, suggests an important role for the ability to utilize erythritol in the tissue tropism of certain *Brucella* spp. Moreover, extracts of fetal fluids, placenta, and chorion have been shown to stimulate growth of *B. abortus*, *B. melitensis,* and *B. suis* (Keppie et al. 1965). Localization of brucellae in the reproductive tract of both sexes may be due to a high concentration of erythritol but definitive information is not available (Keppie et al. 1965).

Brucella abortus

Brucella abortus infection in pregnant cattle involves the placenta and fetus. One investigation indicated initial lesions occurred within the placenta as early as 4 weeks after challenge; at 5 weeks, a more extensive endometritis with erosion of endometrial epithelium was evident (Payne 1959; Anderson and Cheville 1986). Following experimental infection of pregnant cattle via the conjunctival route, bacteria were first detected in the uterus 4-5 weeks postinoculation.

Chorionic epithelium becomes infected and infection extends into the placenta. Considerable variation occurs in distribution and severity of placental lesions in natural and experimental *B. abortus* infections. Seldom are all placentomes involved and frequently only portions of individual placentomas develop lesions. Moreover, fetuses that become infected during late gestation may be aborted, with an absence of grossly recognizable placental lesions. In such cases, microscopic examination of placental tissues often reveals mild focal inflammation (Enright et al. 1984). Other investigators report that extensive fetal inflammatory disease involving multiple organs leads to fetal stress, which results in premature deliveries. Bovine fetuses of a known gestational age were inoculated per os or intramuscularly with low doses of *B. abortus* strain 2308 and examined after abortion or delivery by caesarean section at intervals of 6-16 days postinoculation (Enright et al. 1984). Infected fetuses developed lymphoreticular hyperplasia of lymph nodes as early as 6 days postinoculation. Progression of lymph node changes to primary-follicle formation and germinal-center formation was noted at 9 and 12 days postinoculation, respectively. Morphologic alterations of lymph nodes correlated with elevations of IgM and IgG immunoglobin levels by day 9 postinoculation. Granulomatous responses, composed of macrophages, epithelioid cells, and giant cells, were present in lymphatic tissues, liver, and occasionally within the kidney interstitium as early as 6 and 9 days post-inoculation. Inflammatory lesions were extensive in fetal lungs characterized by thickened interlobular septae and focal-to-diffuse, exudative interstitial pneumonia. Pulmonary lesions were composed primarily of macrophages and a few neutrophils.

Fetal cortisol levels were elevated substantially above normal values and were present as early as 6 days postinoculation. Experimentally infected bovine fetuses were aborted between 7-19 days postinoculation. This relatively short and consistent interval between infection and abortion suggests that the bovine fetus is a sensitive indicator of in utero infection with *B. abortus*. The character and early development of granulomatous inflammatory reaction and a rapid development of the fetal immune response demonstrate that the fetus is capable of reaction to *B. abortus* infection similar to that of mature animals.

Mammary gland infection is often clinically inapparent, and gross lesions are not present. Histologically, there is a lobular and periductal, lymphoplasmacytic and histiocytic interstitial mastitis with leukocytes in alveoli and ducts. Brucellae localize and replicate primarily in macrophages in mammary secretions or in phagocytes in the interstitium. Lesions are not seen in all infected mammals (Meador et al. 1989a). Supramammary lymph nodes are typically enlarged and are characterized by edema, lymphofollicular hyperplasia, medullary plasmacytosis, and sinus histiocytosis. Localization of *B. abortus* in the mammary gland is markedly influenced by nursing; e.g., milk retention in the mammary gland correlates with the degree of infection. Studies in goats have shown that failure to nurse or release milk enhances localization and replication of bacteria in mammary glands after parturition (Meador et al. 1989b). In turn, mammary infection may result in increased systemic spread and persistence of brucellae in the host.

Brucella melitensis

B. melitensis is the principal cause of brucellosis in sheep and goats. The pathogenesis of *B. melitensis* infection in goats and early localization within the mammary gland and pregnant uterus of sheep is similar to that of *B. abortus* in cattle. However, *B. melitensis* is considered to cause more necrosis and less exudation within placental tissue than *B. abortus*.

Brucella ovis

B. ovis is an important cause of epididymitis in rams, but appears to be the least pathogenic brucella that affects animals. Venereal exposure is probably the most frequent route of transmission (Buddle 1955). Lesions in rams are most often located in the tail of the epididymis. Initial localization in the epididymis is accompanied by perivascular edema and infiltration of peritubular tissue by lymphocytes and monocytes; subsequently, neutrophils infiltrate the exudate. Previously inflamed tubular epithelium develops papillary hyperplasia and local hydropic degeneration, with subsequent formation of inflammatory reaction leading to an extravasation of spermatozoa. Host responses to extravasated spermatozoa lead to the formation of large spermatic granulomas, which may result in complete blockage of the epididymis; testicular degeneration and fibrosis are secondary to this blockage. *B. ovis* can be cultured from spleens of infected ewes in which no lesions are observed. Placental pathology experimentally induced by *B. ovis* tends to localize in the intercotyledonary placenta and is often less severe than the placentitis caused by *B. melitensis* or *B. abortus*.

Pregnant ewes inoculated with *B. ovis* in the conjunctiva on day 30-90 of gestation usually develop uterine infections and abort or deliver infected lambs. In contrast to the limited period of susceptibility of fetuses in the intraconjunctivally infected ewes, fetuses exposed to *B. ovis* in utero are susceptible to infection throughout pregnancy. Intervals between in utero infection with *B. ovis* and abortion range from 23 to 80 days postinoculation; by comparison, bovine fetuses infected in utero by *B. abortus* usually abort by 12 days postinoculation. This difference may be attributed to the low pathogenicity of *B. ovis*. The immunologic and granulomatous inflammatory responses of *B. ovis*-infected bovine fetuses are similar to those observed in *B. abortus*-infected bovine fetuses.

Brucella suis

Brucellosis in swine is usually caused by *B. suis*, which may be transmitted by ingestion; venereal transmission occurs (MacMillan 1992). Boars are as susceptible to infection as sows, and many infected boars develop lesions in the testicles and accessory reproductive organs and shed *B. suis* organisms in semen for extended periods. There is a tendency for focal granulomatous inflammation to develop in the endometrium and extend to the entire nongravid uterus. *B. suis* also tends to secondarily localize in a greater variety of tissues than does *B. abortus*. The organism demonstrates a predilection for localization in bone and joints, spleen, liver, kidney, and brain. Prolonged bacteremia often occurs with *B. suis* infection and may contribute to localization in a wide range of tissues.

The early stages in pathogenesis of *B. suis* infection in swine are comparable to the early stages of *B. abortus* infection in cattle. However, the character of the response of swine to *B. suis* differs slightly from the response of cattle to *B. abortus*. *B. suis* multiplies in mononuclear phagocytes and produces granulomatous lesions composed of macrophages and epithelioid cells. The granulomas tend to undergo caseous necrosis and become encapsulated by fibrous connective tissue.

Brucella canis

Brucella canis causes infectious abortions in female dogs and epididymitis, testicular atrophy, and sterility in infected males. Both oral and venereal transmission are important in *B. canis* infections. Depending on the route of infection, dogs typically develop enlarged retropharyngeal lymph nodes or superficial inguinal and external iliac lymph nodes. Susceptible dogs often exhibit bacteremia within 1-3 weeks following exposure to infection; this usually persists for several months to several years.

Tissue responses to *B. canis* consist of lymphoid and reticular cell hyperplasia and multiple granuloma formation. Localization in the epididymis and prostate results in chronic epididymitis and prostatitis; testicular degeneration and atrophy may occur (Carmichael and Kenney 1970). *B. canis* can be detected by immunofluorescence within the cytoplasm of macrophages in testicles, epididymis, and prostate and in epithelial cells lining these organs. Antisperm antibodies are formed in response to spermatozoa. An autoantibody of the IgA class has also been demonstrated on head-agglutinated spermatozoa within accessory sex organs. Autoantibody to spermatozoa may explain infertility observed in infected male dogs.

HOST RESPONSE

The alimentary tract is the major route in the transmission of *B. abortus* in cattle. Licking aborted fetuses and placental membranes or ingesting contaminated milk by calves introduces brucellae to the oral mucosa, tonsils, and gastrointestinal mucosa. Passage of *B. abortus* through epithelial barriers results in acute regional lymphadenitis and bacteremia.

Epithelium covering domes of ileal Peyer's patches, an important site of entry for several bovine pathogens that traverse the intestinal mucosa, is important in uptake of brucellae. Studies of infected ligated ileal loops in calves have shown that transepithelial migration of *B. abortus* occurs chiefly by dome lymphoepithelial cell endocytosis and transport, and that bacteria are degraded by macrophages and neutrophils of the gut-associated lymphoid tissue (Ackermann et al. 1988).

The interaction of bacteria with serum components (i.e., antibody or complement), neutrophils, mononuclear macrophages, fibroblasts, and epithelial cells results in the production of a variety of biologically active substances that activate macrophages, expand clones of antigen-recognizing T lymphocytes, stimulate lymphocytes to secrete lymphokines, stimulate hematopoiesis, and induce inflammation.

Several fractions have been isolated from *B. abortus* that generate chemotactic factors derived from serum (Bertram et al. 1986). A carbohydrate-rich, aqueous methanol fraction of *B. abortus* inhibited chemotactic activity at high concentration; however, a nondialyzable component of this fraction contained a potent stimulator of chemotaxis. Preheating the serum at 56°C for 30 minutes prevented generation of chemotactic activity by different fractions. Protein-rich fractions of *B. abortus* strain 2308 or *B. abortus* strain 19 failed to stimulate chemotaxis.

Neutrophils are considered an important line of defense against infection with the *Brucella* spp. (Riley and Robertson 1984). A component of *B. abortus* is capable of inhibiting release of myeloperoxidase by dose-dependent preferential inhibition of primary granule release from bovine neutrophilic leukocytes (Bertram et al. 1986). Failure of polymorphonuclear leukocytes (PMNs) to kill the organism at the primary site of infection may be responsible in part for the dissemination of *Brucella* spp. to other tissues of the body. Lysates from granules of guinea pig, human or bovine PMNs were highly toxic to smooth *B. abortus* strain 45/0 or to rough *B. abortus* 45/20 (Riley and Robertson 1984). However, an oxygen-dependent killing system appeared to be lethal to both strains. Iodine was more active than chlorine in the presence of H_2O_2 and granule lysate in killing the organism. Ingestion of either strain by PMNs failed to stimulate the hexose monophosphate shunt; therefore, *Brucella* spp. survive possibly because certain surface properties fail to generate a suitable stimulus to activate killing mechanisms during interaction with the plasma membrane.

Cell-mediated responses have been evaluated in cattle infected with virulent *B. abortus* and in cattle vaccinated with *B. abortus* strain 19. Lymphocyte-stimulation responses were significantly greater in cattle infected with virulent *B. abortus* as compared to responses observed in cattle vaccinated with attenuated *B. abortus* strain 19. However, no correlation was found between lymphocyte responses to specific antigen and humoral antibody responses in *B. abortus*-infected cattle. Leukocyte migration-inhibition responses were greater in non-vaccinated heifers experimentally infected with *B. abortus* as compared to vaccinated heifers similarly infected (Wyckoff and Confer 1990).

Differences in subpopulations of lymphocytes and mononuclear macrophages may explain in part how an organism stimulates bactericidal activity of macrophage in one instance and suppresses this activity at other times. The chronic persistence of *B. abortus* infections may be due to intracellular localization of brucellae in macrophages whose bactericidal mechanisms are resistant or refractory to activation. Overcoming bactericidal incompetence in these cells may be necessary for elimination of brucellae. Studies on the nonspecific and specific immunity to brucellae in rodents have provided valuable information on host cell-cell interactions and on host cell-brucellae interactions, which cannot be adequately studied in domestic animals due to lack of genetically identical individuals (Cheers 1984).

The protective effects of submucosal immune responses and inflammatory reactions against invading organisms may substantially alter the ability of *Brucella* spp. to colonize the local lymph nodes. Bacterial and host factors that allow brucellae to penetrate intact mucosa should be examined. Submucosal inflammatory responses composed of macrophages, lymphocytes, plasma cells, and large numbers of eosinophils and neutrophils have been observed as early as 2 and 4 days after conjunctival inoculation of cattle with *B. abortus*. These reactions are present in the submucosa of the conjunctiva, lacrimal duct, and in tonsils. Similar acute inflammatory reactions of mice may actually facilitate the spread of infection. Alteration of submucosal inflammatory reactions by presensitization and their effectiveness in destroying brucellae are unknown. Some pregnancy-associated or pregnancy-specific factors that suppress immune responses may alter the effectiveness of vaccines. The responses of a pregnant host to *Brucella* spp. may represent a useful model to elucidate various mechanisms of immune suppression during pregnancy.

DISEASE CONTROL

Efforts to control and eliminate brucellosis in cattle rely on the (1) use of vaccine prepared from attenuated *B. abortus* strain 19, (2) screening of milk of dairy herds for antibodies using the brucella ring test, (3) detection of infected beef cattle herds by the serologic testing of slaughter cows and bulls and tracing reacting animals to the herd of origin, and (4) identification and removal of *B. abortus*-infected cattle from a herd by use of serologic tests and/or bacteriologic examinations on appropriate tissues and milk. The standard plate and tube agglutination tests have been commonly used for detecting infected animals. The occurrence of agglutinating substances in sera of cattle not known to have been exposed to *Brucella* spp. stimulated the development of tests for differentiating between reactions due to exposure to *Brucella* spp. and those originating from other sources. These tests include acidified plate antigen, heat inactivation, rivanol agglutination, 2-mercaptoethanol tube agglutination, and buffered antigen tests such as the card test and rose bengal plate test. Purified LPS extracted from *B. abortus* has been used in enzyme-linked immunosorbent assays (ELISAs) and immunodiffusion tests for detecting cows infected with field strains of *B. abortus*.

The brucella ring test (BRT), which detects brucellae antibodies in milk, is useful for monitoring dairy herds for possible *B. abortus* infection. More recently, ELISAs have been developed for detecting specific immunoglobulin classes and subclasses in milk (Thoen et al. 1983).

A high percentage of *B. abortus*-infected cows shed large numbers of bacteria in vaginal secretions and fetal tissues, following abortion or delivery of an infected calf; therefore, efforts have been made to detect *Brucella* spp. in these specimens using bacteriologic or serologic procedures. Although bacteriologic examinations are suitably sensitive and highly reliable, there is a need to develop and evaluate more rapid procedures. Application of an ELISA for detecting *Brucella* spp. antigens in vaginal secretions may be useful for the rapid detection of animals shedding organisms, so they can be removed immediately from the herd (Chen et al. 1984). Recently it has been shown the use of polymerase chain reaction (PCR) provides for improved detection of brucellae in tissues (Fekete et al. 1992).

Vaccination of calves 4-8 months of age with *B. abortus* strain 19 provides protection in 65-85% of mature animals challenged by natural exposure. Killed *B. abortus* strain 45/20 suspended in adjuvant has been used for vaccinating cattle of all ages. A major advantage of this vaccine is that it does not induce levels of agglutinating antibodies that interfere with. interpretation of agglutination tests. However, it does stimulate the formation of antibodies detectable by complement-fixation test. Although vaccines are usually administered subcutaneously, other routes, including conjunctival and oral exposure, have been reported (Nicoletti and Milward 1983).

A live attenuated *B. melitensis* Rev 1 has been used for vaccinating sheep and goats. The vaccine is capable of inducing protection and does not cause abortion or stimulate development of humoral antibodies, which interfere with serologic tests; therefore, serologic tests can be utilized for detecting infected animals.

Polysaccharide components of *B. abortus* or *B. melitensis* that are not agglutinogenic provide enhanced protection. The use of tetramisole, a phenyl-imidothiazole salt, reportedly enhanced the immune response in mice vaccinated with killed *B. melitensis* cells suspended in incomplete adjuvant to subsequent challenge with *B. abortus*; however, no significant increase in antibody response was observed. The immune responses in cattle immunized with whole-cell outer membrane or with protein fractions of outer membrane of *B. abortus* 45/20 in combination with trehalose dimycolate (TDM) and muramyl dipeptide (MDP) has been investigated (Dzata et al. 1991). The magnitude of delayed-type hypersensitivity responses and lymphocyte transformation responses induced by vaccination was greater than responses observed in naturally or experimentally infected cattle. These findings suggest that TDM and MDP adjuvants induce elevated immune responses without undesirable side effects. New approaches utilizing genetic engineering techniques in producing vaccines with increased immunogenicity are a subject of current research.

Alum-precipitated bacteria prepared from *B. ovis* have been used to stimulate immune responses in sheep. Vaccines against *B. canis* or *B. suis* have not been shown to adequately stimulate host immunity to provide protection against challenge by natural exposure. Serologic tests and cultural examinations should be used for identifying infected dogs and swine so they can be removed from a population.

The major cycle of tranmission of *B. abortus* is from an aborted fetus to mature females. *B. abortus* infection of farm dogs has been reported; clinical signs are uncommon but abortion, epididymitis, and arthritis occur. Duration of infection varies but may exceed 150 days. Lymph nodes associated with the alimentary canal have the highest incidence of infection (Forbes 1990). Infected dogs shed bacteria via urine, vaginal exudates, feces, or aborted fetuses. Cattle-to-dog transmission occurs through ingestion of infected bovine placental tissue. Dog-to-cattle transmission has been reported, but only experimentally (there is circumstantial evidence that this occurs naturally). Aborted placentas or vaginal exudates probably have the greatest potential for transmitting *B. abortus* from dog to cow. Infected vaginal discharges have been reported up to 42 days after abortion in dogs. Coyotes and timber wolves may also be infected and transmit the disease in nature.

Insects that suck blood or feed on body fluids are capable of transmitting pathogenic *B. abortus*. Life cycles of the face fly *Musca autumnalis* are closely tied to cattle and other ruminants; this fly requires semifluid bovine feces for ova deposition and selectively feeds on bovine body fluids including blood, tears, and placental exudates. Experimentally, face flies will take up, hold, and excrete brucellae in their feces (Cheville et al. 1989). Although this mechanical transmission is probable in nature, it is unlikely to play a major role in transmission and maintenance of the disease in cattle.

REFERENCES

Ackermann, M. R.; Cheville, N. F.; and Deyoe, B. L. 1986. Bovine ileal dome lymphoepithelial cells: Endocytosis and transport of *Brucella abortus* strain 19. Vet Pathol 25:226-37.

Anderson, T. D., and Cheville, N. F. 1986. Ultrastructural morphometric analysis of *Brucella abortus*infected trophoblasts in experimental placentitits. Am J Pathol 124:226-37.

Bertram, T. A.; Canning, P. C.; and Roth, J. A. 1986. Preferential inhibition of primary granule release from bovine neutrophils by a *Brucella abortus* extract. Infect Immun 52:285-92.

Bounous, D. I.; Enright, F. M.; Gossett, K. A.; Berry, C. M.; and Kearney, M. T. 1992. Comparison of oxidant production by bovine neutrophils and monocyte-derived macrophages stimulated with *Brucella abortus* strain 2308. Inflam 16:215-18.

Bricker, B. J., and Halling, S. M. 1992. Identification of *Brucella* species by the polymerase chain reaction. Fed Am Soc Exp Biol J 6:A1624.

Buddle, M. B. 1955. Observations on the transmission of *Brucella* infection in sheep. NZ Vet J 3:10-19.

Canning, P. C.; Roth, J. A.; Tabatabai, L. B.; and Deyoe, B. L. 1985. Isolation of components of *Brucella abortus* responsible for inhibition of function in bovine neutrophils. J Infect Dis 152:913-21.

Carmichael, L. E., and Kenney, R. M. 1970. Canine brucellosis: The clinical disease, pathogenesis and immune response. J Am Vet Med Assoc 156:1726-34.

Cheers, C. 1984. Pathogenesis and cellular immunity in experimental murine brucellosis. In Biological Standardization, Vol. 56. Third International Symposium on Brucellosis. Ed. L. Vallette and W. Heunessen, pp. 237-46. Basel: S. Karger.

Chen, I.; Thoen, C. O.; Pietz, D. E.; and Harrington, R. Application of an enzyme-linked immunosorbent assay for detection of *Brucella* antigens in vaginal discharge of cows. Am J Vet Res 45:32-34.

Cheville, N. F. 1994. Ultrastructure Pathology: An Introduction to Interpretation. Ames: Iowa State University Press. Accepted for publication.

Cheville, N. F.; Rogers, D. G.; Deyoe, W. L.; Drafsur, E. S.; and Cheville, J. C. 1989. Uptake and excretion of *Brucella abortus* in tissues of the face fly (*Musca autumnalis*). Am J Vet Res 50:1302-6.

Crawford, R. P.; Adams, L. G.; and Williams, J. D. 1987. Relationship of fetal age at conjunctival exposure of pregnant heifers and *Brucella abortus* isolation. Am J Vet Res 48:755-57.

Detilleux, P. G.; Deyoe, B. L.; and Cheville, N. F. 1990. Penetration and intracellular growth of *Brucella abortus* in nonphagocytic cells in vitro. Infect Immun 58:2320-28.

Dzata, G. K.; Confer, A. W.; and Wyckoff, J. H., III. 1991. The effects of adjuvants of immune responses in cattle injected with a *Brucella abortus* soluble antigen. Vet Microbiol 29:27-42.

Enright, F. M.; Walker, J. V.; Jeffers, G. W.; and Deyoe, B. L. 1984. Cellular and humoral responses of *Brucella abortus* infected bovine fetuses. Am J Vet Res 45:424-30.

Fekete, A.; Bantle, J. A.; and Halling, S. M. 1992. Detection of *Brucella* by polymerase chain reaction in bovine fetal and maternal tissues. J Vet Diagn Invest 4:79-83.

Forbes, L. B. 1990. *Brucella abortus* infection in 14 farm dogs. J Am Vet Med Assoc 196:911-16.

Halling, S. M.; Detilleux, P. G.; Tatum, F. M.; Judge, B. A.; and Mayfield, J. E. 1991. Deletion of the BCSP31 gene of *Brucella abortus* by replacement. Infect Immun 59:3863-68.

Halling, S. M., and Zehrs, M. 1990. Polymorphism in *Brucella* spp. J Bacteriol 172:6637-40.

Keppie, J.; Williams, A. E.; Witt, K.; and Smith, H. 1965. The role of erythritol in tissue localization of the *Brucellae*. Br J Exp Pathol 46:104-8.

Kreutzer, D. L.; Dreyfus, L. A.; and Robertson, D. C. 1979. Interaction of polymorphonuclear leukocytes with smooth and rough strains of *Brucella abortus*. Infect Immun 23:737-42.

Lin, J.; Adams, L. G.; and Ficht, T. A. 1992. Characterization of the heatshock response in *Brucella abortus* and isolation of the genes encoding the GroE heatshock proteins. Infect Immun 60:2425-31.

MacMillan, A. P. 1992. Brucellosis. In Diseases of Swine, 7th Ed. Eds. A. D. Leman, B. E. Straw, W. L. Mengeling, S. D'Allaire, and D. J. Taylor, pp. 446-53. Ames: Iowa State University Press.

Mayfield, J. E.; Bricker, B. J.; Godfrey, H.; Crosby, R. M.; Knight, D. J.; Halling, S. M.; Balinsky, D.; and Tabatabai, L. B. 1988. The cloning, expression, and nucleotide sequence of a gene coding for an immunogenic *Brucella abortus* protein. Gene 63:1-9.

Meador, V. P.; Deyoe, B. L.; and Cheville, N. F. 1989a. Effect of nursing on *Brucella abortus* infection of mammary blands of goats. Vet Pathol 26:369-75.

__________. 1989b. Pathogenesis of *Brucella abortus* infection of the mammary gland and supramammary lymph node of the goat. Vet Pathol 26:357-68.

Nicoletti, P., and Milward, F. W. 1983. Protection by oral administration of *Brucella abortus* strain 19 against oral challenge with a pathogenic strain of *Brucella*. Am J Vet Res 44:1641-43.

Payne, J. M. 1959. The pathogenesis of experimental brucellosis in the pregnant cow. J Pathol Bacteriol 78:447-63.

Price, R. E.; Templeton, J. W.; Smith, R., III.; and Adams, L. G. 1990. Ability of mononuclear phagocytes from cattle naturally resistant or susceptible to brucellosis to control in vitro intracellular survival of *Brucella abortus*. Infect Immun 58:879-86.

Riley, L. K., and Robertson, D. C. 1984. Ingestion and intracellular survival of *Brucella abortus* in human and bovine polymorphonuclear leukocytes. Infect Immun 46:224-30.

Roop, R. M., II.; Jeffers, G.; Bagchi, T.; Walker, J.; Enright, F. M.; and Schurig, G. G. 1991. Experimental infection of goat fetuses in utero with a stable, rough mutant of *Brucella abortus*. Res Vet Sci 51:123-27.

Roop, R. M., II.; Price, M. L.; Dunn, B. E.; Boyle, S. M.; Sriranganathan, N.; and Schurig, G. G. 1992. Molecular cloning and nucleotide sequence analysis of the gene encoding the immunoreactive *Brucella abortus* Hsp60 protein, BA60K. Microb Pathog 12:47-62.

Samartino, L. E., and Enright, F. M. 1992. Interaction of bovine chorioallantoic membrane explants with three strains of *Brucella abortus*. Am J Vet Res 53:359-63.

Sriranganathan, N.; Boyle, S. M.; Schurig, G.; and Misra, H. 1991. Superoxide dismutases of virulent and avirulent strains of *Brucella abortus*. Vet Microbiol 26:359-66.

Tatum, F. M.; Detilleux, P. G.; Sacks, J. M.; and Halling, S. M. 1992. Construction of CuZn superoxide dismutase deletion mutants of *Brucella abortus*: Analysis of survival in vitro in epithelial and phagocytic cells and in vivo in mice. Infect Immun 60:2863-69.

Thoen, C. O.; Bruner, J. A.; Luchsinger, D. W.; and Pietz, D. E. 1983. Detection of brucella antibodies of different immunoglobulin classes in cow milk by enzymelinked immunosorbent assay. Am J Vet Res 44:306-8.

Wyckoff, J. H., III., and Confer, A. W. 1990. Immunomodulation in cattle immunized with *Brucella abortus* strain 19. Vet Immunol Immunopathol 26:367-83.

21 / *Pseudomonas* and *Moraxella*

BY C. L. GYLES

THE GENUS *PSEUDOMONAS* consists of gram-negative, nonfermentative, rod-shaped bacteria that are motile by means of polar flagella. Members of the genus are obligate aerobes, which are oxidative in their metabolism and produce indophenol oxidase (Palleroni 1984). *Pseudomonas aeruginosa*, *P. pseudomallei*, and *P. mallei* are three species of *Pseudomonas* of importance in animal diseases. All three species cause disease in humans as well, and in the case of *P. aeruginosa,* much of our knowledge of pathogenic mechanisms is drawn from studies of the human diseases.

PSEUDOMONAS AERUGINOSA

Pseudomonas aeruginosa is ubiquitous in the environment and is found in water, soil, and on plants, as well as on skin and mucous membranes of healthy animals (Sabath 1980). The blue-green, water-soluble pigment pyocyanin is unique to this organism and is produced by most strains. *P. aeruginosa* is physiologically versatile, is able to survive and grow on a wide variety of carbon sources over a temperature range of 20-42°C, and is extremely resistant to antibacterial drugs and disinfectants. Differentiation of strains is important for epidemiologic studies and for evaluation of vaccines and protective immunity. Serotyping, immunotyping, pyocin-typing (or aeruginocin-typing), and phage-typing are all used for characterization of isolates (Palleroni 1984). Serotyping is based on recognition of surface antigens of the organism, and there are several schemes in use. Pyocin type is determined either by the sensitivity of the isolate to pyocins produced by a standard set of strains or by the activity of pyocin produced by the isolate against a set of indicator strains. Seven immunotypes have been described on the basis of cross-protection studies in mice. Phage type is identified by the pattern of lysis of an isolate determined by a standard set of bacteriophages.

DISEASES

P. aeruginosa is an opportunistic pathogen (Table 21.1) and depends on a defect in normal host defense mechanisms to produce disease. In humans, this organism is a major nosocomial pathogen (Sabath 1980), which usually establishes in moist environments in hospitals and infects patients with debilitating disease or with deficient immune systems. Thus, when the skin is burned or traumatized, infection with *P. aeruginosa* may occur as a consequence of the defect in this epithelial covering. Wound infections in all species, chronic purulent otitis externa in dogs, ulcerative keratitis in dogs and horses, and dermatitis (fleecerot) in sheep are examples of diseases in which *P. aeruginosa* establishes a local infection following a break in an external epithelial barrier (Sabath 1980; Atherton and Pitt 1982; Bush 1983). There are also localized infections by *P. aeruginosa* in which the breach in normal host defense has not been identified: mastitis in cattle and genital tract infections in horses (Atherton and Pitt 1982).

TABLE 21.1. Diseases in animals caused by *P. aeruginosa*

Animal Species	Diseases
All species	Wound infection
Dogs, cats	Otitis externa, urinary tract infection, ulcerative keratitis, pneumonia
Cattle	Respiratory infections, mastitis, enteritis
Horses	Genital tract infection, abortion, ulcerative keratitis
Swine	Respiratory infections
Poultry	Septicemia, keratitis
Sheep	Pneumonia, mastitis, fleecerot
Mink	Hemorrhagic pneumonia, septicemia
Chinchillas	Pneumonia, septicemia
Laboratory rodents	Septicemia, enteritis

In certain diseases caused by *P. aeruginosa,* the external barriers to entry of the organisms are bypassed by manipulations; urinary tract infection in which the bacteria enter the bladder as a result of catheterization and systemic infection in which the organisms are introduced by injection of contaminated material are examples.

Prolonged antibacterial therapy administered orally to calves may predispose to enteric infection with *P. aeruginosa,* and respiratory viral infection may be a forerunner to pneumonia due to *P. aeruginosa*. In most animal species, diseases due to *P. aeruginosa* occur sporadically, as dictated by individual defects in host defense mechanisms. However, in poultry, mink, and chinchillas, outbreaks of septicemic disease due to *P. aeruginosa* occur. The specific predisposing factors have not been identified, but viral agents, notably calicivirus, have been isolated from mink with pneumonia caused by *P. aeruginosa*. Mink appear to be highly susceptible and have an intratracheal LD_{50} of 10^3-10^4, compared to values of 10^6-10^9 in other species (Long et al. 1980).

VIRULENCE FACTORS

Glycolipoprotein

P. aeruginosa produces an extracellular slime, which is produced by most environmental and tissue isolates (Knirel 1990). The slime contains polysaccharide of the cell wall lipopolysaccharide and a glycolipoprotein (GLP), which may be significant in disease. The biologic activity of GLP resides principally in the glycolipid component of the molecule (Orret al. 1982). Purified slime GLP is lethal for mice, and antibody against GLP protects mice against lethal challenge with the homologous strain. Leukopenia observed in disease in mice is associated with binding of GLP to blood leukocytes and sequestration of the GLP-leukocyte complex in the liver. GLP is also antiphagocytic.

Alginate

Mucoid strains of *P. aeruginosa* produce an exopolysaccharide, which is different from *Pseudomonas* slime. The exopolysaccharide is an acetylated alginate (a heteropolymer of mannuronic and guluronic acids) and is antiphagocytic in vitro, but nontoxic. Mucoid and nonmucoid forms of the same isolate are readily obtained and the extent and type of polysaccharide produced are affected by environmental factors, including temperature (Knirel 1990; May et al. 1991). The mucoid form of the bacterium is found almost exclusively in the lungs of humans with cystic fibrosis (CF).

In CF, infection is initiated by nonmucoid organisms, but as the disease progresses, there is a switch to the mucoid form, which then predominates. Alginate constitutes a surface layer that allows *P. aeruginosa* to resist dehydration in the CF-affected lung; interferes with complement-mediated chemotaxis of polymorphonuclear leukocytes (PMNs); inhibits non-opsonic phagocytosis by PMNs; scavenges hypochlorite and thereby effects resistance to myeloperoxidase-driven bacterial killing; and attaches bacteria to the tracheal epithelium. Exopolysaccharide also contributes to the formation of microcolonies of the organism in pulmonary infections and to the enhanced resistance to phagocytosis and antibacterial drugs observed in this mode of growth. It has also been proposed that alginate potentiates the effects of extracellular products of the bacteria by allowing concentration of these products within the extracellular gel that it forms.

Lipopolysaccharide

The lipopolysaccharide (LPS) from *P. aeruginosa* is weak in its biologic activity; its role in disease is not known. However, the presence and the amount of O polysaccharide as well as the length of the O-antigen chain influence virulence and susceptibility to serum. Mutants that lack the O-antigen moiety are less virulent than their wild-type parents and antibodies against LPS are protective. The LPS appears to protect against opsonization by complement in the nonimmmune host.

Original endotoxic protein (OEP) is a common protein antigen normally complexed with LPS in the cell wall and is capable of stimulating protective immunity.

Exotoxin A

Exotoxin A is the most toxic extracellular product of *P. aeruginosa* and is elaborated by most strains of the organism recovered from disease. The toxin is excreted as a protein of molecular weight 68 kDa and has a mechanism of action like that of diphtheria toxin; it catalyzes the transfer of the adenosine diphosphate-ribosyl (ADPR) moiety from nicotinamide adenine dinucleotide (NAD) to elongation factor 2 (EF2), a GTP-binding protein involved in protein synthesis by eukaryotic cells. ADP-ribosylated EF2 is unable to mediate elongation of polypeptide chains and protein synthesis stops. The end result is a marked cytotoxicity for a variety of cells and lethality for a variety of animals. The LD_{50} for mice is 25 μg/kg body weight by the intraperitoneal route and 3 μg/ kg body weight by the intravenous route. Studies of toxin crystals by X-ray diffraction have identified three structural domains: an amino-terminal domain is involved in binding to cell surface receptors; a second domain is responsible for translocation from endocytic vesicles to the cytoplasm; and the C-terminal third domain contains the active site of the enzyme (Wick et al. 1990).

Exotoxin A is produced in vivo, and specific antibodies are detected in patients recovering from infection. In experimental models of infection, strains that fail to produce the exotoxin have reduced virulence; protection is conferred by antibodies against the toxin. Toxin synthesis is regulated by iron concentration, production being maximal when iron concentration is low. The toxin inhibits the function of phagocytes in vitro, possibly because of its cytotoxicity. The immunosuppressive activity of exotoxin A may aid in long-term residence of the bacteria in the lung of CF patients (Staugas et al. 1992). There is evidence that the toxin may be important in some infections and not in others. Studies in experimental infections in animals indicate that exotoxin A contributes to corneal infections by interfering with host clearance of the bacteria and also causes tissue damage.

Proteases

Three proteases with broad substrate specificities are produced by most strains of *P. aeruginosa:* elastase, the LasA protein, and alkaline protease. Elastase, a zinc metalloprotease, is considered to be a major virulence factor. Despite its name, elastase has only weak elastolytic activity and has strong proteolytic activity (Galloway 1991). It is able to degrade certain plasma proteins such as immunoglobulins, complement factors, and alpha-proteinase inhibitor. The weak elastolytic activity is important in pathogenesis; it is capable of causing tissue damage by its attack on elastin, an important component of lung tissue and blood vessels.

Elastase, the product of the *las*B gene, is a neutral protease that is secreted as a 33-kDa protein. Elastase appears to act in concert with the LasA protein, which is a second elastase. Damage to elastin is initiated by LasA, and the nicked elastin is then highly susceptible to attack by elastase and other proteases (Galloway 1991; Tamura et al. 1992).

Exoenzyme S

Exoenzyme S is a second ADPR transferase produced by *P. aeruginosa* that is relatively heat stable. It preferentially ADP-ribosylates vimentin, a structural cell component, and a number of low-molecular weight, GTP-binding, membrane-associated proteins of eukaryotic cells. Studies of transposon mutants suggest that this toxin plays a role in chronic lung infection (Woods and Sokol 1985). Coburn (1992) noted that in the burned mouse model, mutants that are unable to produce exoenzyme S have an LD_{50} approximately 10^4 times higher than that of their parents, and that these mutants establish at the site of inoculation but fail to become disseminated. Antibodies against exoenzyme S protect against spread of infection. It appears, therefore, that exoenzyme S is involved in overcoming host defenses against spread from the site of infection. A recent finding is that exoenzyme S is an important adhesin (Baker et al. 1990) that mediates binding of the bacteria to glycosphingolipids.

Hemolysins

P. aeruginosa produces both a phospholipase C and a detergentlike glycolipid. Both products are hemolytic, and neither one has been ascribed a major role in disease. The phospholipase C does induce release of inflammatory mediators from granulocytes and mast cells, and the glycolipid results in release of histamine, serotonin, and 12-hydroxyeicosatetraenoic acid from platelets.

Leukocidin (Cytotoxin)

Leukocidin is a third ADPR transferase produced by certain strains of *P. aeruginosa*. Leukocidin damages the cytoplasmic membrane of PMNs of humans and animals and impairs the bactericidal function of PMNs, which undergo lysis subsequent to membrane damage

(Aldona et al. 1985). Leukocidin has been renamed cytotoxin because of its destructive effects on most eucaryotic cells studied. The gene (*ctx*) for cytotoxin was carried by a temperate phage in one strain that was examined by Hayashi et al. (1990).

Pili

P. aeruginosa produce type 4 pili that are polar and antigenically heterogeneous. These pili mediate adherence to upper respiratory tract epithelium, thereby contributing to colonization of respiratory epithelium. Mucins may serve as specific receptors for the organism at epithelial surfaces. Fibronectin ordinarily protects epithelial cells from attachment, and illnesses that result in a loss of the fibronectin coating predispose to epithelial-cell colonization. Cellular injury may also permit attachment of *P. aeruginosa* to epithelial cells; this opportunistic adherence occurs in pseudomonas keratitis and urinary tract infections.

Flagella

Montie et al. (1987) have suggested that flagella may be a virulence factor. Their studies showed that mutants that were either Fla^-, Mot^- or Fla^+, Mot^- had LD_{50} values for burned mice that were 10^3->10^6 times higher than the LD_{50} values for their Fla^+, Mot^+ parents. Furthermore, these researchers demonstrated that anti-flagellar antibodies protected burned mice from lethal invasion by *P. aeruginosa* but not local lesions. They concluded that the antibodies arrested motility, which was necessary for invasion of the bloodstream.

PATHOGENESIS

The source of organisms in disease due to *P. aeruginosa* may be either endogenous or exogenous. Some defect in local (skin or mucosa) defense or generalized defense (immune system) is necessary for this opportunistic pathogen to cause disease. The first phase of disease consists of establishment of infection; this is followed by production of toxic metabolites.

Adherence is the first step in infection, and both pili and exoenzyme S have been shown to mediate attachment of the organisms to epithelial surfaces. Pili can mediate attachment to buccal epithelial cells, damaged tracheal epithelial cells and mucous proteins, but antipilin antibodies only partially inhibit binding of whole bacteria to cells. Recent studies have demonstrated that exoenzyme S is a major adhesin of *P. aeruginosa* and that antibodies to this surface protein almost completely inhibit the binding of bacteria to buccal epithelial cells (Baker et al. 1992). These researchers have suggested that pili may be responsible for initial contact between bacteria and cells and that, subsequently, surface proteins such as exoenzyme S interact with receptors on the cells to effect an irreversible attachment. It is likely that alginate also functions in adherence, but not in the initial phases of infection; it is believed to attach mucoid organisms to tracheal epithelium.

Following adherence, the bacteria must be able to grow in the low-iron environment of the host. *P. aeruginosa* has effective mechanisms for scavenging iron from the host and is able to multiply in vivo. In low-iron environments the siderophores pyochelin and pyoverdin are produced by the bacterium, and iron-regulated membrane proteins are induced to transport iron-siderophore complexes into the cell. Interestingly, synthesis of exotoxin A, elastase, and alkaline protease is also derepressed by low iron concentrations. Damage to host tissue is mediated by bacterial products in acute disease and by the host immune response in chronic disease, as seen in cystic fibrosis.

Several bacterial factors have been identified and shown to be either definite or probable contributors to the disease process. However, pathogenesis of diseases due to *P. aeruginosa* is complex because the organism is invasive, toxinogenic, and endowed with a variety of potential virulence factors, none of which is of preeminent importance (Pollack 1984). Local lesions caused by *P. aeruginosa* show inflammation and necrosis, and may involve invasion of vascular walls and degenerative changes in epithelial and endothelial cells, as well as leukocytes. Changes in pulmonary endothelial and epithelial cells in experimentally infected mink are similar to changes caused by purified protease in rabbit lungs. The organism induces production of substantial amounts of tumor necrosis factor (TNF) in cultured peripheral blood mononuclear leukocytes, and high concentrations of this proinflammatory cytokine in the lung of CF patients could contribute to the chronic inflammation (Staugas et al. 1992).

Current concepts of the pathogenesis of infection due to *P. aeruginosa* are summarized in Figure 21.1. The bacterium colonizes damaged epithelium by attachment mediated by pili, exoenzyme S, and possibly other surface proteins. The organisms multiply at the site, aided by the antiphagocytic effects of LPS and slime, resistance to killing by complement, and the ability to obtain iron from the low-iron environment of the tissue. The infected tissue is then damaged by extracellular enzymes, notably proteases, phospholipase C, and exotoxin A, which contribute to the local lesion. If there is dissemination throughout the body, exoenzyme S contributes to the process, and endotoxin and exotoxin A play a role in systemic toxicity. In persistent infections, as seen in the lung of human patients with CF, tissue damage is likely due to immune-complex-mediated inflammation, involving release of proteases through complement activation and release of serotonin from thrombocytes.

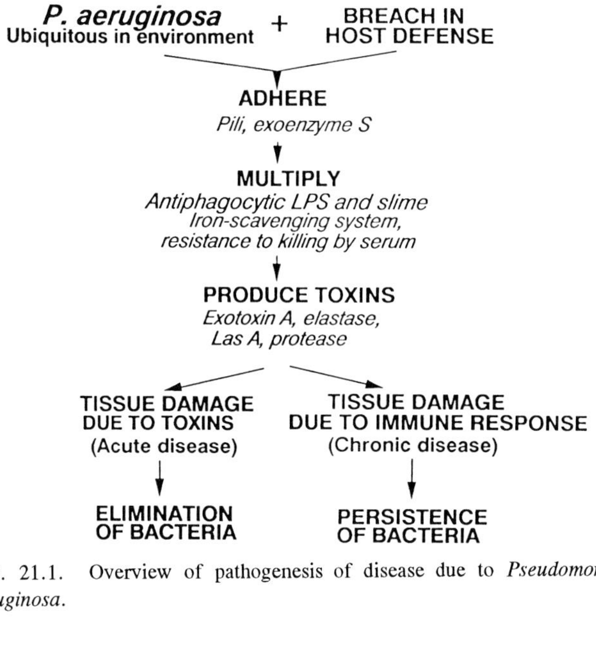

FIG. 21.1. Overview of pathogenesis of disease due to *Pseudomonas aeruginosa*.

HOST RESISTANCE

Phagocytosis by PMNs is the usual mechanism by which *P. aeruginosa* is eliminated from the host. The critical importance of phagocytosis is due to the organism's ability to resist the killing action of serum. Effective phagocytosis occurs in normal serum but is enhanced in the presence of opsonizing antibodies. It appears that adequate numbers of lung phagocytes, plus type-specific opsonic antibody represent the best circumstances for lung defense against *P.*

aeruginosa. Cellular immunity does not seem to be of much significance in resistance to disease caused by *P. aeruginosa* (Sabath 1980).

VACCINES

Vaccines against animal diseases due to *P. aeruginosa* are of practical relevance only in mink and chinchilla. Vaccines have been designed to suppress multiplication of the bacteria by enhancing the ability of PMNs to ingest and kill them, and/or neutralize toxic metabolites. The major difficulty is to ensure that protection extends to a sufficiently wide variety of immunotypes. Humoral antibody against LPS and exotoxin A appear to be most important in providing protective immunity (Cryz et al. 1987). Anti-LPS antibodies promote killing of the organisms by PMNs and antiexotoxin A antibodies likely protect against damage to cells and the immune system.

A formalin-killed culture of *P. aeruginosa* isolated from an outbreak of disease was effective in protecting mink. A multicomponent vaccine consisting of OEP, protease toxoid, and elastase toxoid has been tested in mink and has been reported to be effective in prevention and therapy and to be more effective than an autogenous formalin-killed bacterin preparation (Homma et al. 1983).

A heptavalent LPS vaccine containing O polysaccharide from immunotypes 1-7 coupled to exotoxin A has shown considerable promise in animal models and has evoked neutralizing antitoxic antibody and anti-LPS antibody in healthy humans (Cryz et al. 1987).

DRUG RESISTANCE

Strains of *P. aeruginosa* are resistant to a wide range of antimicrobial drugs and disinfectants. Much of this intrinsic resistance is attributed to the outer-membrane porins, which restrict passage of many antimicrobial agents into the periplasm. Thus, the organism is usually resistant to penicillin, ampicillin, tetracycline, first- and second-generation cephalosporins, sulfonamides, neomycin, streptomycin, kanamycin, chloramphenicol, nitrofurans, and trimethoprim-sulfonamide. Strains of *P. aeruginosa* are often susceptible in vitro to gentamicin, amikacin, tobramycin, colistin, polymyxin B, carbenicillin, ticarcillin, third-generation cephalosporins, and ciprofloxacin.

PSEUDOMONAS MALLEI

Pseudomonas mallei is an oxidase-negative, nonmotile organism that causes "glanders" of horses, mules, and donkeys. The host specificity of the organism has facilitated control and eradication, and the disease no longer occurs in North America but is still encountered in parts of Asia. Humans and carnivorous animals may occasionally become infected by the organism; the disease is often fatal for humans.

Disease may result from ingestion, inhalation, or skin contamination (Minett 1959). The bacteria are most commonly ingested and there is invasion of the gut wall and a subsequent septicemia. The septicemic disease may be accompanied by bronchopneumonia and may give rise to cutaneous lesions. Inhalation usually leads to a bronchopneumonia and lesions in the nasal mucosa. Infection of abraded skin gives rise to cutaneous glanders or "farcy." The characteristic lesions are nodules, which may develop in the lung, nasal mucosa, or skin. Little is known about the bacterial and host factors in the disease process. Recovered horses have some degree of immunity to reinfection but are not solidly immune.

Over the years, efforts have been concentrated on tests for detection of infected animals so that these animals may be slaughtered. The mallein test, which is based on delayed hypersensitivity and involves intrapalpebral inoculation of an extract of the bacterium, is commonly used for this purpose. An indirect hemagglutination test, a complement-fixation test, and counter immunoelectrophoresis are also useful for detection of infection. Affected animals should be slaughtered and treatment is not recommended.

PSEUDOMONAS PSEUDOMALLEI

Pseudomonas pseudomallei is ubiquitous in tropic and subtropic environments, particularly in Southeast Asia (Dance 1991). It is considered to be a soil organism, which occasionally causes disease (Leelarasamme and Bovornkitti 1989; Dance 1990). In Australia, high concentrations of the organism have been recovered from soil during the summer. The bacteria appear to persist in moist clay soils during the dry season and in the surface layers during the rainy season. Wild rodents are an important reservoir of the organism and serve to contaminate soil and water. Although infected animals may disseminate the organisms, they are not usually a source of infection; humans and animals become infected through contact with bacteria in the environment.

The disease melioidosis has been reported in a wide variety of hosts, including cattle, sheep, goats, horses, pigs, dogs, cats, camels, monkeys, and humans (Minett 1959; Stedham 1971; Narita et al. 1982; Mutalib et al. 1984). Infection may occur as a result of ingestion, inhalation, or skin contamination; there may be a long incubation period before disease is evident.

In adult domestic animals, the disease typically runs a chronic course with progressive emaciation leading to death. There is widespread dissemination of organisms throughout the body and abscesses usually develop in infected organs, which often include the lungs, spleen, and liver. *P. pseudomallei* is a major cause of abscess formation in pigs in some areas of Southeast Asia, resulting in substantial losses due to condemnation of carcasses.

There has been little recent interest in the study of melioidosis and almost nothing is known of mechanisms of pathogenicity. However, the organism has been shown to produce a very potent 36-kDa exotoxin, whose mechanism of action is the same as that of exotoxin A of *P. aeruginosa* (Ismail et al. 1991). This toxin is believed to cause cellular damage during the course of disease; antibody to the toxin is produced by infected animals. Other extracellular enzymes produced by this organism that could cause tissue damage are a dermonecrotic protease, a lipase, and a lecithinase (Dance 1990). In humans, melioidosis is often associated with underlying disease or a compromised immune system. This observation, taken with the prolonged incubation time in some cases, and the tendency to have recrudescence have suggested that the organism may be a facultative intracellular parasite.

There is marked variation in virulence of strains. When 500 or more organisms of one strain were injected subcutaneously into goats, all animals developed a severe acute disease, which lead to death (Thomas et al. 1988). However, when more than 10^7 organisms of another strain were inoculated subcutaneously into goats, mild disease or no disease occurred (Narita et al. 1982).

Most studies of the epidemiology of the disease and of experimental reproduction of the disease are found in the older literature (Minett 1959). Although infected animals may shed large numbers of the bacteria, spread to other animals or to humans is extremely rare, suggesting that the form of the organism growing in soil is infectious whereas the form growing in tissues is not. This is reminiscent of the situation with certain dimorphic fungi which have an infective saprophytic mycelial form and a noninfective yeast form found in tissue.

MORAXELLA BOVIS

Moraxella bovis is the causative agent of the disease known as infectious bovine keratoconjunctivitis (IBK), or pink eye of cattle (Baptista 1979; Chandler et al. 1979; Punch and Slatter 1984). This organism is a short, plump, gram-negative rod, which characteristically appears as a diplobacillus. The bacterium is oxidative, positive for oxidase and proteolytic activities, and negative for carbohydrate fermentation and growth on MacConkey agar. Typically, colonies of the organism recovered from diseased animals produce hemolysis on blood agar, are flat and friable, and produce a pitting of the agar.

IBK is an economically very important disease of cattle throughout the world and is noteworthy for its rapid spread within the herd. In the individual animal, one eye is usually affected at the outset, but the other eye may become infected later. Keratitis and conjunctivitis develop rapidly and an initially serous discharge later becomes purulent. Corneal ulceration may occur. Infrequently, the infection may spread to produce meningitis. Affected animals suffer weight loss, and those that are lactating suffer decline in milk production. The lesions in the infected eyes usually heal in 2-6 weeks.

Cattle can carry *M. bovis* in the eye, nasal cavity, or vagina. The source of organisms in an outbreak is usually the carrier animal, which may have had clinically manifest disease and recovered, or may never have had disease resulting from infection.

PATHOGENESIS

Pili (fimbriae) and hemolysin are two bacterial factors that are required for virulence of *M. bovis* (Pedersen et al. 1972; Punch and Slatter 1984). Animals that carry the organisms but show no ill effects often have a high percentage of nonhemolytic, nonpiliated organisms. Loss of ability to produce hemolysin and/or pili also occurs in vitro.

Pili produced by *M. bovis* are classified as type 4 pili and are characteristic of pili found on *Pseudomonas aeruginosa*, *Neisseria*, and *Bacteroides nodosus*. Features of these pili include being thin and flexible, being associated with twitching motility, having N-methylphenylalanine as the first amino acid of the mature pilin subunit, and having a highly homologous hydrophobic N-terminal. These pili also have a mainly polar distribution on the cell. Different strains produce pilin with different molecular weights, and a single strain may encode more than one type of pilin protein. In one strain, inversion of a 2-kb region of DNA was responsible for determining which of the two pilin proteins was produced. Piliated *M. bovis* corrode and pit the agar, autoagglutinate, mediate hemagglutination, and form a pellicle on the surface of broth cultures. These culture characteristics are associated with virulence and are presumed to be due to the presence of pili. Transitions between piliated (P^+) and nonpiliated (P^-) forms occur when the bacteria are subcultured. Loss of piliation results in changes in colony morphology as well as in loss of virulence.

Pili are required for adherence of the bacteria to bovine conjunctival and corneal epithelium, and this adherence is essential for colonization, which precedes development of pathologic changes (Lepper and Power 1988; Moore and Rutter 1989). Adherence is required to overcome the flushing effects of lacrimal secretion and blinking. Pili are not only essential for virulence, but they are also important for stimulation of protective immunity. Pili on isolates collected from several countries have been placed in seven serogroups (A-G) on the basis of an enzyme-linked immunosorbent assay, slide agglutination, and tandem crossed immunoelectrophoresis (Moore and Lepper 1991). Immunity is type-specific.

Hemolysin is a heat-labile, Ca^{++}-dependent protein, which produces transmembrane pores in red blood cells (Clinkenbeard and Thiessen 1991). Colonies of the organism that no longer express hemolysis are avirulent, but the role of hemolysin in the disease process is not clear.

Hemolysin probably plays a role in establishment of infection and in tissue injury and may assist the nutritionally fastidious *M. bovis* to proliferate in vivo by supplying iron from lysed red blood cells. Recently, the hemolysin has been shown to be a leukotoxin, which damages the membranes of neutrophils (Hoien-Dalen et al. 1990). Incubation of living *M. bovis* bacteria with bovine neutrophils results in rapid lysis of the neutrophils, but this effect is not observed when human neutrophils are used. Indeed, human neutrophils phagocytose the bacteria and remain intact (Kagonyera et al. 1989). *M. bovis* cells and culture supernatant both are capable of killing bovine corneal epithelial cells in vitro. These activities are attributed to hemolysin, which is a pore-forming cytolysin rather than a mere hemolysin.

Ocular changes in infected cattle include necrosis and pitting of corneal epithelium. The cytolysin may be responsible for much or all of the changes. It is capable of causing direct damage to corneal epithelial cells and of protecting *M. bovis* from phagocytosis by destroying neutrophils. The cytolysin can also cause indirect damage through the release of neutrophil granules into the tissue; this contributes to corneal damage through the action of collagenases and other proteolytic enzymes.

A number of additional toxic activities have been described, but not well characterized (Pugh et al. 1973; Arora 1982). Dermonecrotic activity has been detected in cell-wall fractions of the organism, and an exotoxin has been proposed to account for certain heat-labile activities produced by cell sap preparations (Pugh et al. 1973). These activities include a lowering of body temperature, ocular edema, and local edema in cattle inoculated intravenously with a cell sap fraction of *M. bovis*. Hyaluronidase and collagenases produced by the organism may also play a role in disease (Arora 1982). An overview of pathogenesis of IBK is shown in Figure 21.2.

Although host, environmental, and bacterial factors important for development of disease have been identified, there are many gaps in our understanding of pathogenesis of the disease. A mouse model of disease in cattle has been developed. The model involves inoculation of *M. bovis* on to the corneal surface of steroid-treated mice. This model holds promise for increasing our understanding of the disease process (Lepper and Power 1988).

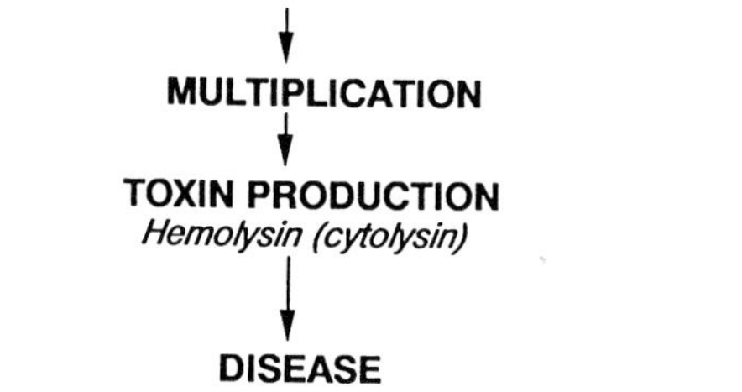

FIG. 21.2. Overview of the pathogenesis of infectious bovine keratoconjunctivitis (IBK) due to *Moraxella bovis*.

HOST RESISTANCE

Several factors that contribute to development and spread of disease have been recognized, and species and breed differences in susceptibility have been documented (Punch and Slatter 1984). Thus, *Bos indicus* breeds are known to be more resistant than *Bos taurus* species; degree of pigmentation of the eyelids associated with different breeds accounts, in part, for differences in breed susceptibility. White-faced breeds, or animals that lack pigmentation, tend to be more susceptible. The presence of large numbers of flies, especially the face fly (*Musca autumnalis*) increases the chances of development and spread of disease. Flies produce mechanical irritation of the conjunctiva and may act as mechanical vectors in the spread of organisms.

Several infectious agents that cause eye damage contribute to the development of IBK: infectious bovine rhinotracheitis virus, *Mycoplasma bovoculi,* and the nematode parasite *Thelazia*.

Solar ultraviolet irradiation predisposes to disease and is one of several factors that contribute to a greater frequency of disease in the warmer months. Mechanical irritants such as dust, grass, and wind also contribute to development of disease. Vitamin deficiency has been noted as a possible factor in some outbreaks of IBK.

The disease occurs more frequently in calves than in adults, and it is possible that this is due to development of immunity following exposure to the organisms. Emergence of new pilus types is sometimes responsible for outbreaks of disease in adults previously exposed to the organism.

IMMUNITY

Following infection of cattle with hemolytic, piliated *M. bovis*, the cattle develop antibodies that neutralize hemolytic activity of a wide variety of strains. However, antibodies that neutralize the pilus-mediated adherence are pilus-type specific (Lepper 1988). Interest in artificial stimulation of protective immunity has concentrated on pilus antigens. Antibody to pili inhibit adherence of *M. bovis* of the same serogroup to corneal epithelial cells in vitro (Moore and Rutter 1989).

The immunology of IBK is poorly understood. Both local and systemic antibodies have been demonstrated following natural and experimental disease (Kopecky et al. 1983; Weech and Renshaw 1983). When antibodies specific for *M. bovis* are measured, high levels of secretory IgA and low levels of IgG and IgM are detected in lacrimal fluid from infected cattle. There is debate about the relationship between serum antibody levels and protection against disease; it has been suggested that local cell-mediated immunity is the important factor in providing protection against the disease.

VACCINES

Although much work has been done on evaluation of vaccines, there is still no suitable vaccine. Heat-killed organisms, ribosomes, and disrupted cells all failed to stimulate protective immunity. Live vaccines or formalin-killed cells have induced some protection against challenge with the homologous organism, but protection against heterologous strains has been weak and variable. These results are undoubtedly associated with the existence of a variety of antigenic types of pili produced by the organism. It is interesting that a nonhemolytic, live, piliated strain of *M. bovis* elicited protection comparable to that obtained with the hemolytic organisms (Pugh et al. 1982), indicating that antibodies against the hemolysin are not needed for protection. Pilus preparations have also been tested as vaccines and have induced some protection against the homologous organism. Administration of a pilus vaccine into the subconjunctiva resulted in enhanced protection, compared with injection into the dewlap of calves (Pugh et al. 1985). Some passive protection is provided to calves that receive colostrum from dams vaccinated with a pilus vaccine or formalin-killed cells. Commercial vaccines are available but independent tests of two of these vaccines demonstrated that neither was effective in protecting calves against experimental or natural challenge infection (Smith et al. 1990). It is likely that a successful vaccine will need to contain pili of all 7 antigenic types and to be administered so as to elicit a strong secretory IgA response in lacrimal secretions.

REFERENCES

Pseudomonas

Aldona, L. B.; Hammer, M. C.; Smith, R. C.; Obrig, T. G.; Conroy, J. V.; Bishop, M. B.; Egy, M. A.; and Lutz, F. 1985. Effects of *Pseudomonas aeruginosa* cytotoxin on human serum and granulocytes and their microbicidal, phagocytic, and chemotactic functions. Infect Immun 48:498-506.

Atherton, J. G., and Pitt, L. T. 1982. Types of *Pseudomonas aeruginosa* isolated from horses. Equine Vet J 14:329-32.

Baker, N. R.; Minor, V.; Deal, C.; Shahrabadi, M. S.; Simpson, D. A.; and Woods, D. E. 1992. *Pseudomonas aeruginosa* exoenzyme S is an adhesin. Infect Immun 60:2859-63.

Bush, B. M. 1983. *Pseudomonas* infections in small animals. Vet Annu 23:162-67.

Coburn, J. 1992. *Pseudomonas aeruginosa* exoenzyme S. Curr Top Microbiol Immunol 175:133-43.

Cryz, S. J.; Furer, E.; Sadoff, J. C.; and Germanier, R. 1987. A polyvalent *Pseudomonas aeruginosa* O-polysaccharide-Toxin A conjugate vaccine. Antibiot Chemother 39:249-55.

Dance, D. A. B. 1990. Melioidosis. Rev Med Microbiol 1:143-50.

__________. 1991. Melioidosis: The tip of the iceberg. Clin Microbiol Rev 4:52-60.

Galloway, D. R. 1991. *Pseudomonas aeruginosa* elastase and elastolysis revisited: recent developments. Mol Microbiol 5:2315-21.

Hayashi, T.; Baba, T.; Matsumoto, H.; and Terawaki, Y. 1990. Phage-conversion of cytotoxin production in *Pseudomonas aeruginosa*. Mol Microbiol 4:1703-09.

Homma, J. Y.; Abe, C.; Yanagawa, R.; and Noda, H. 1983. Effectiveness of immunization with multicomponent vaccines in protection against hemorrhagic pneumonia due to *Pseudomonas aeruginosa* infection in mink. Rev Infect Dis 5S:858-66.

Ismail, G.; Mohamed, R.; Rohana, S.; Sharifah, H. S. M.; and Embi, N. 1991. Antibody to *Pseudomonas pseudomallei* exotoxin in sheep exposed to natural infection. Vet Microbiol 27:277-82.

Knirel, Y. A. 1990. Polysaccharide antigens of *Pseudomonas aeruginosa*. Rev Microbiol 17:273-304.

Leelarasamme, A., and Bovornkitti, S. 1989. Melioidosis: review and update. Rev Infect Dis 11:413-25.

Long, G. G.; Gallina, A. M.; and Gorham, J. R. 1980. *Pseudomonas* pneumonia of mink: Pathogenesis, vaccination, and serologic studies. Am J Vet Res 41:1720-25.

May, T. B.; Shinabarger, D.; Maharaj, R.; Kato, J.; Chu, L.; DeVault, J. D.; Roychoudhury, S.; Zielenski, N. A.; Berry, A.; Rothmel, R. K.; Misra, T. K.; and Chakabarty, A. M. 1991. Alginate synthesis by *Pseudomonas aeruginosa*: A key pathogenic factor in chronic pulmonary infections of cystic fibrosis patients. Clin Microbiol Rev 4:191-206.

Minett, F. C. 1959. Glanders (and melioidosis). In Diseases Due to Bacteria, Vol. 2. Ed. A. W. Stableforth and I. A. Galloway, pp. 296-309. London: Butterworth Scientific.

Montie, T. C.; Drake, D.; Sellin, H.; Slater, O.; and Edmonds, S. 1987. Motility, virulence, and protection with a flagella vaccine against *Pseudomonas aeruginosa* infection. Antibiot Chemother 39:233-48.

Mutalib, A. R.; Sheikh-Omar, A. R.; and Zamri, M. 1984. Melioidosis in a banded leaf monkey (*Presbytis melalophos*). Vet Rec 115:438-39.

Narita, M.; Loganathan, P.; Hussein, A.; Jamalludin, A.; and Joseph, G. 1982. Pathological changes in goats experimentally inoculated with *Pseudomonas pseudomallei*. Natl Inst Anim Health Q 22:170-79.

Orr, T.; Koepp, L. H.; and Bartell, P. F. 1982. Carbohydrate mediation of the glycolipoprotein of *Pseudomonas aeruginosa*. J Gen Microbiol 128:2631-38.

Palleroni, N. J. 1984. Family *Pseudomonadaceae*. In Bergey's Manual of Systematic Bacteriology, Vol. 1. Ed. N. R. Krieg and J. G. Holt. Baltimore: Williams and Wilkins.

Pollack, M. 1984. The virulence of *Pseudomonas aeruginosa*. Rev Infect Dis 6S:617-26.

Sabath, L. D. 1980. *Pseudomonas aeruginosa*: The organism, Diseases it Causes, and Their Treatment. Bern, Switz.: Hans Huber.

Staugas, R. E.; Harvey, D. P.; Ferrante, A.; Nandoskar, M.; and Allison, A. C. 1992. Induction of tumour necrosis factor (TNF) and interleukin-1 (IL-1) by *Pseudomonas aeruginosa* and exotoxin A-induced supression of lymphoproliferation and TNF, lymphotoxin, gamma interferon, and IL-1 production in human leukocytes. Infect Immun 60:3162-68.

Stedham, M. A. 1971. Melioidosis in dogs in Vietnam. J Am Vet Med Assoc 158:1948-50.

Tamura, Y.; Suzuki, S.; and Sawada, T. 1992. Role of elastase as a virulence factor in experimental *Pseudomonas aeruginosa* infection in mice. Microb Pathog 12:237-44.

Thomas, A. D.; Forbes-Faulkner, J. C.; Norton, J. H.; and Trueman, K. F. 1988. Clinical and pathological observations on goats experimentally infected with *Pseudomonas pseudomallei*. Aust Vet J 65:43-46.

Wick, M. J.; Frank, D. W.; Storey, D. G.; and Iglewski, B. H. 1990. Structure, function, and regulation of *Pseudomonas aeruginosa* exotoxin A. Annu Rev Microbiol 44:335-63.

Woods, D. E., and Sokol, P. A. 1985. Use of transposon mutagenesis to assess the role of exoenzyme S in chronic pulmonary disease due to *Pseudomonas aeruginosa*. Eur J Clin Microbiol 4:163-69.

Moraxella

Arora, A. K. 1982. Toxic effects of *Moraxella bovis* and their relationship to pathogenesis of infectious bovine keratoconjunctivitis. Vet Arh 52:175-82.

Baptista, P. J. H. P. 1979. Infectious bovine keratoconjunctivitis. A review. Br Vet J 135:225-42.

Chandler, R. L.; Baptists, P. J. H. P.; and Turfrey, B. 1979. Studies on the pathogenicity of *Moraxella bovis* in relation to infectious bovine keratoconjunctivitis. J Comp Pathol 89:441-48.

Clinkenbeard, K. D., and Thiessen, A. E. 1991. Mechanism of action of *Moraxella bovis* hemolysin. Infect Immun 59:1148-52.

Hoien-Dalen, P. S.; Rosenbusch, R. F.; and Roth, J. A. 1990. Comparative characterization of the leukocidic and hemolytic activity of *Moraxella bovis*. Am J Vet Res. 51:191-96.

Kagonyera, G. M.; George, L.; and Miller, M. 1989. Effects of *Moraxella bovis* and culture filtrates on ^{51}Cr-labeled bovine neutrophils. Am J Vet Res 50:18-21.

Kopecky, K. E.; Pugh, G. W.; and McDonald, T. J. 1983. Infectious bovine keratoconjunctivitis: Evidence for general immunity. Am J Vet Res 44:260-62.

Lepper, A. W. D. 1988. Vaccination against bovine keratoconjunctivitis: Protective efficacy and antibody response induced by pili of homologous and heterologous strains of *Moraxella bovis*. Aust Vet J 65:310-16.

Lepper, A. W. D., and Power, B. E. 1988. Infectivity and virulence of Australian strains of *Moraxella bovis* for the murine and bovine eye in relation to pilus serogroup sub-unit size and degree of piliation. Aust Vet J 65:305-9.

Moore, L. J., and Lepper, A. W. D. 1991. A unified serotyping scheme for *Moraxella bovis*. Vet Microbiol 29:75-83.

Moore, L. J., and Rutter, J. M. 1989. Attachment of *Moraxella bovis* to calf corneal cells and inhibition by antiserum. Aust Vet J 66:39-42.

Pedersen, K. B.; Froholm, L. O.; and Bovre, K. 1972. Fimbriation and colony type of *Moraxella bovis* in relation to conjunctival colonization and development of keratoconjunctivitis in cattle. Acta Pathol Microbiol Scand [B] 80:911-18.

Pugh, G. W.; Hughes, D. E.; and Schulz, V. D. 1973. The pathophysiological effects of *Moraxella bovis* toxins on cattle, mice, and guinea pigs. Can J Comp Med 37:70-78.

Pugh, G. W.; McDonald, T. J.; and Kopecky, K. E. 1982. Experimental infectious bovine keratoconjunctivitis: Efficacy of a vaccine prepared from nonhemolytic strains of *Moraxella bovis*. Am J Vet Res 43:1081-84.

Pugh, G. W.; Kopecky, K. E.; and McDonald, T. J. 1985. Infectious bovine keratoconjunctivitis: Subconjunctival administration of a *Moraxella bovis* pilus preparation enhances immunogenicity. Am J Vet Res 46:811-15.

Punch, P. I., and Slatter, D. H. 1984. Review of infectious bovine keratoconjunctivitis. Vet Bull 54:193-207.

Smith, P. G.; Blankenship, T.; Hoover, T. R.; Powe, T.; and Wright, J. C. 1990. Effectiveness of two commercial infectious bovine keratoconjunctivitis vaccines. Am J Vet Res 51:1147-50.

Weech, G. M., and Renshaw, H. W. 1983. Infectious bovine keratoconjunctivitis: Bacteriologic, immunologic, and clinical responses of cattle to experimental exposure with *Moraxella bovis*. Comp Immun Microbiol Infect Dis 6:81-94.

22 / *Campylobacter*

BY **M. M. GARCIA** AND **B. W. BROOKS**

THE INCREASING recognition of certain species of *Campylobacter* as human pathogens over the last 15 years has reinforced the importance of this genus in veterinary medicine. The zoonotic significance of several species such as *C. jejuni, C. coli,* and *C. fetus* subsp. *fetus* has resulted in accelerated studies of their taxonomy, epidemiology, molecular biology, and pathogenesis. In particular, recent virulence and animal model studies have helped elucidate certain pathogenic mechanisms of these animal campylobacters.

Taxonomic confusion has characterized members of the genus *Campylobacter* for many years. The wide diversity of their habitat and growth characteristics and relative inertness in biochemical tests have proved to be obstacles for proper identification and nomenclature. The interesting background behind this confusion and the practical problems associated with it have been dealt with in previous reviews (Garcia et al. 1983; Penner 1988). With the expanding list of new campylobacters or campylobacter-like organisms implicated in human and animal health, the systematic naming of these bacteria has become the focus of molecular taxonomists. Using DNA-rRNA hybridization, immunotyping, and phenotypic tests, Vandamme et al. (1991) introduced a new taxonomic position of the genus *Campylobacter* within a new rRNA superfamily VI, that is distinct from other rRNA superfamilies within the gram-negative bacterial group. Eleven species of *Campylobacter* are recognized based on the current classification system; most are pathogens or inhabitants of the reproductive, intestinal, and oral tracts. Other previously named campylobacters associated with animal and human infections have been transferred to the new genera *Arcobacter* and *Helicobacter*. An updated list of *Campylobacter* species, their ecology, and the associated diseases are presented in Table 22.1.

CAMPYLOBACTER FETUS

This species contains two subspecies that are almost genetically identical but cause different infections with distinct clinical manifestations. Subspecies *venerealis* and its biovar *intermedius* mainly cause venereal campylobacteriosis in cattle, which is characterized by infertility and abortion. This subspecies is known to have developed a parasitic relationship with the bovine reproductive tract. Biovar *intermedius* can occur in the intestinal tract. On the other hand, subsp. *fetus* originates in the intestines and causes occasional abortion in cattle, enzootic abortion in sheep, and systemic and intestinal infections in humans. Recent analysis of the hypervariable regions of 16S rRNA sequences of both subspecies has indicated a homology of 99.9% and only a single base mismatch in approximately 1400 bases that were sequenced (Wesley et al. 1991). Occasional reports that *C. fetus* subsp. *fetus* has been isolated from bovine infertility cases put in question the reliability of present subspeciation tests, which depend on glycine tolerance tests and disease presentation. More refined methods, including molecular sizing by pulsed-field electrophoresis (Salama et al. 1992), have been recommended as alternatives to the conventional tolerance tests.

TABLE 22.1. Campylobacters significant in animal and human health

Nomenclature	Ecology	Associated Disease States
C. fetus subsp. *venerealis*	Found in bovine semen, preputial smegma, cervical mucus; does not grow in human or animal gastrointestinal (GI) tract	Bovine abortion and infertility; venereal transmission
C. fetus subsp. *fetus*	Found in placentas and gastric contents of aborted fetuses, and in gall bladder of cattle and sheep; grows in animal and human GI tract	Enzootic abortion and enteritis in sheep; sporadic abortion in cattle; enteric and systemic infections in humans; oral transmission
C. jejuni, C. coli, C. lari	Normal GI flora in swine, sheep, cattle, goats, poultry, and wild birds; grows in human GI tract	Abortion in sheep; enteritis in piglets, calves, lambs; human enteritis; occasional sepsis; oral transmission
C. mucosalis	Adenomatous lesions in the intestinal mucosa, oral cavity; grows in swine feces	SPE[a](?); oral transmission
C. hyointestinalis	Lesions of porcine intestinal adenomatosis; grows in cattle and human feces	SPE(?); diarrhea in cattle and humans(?); oral transmission
Campylobacter-like organism(s)	Lesions of proliferative ileitis in swine	Probable cause of swine proliferative ileitis; oral transmission (?).
C. sputorum	Ovine feces; bovine prepuce and vagina; human oral cavity and feces	Experimental enteritis in calves; various abscesses in humans
"*C. upsaliensis*"	Dogs, cats, pets	Enteritis in humans
C. curvus, C. rectus	Human oral cavity	Periodontal disease

Source: Modified from Garcia et al. 1983.
[a] SPE = swine proliferative enteritis.

DISEASES

Venereal Campylobacteriosis and Abortions

Under natural conditions, *C. fetus* subsp. *venerealis* is transmitted to females by coitus with infected bulls or by insemination with contaminated semen. Direct female-to-female transmission is rare, but transmission may occur between bulls where riding behavior is active and large numbers are grouped together. Transmission from infected bulls to cows may approach 100%. Infection in the bovine female mainly causes infertility and sporadic abortion. Infertility is usually indicated by repeat breeding and is more common in young cows; abortion occurs in less than 10% of infected cows. At the onset of the disease, catarrhal vaginitis develops. Some pregnant animals may abort, most often those that are primiparous. Abortions usually occur during the fifth or sixth month of gestation. Other signs may include endometritis, occasional salpingitis, and a long estrous cycle averaging over 1 month. Cows are not permanently sterile and fertility usually returns; however, the organism may survive in the cervix through a full gestation period and infect susceptible bulls at subsequent breedings. Infected bulls are normally asymptomatic; the higher incidence of carrier bulls over 5 years old has been attributed to the increased size and number of crypts in the epithelium of the penis, which provides favorable conditions for the persistence of *C. fetus*.

C. fetus subsp. *fetus* causes sporadic abortion in cattle and epizootic abortion in sheep, in addition to enteric and systemic infections in humans. The infection may be contracted by ingestion of materials contaminated with the organism. During pregnancy, bacteremia may occur spreading the organism from intestine or liver to the uterus, placenta, and fetus; the organism is rarely found in the nongravid uterus. The organism localizes in the placentomes inducing placentitis and, in the second or third trimester of gestation, causes abortion due to inflammation and necrosis of the cotyledons. In sheep, heavy losses from abortions due to *C. fetus* can occur during the lambing season in which the infection is first introduced. Vaccination of sheep is not as widely practised as vaccination in cattle, although acquired immunity does develop.

VIRULENCE FACTORS

Surface Array Proteins (S layer)

Wild-type strains of *C. fetus* are equipped with regular arrays of protein subunits as the outermost component of their cell envelopes. The superficial location of these proteins strongly suggests that this layer may play a role in host-pathogen interactions. The S layer represents at least 10% of the total cellular proteins and may occur as multiple forms with molecular sizes ranging from 90 to 149 kDa (Dubreuil et al. 1988, 1990; Pei et al. 1988; Fujimoto et al. 1991). These paracrystalline proteins are two-dimensional self-assemblies and can assume hexagonal, tetragonal, or oblique patterns; these specific forms may be related to molecular size and antigenicity. Shared determinants may also be present among different forms, suggesting that strains of *C. fetus* produce a family of S-layer proteins with common structural and antigenic characteristics. Multiple forms of S-layer proteins can be expressed by a single cell with one form predominating. Initially reported to be a glycoprotein microcapsule, recent biochemical studies have shown that *C. fetus* S layer is composed of highly acidic and hydrophobic proteins with amino acid terminal sequences unique to the species. Moreover, there has been no confirming evidence of glycosylation or the presence of sugars in *C. fetus* S layer. The noncovalent binding of S layer to the underlying outer-membrane components facilitates the extraction of the proteins in relatively pure form. This implies that the S-layer proteins can easily be lost but may also be reattached to specific underlying cell components. In elegant experiments, Yang et al. (1992) demonstrated that the binding of S layer involves the N-terminal half of the S layer to the O-chain lipopolysaccharide (LPS) and is serotype specific. This binding is mediated by divalent cations such as Ca^{++} which act as a bridge between negatively charged groups on the LPS side chains, thus producing a dimer and aligning the side chains to the proper orientation needed for binding to the S-layer protein dimer.

Little is known about the genetic-regulation mechanisms involving the ability of *C. fetus* to produce multiple forms of S layer, as well as to switch off and on their production. Recently, Blaser and Gotschlich (1990) successfully cloned the structural gene of the 98-kDa S-layer protein using bacteriophage λGt11 as a vector. The clone contains an insert with an open reading frame encoding a 933-amino acid polypeptide, which closely matches the molecular mass of the purified S-layer protein. A ribosomal binding site and a putative transcription terminator sequence have been identified also. A significant observation indicated that the S-layer protein was exported without a leader sequence (a unique untranslated nucleotide segment preceding the initiation codon of mRNA). However, a short internal segment that showed homology to leader sequences of fimbrial proteins of various bacteria was interpreted to function as targeting the S layer to the cell surface. Furthermore, another internal sequence was found to correspond to signal sequences for exported proteins in other species.

The ability of many pathogens to produce surface components with different antigenic specificities that interface between the pathogen and its host affords the organism an effective defense mechanism against host immune factors. McCoy et al. (1975) were the first to suggest the significance of *C. fetus* S layer as a virulence factor by showing that its presence in the bacterial cell imparted antiphagocytic properties. Subsequent in vitro and mouse model studies (Blaser et al. 1987, 1988; Pei and Blaser 1990) have demonstrated that the S layer is a major factor responsible for the ability of *C. fetus* to cause extraintestinal and systemic infections. Its presence inhibits the binding of host cell C3b complement component to wild-type *C. fetus*, preventing efficient C5 convertase formation. Impaired C3 binding, even in the presence of immune serum, suggests that antibody-directed complement binding to cell surface sites is not important for the serum resistance of encapsulated *C. fetus* strains.

Although studies of the S-layer have been based mainly on strains of *C. fetus* subsp. *fetus*, they have direct relevance to the pathogenesis of venereal campylobacteriosis caused by *C. fetus* subsp. *venerealis*. Current studies in our laboratory (Garcia and Brooks, unpublished) indicate that strains of this subspecies also produce high molecular-sized S-layer proteins, which show antigenic shifts in infected female cattle. Shifts in the predominant S-layer form may explain the antigenic variations observed in cattle infected with *C. fetus*. The mechanism underlying the ability of *C. fetus* to alter the gene product during the course of infection, however, is far from clear.

Lipopolysaccharide

Campylobacter abortion is thought to be an allergic response to heat-stable endotoxins. Fatal or sublethal anaphylactic shock has been demonstrated when animals are injected with fresh or boiled *C. fetus* subsp. *venerealis* cells. Although the LPS core may have common antigenic determinants, there is heterogeneity in the antigenic specificities of the O side chains of LPS of *C. fetus* strains that is the basis for serotyping (Perez-Perez et al. 1986). There appears to be a correlation between serum resistance and LPS serotype.

HOST IMMUNITY AND INTERACTIONS

Knowledge of the immunology of bovine campylobacteriosis has been elucidated (Corbeil and Winter 1978). In natural infections, the venereal pathogen is normally introduced into the female genital tract during the ovulatory phase of the estrous cycle when neutrophils are abundant in the secretion. Organisms that escape phagocytosis are able to multiply and invade the uterus during the luteal phase when neutrophil response is low. This provides an antigenic stimulus, which results in the production of IgG antibodies mainly in the uterus and IgA antibodies in the vagina. IgG plays an opsonizing role in the phagocytosis of the pathogen by neutrophils and mononuclear cells but not in complement-mediated killing. IgA, which is not a good opsonizing agent, immobilizes the organism on the cervicovaginal surfaces and limits entry to the uterus or penetration of the mucous coat, facilitating either the disposal of *C.*

fetus, particularly during estrus, or the establishment of a carrier state. In carrier bulls, immunoglobulins against *C. fetus* are usually at low levels, probably because the superficial location of this organism in the prepuce results in minimal stimulation of the immune system (Winter et al. 1981). In addition to antibiotic treatment, vaccination of bulls is widely used. In herds where the incidence of venereal campylobacteriosis is high, vaccination of heifers is also recommended. Good correlation between serum antibody titers and protection against bovine campylobacteriosis has been demonstrated.

Problems

Although much knowledge of the immunology of venereal campylobacteriosis has been gained, there is little information on mechanisms involving host-parasite interactions in the venereal environment, adhesion, chemotaxis, or tissue tropism. The possible role of *C. fetus* flagella in colonization of mucus has not been explored. Another relatively untouched area is the role of hormones in the persistence of *C. fetus* and other venereal infectious agents in the reproductive tract. Future designs of diagnostic antigens and vaccines will rely on increased knowledge of the genetic regulation of S layer and a better understanding of antigenic variation in *C. fetus* infections.

CAMPYLOBACTER JEJUNI, CAMPYLOBACTER COLI, AND *CAMPYLOBACTER LARI*

These three species are thermophilic campylobacters. *C. jejuni* consists of two subspecies, subsp. *jejuni* and subsp. *doylei*. *C. jejuni* subsp. *jejuni* commonly lives as a commensal in the intestinal tract of a wide range of birds and mammals, including the domestic animals most widely used for food production and pets. Animals, especially poultry, are the most frequent sources of *C. jejuni* subsp. *jejuni* infections in humans. *C. jejuni* subsp. *doylei* has been mainly isolated from adult human gastric biopsy specimens and from the feces of children with diarrhea. *C. coli* is frequently recovered from the intestines of swine and has also been isolated from cattle, poultry, humans, and other sources. *C. lari* was initially isolated from sea gulls but has also been recovered from cattle, dogs, chickens, and other species.

C. jejuni is an important cause of abortion in sheep, showing similar symptoms to those observed in *C. fetus* subsp. *fetus* abortions and an infrequent cause of abortion in dogs, goats, and other animals. This organism is also a cause of enteritis in a wide variety of animal species including cattle, chickens, pigs, cats, dogs, ferrets, and nonhuman primates, particularly in young animals (Shane 1992). It has been associated with bovine mastitis, avian "vibrionic" hepatitis, and abortion in mink. It is best recognized as a cause of enteritis in humans, and worldwide it is the major bacterial cause of human enteritis. *C. coli* and *C. lari* account for only approximately 5% of the *Campylobacter* isolates from cases of human enteritis.

C. jejuni enteritis in animals is generally mild and typically associated with the passage of soft or watery feces containing mucus and flecks of blood. Under farm conditions, most animals will encounter *C. jejuni* early in life when passive immunity is present; disease will be subclinical. Clinically recognizable disease is more likely to be observed in pets and other animals that are kept under good hygiene and become infected in the absence of active or passive immunity.

VIRULENCE FACTORS

Colonization and Invasiveness

Experimental infection studies with chicks and mice indicate that *C. jejuni* localizes in the mucus-filled crypts of the ceca and appears to colonize the mucus without attaching to the epithelial cells of the microvilli (Lee et al. 1986). Rapid, darting motility and the spiral shape of the organism may facilitate colonization of mucus. *C. jejuni* is chemoattracted to mucin, the principal constituent of mucus, and can utilize mucin as a sole substrate for growth. Colonization of the mucus in chick cecal crypts with *C. jejuni* is reduced or prevented by prior colonization with *Escherichia coli*, *Klebsiella pneumoniae*, and *Citrobacter diversus*, which occupy the same niche and produce metabolites antagonistic to *C. jejuni* (Schoeni and Doyle 1992). A pair of congenic strains of *C. jejuni* has been developed that differ in their ability to colonize the intestinal tract of chicks. A 69-kDa component, which is present in the colonizing strain but not detected in the noncolonizing strain, may be involved in colonization. The ability of *C. jejuni* to invade intestinal epithelial cells has been confirmed in studies with chicken embryos, mice and hamsters, and mammalian cell lines. *C. jejuni* proteins with molecular masses in the 28- to 42-kDa range may be invasive factors (De Melo and Pechère 1990).

Campylobacters are actively motile by means of a single polar flagellum at one or both ends of the cell, and motility appears to play a major role in colonization. Nonmotile variants of *C. jejuni* with complete, incomplete, or no flagella fail to colonize or require a larger inoculum than motile strains with complete flagella to colonize the gastrointestinal tract of infant mice. Some *C. jejuni* strains undergo a bidirectional transition, termed phase variation, between flagellated and nonflagellated phenotypes. When nonflagellated *C. jejuni* variants were used to infect rabbits, only flagellated cells could be recovered from fecal samples (Caldwell et al. 1985). Other campylobacter strains reversibly express two flagellar types that are distinct antigenically and in the apparent relative mobility (M_r) in sodium dodecyl sulfate polyacrylamide gel electrophoresis of their flagellin subunits. *C. coli* produces an antigenic phase 1 (P1) flagellin of 61.5 kDa and a phase 2 (P2) flagellin of 59.5 kDa. Rabbits experimentally infected with P2 cells shed predominantly P2 cells throughout the infection, while rabbits infected with P1 cells initially shed P1 cells but later (approximately day 7) shed only P2 cells (Logan et al. 1989). Antigenic variation of *C. coli* flagella is accompanied by a reversible DNA rearrangement and involves two flagellin genes (Guerry et al. 1991). Two flagellin genes (*fla*A and *fla*B) are also present in *C. jejuni* strains and are organized in a similar way as in *C. coli* (Nuitjen et al. 1992).

Lipopolysaccharide

All *C. coli* reference strains in a serotyping scheme based on thermostable antigens have smooth-type LPS with high M_r O side chains in addition to low M_r components. In contrast, approximately two-thirds of the *C. jejuni* reference strains have only low M_r components while the other one-third have smooth-type LPS. The lipid A portion of the LPS molecule is similar chemically and antigenically to that of other gram-negative bacteria. Changes in the somatic O LPS antigenic specificities have been observed in two serostrains of *C. coli* after continuous laboratory subculture and may be due to genomic rearrangements (Mills et al. 1991). In addition, the LPS of some *C. jejuni* strains contain N-acetylneuraminic acid (NeuAc), a component rarely found in prokaryotes but commonly encountered in mammalian glycolipids and glycoproteins. LPS that contains NeuAc may be nonimmunogenic or only poorly immunogenic in the host, and its presence may confer serum resistance on these campylobacters (Moran et al. 1991).

Toxins

C. jejuni, C. coli, and *C. lari* produce a cytotonic factor or enterotoxin that is similar to the heat-labile, adenylate cyclase-activating enterotoxins of *Vibrio cholera* (cholera toxin, CT) and *E. coli* (heat-labile enterotoxin, LT) (Ruis-Palacios et al. 1983). The enterotoxin is trypsin sensitive, has a molecular mass in the range of 60 to 70 kDa, and is completely inactivated at pH 2 and 8. There is disagreement as to the conditions required for the heat inactivation of this toxin. The enterotoxin produces a cytotonic response in mouse adrenal tumor (Y-1) and Chinese hamster ovary (CHO) cells, increases intracellular cAMP levels in CHO and green monkey kidney (Vero) cells, evokes fluid secretion in ligated rat and rabbit ileal loops, and increases permeability in rabbit skin tests. The level of enterotoxin production in campylobacter strains is 20 - 2000 times lower than that for LT or CT (Johnson and Lior 1986). The enterotoxin is partially neutralized by antibody to CT and LT, and preincubation with antiserum to CT abolishes the cytotonic effect in tissue cultures and rat ileal loops. Like the B subunits of CT and LT, the enterotoxin attaches to the GM_1 ganglioside tissue receptor, reacts in enzyme-linked immunosorbent assays (ELISAs) with GM_1 as the solid phase, and can be purified by methods based on its affinity to the galactose component of GM_1 ganglioside; preincubation with GM_1 inhibits its cytotonic response in tissue culture assay.

C. jejuni, *C. coli,* and *C. lari* also produce a trypsin-sensitive cytotoxin that is toxic for bovine kidney, Vero, and human cervix epithelial carcinoma (HeLa) cells. The cytotoxin is heat labile at 70°C for 30 minutes, stable at 60°C for 30 minutes, and is not neutralized by antisera to *Shigella dysenteriae* Shiga toxin, *Clostridium difficile* cytotoxin, or *E. coli* Verotoxin (Johnson and Lior 1986). Both an enterotoxin and a cytotoxin may be produced by these thermophilic campylobacters. In addition to an enterotoxin and a cytotoxin, a heat-labile, trypsin-sensitive, nondialyzable, cytolethal distending toxin (CLDT), with a molecular mass of over 30 kDa, is present in culture filtrates of many *C. jejuni*, *C. coli*, and *C. lari* strains (Johnson and Lior 1988). It is cytolethal to CHO, Vero, HeLa, and human epithelial carcinoma (HEp-2) cells and negative in Y-1 cells. CLDT is negative in adult rabbit ligated ileal loop, suckling mouse, and rabbit skin tests. CLDT is neutralized only by homologous rabbit antitoxin. An adequate amount of available iron may be important for expression of these toxins.

HOST IMMUNITY AND INTERACTIONS

Flagella facilitate attachment of *C. jejuni* cells to human intestinal epitheliallike (INT 407) cells; binding is not reduced in the presence of fucose. LPS also mediates adherence of *C. jejuni* to INT 407 cells, but the binding of purified LPS and whole cells is reduced in the presence of fucose (McSweegan and Walker 1986). LPS, but not flagella, binds to intestinal mucus (McSweegan and Walker 1986). In addition, an outer-membrane protein of 27 kDa appears to be involved in adherence of enteric *Campylobacter* spp. to epithelial cells. These studies suggest that host epithelial cells may have receptors for several *Campylobacter* components. The resistance of certain *C. jejuni* and *C. coli* strains to phagocytosis has also been demonstrated and may contribute to the ability of these organisms to persist in an infected host.

In an infant mouse model, vaccination of dams prevented *C. jejuni* colonization in 85% of offspring challenged with the vaccine strain and 57% of offspring challenged with different strains having the same heat-labile antigens detected by the Lior typing scheme. However, vaccination failed to prevent colonization when challenge strains were from different Lior serogroups. Protection was associated with high concentrations of IgG antibodies in the mammary secretion of the vaccinated dam and may involve antibody response to flagella and other heat-labile antigens. No protection was observed with strains having the same heat-stable antigens in the Penner serotyping scheme but belonging to different Lior serogoups (Abimiku and Dolby 1988). Studies with other models have shown that animals develop both local and systemic antibody responses after challenge with *C. jejuni*. IgA, which can persist longer than IgG, is induced by flagellar and other surface antigens and appears to be important in the establishment of *C. jejuni* in the gut.

PROBLEMS

Although a number of potential virulence factors of the thermophilic campylobacters have been identified and partially characterized, their role in the pathogenesis of disease is not clearly understood. Enterotoxin production has not been consistently demonstrated in *C. jejuni* isolates from humans with diarrhea, and it is essential that discrepancies in identifying enterotoxigenic strains be clarified. Further studies are needed to establish the importance of enterotoxins, cytotoxins, adhesins, invasins, and other factors in the biology of these organisms (Tompkins 1992) and to define mucosal immunity to *C. jejuni*. Such studies will be facilitated by an improved understanding of the molecular genetics of campylobacters.

CAMPYLOBACTER MUCOSALIS, CAMPYLOBACTER HYOINTESTINALIS, AND *CAMPYLOBACTER*-LIKE ORGANISMS

Campylobacter mucosalis and *C. hyointestinalis* have been isolated from the intestines of pigs with swine proliferative enteritis (SPE), a complex disease of the lower small intestine and occasionally of the cecum and colon (Lawson and Rowland 1974; Gebhart et al. 1985). SPE includes four enteric disorders: intestinal adenomatosis, necrotic enteritis, regional ileitis, and proliferative hemorrhagic enteropathy. These conditions occur worldwide and cause severe economic losses in recently weaned pigs. An initial phase of hyperplasia of the crypt cells of the intestinal epithelium and the presence of curve-shaped, *Campylobacter*-like bacteria located intracellularly in the proliferating cells are among the consistent pathological features of SPE. Electron microscopy has revealed that the intracellular bacteria are not bound by a membrane but lie free in the host cell cytoplasm. *C. mucosalis* (formerly *C. sputorum* subsp. *mucosalis*) was first isolated from lesions of porcine intestinal adenomatosis, a benign adenoma of the intestinal mucous membrane of weaned pigs. The organism was subsequently isolated from other conditions of SPE and the oral cavity of some healthy pigs, where intestinal adenomatosis occurs. *C. hyointestinalis* has been isolated from lesions of SPE as well as from the intestines of healthy cattle and, more recently, from stools of human gastrointestinal cases. Although either or both *C. mucosalis* and *C. hyointestinalis* may be isolated consistently from lesions of SPE, attempts to reproduce the disease by experimental inoculation of both species in conventional and gnotobiotic pigs have not succeeded. Recently, a *Campylobacter*-like organism (CLO) isolated from the terminal ileum of SPE lesions was demonstrated to produce gross pathological lesions consistent with SPE in commercial crossbred pigs (Alderton et al. 1992). Designated as RMIT 32A, this CLO was shown to have different phenotypic characteristics as well as the protein and LPS patterns of *C. mucosalis* and *C. hyointestinalis*. The consistent presence in SPE lesions of a nonculturable intracytoplasmic CLO that is antigenically and genetically different from either *C. mucosalis* or *C. hyointestinalis* has also been reported (McOrist et al. 1989; Gebhart et al. 1991). This suggests that CLOs could play an important role in SPE.

PROBLEMS

Improved methods to recover the etiological agents of SPE as well as comparative antigenic and genetic analyses of culturable and nonculturable CLOs are important from the standpoint of diagnosis and understanding the pathogenesis of this increasingly important disease complex.

CAMPYLOBACTER SPUTORUM AND "*CAMPYLOBACTER UPSALIENSIS*"

The three biovars of *C. sputorum* have been implicated in animal or human infections. Biovar *fecalis* is generally regarded as nonpathogenic, although experimental inoculation of calves with pure cultures of the organism can induce enteric infection. Biovar *bubulus* is a saprophytic organism that is commonly encountered in the bull prepuce (Garcia et al. 1983). Both biovar *bubulus* and biovar *sputorum* strains have been isolated from human abscesses. *C. upsaliensis* causes diarrhea and bacteremia in both healthy and immunocompromised persons (Mishu et al. 1992). Most human isolations have been from stools of children with self-limited diarrheal illness. This organism was first reported in dogs with and without diarrhea that were suspected of transmitting the pathogen to humans (Sandsted et al. 1983). The presence of identical plasmids in strains of *C. upsaliensis* isolated from dogs and humans suggests these infections are zoonotic. Thus far, the pathogenic mechanisms involved in *C. upsaliensis* infections are not known.

REFERENCES

Abimiku, A. G., and Dolby, J. M. 1988. Cross-protection of infant mice against intestinal colonisation by *Campylobacter jejuni*: Importance of heat-labile serotyping (Lior) antigens. J Med Microbiol 26:265-68.

Alderton, M. R.; Borland, R.; Coloe, P. J. 1992. Experimental reproduction of porcine proliferative enteritis. J Comp Pathol 106:159-67.

Blaser, M. J., and Gotschlich, E. C. 1990. Surface array protein of *Campylobacter fetus*. J Biol Chem 265: 14529-35.

Blaser, M. J.; Smith, P. F.; Hopkins, J. A.; Heinzer, I.; Bryner, J. H.; and Wang, W. L. 1987. Pathogenesis of *Campylobacter fetus* infection: Serum resistance associated with high-molecular-weight surface proteins. J Infect Dis 155:696-706.

Blaser, M. J.; Smith, P. F.; Repine, J. E.; and Joiner, K. A. 1988. Pathogenesis of *Campylobacter fetus* infections. Failure of encapsulated *Campylobacter fetus* to bind to C3b explains serum and phagocytosis resistance. J Clin Invest 81:1434-44.

Caldwell, M. B.; Guerry, P.; Lee, E. C.; Burans J. P.; and Walker R. I. 1985. Reversible expression of flagella in *Campylobacter jejuni*. Infect Immun 50:941-43.

Corbeil L. B., and Winter, A. J. 1978. Animal model for the study of genital secretory immune mechanisms: Venereal vibriosis in cattle. In Immunobiology of *Neisseria gonorrhoeae*. Ed. G. F. Brooks, E. C. Gotschlich, K. K. Holmes, W. D. Sawyer, and F. E. Young, pp. 293-302. Washington, D.C.: American Society for Microbiology.

De Melo, M. A., and Pechère, J. C. 1990. Identification of *Campylobacter jejuni* surface proteins that bind to eucaryotic cells in vitro. Infect Immun 58:1749-56.

Dubreuil, J. D.; Logan, S. M.; Cubbage, S.; Eidhin, D. N.; McCubbin, W. D.; Kay, C. M.; Beveridge, T. J.; Ferris, F. G.; and Trust, T. J. 1988. Structural and biochemical analyses of surface array protein of *Campylobacter fetus*. J Bacteriol 170:4165-73.

Dubreuil, J. D.; Kostrzynska, M.; Austin, J. W.; and Trust, T. J. 1990. Antigenic differences among *Campylobacter fetus* S-layer proteins. J Bacteriol 172:5035-43.

Fujimoto, S.; Takade, A.; Amako, K.; Blaser, M. J. 1991. Correlation between molecular size of the surface array protein and morphology and antigenicity of the *Campylobacter fetus* S layer. Infect Immun 59:2017-22.

Garcia, M. M., and Brooks, B. W. 1993. Animal Diseases Research Institute, Nepean, Ontario. Unpublished.

Garcia, M. M.; Eaglesome, M. D.; and Rigby, C. 1983. *Campylobacters* important in veterinary medicine. Vet Bull 53:793-818.

Gebhart, C. J.; Edmonds, P.; Ward, G. E.; Kurtz, H. J.; and Brenner, D. J. 1985. "*Campylobacter hyointestinalis*" sp. nov.: A new species of *Campylobacter* found in the intestines of pigs and other animals. J Clin Microbiol 21:715-20.

Gebhart, C. J.; Lin, G.-F.; McOrist, S. M.; Lawson, G. H. K.; and Murtaugh, M. P. 1991. Cloned DNA probes specific for the intracellular *Campylobacter*-like organism of porcine proliferative enteritis. J Clin Microbiol 29:1011-15.

Guerry, P.; Alm, R. A.; Power, M. E.; Logan, S. M.; and Trust, T. J. 1991. Role of two flagellar genes in *Campylobacter* motility. J Bacteriol 173:4757-64.

Johnson, W. M., and Lior, H. 1986. Cytotoxic and cytotonic factors produced by *Campylobacter jejuni*, *Campyobacter coli*, and *Campylobacter laridis*. J Clin Microbiol 24:275-81.

__________. 1988. A new heat-labile cytolethal distending toxin (CLDT) produced by *Campylobacter* spp. Microb Pathog 4:115-26.

Lawson, G. H. K., and Rowland, A. C. 1974. Intestinal adenomatosis in the pig: A bacteriological study. Res Vet Sci 17:331-36.

Lee, A.; O'Rourke, J. L.; Barrington, P. J.; and Trust, T. J. 1986. Mucus colonization as a determinant of pathogenicity in intestinal infection by *Campylobacter jejuni*: A mouse cecal model. Infect and Immun 51:536-46.

Logan, S. M.; Guerry P.; Rollins D. M.; Burr D. H.; and Trust T. J. 1989. In vivo antigenic variation of *Campylobacter* flagellin. Infect Immun 57:2583-85.

McCoy, E. C.; Doyle, D.; Burda, K.; Corbeil, L. B.; and Winter, A. J. 1975. Superficial antigens of *Campylobacter (Vibrio) fetus*: Characterization of an antiphagocytic component. Infect Immun 11:517-25.

McOrist, S.; Boid, R.; and Lawson, G. H. K. 1989. Antigenic analysis of *Campylobacter* species and an intracellular *Campylobacter*-like organism associated with porcine proliferative enteropathies. Infect Immun 57:957-62.

McSweegan, E., and Walker, R. I. 1986. Identification and characterization of two *Campylobacter jejuni* adhesins for cellular and mucous substrates. Infect Immun 53:141-48.

Mills, S. D.; Kurjanczyk, L. A.; Shames, B.; Hennessy, J. N.; and Penner, J. L. 1991. Antigenic shifts in serotype determinants of *Campylobacter coli* are accompanied by changes in the chromosomal DNA restriction endonuclease digestion pattern. J Med Microbiol 35:168-73.

Mishu, B.; Patton, C. M.; and Tauxe, R. V. 1992. Clinical and epidemiologic features of non-*jejuni*, non-*coli Campylobacter* species. In *Campylobacter jejuni:* Current Status and Future Trends. Ed. I. Nachamkin, M. J. Blaser, and L. S. Tomkins, pp. 31-41. Washington D.C.: American Society for Microbiology.

Moran, A. P.; Rietschel, E. T.; Kosunen, T. U.; and Zahringer, U. 1991. Chemical characterization of *Campylobacter jejuni* lipopolysaccharides containing N-acetylneuraminic acid and 2,3-diamino-2,3-dideoxy-D-glucose. J Bacteriol 173:618-26.

Nuijten, P. J. M.; Wassenaar, T. M.; Newell, D. G.; and Van Der Zeijst, B. A. M. 1992. Molecular characterization and analysis of *Campylobacter jejuni* flagellin genes and proteins. In *Campylobacter jejuni*: Current Status and Future Trends. Ed. I. Nachamkin, M. J. Blaser, and L. S. Tompkins, pp. 282-96. Washington, D.C.: American Society for Microbiology.

Pei, Z., and Blaser, M. J. 1990. Pathogenesis of *Campylobacter fetus* infections. Role of surface array proteins in virulence in a mouse model. J Clin Invest 85:1036-43.

Pei, Z.; Ellison, R. T.; Lewis, R. V.; and Blaser, M. J. 1988. Purification and characterization of a family of high molecular weight surface-array proteins from *Campylobacter fetus*. J Biol Chem 263:6416-20.

Penner, J. L. 1988. The genus *Campylobacter*: A decade of progress. Clin Microbiol Rev 1:157-72.

Perez-Perez, G. I.; Blaser, M. J.; and Bryner, J. H. 1986. Lipopolysaccharide structures of *Campylobacter fetus* are related to heat-stable serogroups. Infect Immun 51:209-12.

Ruiz-Palacios, G. M.; Torres, J.; Torres, N. I.; Escamilla, E.; Ruiz-Palacios, B. R.; and Tamayo, J. 1983. Cholera-like enterotoxins produced by *Campylobacter jejuni*: Characterization and clinical significance. Lancet 2:250-52.

Salama, S.; Garcia, M. M.; and Taylor, D. E. 1992. Differentiation of subspecies of *Campylobacter fetus* by genomic sizing. Int J Syst Bacteriol 42:446-50.

Sandstedt, K.; Ursing, J.; Walder, M. 1983. Thermotolerant *Campylobacter* with no or weak catalase activity isolated from dogs. Curr Microbiol 8:209-13.

Schoeni, J. L., and Doyle, M. P. 1992. Reduction of *Campylobacter jejuni* colonization of chicks by cecum-colonizing bacteria producing anti-*C. jejuni* metabolites. Appl Environ Microbiol 58:664-70.

Shane, S. M. 1992. The significance of *Campylobacter jejuni* infections in poultry: A review. Avian Pathol 21:189-213.

Thompkins, L. S. 1992. Genetic and molecular approach to *Campylobacter* pathogenesis. In *Campylobacter jejuni:* Current Status and Future Trends. Ed. I. Nachamkin, M. J. Blaser, and L. S. Tompkins, pp. 241-54. Washington D.C.: American Society for Microbiology.

Vandamme, P.; Falsen, E.; Rossau, R.; Hoste, B.; Segers, P.; Tytgat, R.; and de Ley, J. 1991. Revision of *Campylobacter, Helicobacter* and *Wolinella* Taxonomy: Emendation of generic descriptions and proposal of *Arcobacter* gen. nov. Int J Syst Bacteriol 41:88-103.

Wesley, I. V.; Wesley, R. D.; Cardella, M.; Dewhirst, F. E.; and Paster, B. J. 1991. Oligodeoxynucleotide probes for *Campylobacter fetus* and *Campylobacter hyointestinalis* based on 16S rRNA sequences. J Clin Microbiol 29:1812-17.

Winter, A. J.; Clark, B. L.; Parsonoson, I. M.; Duncan, J. R.; and Bier, P. J. 1981. Nature of immunity in the male bovine reproductive tract based upon responses to *Campylobacter fetus* and *Trichomonas fetus*. In the Ruminant Immune System. Ed. J. E. Butler, J. R. Duncan, and K. Nielsen, pp. 744-52. New York: Plenum Press.

Yang, L.; Pei, Z.; Fujimoto, S.; and Blaser, M. J. 1992. Reattachment of surface array proteins to *Campylobacter fetus* cells. J Bacteriol 174:1258-67.

23 / Gram-Negative Anaerobes

BY J. F. PRESCOTT

THE GRAM-NEGATIVE, anaerobic bacteria of humans and animals are an important frontier in the study of infectious disease. Understanding their role in health and disease is hampered by difficulties in isolation and identification, the large number of species involved, and the complexity of their interaction with the host and with each other. Although remarkable advances have been made in the isolation and taxonomy of important human gram-negative pathogens, detailed knowledge of their pathogenic mechanisms is relatively slight, but recombinant DNA techniques are increasingly being used to isolate genes responsible for virulence. With notable exceptions, similar efforts in veterinary medicine have been rudimentary.

This chapter summarizes the types of disease caused by gram-negative, anaerobic bacteria; the identity of the agents involved; and mentions identified virulence factors before discussing in detail some specific and well-understood, important anaerobic infections of animals (*Fusobacterium necrophorum*, *Dichelobacter nodosus*, *Serpulina hyodysenteriae*, and *Bacteroides fragilis*).

NONSPECIFIC INFECTIONS

CHARACTERISTICS AND SOURCES OF THE BACTERIA

The main genera and common pathogenic species of anaerobic bacteria isolated from nonspecific infections in animals are shown in Table 23.1. *Bacteroides* and *Fusobacterium* species account for over half the isolates from these infections. The sources of infection are mucosal surfaces or the intestinal content. Of the hundreds of species of anaerobic bacteria found in the large bowel and mouth of humans, perhaps 15-20 species are significant opportunist pathogens in humans. Some of these same species occur in animals (Jang and Hirsh 1991), but other animal isolates are unique (Love et al. 1987). Apart from defined virulence determinants, one feature common to these nonspecific pathogens may be their relative insensitivity to oxygen compared to the strictest anaerobes in the intestine, which are usually not pathogenic.

TABLE 23.1. Anaerobic bacteria isolated from nonspecific infections in animals

Genus	Range of incidence (%)	Common species isolated
Actinomyces	1-9	Varies with host species
Bacteroides	20-30	*B. fragilis*
		B. ruminicola
Bifidobacterium	0-2	
Clostridium	4-30	*C. perfringens*
Eubacterium	2-3	
Fusobacterium	6-20	*F. necrophorum*
		F. nucleatum
Peptococcus	2-12	*P. indolicus*
Peptostreptococcus	6-15	*P. anaerobius*
Porphyromonas	10-20	*P. asaccharolytica*
		P. gingivalis

DISEASE

Gram-negative, anaerobic bacteria cause both nonspecific and specific infections. Nonspecific infections in which such anaerobic bacteria are commonly involved include abscesses, dental and oral infections, chronic pleuropneumonic conditions, peritonitis, chronic sinusitis, infections of the female genital tract, mastitis, bite wound infections, and cellulitis (Finegold et al. 1986). The common feature of these infections is that they involve tissue necrosis near a mucosal surface; many of the anaerobes involved are opportunist pathogens, i.e. the more virulent members of the commensal flora. These infections are often polymicrobial. Such "mixed" infections occur because of the requirement for aerobic bacteria to remove oxygen and reduce the local environment sufficiently for anaerobes to establish, with the resultant selection of the anaerobic species resistant to nonspecific host defenses from a larger inoculum of bacteria that may have breached the mucosal barrier. The synergistic associations of certain anaerobes and aerobes, because of the presence of growth factors, are well established but generally not well characterized (Smith et al. 1991). One example of such synergism is that the vitamin K requirement of *Porphyromonas (Bacteroides) asaccharolytica* is provided by other bacteria, not by the host, so that these organisms never occur on their own. Associations between the commensal flora and the host are complex. An interesting example is that *Porphyromonas gingivalis* can substitute progesterone and estradiol for vitamin K, a substitution likely responsible for the enhanced susceptibility to periodontal disease in pregnant women (Kornman and Loesche 1982).

There is increasing interest in the role of *Bacteroides fragilis* in diarrhea in humans and animals (see below under *B. fragilis*). Our understanding of the involvement of this organism in enteric disease may be enhanced as genetic methods of identifying enterotoxigenic strains help define the role of this and perhaps of other gram-negative anaerobic bacteria in diarrheal illness.

BACTERIAL VIRULENCE FACTORS

Virulence in pathogenic bacteria is often multifactorial; grám-negative anaerobes are no exception. Although the virulence determinants have not been fully defined for any species, several factors discussed below are thought to be important in the pathogenesis of nonspecific infections. Those familiar with the characteristic stink of anaerobic infections will not be surprised by the cytotoxic action of butyric acid and ammonia, which are partially responsible for such smells. The polymicrobial nature of many nonspecific anaerobic infections means that many unique combinations of virulence factors will be operating in individual infections.

Adherence

Most pathogenic *Porphyromonas* spp. produce fimbriae, which are thought to be important in adherence to host epithelial cells or to other bacteria (Mayrand and Holt 1988).

Capsules

Capsules may promote resistance to the bactericidal effects of normal serum (i.e., complement) and to phagocytosis, and there is evidence that capsular material from *B. fragilis* or *Porphyromonas* spp. may also inhibit phagocytic killing of facultative anaerobes with which the capsular material may be associated. Highly purified polysaccharide capsule from *B. fragilis* provoked abscess formation experimentally (Kasper et al. 1984).

Extracellular or Cell-bound Enzymes

The pathogenic role of the collagenase produced by pathogenic *Porphyromonas* spp. has been clearly shown to contribute to virulence but the role of the IgA and IgG protease, neuraminidase, DNAase, fibrinolysin, and other enzymes produced by these and by some *Bacteroides* spp. are less well established (Hofstad 1984).

Lipopolysaccharide

Activation of complement by direct action of lipopolysaccharide (LPS) promotes neutrophil chemotaxis, and tissue damage may result when phagocytic cells release their hydrolytic enzymes. Complement-dependent opsonization of bacteria and their subsequent killing by neutrophils in the nonimmune host appear critical for effective clearing of mixed anaerobic infections. Thus, interference in opsonization, because of depletion of opsonins, is an important virulence contribution by pathogenic anaerobes (Finlay-Jones et al. 1991). However, several studies have shown that the LPS of gram-negative anaerobes has relatively little chemotactic activity and is relatively nontoxic compared to LPS of *Salmonella* and other *Enterobacteriaceae* (Lindberg et al. 1990). LPS-evoked macrophage release of inflammatory mediators such as interleukin-1 (IL-1) or tumor necrosis factor alpha may contribute to local inflammation.

FUSOBACTERIUM NECROPHORUM

CHARACTERISTICS OF THE BACTERIUM

Fusobacterium necrophorum is an anaerobic, gram-negative, often highly filamentous rod. It is divided into three biovars largely on the basis of hemagglutinating activity, but there are other differences. Two biovars (A, B) have recently been proposed as subspecies *necrophorum* and *funduliforme*, respectively (Shinjo et al. 1991). An intermediate (AB) biovar also appears to be recognized and to be virulent. Subspecies *necrophorum* is both more hemolytic and more virulent than subspecies *funduliforme*, which is an essentially nonpathogenic member of the commensal flora. Biovar C has been reclassified as *F. pseudonecrophorum* (Shinjo et al. 1990).

SOURCES

This organism is found normally in the intestine of cattle, sheep, swine, and sometimes other species, including humans. It survives well in wet soil that has a high content of manure.

DISEASE

F. necrophorum is an important cause of liver abscessation in feedlot cattle fed grain-rich diets. The disease is of economic significance because it leads to condemnation of livers at slaughter and appears to cause reduced growth of affected animals. Death is uncommon but may result if an abscess erodes through a portal vein and suddenly showers the body with organisms. Infection of livers arises from organisms that are part of the normal rumen flora: rumen acidosis in grain-fed cattle produces ruminal epithelial ulceration with subsequent colonization by *F. necrophorum* and translocation of the bacteria to the liver.

F. necrophorum is a common opportunist pathogen in nonspecific anaerobic infections in cattle, sheep, swine, and wallabies. In cattle, infection is often associated with *Actinomyces pyogenes* and other anaerobes, but sometimes *F. necrophorum* is found on its own. It is of central importance in footrot in cattle, sheep, goats, and deer, in which it is synergistic with *Dichelobacter nodosus* and in heel abscess in cattle, in which it is sometimes synergistic with *P. asaccharolytica*. The minimum infective dose in experimental infections in mice was reduced from 10^6 to less than 10 organisms when *F. necrophorum* was mixed with a variety of aerobic opportunist pathogens (Smith et al. 1991).

BACTERIAL VIRULENCE FACTORS

Relatively little is known in detail about virulence in *F. necrophorum*. There is a correlation between the pathogenicity of *F. necrophorum* and its production of an extracellular leukocidin. Because of its lability, there are conflicting reports of the size of the leukocidin, but it appears to be a 103-kDa dimeric protein (with 13- and 14-kDa components) that is antigenically common among isolates from different sources (Emery et al. 1986) and lethal to peripheral blood leukocytes of cattle, sheep, rabbits, and humans (in order of susceptibility) (Emery et al. 1984). The leukocidin may be identical to the hemolysin (Kanoe et al. 1984); the latter has phospholipase A activity. An extracellular protease from *F. necrophorum* has been partially characterized (Nakagaki et al. 1991) but its role in virulence is undetermined. The role in pathogenesis of the biologically active LPS of *F. necrophorum* has also not been defined but it may be significant in abscess formation. Other surface characteristics of the organism (fimbriae, capsule) have not been described in detail.

IMMUNITY

The cell wall antigens of *F. necrophorum* are serologically heterogeneous, with some common antigens described. Serogrouping has unfortunately not been developed. The basis of immunity remains to be defined, but circulating antibody appears to offer little protection against disease. Extensive experimental vaccination studies have used undefined antigens and have produced variable results. While it has been suggested that the leukocidin is an important protective antigen, experimental studies in mice do not support a role for either the leukocidin or the LPS as protective antigens (Emery and Vaughan 1986).

Curiously, vaccination of wallabies against lumpy jaw, an infection associated with *F. necrophorum*, using *D. nodosus* was effective in preventing infection (Blanden et al. 1987). The reason for this is unexplained, since *D. nodosus* has never been isolated from such lesions.

PROBLEMS

Most aspects of virulence and immunity need to be understood.

DICHELOBACTER NODOSUS

Many aspects of *Dichelobacter nodosus* have been recently reviewed (Egerton et al. 1989).

CHARACTERISTICS OF THE BACTERIUM

D. nodosus is an anaerobic, gram-negative, rod-shaped organism, with characteristic knobs at each end, and is often heavily fimbriated, with several hundred 10-μm fimbriae arranged in a polar distribution (Fig. 23.1). It has recently been reclassified from *Bacteroides* to *Dichelobacter* (Dewhirst et al. 1990).

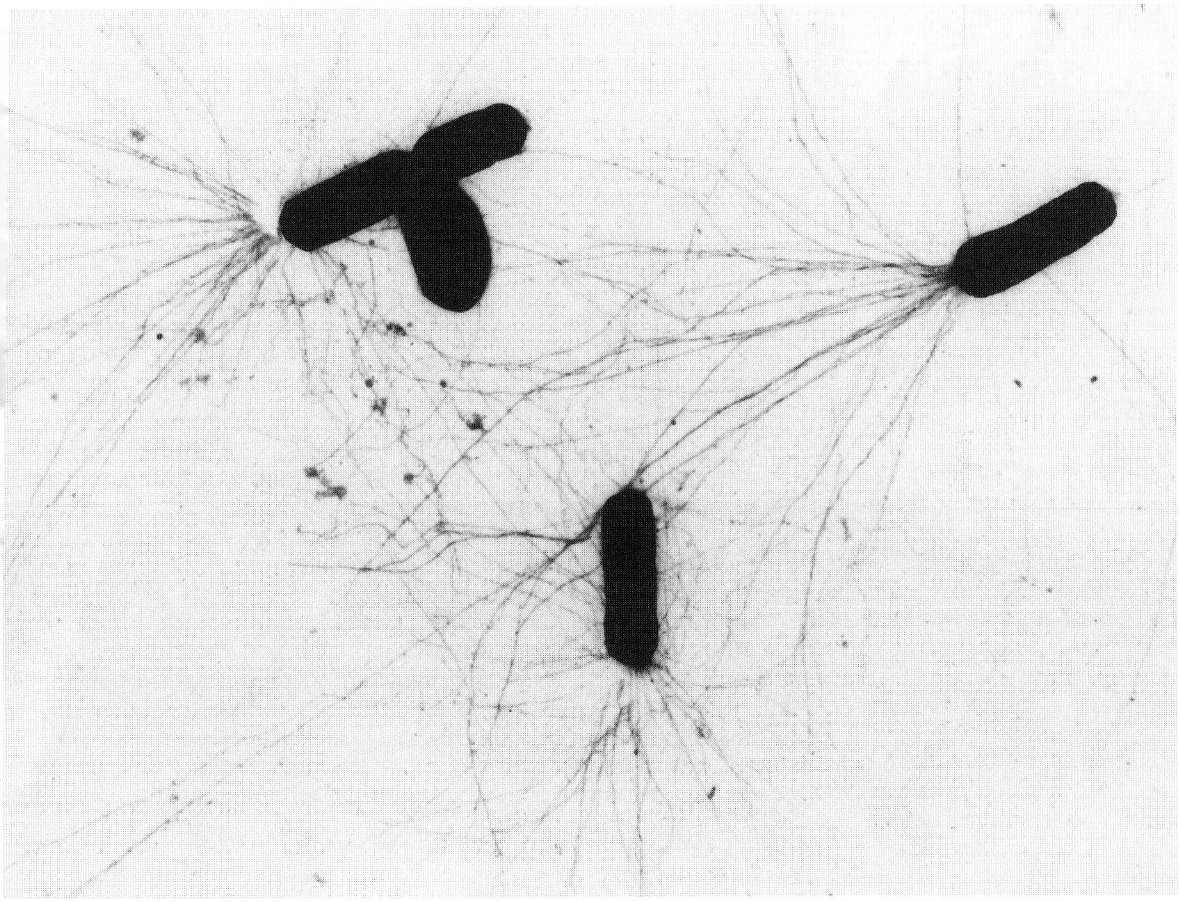

FIG. 23.1. *Dichelobacter nodosus* (virulent foot rot strain) grown in liquid culture, showing abundant pili. Negative uranyl acetate stain. (Courtesy of Dr. T. M. Skerman, Wallaceville Animal Research Station, Ministry of Agriculture and Fisheries, New Zealand.)

SOURCES

This organism is an obligate parasite of the epidermal tissues of the hoof of sheep, goats, and cattle. It is not normally isolated from healthy animals but it may be isolated from chronic lesions in clinically healthy animals, in which it persists until pasture conditions predispose to the development of disease. *D. nodosus* is often found in association with poorly characterized fusiform and spirochetal bacteria, which may contribute to infection. The bacteria can survive in the environment (e.g., soil, bedding) for no longer than several days under optimal conditions of high humidity and temperatures over 10°C. Bovine isolates are essentially avirulent for sheep; some goat isolates may be virulent for sheep. The basis of the well-established difference in virulence between strains is described below.

DISEASE

Footrot is an economically important contagious disease of sheep, cattle, goats, and deer. In sheep, the germinal layers of the epidermis of the interdigital skin and horn matrix are invaded by *D. nodosus* and *F. necrophorum*, resulting in progressive destruction and chronic lameness. In the other animal species, infection is more usually confined to the interdigital skin. Rainfall predisposes to disease in temperate climates. In sheep, following prolonged wetting of the stratum corneum, the interdigital epidermis is colonized by diphtheroids and cocci, followed by *F. necrophorum* from a manured environment. Acute interdigital dermatitis results. If virulent *D. nodosus* are present, synergistic interaction with *F. necrophorum* produces progressive invasion of the soft epidermal tissues of the horn laminae of the sole and heel. Eventually this may be so severe that the hoof may separate. Other bacteria may also have a synergistic effect in enhancing the severity of the infectious process but are not essential. There is, however, an essential and reciprocal synergism in foot rot between *D. nodosus* and *F. necrophorum*.

In cattle, goats, and deer, foot rot is an interdigital dermatitis in which *D. nodosus* plays an essential role, usually in conjunction with *F. necrophorum* or *P. asaccharolytica*, under environmental conditions similar to those that promote infection in sheep. The severely inflamed interdigital lesions are characterized by superficial ulceration, necrosis, fissuring, and hyperkeratosis. Although infection is usually confined to the interdigital skin, serious results of chronic infection include underrunning of the heel, sole ulcers, interdigital fibroma, and hoof overgrowth (Egerton et al. 1989).

BACTERIAL VIRULENCE FACTORS

There are substantial and clinically relevant differences in virulence among strains isolated from sheep (Skerman 1989), with a continuum of clinical disease ranging from virulent foot rot, through an intermediate stage, to benign foot rot. Phenotypic differences that divide strains into broad categories of virulent, intermediate, and benign include colony morphology, degree of motility, proteolytic activity, elastase activity, and thermostability and electrophoretic zymogram patterns of the different proteases (Skerman 1989). Genetic characterization of virulent strains is needed for more definitive and rapid classification, and as a start to this characterization, gene regions associated with virulence have recently been described (Katz et al. 1991).

Virulence correlates remarkably with the production of distinct extracellular proteases, which are likely serine proteases that require divalent cations for activity. The genetic organization of fimbrial genes is well understood and is discussed under Immunity, below.

HOST INTERACTION WITH BACTERIUM

Why the pathogenicity of *D. nodosus* is expressed restrictively in the hoof epidermis (with or without interaction with accessory bacteria) of a limited number of ruminant species (Skerman 1989) has not been determined. It is possible, but not proven, that heavy piliation enhances virulence in cases where interdigital skin is undamaged (damage is usually by wetting or other means). It is likely, but also unproven, that the thermostable keratolytic protease plays a major role in destruction of the germinal layers of the interdigital skin and horn matrix (Depiazzi et al. 1991). The precise basis of synergism with *F. necrophorum* is not understood. Host genetic factors are increasingly recognized as determining the outcome of infection (Skerman 1989). A positive association between resistance to foot rot in sheep and particular lymphocyte antigen types has been described (Outteridge et al. 1989).

IMMUNITY

Sheep vaccinated with *D. nodosus* become highly resistant to infection with homologous organisms. Protection is proportional to the quantity of fimbriae in the vaccine as well as to the type of adjuvant, but lasts only several months. Research in the development of vaccines has continued to emphasize fimbriae, and remarkable advances have been made in understanding the pilin genes of *D. nodosus* and the basis of the antigenic heterogeneity of this pilin (Elleman 1988).

There is homology between pilins from a diverse group of organisms (eg, *Neisseria gonorrhoeae, Pseudomonas aeruginosa*) that show "twitching motility," a colony surface-translocation effect associated with fimbriation. These pilins are similar in size (16-18 kDa) and show extensive similarity in the amino- and carboxy-terminal regions of their amino acid sequence. The first 32 amino acid residues are nearly identical and highly hydrophobic. These pilins are termed mePhe (type 4) pilins because they possess a common N-methylphenyalanine as a modified amino-terminal residue. Considerable detail of the molecular biology of *D. nodosus* pilin is available (Elleman 1988; Mattick 1989).

Isolates have been characterized into at least nine serogroups (A-I), with different serotypes identifiable within these groups. Protective immunity is to the serogroup, at best. Vaccines enriched in pilin from *D. nodosus* are expensive to produce, but these genes have been cloned and expressed in *P. aeruginosa*, producing fimbriae indistinguishable from indigenous *D. nodosus* fimbriae; production of such vaccines is considerably less expensive (Elleman 1988).

Based on the extent of conservation of pilin amino acid sequence outside the major conserved N-terminal region, serogroups A-H can be divided into two distinct categories, an A set (six serogroups) and a B set (two serogroups). The degree of amino acid conservation within the sets corresponds to the degree of serological cross-reaction and cross-protection (Elleman 1988). The pilin genes of *D. nodosus* and *P. aeruginosa* are present only as single copies in the genome, whereas *Moraxella bovis* and *N. gonorrhoea*, for example, have multiple copies, which allow them to undergo antigen variation of considerable importance to persistence in the host. The single pilin gene in *D. nodosus* rules out homologous recombination within the chromosome as a way of producing antigenic variants.

The nucleotide sequences of genes encoding the fimbrial subunits of known serogroups have been determined (Mattick et al. 1991). These genes are preceded by highly conserved regions of the promoter and transcriptional regulator sites, and ribosomal-binding site. The fimbrial subunit genes fall into two major classes (I and II), based on the sequence of the major subunit gene *fimA* corresponding to the A and B subgroups, with class II having more in common with other mePhe-fimbriated organisms (Mattick et al.1991). Comparison of the gene sequences showed the surprising feature of absolute nucleotide conservation of the 5′ coding sequences, which contrasted with the high frequency of silent base changes in the rest of the gene; this suggests that the former sequences might be involved in regulating gene expression or in mediating site-specific recombination. Mattick et al. (1991) made the intriguing speculation that structural variation in the fimbrial subunits might be the means by which *D. nodosus* evaded infection with fimbrial serotype bacteriophage rather than evading the host immune response (which is virtually nonexistent). There is distinct genetic organization of the fimbrial gene region in the two classes of *fimA* genes. Comparisons of nucleotide sequences suggest that class II sequences that may have evolved in another mePhe-fimbriated bacterial species were subsequently substituted in the *D. nodosus* genome by lateral transfer (Hobbs et al. 1991). Analysis of the sequences flanking *fimA* suggests that recombinational exchange of both *fimA* and the entire operon has occurred between different strains, perhaps accounting for antigenic diversity of the serogroups (Hobbs et al. 1991).

PROBLEMS

Defining the genetic basis for virulence in *D. nodosus* will be an interesting and rewarding study over the next few years. This will likely be done by the Australian research workers cited, who have contributed so much to our detailed knowledge about *D. nodosus*.

SERPULINA HYODYSENTERIAE

CHARACTERISTICS OF THE BACTERIUM

Serpulina hyodysenteriae is a gram-negative, anaerobic spirochete, with typical spirochetal ultrastructure of a protoplasmic cylinder and endoflagella surrounded by an outer sheath. It has recently been reclassified from *Treponema* and subsequently *Serpula hyodysenteriae* (Stanton et al. 1991; Stanton 1992). It has been differentiated from the nonpathogenic *S. innocens* by the absence of significant hemolysis by *S. innocens*, but DNA probes are a more useful approach (Jensen et al. 1990; Combs and Hampson 1991). *S. hyodysenteriae* has been distinguished from other porcine spirochetes by multilocus enzyme electrophoresis (Lymbery et al. 1990). At least nine serogroups of *S. hyodysenteriae* have been described, based on LPS antigens (Hampson 1990). Within *S. hyodysenteriae*, restriction enzyme analysis of chromosomal DNA was found to discriminate strains within serotypes (Combs et al. 1989). Classification of other intestinal spirochetes from pigs and from other species is in a rudimentary stage.

SOURCES

Pigs and mice on infected farms are reservoirs, although mice are probably unimportant in transmission of infection. The major reservoir is recovered carrier pigs, which may shed the organisms for at least 2 months following recovery. The organism may survive in pig feces or farm effluent for several months at low environmental temperatures.

DISEASE

Serpulina hyodysenteriae is the cause of swine dysentery, an acute-to-chronic disease of weaned pigs characterized by hemorrhagic to fibrinonecrotic colitis and mucohemorrhagic diarrhea, which may be fatal. Husbandry factors, probably including diet, are important in determining the severity of infection. In experimental infections in gnotobiotic pigs, other bacterial species often enhance colonization of the large bowel by *S. hyodysenteriae*, either by providing nutrients or appropriately anaerobic conditions for this strict anaerobe. Only certain anaerobic bacteria, however, appear capable of such synergism (Hayashi et al. 1990).

BACTERIAL VIRULENCE FACTORS

Avid and irreversible attachment to a variety of cultured cells in vitro is a feature of *S. hyodysenteriae* and may be the result of a bacterial ligand containing sialic acid residues (Bowden et al. 1989). Such binding to cells or to mucus may be important in vivo. The characteristic hemolysin of *S. hyodysenteriae* appears to be a critical virulence factor. The purified, 19-kDa hemolytic toxin caused severe epithelial damage in ligated loops of the ileum and colon of gnotobiotic pigs but did not cause fluid accumulation (Lysons et al. 1991). The hemolysin gene, *tly*, has been cloned and sequenced (Muir et al. 1992). It contains an open reading frame capable of encoding a 26.9-kDa protein. Production of mutants with an inactivated hemolysin gene by homologous recombination reduced, but did not eliminate, cecal lesions in experimental infections in mice (ter Huurne et al. 1992) suggesting that other virulence determinants are also important. These mutants had reduced hemolysis, which suggests that the *tly*-encoded hemolysin is not the only hemolysin produced by *S. hyodysenteriae*. The organism possesses a biologically active LPS in the cell wall, the effects of which have been correlated with the severity of disease in mouse models (Nuessen et al. 1983).

HOST INTERACTION WITH BACTERIUM

The pathogenetic mechanisms of disease are not well understood. The first stage is spirochetal colonization and proliferation deep in the crypts of Lieberkuhn in the colon (Figs. 23.2 and 23.3). The next stage is active invasion of goblet cells by the *Serpulina*, which may subsequently penetrate intercellular spaces through the junctional complex into the lamina propria, penetrate degenerating enterocytes, and be taken into cells by endocytosis (Sueyoshi and Adachi 1990). Mucus production increases markedly as goblet cells discharge and large numbers of spirochetes colonize the increased mucus that covers the colonic surface epithelium. Adherence to, or invasion of, mucosal surface epithelium does not seem to be an important feature of disease. The surface epithelial necrosis and erosion suggest the action of a locally active cytotoxin, probably the hemolysin. Secondary bacterial infection of the eroded surface epithelium appears to enhance the severity of the disease under field conditions. Diarrhea observed in pigs with dysentery results from a failure to absorb fluid and electrolytes in the colon rather than from enhanced secretion. Since half the extracellular fluid volume of the pig is presented daily for reabsorption in the large bowel, loss of absorptive capacity causes severe dehydration. Accumulation of fluid in ligated loops in the rabbit does, however, suggest that active fluid secretion occurs.

IMMUNITY

Solid immunity develops in pigs recovered from clinical disease and seems to be directed, in part, against the LPS antigens and thus to be, in part, serotype specific (Joens et al. 1983). The basis of such immunity has not been determined but is likely to be at least, in part, a local, IgA-based immunity. Attempts to induce immunity to swine dysentery using whole-cell bacterins have, however, been generally only partially successful in pigs, suggesting either that some component is absent from the preparations or that antigen presentation should be by a different route or method. There has been interest in defining the protective antigens, looking particularly at the outer membrane for candidates (Sellwood et al. 1989; Smith et al. 1990). A 16-kDa lipoprotein has been identified as an antigenically common, surface-expressed antigen (Sellwood et al. 1989; Thomas et al. 1992), but there are other proteins (29-45 kDa) that are conserved and apparently important in the immune response (Wannemuehler et al. 1988).

Recently, *S. hyodysenteriae* antigens have been cloned in *E. coli* and characterized (Boyden et al. 1989). One cloned antigen that was serologically related to the endoflagellar proteins induced bactericidal antibody and protected mice against oral challenge with both a homologous and a heterologous serotype (Boyden et al. 1989). The surprising observation that an endoflagellar protein might be a protective immunogen is hard to explain, since it would not likely be surface expressed.

PROBLEMS

Current problems include defining protective antigens and using recombinant techniques for their production; differentiating pathogenic spirochetes from intestinal disease in animals and in humans; developing rapid, genetically based diagnostic systems for spirochetal diarrheas in animals and in humans; and defining the virulence factors of *S. hyodysenteriae*.

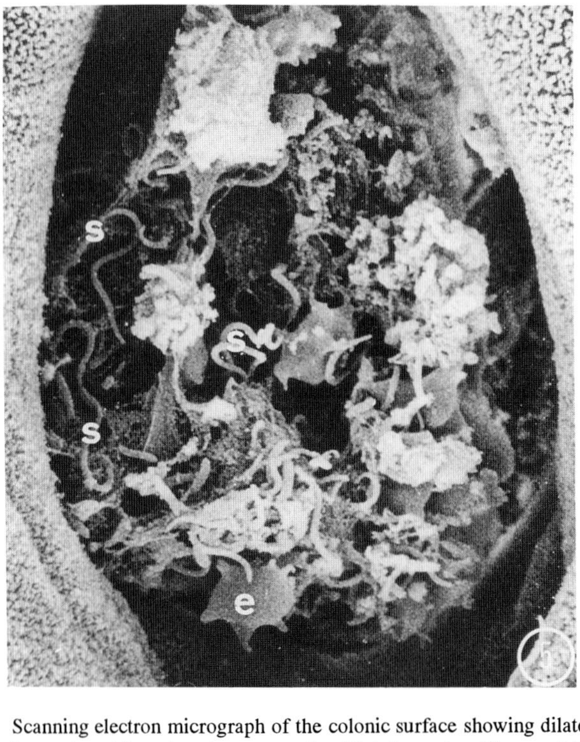

FIG. 23.2. Scanning electron micrograph of the colonic surface showing dilated crypt of Lieberkuhn in a pig with swine dysentery; the dilated crypt orifice contains mucus, erythrocytes (e), and spirochetes (s). x3000. (Teige et al. 1981; reprinted with permission, Acta Vet Scand 22: 218-25.)

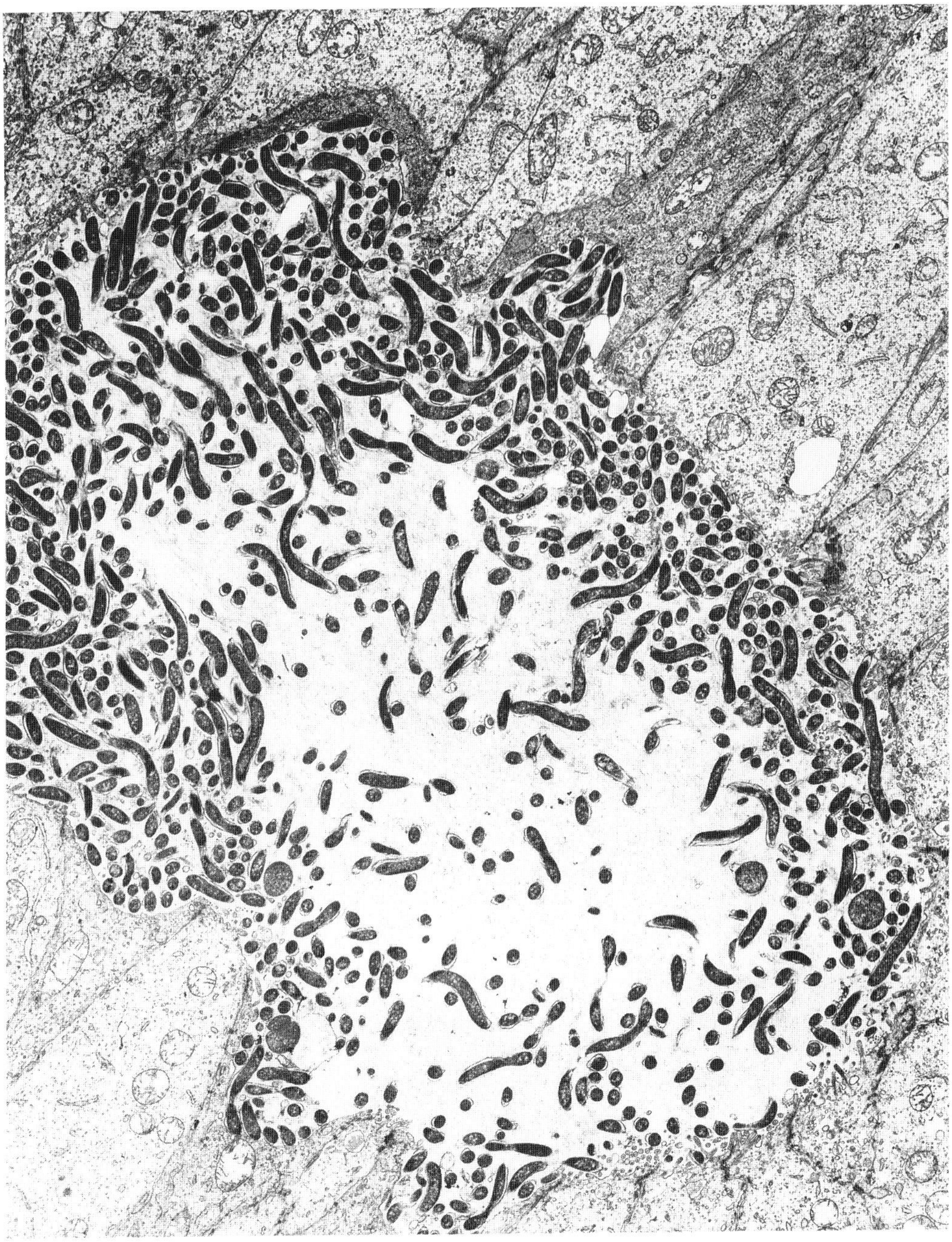

FIG. 23.3. Transmission electron micrograph showing numerous spirochetes in the crypt lumen of the proximal colon of a pig with early swine dysentery; spirochetes are present in the apical part of disgorged goblet cells. x7980. (Courtesy of Dr. J. F. Pohlenz, Iowa State University.)

BACTEROIDES FRAGILIS

Bacteroides fragilis is the major pathogenic *Bacteroides* spp. of humans and is isolated frequently from non-spore-forming anaerobic infections in animals. It is by contrast only a minor member of the normal intestinal flora. An antiphagocytic, polysaccharide capsule and lipopolysaccharide are virulence determinants identified as playing a role in nonspecific infections. Encapsulation has been used to explain the particular virulence of the species (Kasper et al. 1984).

Apart from interest in *B. fragilis* as an important opportunistic pathogen in nonspecific infections, there is now considerable interest in this species as a possible cause of diarrhea in humans and in several species of domestic animals (Myers et al. 1989). Diarrheal illness has been produced by oral inoculation of cultures in rabbits and gnotobiotic pigs. A 19-kDa molecular weight heat-labile enterotoxin has been isolated from toxigenic strains of *B. fragilis* that causes intestinal fluid accumulation and sloughing of intestinal epithelial cells (Van Tassell et al. 1992).

REFERENCES

Blanden, D. R.; Lewis, P. R.; and Ferrier, G. R. 1987. Vaccination against lumpy jaw and measurement of antibody response in wallabies *(Macropus eugenii)*. Vet Rec 121:60-62.

Boyden, D. A.; Albert, F. G.; and Robinson, C. S. 1989. Cloning and characterization of *Treponema hyodysenteriae* antigens and protection in a CF-1 mouse model by immunization with a cloned endoflagellar antigen. Infect Immun 57:3808-15.

Bowden, C. A.; Joens, L. A.; and Kelley, L. M. 1989. Characterization of the attachment of *Treponema hyodysenteriae* to Henle intestinal epithelial cells in vitro. Am J Vet Res 50:1481-85.

Combs, B.; Hampson, D. J.; Mhoma, J. R. L.; and Buddle, J. R. 1989. Typing of *Treponema hyodysenteriae* by restriction endonuclease activity. Vet Microbiol 19:351-59.

Combs, B. G.; and Hampson, D. J. 1991. Use of a whole chromosomal probe for identification of *Treponema hyodysenteriae*. Res Vet Sci 50:286-89.

Depiazzi, L. J.; Richards, R. B.; Henderson, J.; Rood, J. I.; Palmer, M.; and Penhale, W. J. 1991. Characterization of virulent and benign strains of *Bacteroides nodosus*. Vet Microbiol 26:151-60.

Dewhirst, F. E.; Paster, B. J.; La Fontaine, S.; and Rood, J. I. 1990. Transfer of *Kingella indologenes* (Snell and Lapage 1976) to the genus *Suttonella* gen. nov. as *Suttonella indologenes* comb. nov.; transfer of *Bacteroides nodosus* (Beveridge 1941) to the genus *Dichelobacter* gen. nov. as *Dichelobacter nodosus* com. nov.; and assignment of the genera *Cardiobacterium*, *Dichelobacter*, and *Suttonella* to *Cardiobacteriaceae* fam. nov. in the gamma division of *Proteobacteria* based on 16S ribosomal ribonucleic acid sequence comparisons. Int J Syst Bacteriol 40:426-33.

Egerton, J. R.; Yong, W. K.; and Riffkin, G. G., (eds). 1989. Footrot and Foot Abscess of Ruminants. Boca Raton, Fla: CRC Press.

Elleman, T. C. 1988. Pilins of *Bacteroides nodosus*: Molecular basis of serotypic variation and relationships to other bacterial pilins. Microbiol Rev 52:233-47.

Emery, D. L.; Dufty, J. H.; and Clark, B. L. 1984. Biocehmical and functional properties of a leucocidin produced by several strains of *Fusobacterium necrophorum*. Aust Vet J 61:382-87.

Emery, D. L., and Vaughan, J. A. 1986. Generation of immunity against *Fusobacterium necrophorum* in mice inoculated with extracts containing leucocidin. Vet Microbiol 12:255-68.

Emery, D. L.; Edwards, R. D.; and Rothel, J. S. 1986. Studies on the purification of the leucocidin of *Fusobacterium necrophorum* and its neutralization by specific antisera. Vet Microbiol 11:357-72.

Finegold, S. M.; George, W. L.; and Mulligan, M. E. 1986. Anaerobic Infections. Chicago: Yearbook Medical Publishers.

Finlay-Jones, J. J.; Kenny, P. A.; Nulsen, M. F.; Spencer, L. K.; Hill, N. L.; and McDonald, P. J. 1991. Pathogenesis of intraabdominal abscess formation abscess-potentiating agents and inhibition of complement-dependent opsonization of abscess-inducing bacteria. J Infect Dis 164:1173-79.

Hampson, D. J. 1990. New serogroups of *Treponema hyodysenteriae* (G, H, and I). Vet Rec 127:524.

Hayashi, T.; Suenaga, I.; Omeda, T.; and Yamazaki, T. 1990. Role of *Bacteroides uniformis* in susceptibility of Ta:CF#1 mice to infection with *Treponema hyodysenteriae*. Zentralbl Bakteriol 274:118-25.

Hobbs, M.; Dalrymple, B. P.; Cox, P. T.; Livingstone, S. P.; Delaney, S. F.; and Mattick, J. S. 1991. Organization of the fimbrial gene region of *Bacteroides nodosus*: Class I and class II strains. Mol Microbiol 5:543-60.

Hofstad, T. 1984. Pathogenicity of anaerobic Gram-negative rods: Possible mechanisms. Rev Infect Dis 6:189-99.

Jang, S. S., and Hirsh, D. C. 1991. Identity of *Bacteroides* and previously named *Bacteroides* spp. in clinical specimens of animal origin. Am J Vet Res 52:738-41.

Jensen, N. S.; Casey, T. A.; and Stanton, T. B. 1990. Detection and identification of *Treponema hyodysenteriae* by using oligonucleotide probes complementary to 16S rRNA. J Clin Microbiol 28:2717-21.

Joens, L. A.; Whipp, S. C.; Block, R. D.; and Neussen, M. E. 1983. Serotype-specific protection against *treponema hyodysenteriae* infection in ligated colonic loops of pigs recovered from swine dysentery. Infect Immun 39:460-62.

Kanoe, M.; Kitamoto, N.; Toda, M.; and Uchida, K. 1984. Purification and partial characterization of *Fusobacterium necrophorum* haemolysin. Fed Eur Microbiol Soc Microbiol Lett 25:237-42.

Kasper, D. L.; Lindberg, A. A.; Weintraub, A.; Onderdonk, A. B.; and Lonngren, J. 1984. Capsular polysaccharides and lipopolysaccharides from two strains of *Bacteroides fragilis*. Rev Infect Dis 6S:25-29.

Katz, M. E.; Howarth, P. M.; Yong, W. K.; Riffkin, G. G.; Depiazzi, L. J.; and Rood, J. I. 1991. Identification of three gene regions assoicated with virulence in *Dichelobacter nodosus*, the causative agent of footrot. J Gen Microbiol 137:2117-24.

Kornman, K. S., and Loesche, W. J. 1982. Effects of estradiol and progesterone on *Bacteroides melaninogenicus* and *Bacteroides gingivalis*. Infect Immun 35:256-63.

Lindberg, A. A.; Weintraub, A.; Zahringer, J.; and Rietschel, E. T. 1990. Structure-activity relationships in lipopolysaccharides of *Bacteroides fragilis*. Rev Infect Dis 12:S133-S141.

Love, D. N.; Johnson, J. L.; Jones, R. F.; and Calverley, A. 1987. *Bacteroides salivosus* sp. nov., an asaccharolytic, black-pigmented species from cats. Int J Syst Bacteriol 37:307-9.

Lymbery, A. J.; Hampson, D. J.; Hopkins, R. M.; Combs, B.; and Mhoma, J. R. L. 1990. Multilocus enzyme electrophoresis for identification and typing of *Treponema hyodysenteriae* and related spirochetes. Vet Microbiol 22:89-99.

Lysons, R. J.; Kent, K. A.; Bland, A. P.; Robinson, W. F.; and Frost, A. J. 1991. A cytotoxic haemolysin from *Treponema hyodysenteriae* - a probable virulence determinant. J Med Microbiol 34:97-102.

Mattick, J. S. 1989. The molecular biology of the fimbriae (pili) of *Bacteroides nodosus* and the development of a recombinant-NA-based vaccine. In Footrot and Foot Abscesses of Ruminants. Ed. J. R. Egerton, W. K. Yong, and G. G. Riffkin. Boca Raton, Fla: CRC Press.

Mattick, J. S.; Anderson, B. J.; Cox, P. T.; Dalrymple, B. P.; Bills, M. M.; Hobbs, M.; and Egerton, J. R. 1991. Gene sequences and comparison of the fimbrial subunits representative of *Bacteroides nodosus* serotypes A to I class I and class II strains. Mol Microbiol 5:561-73.

Mayrand, D., and Holt, S. C. 1988. Biology of asaccharolytic black-pigmented *Bacteroides* species. Microbiol Rev 52:134-52.

Muir, S.; Koopman, M. B. H.; Libby, S. J.; Joens, L. A.; Heffron, F.; and Kusters, J. G. 1992. Cloning and expression of a *Serpula (Treponema) hyodysenteriae* hemolysin gene. Infect Immun 60:529-35.

Myers, L. L.; Shoop, D. S.; Collins, J. E.; and Bradbury, W. C. 1989. Diarrheal disease caused by enterotoxigenic *Bacteroides fragilis* in infant rabbits. Clin Microbiol 27:2025-30.

Nakagaki, M.; Fukuchi, M.; and Kanoe, M. 1991. Partial characterization of *Fusobacterium necrophorum* protease. Microbios 66:117-23.

Nuessen, M. E.; Birmingham, J. R.; and Joens, L. A. 1983. Involvement of lipopolysacchraide in the pathogenicity of *Treponema hyodysenteriae*. J Immunol 131:997-99.

Outteridge, P. M.; Stewart, D. J.; Skerman, T. M.; Dufty, L. H.; Egerton, J. R.; Ferrier, G.; and Marshall, D. 1989. A positive association between resistance to ovine footrot and particular lymphocyte antigen types. Aust Vet J 66:175-79.

Sellwood, R.; Kent, K. A.; Burrows, M. R.; Lysons, R. J.; and Bland, A. P. 1989. Antibodies to a common outer envelope antigen of *Treponema hyodysenteriae* with antibacterial activity. J Gen Microbiol 135:2249-57.

Shinjo, T.; Hiraiwa, K.; and Miyazato, S. 1990. Recognition of biovar C of *Fusobacterium necrophorum* (Flugge) Moore and Holdeman as *Fusobacterium pseudonecrophorum* sp. nov., nom. rev. (ex Prevot 1940). Int J Syst Bacteriol 40:71-73.

Shinjo, T.; Fujisawa, T.; and Mitsuoka, T. 1991. Proposal of two subspecies of *Fusobacterium necrophorum* (Flugge) Moore and Holdeman: *Fusobacterium necrophorum* subsp. *necrophorum* subsp. nov., nom. rev. (ex Flugge 1886), and *Fusobacterium necrophorum* subsp. *funduliforme* subsp. nov., nom. rev. (ex Halle 1898). Int J Syst Bacteriol 41:395-97.

Skerman, T. M. 1989. Isolation and identification of *Bacteroides nodosus*. In Footrot and Foot Abscess of Ruminants. Ed. J. R. Egerton, W. K. Yong, and G. G. Riffkin. Boca Raton, Fla: CRC Press.

Smith, G. R.; Barton, S. A.; and Wallace, L. M. 1991. Further observations on enhancement of the infectivity of *Fusobacterium necrophorum* by other bacteria. Epidemiol Infect 106:305-10.

Smith, S. C.; Roddick, F.; Ling, S.; Gerraty, N. L.; and Coloe, P. J. 1990. Biochemical and immunochemical characterization of strains of *Treponema hyodysenteriae*. Vet Microbiol 24:29-41.

Stanton, T. B.; Jensen, N. S.; Casey, T. A.; Tordoff, L. A.; Dewhirst, F. E.; and Paster, B. J. 1991. Reclassification of *Treponema hyodysenteriae* and *Treponema innocens* in a new genus, *Serpula* gen. nov., as *Serpula hyodysenteriae* comb. nov. and *Serpula innocens* comb. nov. Int J Syst Bacteriol 41:50-58.

Stanton, T. B. 1992. Proposal to change the genus designation *Serpula* to *Serpulina* gen. nov. containing the species *Serpulia hyodysenteriae* comb. nov. and *Serpulia innocens* comb. nov. Int J Syst Bacteriol 42:189-90.

Sueyoshi, M., and Adachi, Y. 1990. Diarrhea induced by *Treponema hyodysenteriae*: A young chick cecal model for swine dysentery. Infect Immun 58:3348-62.

ter Huurne, A. A. H. M.; van Houten, M.; Muir, S.; Kusters, J. G.; van der Zeijst, B. A. M.; and Gaastra, W. 1992. Inactivation of a *Serpula (Treponema) hyodysenteriae* hemolysin gene by homologous recombination; importance of this hemolysin in pathogenesis of *S. hyodysenteriae* in mice. Fed Eur Microbiol Soc Microbiol Lett 92: 109-14.

Thomas, W.,; Sellwood, R.; and Lysons, R. J. 1992. A 16-kilodalton lipoprotein of the outer membrane of *Serpulina (Treponema) hyodysenteriae*. Infect Immun 60:3111-16.

Van Tassell, R. L.; Lyerly, D. M.; and Wilkins, T. D. 1992. Purification and characterization of an enterotoxin from *Bacteroides fragilis*. Infect Immun 60:1343-50.

Wannemuehler, M. J.; Hubbard, R. D.; and Greer, J. M. 1988. Characterization of the major outer membrane proteins of *Treponema hyodysenteriae*. Infect Immun 56:3032-39.

24 / *Leptospira*

BY **J. F PRESCOTT** AND **R. L. ZUERNER**

LEPTOSPIROSIS is a fascinating, important, and complex disease of animals and people, caused by pathogenic members of the genus *Leptospira*. It causes loss in animals through acute (septicemia, hepatitis, nephritis, meningitis) or chronic (abortion, stillbirth, nephritis) disease. Frequently, infection is not recognized clinically. Leptospirosis is an important zoonotic infection, with humans acquiring infection from a wide variety of animal reservoirs, often from urine-contaminated water. Many aspects of leptospirosis are poorly understood. This is due, in part, because of the insensitivity of traditional diagnostic methods, difficulties associated with taxonomic classification, and the complexities of leptospiral-host parasite relationships. Changing patterns of infection associated with vaccination and antibiotic usage may also contribute to confusion regarding the disease. Use of molecular genetic techniques to study leptospirosis is revolutionizing our understanding of a generally neglected, frequently confusing, and surprisingly backward subject. While considerable advances have been made in taxonomy, understanding of the molecular basis of pathogenicity of leptospires is rudimentary.

CHARACTERISTICS OF THE LEPTOSPIRES

Leptospires are slender, helically shaped, motile, spirochetes 0.2-0.3 μm in diameter and 20-30 μm long. The family *Leptospiraceae* contains the genera *Leptospira* and *Leptonema*. The genus *Leptospira* has traditionally been divided into the pathogenic species *L. interrogans* and the free-living nonpathogenic species *L. biflexa* and *L. parva*. Characteristically, *Leptospira* have hooked ends and two periplasmic flagella (PF), also known as axial filaments and endoflagella. These flagella are inserted near the two ends of the cell and wrap around the cytoplasmic membrane in the periplasmic space along the length of the cell. Rotation of the PF results in the distinct spinning mobility of *Leptospira* (Bromley and Charon 1979). The PF are complex structures consisting of a solid inner core surrounded by two tubular sheaths (Trueba et al. 1992). The cytoplasmic cylinder, into which the flagella are inserted, has a peptidoglycan layer and cytoplasmic membrane. The outer envelope contains mainly lipooligosaccharide, proteins, and phospholipids. *Leptospira* divide by binary fission.

Pathogenic leptospirae are obligate aerobes that use long-chain (> 15 carbon atoms) fatty acids or fatty alcohols, rather than carbohydrates and amino acids, as their energy and carbon sources. Ammonium salts are used as the major source of nitrogen but parasitic leptospires can also use urea. Other nutritional requirements include vitamins B_1 and B_{12} and purines (Johnson and Faine 1984). The most widely used culture medium is EMJH (named for the individuals who developed it: Ellinghausen and McCullough 1965, and Johnson and Harris 1967). This medium has a bovine serum-albumin base, which complexes with (i.e., detoxifies) the free fatty acids used by the bacteria for growth. In addition, EMJH contains several other medium components that are better defined than in the rabbit serum-based medium used previously. It is important to use a 0.2% semisolid agar medium for initial isolation from tissue, possibly allowing organisms to use their motility to find optimal conditions. Despite the improvements in leptospiral media, some serovars may take weeks or months to isolate from tissues because appropriate but unknown media constituents are absent. The long generation time of even the fast-growing serovars (6-16 hours) means that isolation usually takes a minimum of 7-10 days. Surprisingly, considering their pathogenicity for mammals, the optimal temperature for growth in vitro is 30°C.

CLASSIFICATION AND NOMENCLATURE

The taxonomy of pathogenic leptospires is in transition from an antigenic to a genetic classification because genetic classification has considerably greater power to discriminate among isolates. This attribute of genetic classification contributes to improving our understanding of the epidemiology of leptospirosis.

Antigenic classification is based on the microscopic agglutination test and uses polyclonal antiserum and a cumbersome process of cross-agglutination and agglutination-absorption to identify serovars. Recently, this classification process has been simplified by typing with monoclonal antibodies (Terpstra et al. 1987). Antigenic composition has traditionally been used for taxonomy of leptospires. In this approach, *L. interrogans* was regarded as the single pathogenic species within the genus and was divided into about 190 different serovars, arranged into 25 major antigenically related serogroups (Waitkins 1986).

Recent genetic studies have revealed considerable heterogeneity among pathogenic leptospires. The former species *L. interrogans* is now divided into 7 species: *L. borgpetersenii*, *L. inadai*, *L. interrogans*, *L. kirschneri (L. alstonii)*, *L. noguchii*, *L. santarosai*, and *L. weilii* (Yasuda et al. 1987; Ramadass et al. 1992). Likewise, saprophytic leptospires that belonged to the former species *L. biflexa* now comprise at least three new species: *L. biflexa*, *L. meyeri*, and *L. wolbachii* (Yasuda et al. 1987).

There is limited correlation between the genetic and antigenic typing schemes for pathogenic leptospires. Restriction enzyme analysis (REA) of chromosomal DNA isolated from different pathogenic leptospires is often used as a measure of genetic relatedness. Use of REA has shown the validity of the genetic classification scheme by detecting differences between antigenically identical isolates belonging to different species (Marshall et al. 1981; Thiermann and Ellis 1986). For example, isolates characterized antigenically as *L. interrogans* serovar *hardjo* were divided by REA into 2 "genotypes" (later shortened to "types"), hardjoprajitno and hardjo-bovis (Thiermann and Ellis 1986). Differences detected by REA correlate with the different species to which these antigenically similar isolates belong (*L. interrogans* for *hardjoprajitno* and *L. borgpetersenii* for *hardjo-bovis*), and to differences in geographic distribution. In serovar *bratislava* infection in swine, REA has also detected differences that may be important in understanding the epidemiology and pathogenesis of infection. For example, genotypes B2b and M2 accounted for the majority of isolates from stillborn or aborted pigs in Northern Ireland, whereas genotype B2a was recovered from the brains of pigs suffering from meningitis (Ellis et al. 1991). Hybridization with DNA probes prepared against repetitive sequences or ribosomal genes have supported the genetic classification, and in some cases further differentiated related organisms (Perolat et al. 1990; Zuerner and Bolin 1990).

Within the different species of pathogenic leptospires, isolates are further antigenically classified as serovars, within which further genetic differences may be recognized. For ease of description, they will be described below as serovars without the preceding species epithet.

SOURCES OF PATHOGENIC LEPTOSPIRES

Leptospirosis is an important zoonotic infection affecting most mammals worldwide. In addition, leptospires have been isolated from amphibians, arthropods, birds, and reptiles (Thiermann 1984). Usually, only a small number of serovars are endemic to a particular region or country, but there are considerable differences between regions.

The natural habitat of pathogenic leptospires is the proximal convoluted tubule of the kidney (and sometimes the genital tract) of specific maintenance (reservoir) hosts. Often, there is a clear adaptation of particular serovars for long-term persistence in particular hosts. Incidental (accidental) hosts are infected by direct transmission through infected urine, placental or fetal tissues, or indirectly through contact with a contaminated environment. Venereal transmission of serovar *bratislava* may be an important means of transmission in pigs. The potential for venereal transmission in other species may be limited. Vertical transmission from the mother to the fetus may also occur in several species (e.g., *hardjo-bovis* in cattle).

Infection occurs through mucous membranes (nasal, genital, ocular, intestinal) or through abraded or water-softened skin. A warm moist, environment with the presence of groundwater with a neutral or slightly basic pH favors the survival of leptospires outside the host. There is thus a high incidence of leptospirosis in tropical and subtropical regions, but seasonal infections also occur in temperate regions. Under optimal conditions, leptospires may survive outside the host for about 6 weeks, although survival for considerably longer has been described. Common leptospiral serovars in domestic animals and their maintenance hosts are shown in Table 24.1.

DISEASES

The leptospire-host relationship is complex. In some cases the leptospira-maintenance host pairing is unique to an individual serovar and host species. In contrast, there is no strict serovar-host specificity for infection of incidental hosts. For some maintenance hosts, leptospirosis can be thought of not as a single disease but rather as a series of distinct diseases caused by different serovars. Some generalizations can nevertheless be made.

Leptospirosis can occur as an acute and severe disease (septicemia with evidence of endotoxemia such as hemorrhages, hepatitis, nephritis, meningitis), as a subacute moderately severe disease (nephritis, hepatitis, agalactia, meningitis), as a chronic disease (iridocyclitis, abortion, stillbirth, infertility), or in subclinical form recognized only by the development of antibody. The basic lesion in acute disease is damage to the endothelium of small blood vessels, leading to extravasation of blood and of leptospires into tissues; reduced oxygen supply and secondary ischemia result in damage to organs, particularly the kidneys, liver, and adrenals (Faine 1982). Severe clinical signs associated with bacteremic or septicemic spread and localization and multiplication in tissues (especially liver, kidney, and cerebrospinal fluid) include fever, marked malaise, severe muscle pain, hemoglobinuria, jaundice, anemia, anuria or other evidence of acute nephritis, severe headache and other meningeal signs, gastrointestinal signs, skin rashes, and photophobia.

TABLE 24.1. Common leptospiral serovars of significance in disease in domestic animals

Serovar	Species	Maintenance Host	Distribution	Diseases in Major Hosts
bratislava	*interrogans*	Pigs, horses, (dogs?)	Global	Pigs: abortion, stillbirth, infertility
canicola	*interrogans*	Dogs	Global	Dogs: acute, chronic interstitial nephritis Other species: acute, chronic leptospirosis
grippotyphosa	*kirschneri* (*alstonii*)	Racoons, skunks, small rodents	Global	?
hardjo-bovis	*borgpetersenii*	Cattle	Global	Agalactia, abortion
hardjoprajitno	*interrogans*	Cattle	Europe	Milk loss syndrome, agalactia, abortion
icterohaemorrhagiae	*interrogans*	Rodents	Global	Mild disease in rodents?; fulminant leptospirosis in incidental hosts
pomona (type kennewicki)	*interrogans*	Pigs, wildlife,	Americas	Acute leptospirosis in piglets; abortion, chronic nephritis in adults Acute to chronic lepto-spirosis in other species
tarrasovi	*borgpetersenii*	Pigs	Europe, Australia	Abortion

In maintenance hosts to which serovars appear well adapted (e.g., *bratislava* in pigs, *hardjo-bovis* in cattle), serovars tend to be of relatively low pathogenicity and to cause chronic (i.e., abortion, stillbirth, and infertility) rather than acute disease. Typically, the host has a relatively weak antibody response, and the bacteria can persist in the kidney and sometimes genital tract for months or years. By contrast, in accidental hosts (e.g., *icterohaemorrhagiae* in humans), serovars tend to cause acute, severe disease, which is associated with a marked antibody response and rapid elimination from the kidney. These distinctions in general behavior between infection in maintenance or incidental hosts are generalizations to which there are clear exceptions. Below are some typical examples of the disease caused by adapted or incidental serovars.

In cattle, the most severe but uncommon manifestation of acute infection occurs in calves infected with incidental serovars, especially *pomona* type kennewicki. The more common form of acute disease, noted particularly with *hardjoprajitno*, is a marked drop in milk production ("flabby udder mastitis," "milk drop syndrome") associated with pyrexia. The chronic form of the disease, associated with these serovars and with *hardjo-bovis*, is abortion, stillbirth, or the production of weak and premature calves. Infected but apparently healthy calves may also be born. Subclinical carriage in the kidneys and genital tract of cows and bulls is also common (Ellis 1986).

In pigs, acute illness is uncommon, but incidental infection with serovar *pomona* may produce hemoglobinuria and jaundice in piglets. More commonly, serovars *pomona* and *tarrasovi* cause abortion in the last third of pregnancy, stillbirth in piglets, or variable lesions of interstitial nephritis noted at slaughter. By contrast, infection with adapted serovars *bratislava* or *muenchen* is associated with abortion and especially with stillbirth and reduced litter size, as well as with infertility ("repeat breeder syndrome") (Ellis 1989).

In dogs, serovar *icterohaemorrhagiae* tends to cause acute, fulminant leptospirosis characterized by fever, a marked tendency to hemorrhage, anaemia, and jaundice, whereas serovar *canicola* is more likely to cause a disease characterized by acute diffuse nephritis or by chronic interstitial nephritis. However, in dogs it is not possible to distinguish, on the basis of clinical illness, acute infections caused by different serovars.

IMMUNITY

Recovery from acute leptospiral infection is associated with a humoral immune response that is serovar (and perhaps to a lesser extent serogroup) specific. The initial short-lasting IgM antibody response can be detected using the microscopic agglutination test (MAT). However, this commonly used test is less sensitive in detecting IgG, so that animals may be immune to reinfection in the absence of antibody detected by the MAT. Immunity is associated with either agglutinating or opsonizing antibodies, of which the latter are not detected by the MAT but can be demonstrated by passive immunization or other means. The IgG antibody response confers greater protection than the IgM. Immunity can last about 1 year following vaccination with bacterins, although the MAT response to vaccination is usually transient and minor. The role of the cellular immune response in protective immunity to leptospirosis is unclear.

Localization of pathogenic leptospires to the proximal convoluted tubules of the kidney and their survival in the cerebrospinal fluid and vitreous humor of the eye of some infected animals, in part, reflects the inability of antibody to penetrate these sites in the absence of inflammation. The basis of the persistence of host-adapted serovars such as *bratislava* and *hardjo-bovis* in the porcine and bovine genital tracts, respectively, has not been determined. It is possible that these organisms are protected from opsonization by IgG by a local IgA response in these sites, a mechanism responsible for the similar persistence of *Campylobacter fetus* in the genital tract of cattle. Possible alternative explanations for the persistence of these host-adapted serovars include an apparently low antigenicity of such serovars in their hosts or antigenic variation of leptospiral lipooligosaccharide (LOS) (Ito et al. 1988). Antibody-selected LOS-variants show a stable change to different serovars (Shimono et al. 1980; Yamaguchi et al. 1988). These findings might also explain the paradoxical serological reactions observed in acute leptospirosis. The failure of vaccination with serovar *hardjo-bovis* to protect against experimental and possibly natural infection caused by the same serovar (Bolin et al. 1989a,b) demonstrates the importance of determining the basis of both systemic and local immunity against leptospires.

In some cases, protective immunity is associated with development of opsonizing and bactericidal antibodies directed against the LOS (Farrelly et al. 1987). The LOS of pathogenic leptospires does not have the classical electrophoretic ladderlike appearance but rather consists of a number of distinct fractions ranging from 14 to 30 kDa, which can be recognized as either serovar- or serogroup-specific components (Adler et al. 1989). Since development of high antibody titers to LOS may not protect cattle against hardjo-bovis infection (Bolin et al. 1989b), antibodies to protein antigens may also be required for protective immunity (Alexander et al. 1971).

The antigenically common, surface-exposed outer-membrane proteins (OMPs) of pathogenic leptospires have attracted interest because of their potential for use as heterologous protective antigens. Evidence for protective immunity is seen in cross-protection experiments with guinea pigs after reciprocal superinfection with antigenically unrelated serovars (Kemenes 1964). Molecular characterization of protein antigens is in the initial stage. A number of surface-exposed proteins of differing molecular size have been identified in different studies (Haake et al. 1991; Zuerner et al. 1991). Quantitative and qualitative differences between OMPs of a virulent and an avirulent strain of serovar *grippotyphosa* was noted (Haake et al. 1991) but the function of these proteins is unknown. In serovar *pomona*, the major outer-membrane protein antigen is cleaved by an endogenous protease at 37°C, but not at 30°C (Zuerner et al. 1991). Presumably, this cleavage alters the antigenic composition of the bacterial surface during growth in the host. Two serovar *canicola* genes thought to encode OMPs have been reported (Doherty et al. 1989). Cloning and characterization of these genes should be a useful approach in determining the role of surface expressed OMPs in protective immunity to leptospirosis.

BACTERIAL VIRULENCE FACTORS

Details of the molecular pathogenesis of leptospirosis and of the basis of virulence in leptospires are sketchy. Little is known about the factors that allow leptospires to survive in serum or to persist for years in the body. The factors responsible for causing monocytes to release host-damaging cytokines are unknown. The bacterial phospholipases and cytotoxic factors are poorly understood, particularly how these factors affect pathogenicity.

Leptospires invade the host across mucosal surfaces or softened skin. Virulent strains bind to epithelial cell lines in vitro more efficiently than nonvirulent strains (Ballard et al. 1986) and attach to the constituents of the extracellular matrix (e.g., fibronectin and collagen) (Ito and Yanagawa 1987). Adhesion appears to be an active process involving unidentified leptospiral surface proteins (Thomas and Higbie 1990). While pathogenic leptospires may also invade epithelial and other cell types (Miller and Wilson 1967), they are found extensively extracellularly, between cells of the liver and the kidney (Van den Ingh and Hartman 1986).

Release of lymphokines such as tumor necrosis factor alpha (TNF-alpha) from monocytes through the endotoxic activity of the leptospiral LOS may be an important virulence mechanism. Induction of TNF-alpha release may help explain the damage to endothelial cells with resultant hemorrhage and bleeding abnormalities typical of disseminated intravascular coagulation that are seen in acute, severe leptospirosis. A virulent strain of serovar *icterohaemorrhagiae* was shown to produce greater procoagulant effect than a nonvirulent strain incubated with human monocytes *in vitro* (Miragliotta and Fumarola 1983). Isogai et al. (1989) have detected differences in the effect of LOS from different serovars that cause thrombocytopenia and platelet aggregation. The LOS content of virulent and avirulent strains of particular serovars may differ (Niikura et al. 1987; Haake et al. 1991). The toxic and mitogenic effect of lipid A isolated from leptospiral LOS has been reported to be much lower than that of the lipopolysaccharide of *E. coli* or *Salmonella* (Tu et al. 1986; Shimizu et al. 1987). Yet, leptospiral LOS has been shown to be a powerful inducer of cytokines from monocytes (Isogai et al. 1990). In the absence of specific antibody, the LOS content of the cell wall is thought to be responsible for resistance of leptospires to the bactericidal effect of normal serum (Isogai et al. 1986).

Leptospiral hemolysins may contribute to the hemoglobinuria, hemolytic anemia, and tissue damage observed in acute leptospirosis (Thompson and Manktelow 1989). Sphingomyelinase C has been recognized only in pathogenic strains (Bazovska 1978), and different but related multiple sphingomyelinase genes that code for extracellular sphingomyelinase have been identified in different leptospiral strains, suggesting the importance of these genes (Del Real et al. 1989; Segers et al. 1990a,b). The various sphingomyelinase genes identified correspond with the genetic grouping of pathogenic leptospires (Segers 1991). For strains of serogroup Pomona, a correlation was reported between virulence and sphingomyelinase production (Stallheim 1971) as well as phospholipase A activity (Volina et al. 1986). While the role of sphingomyelinases C and phospholipases A in pathogenesis is still speculative, these enzymes may have a cytolytic effect on host cells, and provide nutritional benefits to the organism by making iron and fatty acids available. Cytotoxic activity, not associated with hemolytic activity but closely associated with whole cells, was associated with a virulent but not an avirulent strain of serovar *pomona* (Miller et al. 1970).

GENOMIC STRUCTURE AND ORGANIZATION

Using nucleotide sequence analysis of ribosomal RNA as a molecular chronometer, the leptospires are thought to be one of the oldest branches in eubacterial evolution. Little is known about *Leptospira* genetics, and because of their unique position in evolution, little can be extrapolated from other bacterial species. Lack of any known system for genetic exchange has prevented use of classical genetic techniques for genetic analysis. In contrast, knowledge of genetic-exchange systems has made possible the analysis of a wide variety of other bacteria, notably *Bacillus*, *Escherichia*, and *Salmonella*. Genetic analysis of *Leptospira* has relied on molecular cloning and analytical techniques. Still, few leptospiral genes have been cloned, and fewer still from pathogenic serovars. As mentioned above, among the leptospiral genes that have been cloned are genes encoding hemolysin, outer membrane proteins, and, more recently, a flagella subunit protein.

Since few genes have been analyzed, little is known about *Leptospira* gene organization. Initial studies suggest that the organization is significantly different from other, more intensively studied bacteria. For example, genes encoding a putative RNA polymerase subunit and two ribosomal proteins are found clustered in a manner more similar to archaebacteria than to eubacteria (Zuerner and Charon 1990). Likewise, genes encoding leptospiral ribosomal RNAs are not clustered together, as they are for most other organisms (Fukunaga et al. 1990).

Several spirochetal genomes have been analyzed by pulsed-field gel electrophoresis, a technique that separates large restriction endonuclease fragments of chromosomal DNA. The genomic content and organization of spirochetes shows substantial diversity. In *Leptospira*, the genome contains two circular DNA molecules: a 4400-kb chromosome and a 350-kb extra-chromosomal DNA (thought to be a plasmid) (Zuerner 1991). In contrast, *Treponema pallidum* contains a 900-kb circular chromosome (Walker et al. 1991). The *Borrelia* genome is unique among all bacteria studied to date, possessing a 100-kb linear chromosome (Baril et al. 1989; Ferdows and Barbour 1989), and a collection of circular and linear plasmid DNA molecules, which can comprise about one-third of the total genetic information.

Repetitive DNA present in the leptospiral genome (Zuerner and Bolin 1988) may contribute to genetic recombination, resulting in the varied REA patterns observed among serovars. These rearrangements could easily affect gene expression and thus alter expression of antigens as well as virulence traits. Development of a combined physical and genetic map of the leptospiral genome should aid in localizing such rearrangements and help identify virulence-associated genes.

PROBLEMS

There is still much left to understand about the pathogenic leptospires. The basic biology, genetics, and virulence attributes of these bacteria are poorly characterized or understood. Items as basic to the field as an improved medium would help greatly improve initial isolation and cultivation of nutritionally fastidious serovars. Although a few toxins elicited from these bacteria have been identified, most are poorly characterized. These toxins need to be purified and the mechanisms of their action determined. New diagnostic tests using the polymerase chain reaction (Hookey 1992) make it easier to identify infected animals and thus obtain more definitive epidemiological data. Use of nucleic acid probes for genetic typing is proving to be more sensitive than REA in detecting subtle genetic differences among serovars. This approach may be useful in identifying genetic changes that result in different levels of virulence, and thus help in identifying genetic determinants of pathogenicity. Likewise, an understanding of how virulence determinants are regulated may provide insight into how disease may be prevented. The mechanisms of leptospira adaptation to a particular host species pose one of the greatest challenges, for many of the features of the persistent species-adapted infection run counter to the findings based on acute, incidental-host infections. In this context, antigens that contribute to development of protective immunity to host-adapted strains need to be identified and characterized. While cell-mediated immunity does not appear to play an important role in incidental-host, acute disease, it may play a critical role in protecting against host-adapted chronic infection. Also, the role of lymphokines in the disease process needs to be elucidated.

REFERENCES

Adler, B.; Ballard, S. A.; Miller, S. J.; and Faine, S. 1989. Monoclonal antibodies reacting with serogroup and serovar specific epitopes on different lipooligosaccharide subunits of *Leptospira interrogans* serovar *pomona*. Fed Eur Microbiol Soc Microbiol Immunol 1:213-18.

Alexander, A. D.; Wood, G.; Yancey, Y.; Byrne, R. J.; and Yager, R. H. 1971. Cross neutralization of leptospiral hemolysins from different serotypes. Infect Immun 4:154-59.

Ballard, S. A.; Williamson, M.; Adler, B.; Tu, V.; and Faine, S. 1986. Interactions of virulent and avirulent leptospires with primary cultures of renal epithelial cells. J Med Microbiol 21:59-67.

Baril, C.; Richaud, C.; Baranton, G.; and Saint Girrons, I. 1989. Linear chromosome of *Borrelia burgdorferi*. Res Microbiol 140:507-16.

Bazovska, S. 1978. Sphingomyelinase activity of leptospira cultures. Czech Epidem Microbiol Immunol 27:137-43.

Bolin, C. A.; Thiermann, A. B.; Handsaker, A. L.; and Foley, J. W. 1989a. Effect of vaccination with a pentavalent vaccine of *Leptospira interrogans* serovar *hardjo* type hardjo-bovis infection of pregnant cattle. Am J Vet Res 50:161-65.

Bolin, C. A.; Zuerner, R. L.; and Trueba, G. 1989b. Effect of vaccination with a pentavalent leptospiral vaccine containing *Leptospira interrogans* serovar *hardjo* type hardjo-bovis on type hardjo-bovis infection of cattle. Am J Vet Res 50:2004-8.

Bromley, D. B., and Charon, N. W. 1979. Axial filament involvement in the motility of *Leptospira interrogans*. J Bacteriol 137:1406-12.

Del Real, G.; Seeger, R. P. A. M.; van der Zeijst, B. A. M.; and Gaastra, W. 1989. Cloning of a hemolysin gene from *Leptospira interrogans* serovar *hardjo*. Infect Immun 57:2588-90.

Doherty, J. P.; Adler, B.; Rood, J. I.; Billington, S. J.; and Faine, S. 1989. Expression of two conserved leptospiral antigens in *Escherichia coli*. J Med Microbiol 28:143-49.

Ellinghausen, H. C., and McCullough, W. G. 1965. Nutrition of *Leptospira pomona* and growth of 13 other serotypes fractions of oleic albumin complex and a medium of bovine albumin and polysorbate 80. Am J Vet Res 26:45-51.

Ellis, W. A. 1986. Effects of leptospirosis on bovine reproduction. In Current Therapy in Theriogenology, 2nd Ed. Ed. D. A. Morrow, pp. 267-71. Philadelphia: W. B. Saunders.

__________. 1989. *Leptospira australis* infection in pigs. Pig Vet J 22:83-92.

Ellis, W. A.; Montgomery, J. M.; and Thiermann, A. B. 1991. Restriction endonuclease analysis as a taxonomic tool in the study of pig isolates belonging to the Australis serogroup of *Leptospira interrogans*. J Clin Microbiol 29:957-61.

Farrelly, H. E.; Adler, B.; and Faine, S. 1987. Opsonic monoclonal antibodies against lipopolysaccharide antigens of *Leptospira interrogans* serovar *hardjo*. J Med Microbiol 23:1-7.

Faine, S., (ed). 1982. Guidelines for the Control of Leptospirosis. Geneva: World Health Organization.

Ferdows, M. S., and Barbour, A. G. 1989. Megabase-sized linear DNA in the bacterium *Borrelia burgdorferi*, the Lyme disease agent. Proc Natl Acad Sci USA 86:5969-73.

Fukunaga, M.; Masuzawa, T.; Okuzako, N.; Mifuchi, I.; and Yanagihara, Y. 1990. Linkage of ribosomal RNA genes in *Leptospira*. Microbiol Immunol 34:565-73.

Haake, D. A.; Walker, E. M.; Blanco, D. R.; Bolin, C. A.; Miller, J. N.; and Lovett, M. A. 1991. Changes in the surface of *Leptospira interrogans* serovar *grippotyphosa* during in vitro cultivation. Infect Immun 59:1131-40.

Hookey, J. V. 1992. Detection of *Leptospiraceae* by amplification of 16S ribosomal DNA. Fed Eur Microbiol Soc Microbiol Lett 90:267-74.

Isogai, E.; Isogai, H.; and Ito, N. 1986. Decreased lipopolysaccharide content and enhanced susceptibility of leptospiras to serum bactericidal action and phagocytosis after treatment with diphenylamine. Zentralbl Bakteriol Hyg A 262:438-47.

Isogai, E.; Kitagawa, H.; Isogai, H.; Matsuzawa, T.; Shimizu, T.; Yanagihara, Y.; and Katami, K. 1989. Efffects of leptospiral lipopolysaccharide on rabbit platelets. Int J Med Microbiol 271:186-96.

Isogai, E.; Isogai, H.; Fujii, N.; and Oguma, K. 1990. Macrophage activation of leptospiral lipopolysaccharide. Int J Med Microbiol 273:200-208.

Ito, T., and Yanagawa, R. 1987. Leptospiral attachment to extracellular matrix of mouse fibroblast (L929) cells. Vet Microbiol 15:89-96.

Ito, T.; Takimoto, T.; and Yanagawa, Y. 1988. Presence of a variety of antigenic variants belonging to at least 4 different serovars in a population of *Leptospira interrogans* serovar *hebdomadis*. Zentralbl Bakteriol Hyg A 269:15-25.

Johnson, R. C., and Faine, S. 1984. Family II. *Leptospiraceae* Hovind-Hougen 1979, 245. In Bergey's Manual of Systematic Bacteriology, Vol. 1. Ed. N. R. Krieg and J. G. Holt. Baltimore: Williams and Wilkins.

Johnson, R. C., and Harris, V. G. 1967. Differentiation of pathogenic and nonpathogenic leptospires. J Bacteriol 94:27-31.

Kemenes, F. 1964. Cross-immunity studies on virulent strains of leptospires belonging to different serotypes. Z Immunitat Alerg 127:209-29.

Marshall, R. B.; Wilton, B. E.; and Robinson, A. J. 1981. Identification of leptospira serovars by restriction-endonuclease analysis. J Med Microbiol 14:163-66.

Miller, N. G., and Wilson, R. B. 1967. Electron microscopic study of the relationship of *Leptospira pomona* to the renal tubules of the hamster during acute and chronic leptospirosis. Am J Vet Res 28:225-35.

Miller, N. G.; Froehling, R. C.; and White, R. A. 1970. Activity of leptospires and their products on L cell monolayers. Am J Vet Res 31:371-77.

Miragliotta, G., and Fumarola, D. 1983. In vitro effect of *Leptospira icterohaemorrhagiae* on human mononuclear leukocytes procoagulant activity: Comparison of virulent with nonvirulent strain. Can J Comp Med 4770-73.

Niikura, M.; Ono, E.; and Yanagawa, R. 1987. Molecular comparison of antigens and proteins of virulent and avirulent clones of *Leptospira interrogans* serovar *copenhageni*, strain Shibaura. Zentralbl Bakteriol Hyg A 266:453-62.

Perolat, P.; Grimont, F.; Regnault, P. F.; Grimont, P. A.; Fournie, E.; Thevenet, H.; and Baranton, G. 1990. rRNA gene restriction patterns of *Leptospira*: A molecular typing system. Res Microbiol 141:159-71.

Ramadass, P.; Jarvis, B. W. D.; Corner, R. J.; Penny, D.; and Marshall, R. B. 1992. Genetic characterization of pathogenic *Leptospira* species by DNA hybridization. Int J Syst Bacteriol 42: 215-19.

Segers, R. P. A. M. 1991. The molecular analysis of sphingomyelinase genes of *Leptospiraceae*. PhD diss: University of Utrecht.

Segers, R. P. A. M.; Van Der Drift, A.; De Nijs, A.; Corcione, A.; Van der Zeijst, B.; and Gaastra, W. 1990a. Molecular analysis of a sphingomyelinase C gene from *Leptospira interrogans* serovar *hardjo*. Infect Immun 58:2177-85.

Segers, R. P. A. M.; Van der Drift, A.; Van Eys, G. J. J. M.; De Nijs, A.; Van der Zeijst, B. A. M.; and Gaastra, W. 1990b. The presence among *Leptospira interrogans* strains of DNA sequences similar to the sphingomyelinase gene of a serovar *hardjo* strain. In Proceedings of the Leptospirosis Research Conference 1990, pp. 388-99. ed. Y. Kobayashi. Matsujama, Japan.

Shimizu, T.; Matsusaka, E.; Takayanagi, T.; Masuzawa, T.; Iwamoto, Y.; Morita, T.; Mifuchi, I.; and Yanagihara, Y. 1987. Biological activities of lipopolysaccharide-like substance (LLS) extracted from *Leptospira interrogans* serovar *canicola* strain Moulton. Microbiol Immunol 31:727-35.

Shimono, E.; Yanagawa, R.; and Torres Barranca, G. 1980. Isolation of revertants from antigenic variants of leptospiras. Zentralbl Bakteriol Hyg (I Abt Orig A) 147:392-99.

Stallheim, O. H. V. 1971. Virulent and avirulent leptospires: Biochemical activities and survival in blood. Am J Vet Res 29:1463-69.

Terpstra, W. J.; Korver, H.; Schoone, G. J.; van Leeuwan, J.; Schonemann, C. E.; De Jong-Agilbut, S.; and Kolk, A. H. 1987. The classification of Sejroe group serovars of *Leptospira interrogans* with monoclonal antibodies. Zentralb Bakteriol (I Abt Orig A) 266:412-21.

Thiermann, A. B. 1984. Leptospirosis: Current developments and trends. J Am Vet Med Assoc 184:722-25.

Thiermann, A. B.; and Ellis, W. A. 1986. Identification of leptospires of veterinary importance by restriction endonucelase analysis. In The Present State of Leptospirosis Diagnosis and Control. Ed. W. A. Ellis and T. W. A. Little. Dordrecht: Martinus Nijhoff.

Thomas, D. D., and Higbie, L. M. 1990. In vitro association of leptospires with host cells. Infect Immun 58:581-85.

Thompson, J. C., and Manktelow, B. W. 1989. Pathogenesis of renal lesions in haemoglobinaemic and non-haemoglobinaemic leptospirosis. J. Comp. Pathol. 101:201-214.

Trueba, G. A.; Bolin, C. A.; and Zuerner, R. L. 1992. Characterization of the peroplasmic flagellum proteins of *Leptospira interrogans*. J Bacteriol 174:4761-68.

Tu, V.; Adler, B.; and Faine, S. 1986. Glycolipoprotein cytotoxin from *Leptospira interrogans* serovar *copenhageni*. J Gen Microbiol 135:2663-73.

Van den Ingh, T. S., and Hartman, E. G. 1986. Pathology of acute *Leptospira interrogans* serotype *icterohaemorrhagiae* in the Syrian hamster. Vet Microbiol 12:367-76.

Volina, E. G.; Levina, L. F.; and Soboleva, G. L. 1986. Phospholipase activity and virulence of pathogenic leptospirae. J Hyg Epidemiol Microbiol Immunol 30:163-69.

Waitkins, S. A. 1986. Leptospirosis as an occupational disease. Br J Ind Med 43:721-25.

Walker, E. D.; Arnett, J. K.; Don Heath, J.; and Norris, S. J. 1991. *Treponema pallidum* has a single, circular chromosome with a size of <900 kilobase pairs. Infect Immun 59:2476-79.

Yamaguchi, T.; Ono, E.; Ito, T.; and Yanagawa, R. 1988. Restriction endonuclease DNA analysis of antigeic variants of leptospires selected by monoclonal antibodies. Microbiol Immunol 32:1007-11.

Yasuda, P. A.; Steigerwalt, A. G.; Sulzer, K. R; Kaufman, A. F.; Rogers, F.; and Brenner, D. J. 1987. Deoxyribonucleic acid relatedness between serogroups,and serovars in the family *Leptospiraceae* with proposals for seven new *Leptospira* species. Int J Syst Bacteriol 37:407-15.

Zuerner, R. L. 1991. Physical map of chromosomal and plasmid DNA comprising the genome of *Leptospira interrogans*. Nucl Acid Res 19:4857-60.

Zuerner. R. L., and Bolin, C. A. 1988. Repetitive sequence element cloned from *Leptospira interrogans* serovar *hardjo* type hardjo-bovis provides a sensitive diagnostic probe for bovine leptospirosis. J Clin Microbiol 26:2495-2500.

Zuerner, R. L., and Charon, N. W.; 1990. Nucleotide sequence analysis of the *Leptospira biflexa* serovar *patoc rps*L and *rps*G genes. J Bacteriol 172:6165-68.

Zuerner, R. L.; Knudtson, W.; Bolin, C. A.; and Trueba, G. 1991. Characterization of outer membrane and secreted proteins of *Leptospira interrogans* serovar *pomona*. Microb Pathog 10:311-22.

25 / *Mycoplasma*

BY S. ROSENDAL

NATURE OF MYCOPLASMAS

The trivial name, mycoplasma, is used for organisms of the following general description, even though a more correct term would be mollicutes, since it is intended to denote all organisms in the class *Mollicutes*. Mycoplasmas are prokaryotic organisms capable of independent existence in either aerobic or anaerobic environments. Ultrastructurally, the enclosing plasma membrane is similar to the structure of cell membranes of eukaryotic cells and is made up of lipids and proteins. There are no internal membrane structures and no cell wall external to the plasma membrane; however, many strains possess surface structures equivalent to a capsule. With the exception of acholeplasmas, mycoplasmas depend on a supply of intact cholesterol, which they incorporate into the membrane, creating sufficient osmotic stability for survival under normal physiological conditions. Mycoplasmas are cultured on highly enriched media, which provide carbohydrates and amino acids for energy metabolism and protein synthesis and lipids for membrane synthesis. The growth is slow, with doubling times varying between 1 and 10 hours; doubling time may occasionally be longer, particularly during primary isolation. The genomic division is semiconserved but not always synchronized with cell division, resulting in multigenomic branching filaments. The smallest reproducing cell is approximately 300 nm in diameter.

The genome of members of *Mycoplasmatales* is approximately 800- to 1200-kb pairs, corresponding to approximately 600-800 genes, which is believed to be the bare minimum for independent life and requires essentially monocistronic gene usage. Mycoplasma DNA is remarkably poor in guanine (G) and cytosine (C) (18-40%), and rich in adenine (A) and thymidine (T). Very little of the nucleotide sequence is shared with other bacteria. As expected, the codon usage is rich in uracil (U) and adenine. It is generally not difficult to isolate and clone genetic sequences into *Escherichia coli* strains but it has proven somewhat difficult to express mycoplasma genes in this host. This is possibly due to the difference in codon usage; most mycoplasmas read UGA as tryptophane, whereas *E. coli* reads it as a stop codon. Transposons and extrachromosomal DNA, which should be useful as tools for gene manipulations in mycoplasmas, have been described. Sequence analyses of the 16S rRNA genes have shown the closest relation to members of the *Clostridium* genus and suggest that mycoplasmas constitute degenerate, evolutionary forms (Weisburg et al. 1989).

Mycoplasmas can be divided into helical and nonhelical organisms. The former belong in the Order *Spiroplasmatales*, which inhabits plants and arthropods; many of them are important plant pathogens. *Spiroplasma mirum*, isolated from ticks, is pathogenic for vertebrates inoculated in the laboratory but is not associated with natural diseases of higher animals. The nonhelical mycoplasmas, which do not require cholesterol, are members of the order *Acholeplasmatales*. Although they are commensals of animals, none are significant pathogens under natural conditions. The cholesterol-dependent, nonhelical mycoplasmas belong to *Mycoplasmataceae*, which comprises the genera *Mycoplasma* (urease negative) and *Ureaplasma* (urease positive). Many of the approximately 95 *Mycoplasma* and *Ureaplasma* species in humans and animals are pathogenic.

Following successful isolation on appropriate media, mycoplasmas are identified on the basis of their biochemical reactivity and antigenic makeup. The latter requires panels of reference antisera raised against representative strains of each species. These sera recognize strains of the homologous species in immunoenzyme, fluorescent-antibody, and growth-inhibition assays. The inhibition of growth likely involves steric interference with certain enzymes in the membrane, because complement appears not to be required. A detailed taxonomic description of mycoplasmas is available (Razin and Freundt 1984).

Mycoplasmas are hostdependent, extracellular parasites. The latter quality may be questionable because recent evidence suggests that certain strains of *M. fermentans* and the new species *M. penetrans* may be intracellular potential pathogens in HIV-infected people.

Mycoplasmas are very fragile outside the host; the organisms are highly susceptible to heat, detergents, disinfectants, and antimicrobial drugs that inhibit transcription, translation, or the DNA gyrase enzyme. They are constitutively resistant to drugs with a primary effect on bacterial cell-wall synthesis or cell-wall integrity. Mycoplasmas inhabit mucosal surfaces and are commonly isolated from conjunctiva, nasal cavity, oropharynx, larynx, and intestinal and genital tracts. Some mycoplasmas may have distinct tropism for one particular anatomical site, others may occur in a variety of sites.

With some exceptions, mycoplasmas are host-specific organisms. This adaptation may be based on specific colonization factors, for which the host provides the receptors, or it may be based on the inability of the natural host to recognize and respond to the parasite.

PATHOGENESIS

From a purely pathogenetic point of view it seems reasonable to categorize mycoplasma diseases as (I) diseases where septicemia is the important feature; (II) diseases that involve a usually unrecognized phase of mycoplasma dissemination through the blood, followed by localization and inflammation in serosal cavities and/or joints; and (III) local diseases of the respiratory tract, genital tract, mammary gland, or conjunctiva (Table 25.1). The group III diseases may result from uncontrolled local proliferation of the mycoplasmas and spread along mucosal surfaces to more sensitive sites of the system or organ. Dissemination via the bloodstream is not the typical feature but may occur on occasion.

The septicemic diseases are acute with fever, often leading to death within a few days. The remaining diseases are generally subacute to chronic and characterized by persistent infection in the presence of an inflammatory disease process.

For many strongly host-adapted microorganisms, disease, as a result of infection, is the exception rather than the rule. However, it is possible that certain disease manifestations may facilitate transmission, and therefore be of survival advantage to the microorganism. Coughing and sneezing conceivably promote aerosolization of respiratory secretions and thereby promote the spread of respiratory pathogens.

Mycoplasma diseases are typical examples of multifactorial diseases, where factors such as intercurrent infections, crowding, inclement climatic conditions, age, genetic constitution, and stress from transportation, handling, and experimentation are important determinants of the final outcome of infection.

TABLE 25.1. Diseases in animals caused by mycoplasmas

Primary Disease	Host(s)	Agents	Other Manifestations
I			
Septicemia	Goats, sheep	*M. mycoides* subsp. *mycoides* LC type[a]	Polyarthritis, pneumonia, mastitis, conjunctivitis
		M. capricolum	Arthritis, mastitis, pneumonia
II			
Polyserositis-arthritis	Swine	*M. hyorhinis*	Pneumonia (?)
Tenosynovitis, arthritis	Chickens, turkeys	*M. synoviae*	Airsacculitis
Arthritis	Cattle	*M. bovis*	Mastitis, pneumonia
	Swine	*M. hyosynoviae*	
	Rats	*M. arthritidis*	
III			
Sinusitis	Chickens, turkeys	*M. gallisepticum*	Arthritis
Airsacculitis	Turkeys	*M. meleagridis*	Osteodystrophy
Pneumonia	Cattle	*M. dispar*	
		M. bovis	Mastitis, arthritis
		Ureaplasma diversum	Reproductive diseases
	Sheep, goats	*M. ovipneumoniae*	
		M. capricolum	Septicemia, arthritis, mastitis
		M. mycoides subsp. *capri*	
	Swine	*M. hyopneumoniae*	
	Dogs	*M. cynos*	
	Mice, rats	*M. pulmonis*	Salpingitis, endometritis
Pleuropneumonia	Cattle (CBPP)[c]	*M. mycoides* subsp. *mycoides* SC type[b]	Arthritis in calves
	Goats (CCPP)[d]	*M.* sp. F-38	
Pleuritis	Horses	*M. felis*	
Vulvitis	Cattle	*Ureaplasma diversum*	Infertility, abortion, pneumonia
Seminal vesiculitis	Cattle	*M. bovigenitalium*	Reduced sperm motility, mastitis
Mastitis	Cattle	*M. bovis*	Pneumonia, arthritis
		M. californicum	
		M. canadense	
		M. alkalescens	
	Sheep, goats	*M. agalactiae*	Septicemia, arthritis, kerato-conjunctivitis
		M. capricolum	As for *M. agalactiae*
		M. mycoides subsp. *mycoides* LC type	As for *M. agalactiae*
	Goats	*M. putrefaciens*	
Conjunctivitis	Cattle	*M. bovoculi*	
	Sheep, goats	*M. conjunctivae*	
	Cats	*M. felis*	
	Mice	*M. neurolyticum*	"Rolling disease"

[a] *Mycoplasma mycoides* subspecies *mycoides*, large colony type. [b] *Mycoplasma mycoides* subspecies *mycoides*, small colony type. [c] CBPP = contagious bovine pleuropneumonia. [d] CCPP = contagious caprine pleuropneumonia.

Septicemia

The large colony (LC) type of *M. mycoides* subsp. *mycoides* (*M. mycoides*) enters the bloodstream when the animal's resistance is compromised or as a sequel to conjunctivitis, pneumonia, or mastitis. Outbreaks of septicemia have been observed in kids exposed to large doses in milk of mastitic does excreting this organism. The disease is usually fatal, with necropsy changes characterized by fibrinopurulent polyarthritis, embolic pneumonia, renal infarcts, and thromboembolic lesions in various tissues resembling disseminated intravascular coagulation. Animals surviving the septicemia may develop chronic destructive arthritis in one or more joints.

M. mycoides, LC type, is recovered from nasopharyngeal secretions, milk, and the external ear canal of susceptible animals during outbreaks. The ear may be an important reservoir and port of entry. Ear mites may play a role as vehicles for transmission of the microorganism and for breaking the barrier to the bloodstream.

M. capricolum is occasionally isolated from outbreaks of septicemia in sheep and goats, similar in clinical and pathological features to septicemia caused by *M. mycoides*, LC type. The external ear canal may also serve as habitat of this mycoplasma, but the organism is rarely isolated from healthy animals.

Polyserositis, Tenosynovitis, and Polyarthritis

M. hyorhinis inhabits the upper airways and is a common secondary agent in porcine pneumonia. Before or around weaning age, this mycoplasma may cross the epithelial barrier and disseminate, provoking acute inflammation in serosal and synovial cavities. The major weight-bearing joints seem to be most severely affected. The disease is rarely fatal but chronic fibrosing pleuritis, pericarditis, and peritonitis are often seen in slaughtered pigs; the disease has a negative effect on growth performance. In acute arthritis, the joints are distended with fibrinous serohemorrhagic fluid containing lymphocytes, neutrophils, and exfoliated synovial cells. In the chronic phase, the synovial membranes undergo nonsuppurative proliferative changes with marked mononuclear infiltration originating around subsynovial capillaries. Pannus formation and erosion of cartilage are common but rarely amount to permanent impaired function. Mycoplasma antigen may persist in degenerated and collapsed cartilage lacunae in the face of high local and systemic antibody titers.

Arthritis caused by *M. hyosynoviae* occurs in growing pigs of heavy breeds between 30- and 100-kg bodyweight and often in association with movement. The inflammation is similar to *M. hyorhinis* arthritis but milder and regresses after a few weeks. *M. hyosynoviae* resides in tonsils and nasal cavity.

Tenosynovitis and arthritis follow localization in the synovial cavities when *M. synoviae* enters the blood from the airsacs in chickens and turkeys. Inflammation is acute and suppurative with destruction of synovial membranes, cartilage, and eventually bone. In cattle, *M. bovis* arthritis may be secondary to infection of the respiratory tract or mammary gland. The infection may persist in affected joints where synovial antibodies can reach high titers.

M. arthritidis causes an acute arthritis in rats affecting the distal joints of the extremities but occasionally also vertebral joints. The inflammation usually subsides after 1-2 months. Experimental *M. arthritidis* arthritis in mice has served as a model for rheumatoid arthritis in humans because of clinical and pathomorphological similarities.

Sinusitis, Airsacculitis

M. gallisepticum causes a severe inflammation of the sinuses, trachea, and lungs and airsacs of chickens and turkeys, often referred to as chronic respiratory disease (CRD). There is an accumulation of mucus and exudate in the airways, epithelial damage, and mononuclear inflammatory-cell infiltration. Sneezing and coughing birds spread the infection horizontally. Infection of the genital tract may result from blood-borne dissemination and lead to egg transmission. Suppurative arthritis is occasionally seen as a sequel to CRD.

M. meleagridis causes a milder respiratory disease, essentially confined to airsacculitis of turkey poults. Mild arthritis, synovitis, and occasionally osteodystrophy of tarsal bones can be seen in turkeys with *M. meleagridis* infection. Transmission is similar to that described for *M. gallisepticum*, but in addition, cloacal infection of both sexes is common and can lead to venereal transmission.

M. synoviae infects both chickens and turkeys causing airsacculitis and sinusitis from which site it spreads to joints, tendon sheets, and bursae.

Pneumonia, Pleuropneumonia, Pleuritis

Bronchitis, bronchiolitis, and alveolitis with predominantly neutrophils and mononuclear cellular response constitute the very early inflammation in mycoplasma pneumonias. Along with the persistent infection of the airways, this pathomorphology progresses toward an interstitial pneumonia. Inflammatory cells in the chronic phase are mainly lymphocytes, plasma cells, and macrophages. The bronchoalveolar lymphoid tissue (BALT) undergoes proliferation forming the so-called "cuffs" around the bronchioles, a characteristic of chronic mycoplasma pneumonias except pneumonia in calves caused by *M. dispar*, which does not seem to have the lympho-stimulatory capacity of other mycoplasmas. The cranio-ventral distribution of mycoplasma pneumonia indicates that gravitation is an important factor in its development. Bronchiectasis is a typical feature of pneumonia in mice and rats caused by *M. pulmonis*.

Contagious bovine pleuropneumonia (CBPP) and contagious caprine pleuropneumonia (CCPP) are characterized by substantial pulmonary necrosis, sequestration, and marked serosanguinous fluid accumulation in interstitia and pleura. Vasculitis appears to be an important component of the pathological changes in these diseases, explaining the marked exudation and pleurisy. Thrombosis can explain ischemic necrosis and infarcts of the lung. Infection of older cattle with *M. mycoides* (SC type) is confined to the respiratory tract and typical CBPP, whereas infection of calves may result in dissemination and arthritis.

M. felis is an occasional cause of a self-limiting pleuritis in horses. The organism is harbored in the upper respiratory tract and in the genital tract and apparently enters the pleural cavity after exhaustive exertion. The immune response to the mycoplasma appears to be responsible for spontaneous recovery.

Vulvitis

A self-limiting inflammation of the vulva occurs in cattle infected with *Ureaplasma diversum*. Vaginal discharge is present for a few weeks and prominent granules composed of hyperplastic lymphoid tissue are observed on the vulvar mucosa. If the organism is introduced into the uterus during insemination it may induce a mild endometritis-salpingitis and hinder fertilization and implantation. This ureaplasma is also a sporadic cause of placentitis, amnionitis, and fetal pneumonia seen in aborted calves or premature weak calves.

Seminal Vesiculitis

M. bovigenitalium is very common in the prepuce and distal part of the urethra of bulls, and it may occasionally spread to the seminal vesicles and epididymes and elicit an inflammatory response. Persistent mycoplasma shedding in semen and chronic fibrosing inflammation seem to characterize the natural cases. Semen contaminated with *M. bovigeni-*

talium may have reduced motility as a result of adhesion to the spermatozoa and seems to have reduced freezeability. *M. bovigenitalium* is also a rare cause of mastitis.

Mastitis

The most common cause of mycoplasma mastitis in cattle is *M. bovis*. Under experimental conditions, only a few organisms are required to induce mastitis. The mastitic glands are swollen, and the milk yield is dramatically decreased while the somatic cell count is high. The inflammatory secretion consists of fibrin clots and flakes in a clear whey. Histologic changes in the acute phase are characterized by subepithelial hemorrhages and neutrophilic alveolitis. In the chronic phase, lymphocytes, plasma cells, and macrophages predominate in the alveoli and the interstitial tissue. Persistent infections are common, and flare-ups may occur in subsequent lactations. The infection spreads in dairy herds by direct and indirect contact; to control requires culling of carriers. Spontaneous mastitis in cattle due to *M. bovigenitalium, M. canadense, M. alkalescens*, and *M. californicum* are similar but generally milder in their clinical course.

Contagious agalactia in sheep and goats is caused by *M. agalactiae*. The organism seems adapted to the mammary gland but can also establish in the intestinal tract. Severe mastitis and agalactia typically occur shortly after parturition, when stress may tip the balance in favor of the mycoplasma. *M. agalactiae* is invasive and enters the circulation in many animals, resulting in arthritis. The pathomorphological changes in the gland or joints depend on the stage of inflammation but seem to be nonspecific.

Mastitis in goats caused by the LC type of *M. mycoides* is very severe, often progressing to sepsis and death. There is vascular thrombosis, resulting in necrosis of lactogenic tissue. *M. putrefaciens* mastitis in goats is mild and selflimiting, although one outbreak with high morbidity and mortality among goats in a dairy herd with poor management practices has been described.

Conjunctivitis

M. bovoculi is part of the etiologic complex of infectious bovine keratoconjunctivitis. Persistent infection of the bovine eye with pure cultures results in a mild conjuntivitis with edema, hyperemia, and serous discharge. The inflammation may go through several peak episodes with intervals of regression. *M. bovoculi* infection predisposes for the more severe infection with *Moraxella bovis*. In cattle, *M. bovoculi* infection does not seem to involve the cornea. This is in contrast to the *M. conjunctivae* infection in sheep and goats, which results in keratitis as well as conjunctivitis. The inflammation is usually selflimiting after 1-2 months and does not permanently damage the eye. *M. felis* is associated with conjunctivitis in cats, but the disease has not been reproduced convincingly.

Conjunctivitis is the natural disease with which *M. neurolyticum* has been associated. The interest in this mycoplasma stems from the fact that it is the only mycoplasma known to secrete an exotoxin. Intravenous injection of this toxin into mice elicits a neurologic syndrome called "rolling disease."

VIRULENCE

Virulence is the capacity of an organism to overcome natural defenses of the host and induce disease. Considering the marked genetic diversity among mycoplasmas, the subject of virulence should perhaps be treated for each species separately. However, some attributes may be general even though they have only been studied in a few.

It is quite well established that virulence varies among strains within the same mycoplasma species. The dose of *M. pulmonis* required to induce gross pneumonic lesions in 50% of challenged mice can vary from 10^3 colony-forming units (CFU) to $>10^7$ (Davidson et al. 1988). A highly variable surface-expressed protein may play an important role in the ability of this respiratory murine mycoplasma to establish and persist in the lung. The capacity to vary antigenic make-up allows the parasite to escape immune surveillance. *M. hyorhinis* has been shown to possess similar variable surface lipoproteins, which are recognized by the immune system of the pig (Rosengarten and Wise 1991). The shift in antigenic repertoire within a dominant surface- expressed molecule is a recently recognized phenomenon and seems to apply to all the mycoplasma species that have been examined for this property.

Virulence attenuation occurs readily among mycoplasmas. Pneumonia in pigs is difficult to reproduce with strains of *M. hyopneumoniae* propagated in the laboratory for more than a few passages. The attenuation seems related to the loss of a surface layer (capsulelike material) and may be regained along with virulence by animal passage.

Many of the mycoplasmas listed in Table 25.1 have the ability to invade the bloodstream. It is possible that invasion occurs frequently from local infections, but in healthy animals and immune animals these mycoplasmas are cleared rapidly. The circumstances permitting the penetration of the epithelial barriers are not well understood, but experience suggests that disease, as a result of invasion, is related to general suppression of host resistance or immature immune response. Immunologic condition may play a role in susceptibility, as aerosol exposure of young gnotobiotic pigs results in dissemination, whereas intranasal exposure of healthy conventional, but *M. hyorhinis*-free, pigs does not. However, if the mucosal barrier is circumvented by either intravenous or intraperitoneal inoculation, mycoplasmas will be present in the blood for several weeks.

The LC type of *M. mycoides* readily invades both the intestinal tract and the respiratory tract. *M. bovis* is able to enter the mucosal surface of the airways into pulmonary tissue and produce suppurative lesions.

In T cell-suppressed mice exposed intranasally to *M. pulmonis*, mycoplasmas are present in tissues outside the lung more frequently than in normal mice. The SC type of *M. mycoides* produces a galactan-polymer that seems, at least in calves, to modulate the immune function and promote dissemination of a primary pulmonary infection.

Mycoplasmas are generally regarded as extracellular parasites or commensals; however, some mycoplasmas isolated from human patients with HIV infection have been located intracellularly. Invasion of cells in vitro has also been shown for these AIDS-associated mycoplasmas. Whether cell invasion occurs and whether it contributes to the virulence of animal mycoplasmas need investigation.

The mucosal or epithelial habitation of animal mycoplasmas necessitates attachment mechanisms. Cytoadsorbtion and cytoagglutination reactions constitute useful in vitro models for attachment studies. Furthermore, it is possible to study attachment to explants of epithelium from the natural site of infection. Such studies have invariably shown that adhesion is mediated by proteins or protein conjugates.

Some mucosal colonizers do not have specialized structures for attachment. Others, such as *M. dispar* and *M. hyopneumoniae,* have anionic surface layers contributing to attachment, and *M. gallisepticum, M. iowae,* and *M. pneumoniae* possess tip structures, which face the host epithelial cells (Fig. 25.1).

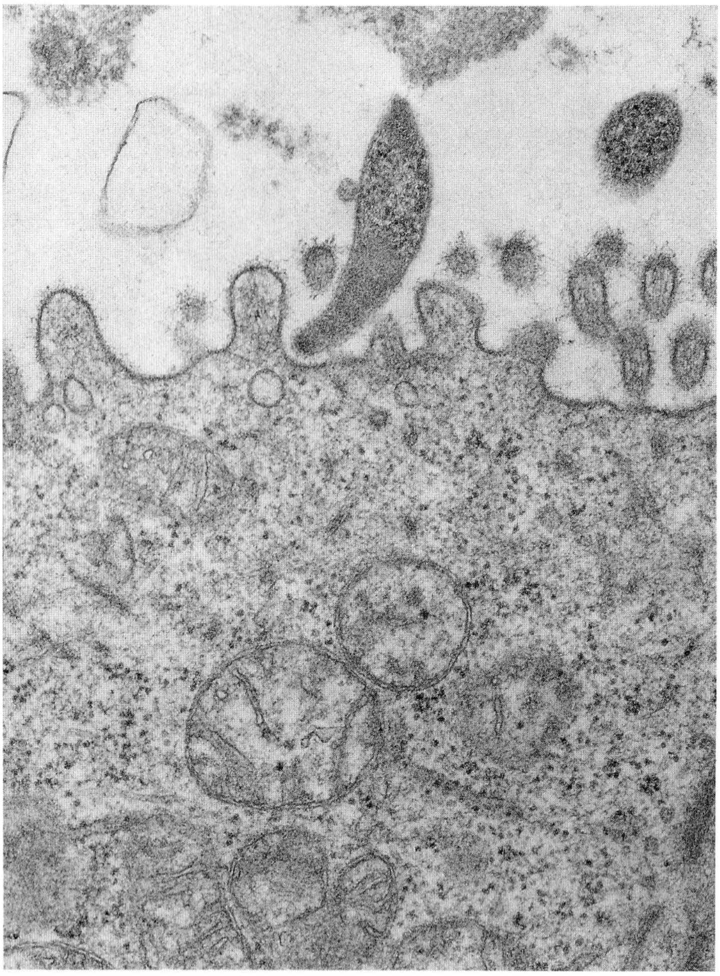

FIG. 25.1. *Mycoplasma iowae* attaching to intestinal cell of chick embryo. The tip structure contains fine granules and is oriented toward the host cell.

The important human species, *M. pneumoniae*, has been studied in detail and found to possess several proteins that are located on the tip structure and are involved in attachment. The most prominent is the P1 protein of 170 kDa molecular mass (Razin and Jacobs 1992). Antibodies to these proteins inhibit attachment and pathogenicity in a hamster pneumonia model. *M. gallisepticum* shares antigenic epitopes on a 155-kDa protein with the P1 adhesin of *M. pneumoniae*. Homologous sequences of the P1 gene have been found in *M. gallisepticum* and *M. genitalium* (human mycoplasma), suggesting a common ancestral adhesin gene.

The receptor counterparts are glycoconjugates and in some cases sialoglycoconjugates. Recently, sulphated glycolipids have been shown to function as receptor moieties for human, bovine, and murine ureaplasmas and mycoplasmas able to colonize the genital tract (Lingwood et al. 1990). The receptor for *M. hyopneumoniae* on cilia of porcine respiratory epithelial cells likewise appears to be sulphated glycolipid.

Toxicity and Toxins

Many of the mycoplasmas that cause pneumonia adhere to ciliated epithelium and induce ciliostasis, loss of cilia, and cytopathic changes of epithelial cells. The basis for ciliostasis may be mycoplasma-induced ciliary membrane alterations and disruption of the ciliary necklace. The contact-cytotoxicity is related to peroxidation but may also be more nonspecific in nature, involving competition for essential nutrients, and effects of mycoplasma proteases, nucleases, and lipases.

Many mycoplasmas produce peroxide and superoxide, which impair host cell-membrane integrity while at the same time inactivating host cell catalase and superoxide dismutase. Interestingly, most epithelial surface colonizers produce their own superoxide dismutase, which may be protective and enhance their ability to survive in the host. *M. equigenitalium* is occasionally associated with abortion in horses and produces ciliary damage and cytotoxicity to Fallopian tube explants. This effect can be reversed by catalase. Although associated with infertility and abortion, *Ureaplasma diversum* does not seem to damage bovine oviductal cilia during colonization, but leakage of calmodulin from colonized cells is taken as evidence for a cytotoxic effect. The role of the urease enzyme of ureaplasmas needs to be investigated. Ammonia produced in the microenvironment of the organism may have detrimental effects on host cells.

M. mycoides, SC-type, was shown 30 years ago to possess endotoxic activity. A phenol-water extract was pyrogenic, chick embryo lethal, and it induced leukopenia (Villemot et al. 1962). This finding needs reevaluation because of the potential risk of bacterial endotoxin contamination of mycoplasma extracts made from organisms grown in very rich and complex media. It is now clear that endotoxicity is primarily mediated by IL-1, tumor necrosis factor (TNF), and IL-6 released from CD14-positive mononuclear cells. Evidence is also mounting that many mycoplasmas can induce one or more of these mediators although the molecules with this activity are not always known. Lipoglycans are found in many mycoplasma species, including the nonpathogenic acholeplasmas, thus stressing the need for more studies on their contribution to virulence. The lipoglycan of *M. mycoides*, both LC and SC type, has been characterized as a beta 1-6 galactofuranose (galactan) with phosphate and lipid substitution. Calves, sheep, and goats respond to this galactan with increased pulmonary resistance, pulmonary edema, and in some cases pulmonary capillary thrombosis. The disease in goats caused by the LC type of *M. mycoides* is acute with high fever, coagulopathy and widespread thrombosis (DIC) and therefore resembles gram-negative bacterial sepsis. Direct endothelial cell cytotoxicity may be part of this effect, and endotoxicity of this organism would seem likely, but is unsubstantiated.

The mouse lethal effect of *M. fermentans* (associated with AIDS), given in large doses either as whole cells or membranes, is similar to endotoxic shock and can be enhanced by drugs known to increase sensitivity to endotoxin. Extracts of *M. fermentans* grown in pyrogen-free media and ascertained to be free of endotoxin contamination have recently been shown to induce IL-1, TNF, and IL-6 from activated macrophages, thus offering a plausible explanation for the mouse lethal effect observed many years earlier (Muhlradt et al. 1991).

A protein (200 kDa) isolated from the supernate of *M. neurolyticum* cultures causes lethality within a few hours of intravenous inoculation of mice. The syndrome has been labelled "rolling disease" because the mice roll around their long axis before they die. There is vascular damage in the brain and dramatic perivascular astrocytic hypertrophy, but the biochemical events underlying these changes are unknown. A similar effect has been observed in turkey poults inoculated intravenously with large numbers of *M. gallisepticum*, but a toxin has not been identified.

A toxin isolated from the membrane of *M. bovis* has been shown to elicit inflammation in the bovine mammary gland and in the skin of guinea pigs. The toxin is a protein-conjugated 73-kDa carbohydrate (Geary et al. 1981).

M. hyopneumoniae is cytotoxic for porcine pulmonary fibroblasts. Two membrane proteins appear to be responsible for the cytotoxicity, which can be neutralized by antibody.

Mycoplasma phospholipases are potentially important in pneumonia for they may reduce surface tension of the alveolar surfactants, thus resulting in atelectasis.

HOST INTERACTION

Nonspecific Resistance

The mucous blanket and the mucosal epithelium form an efficient barrier between the resident microbial flora and the underlying tissues. Many of the respiratory tract pathogens exhibit a peculiar gliding motility, and it has been suggested that this may enable them to penetrate the mucous blanket. The ability of mycoplasmas to adhere to a variety of cells strongly suggests that the normal mycoplasma flora is closely associated with epithelial cell membranes. It is probably safe to assume that breaking the anatomical and physiological integrity of the epithelial barrier is the initial step in disease (Stadtlander et al. 1991).

Pathogenic mycoplasmas readily adhere to neutrophils and macrophages, but without specific antibodies or alternative means of opsonization phagocytosis does not proceed. The interaction of rat alveolar macrophages and *M. pulmonis* appears to be the exception to this rule, since their killing efficiency is stronger than mouse alveolar macrophages in the absence of opsonins. Opsonins, present in bronchoalveolar fluid but not yet fully characterized, appear to facilitate pulmonary clearance in the mouse lung (Davidson et al. 1990). The ability of *M. pulmonis* to resist phagocytosis by mouse macrophages has been linked to a trypsin-sensitive membrane component and correlated with pneumogenic virulence.

Several mycoplasmas are able to impair phagocytic function of neutrophils or macrophages. Peritoneal macrophages from *M. arthritidis*- or *M. hominis*-infected rats had reduced killing capacity when challenged with *E. coli*. Similar results have been noticed when bovine neutrophils were exposed to *M. bovis* or *M. dispar*. *M. bovis* adheres to bovine neutrophils and inhibits the respiratory burst. The inhibition is at the level of or distal to protein kinase C in the signal pathway of the cell (Thomas et al. 1991). The antiphagocytic property of *M. dispar* resides in the capsule, which suppresses IL-1 and TNF release, and glucose-consumption of bovine alveolar macrophages stimulated by endotoxin (Almeida et al. 1992). Phagocytic function of porcine alveolar macrophages is impaired when exposed to *M. hyopneumoniae* and *Actinobacillus pleuropneumoniae*, and susceptibility to *Pasteurella multocida* pneumonia is enhanced in pigs infected with *M. hyopneumoniae*.

Recent studies using the murine mycoplasma pneumonia model has revealed that the natural killer (NK) cells of both spleen and lung mediate natural resistence. The mycoplasma-inhibiting effect is enhanced 3-7 days after infection and seems to be effected both directly by the NK cells as well as indirectly by the production and release of interferon-gamma (IFN-gamma), which stimulates and activates macrophages (Lai et al. 1990). Environmental stress factors and pollutants (NH_3 and NO_2) increase susceptibility to mycoplasma infections, presumably by impairing host defenses, such as macrophage function and NK cell activity (Simecka et al. 1992).

The basis for the host specificity of mycoplasma infections is poorly understood, but may be related to the availability of specific receptors or the inability of the host to recognize the organisms as nonself. Individual differences in susceptibility to colonization may be hormone dependent, as has been demonstrated in mice infected experimentally with a number of mycoplasma species. Genetically determined differences in susceptibility within host species have been recognized in the case of *M. pulmonis*, where Lewis rats develop marked pneumonia but Fisher 344 rats do not. Pigs of the Piney Wood miniature strain are less susceptible to *M. hyorhinis* serositis-arthritis than are Yorkshire pigs.

Inflammatory Response

Direct activation of inflammatory-response mediators is poorly researched in mycoplasmology. Potent tissue-necrotizing toxins are not produced and diseases are often subclinical and chronic. The subtle injury caused by peroxidation undoubtedly initiates local inflammation aimed toward repair. Morphologically, the typical reactions to mycoplasma infections are proliferative, suggesting powerful chemoattractants and mitogenic factors.

Activation of Complement. Mycoplasma-specific antibodies and complement result in lysis of mycoplasmas due to activation of the complement cascade on the membrane surface. Damage may extend to host cells where these are in intimate contact with mycoplasmas.

Arthritogenic mycoplasmas or their antigens may persist in inflamed joints in the presence of immunoglobulin. The perpetuation of inflammation may be based on immune-complex hypersensitivity, since antigen, immunoglobulin, and complement are present. Potential immune-complex-mediated inflammation may also be elicited by mycoplasmas with immunoglobulin Fc receptors, a possible scenario in *M. synoviae* tenosynovitis-arthritis (Lauerman and Reynolds-Vaughn 1991). The histologic similarity and the fact that rheumatoid factor (IgM antibodies against IgG) is demonstrable in some of the chronic mycoplasma arthritides has made them interesting models for rheumatoid arthritis in humans.

The LC type of *M. mycoides* can activate the classical complement pathway in the abscence of detectable antibodies; activation may result in release of anaphylatoxins (C3a and C5a), initiation of inflammation, and damage to bystander cells. Conversely, activation of complement may be a beneficial host reaction leading to opsonization and killing of the mycoplasma. *M. pulmonis* activates complement and attracts macrophages to the murine lung during infection. The inflammatory glycoprotein of *M. bovis* activates complement, which may explain its effect. *M. pneumoniae* activates complement via the classical, as well the alternative, complement pathway, releasing C3a and C5a and potentially mediating inflammation.

Low molecular weight chemoattractants for neutrophils have been recognized in the case of *M. arthritidis*, which also appears to cause neutrophil aggregation. *M. pulmonis* produces a potent membrane protein with direct chemotactic activity for B cells.

It is now clear that many mycoplasmas can induce primary inflammatory-response mediators in stimulated or nonstimulated macrophages. A role for TNF has been established in murine mycoplasma pneumonia, where TNF is recognized locally after experimental infection with *M. pulmonis*, and neutralization with antibodies to TNF abrogates the inflammation (Faulkner et al. 1992). A heat-stable membrane factor of *M. capricolum* induces TNF in murine macrophages in a fashion similar to bacterial endotoxin. Furthermore, macrophages from lipopolysaccharide (LPS) nonresponsive mice (C3H/HeJ) are equally responsive, excluding the possibility of contamination and suggesting an alternative signal mediation. The role of TNF in disease caused by this organism is not known.

M. fermentans stimulates IFN-gamma-activated macrophages to release TNF, IL-1, and IL-6, with subsequent stimulation of T-helper cells (both 1 and 2) and activation of cytotoxic T lymphocytes. The potential for further stimulation of host phospholipase A_2 is obvious and suggested by the simultaneous demonstration of prostaglandin (Muhlradt and Schade 1991). Some mycoplasmas produce phospholipase A_2, but the possible triggering effect on the arachidonic acid prostaglandin cascade needs study (Bhandari and Asnani 1989).

Immune System Response

Stimulation. Mycoplasma infections give rise to specific humoral and cell-mediated immune (CMI) responses. Antibodies are produced both locally and systemically. CMI responses can be demonstrated as skin hypersensitivity to appropriate antigens or by in vitro lymphocyte stimulation and lymphokine production.

Mycoplasmas may adhere intimately to lymphocytes and eventually form a cap similar to lectin-induced capping. The interaction may actually proceed to complete fusion of mycoplasma and lymphocyte (Franzoso et al. 1992). Therefore, it is not surprising, that this intimate relationship can have profound effects on lymphocytic function.

In many mycoplasma diseases, particularly the chronic pneumonias, there is a dramatic lymphoplasmocytic proliferation in the interstitial tissue, which is partly due to antigen-specific humoral and CMI response and partly due to nonspecific mitogenic activity of the inciting mycoplasma. Several mycoplasma species, including those causing pneumonia, are mitogenic for B and T cells, but some nonpathogenic species are active as well. In the case of *M. pulmonis*, the mitogen has been shown to cause alveolitis in the lungs of experimentally challenged mice. In those organisms, where polyclonal B-cell stimulation occurs, antibodies are produced to a variety of irrelevant antigens, including self antigens. *M. pneumoniae* infection in humans gives rise to antibodies against group I erythrocytes, lymphocyte antigens, brain, lung, smooth muscle, and mitotic spindle apparatus. This autoimmunity leads to disease in a small but significant proportion of *M. pneumoniae* patients.

Mitogenic response of lymphocytes from Lewis rats is much higher than the response of Fisher 344 rats and correlates with their susceptibility to *M. pulmonis* pneumonia. However, the difference in mitogenic response is not confined to mycoplasma mitogens, but seems to be a general characteristic of the two strains of rats.

It has long been known that *M. arthritidis* secretes a factor, *M. arthritidis* mitogen (MAM), with marked T-cell mitogenic activity. This factor is now known to be a small peptide with the characteristics of superantigens, i.e., antigens bridging the major histocompatability complex class II (MHC II) molecule on antigen-presenting cells and certain V beta regions of the T-cell receptors (Cole and Atkin 1991). The MAM molecule does not need processing by the presenting cells. The superantigen stimulates a broad section of the T-cell population of both CD4 and CD8 specificity. This results in activation of cytotoxic T cells, stimulation of B cells through the helper activity, and release of biologically active cytokines, with potential for acute shocklike reaction, self-reactivity, and generation of inflammation.

Several mycoplasma species have been shown to upregulate MHC expression of macrophages and B cells, an effect facilitating immune response and with the potential of inadvertant response to self antigens (Stuart et al. 1989).

The galactan produced by LC and SC strains of *M. mycoides* shares epitopes with pneumogalactan of cattle, suggesting that an immune response to the mycoplasma galactan could elicit an autoimmune hypersensitivity reaction in the lung. Stress-response proteins have also been found in mycoplasmas and suggest a possible role in autoreactions.

Suppression. The large number of chronic mycoplasma diseases and persistent infections suggest that the host response somehow is inadequate. It is interesting that many mycoplasmas with an immunostimulatory activity also have a suppressive effect, such as decreased response to standard lymphocyte mitogens. In addition, there is evidence of shared antigens between host and mycoplasma, a phenomenon called "biological mimicry," which may reduce the ability of the host to recognize the organism as foreign, thus permitting it to evade recognition. *M. arthritidis* shares antigens with the rat and *M. mycoides* shares the galactan antigen with a pneumogalactan in cattle. The stress proteins of mycoplasmas and their hosts may fall under this category as well.

The MAM superantigen of *M. arthritidis* is strongly suppressive by suppressing T-H1 function and thereby IL-2 secretion. T-H2 function, on the other hand, is stimulated resulting in increased IL-4 and IL-6 secretion and antibody production.

Immunosuppression has been recognized in vivo in cattle in the terminal stages of CBPP, in chickens immunized with inactivated suspensions of *M. gallisepticum*, and in humans with *M. pneumoniae* disease. The capsule of *M. dispar* suppresses the ability of bovine alveolar macrophages to produce IL-1 and TNF, which presumably slows or reduces T- and B-cell response. Interestingly, the bronchiolitis in calves with *M. dispar* lacks the lymphoproliferative feature of other mycoplasma pneumonias.

The close association between mycoplasma and host lymphocytes may lead to membrane fusion and altered lymphocytic function. Another result of this interaction may be acquisition of host antigens. It has been shown that *M. hyorhinis* acquired T-cell marker antigen from lymphocytes grown in vitro. Whether this occurs in vivo is not known, but it would have the potential of allowing the mycoplasma to escape immune recognition. Conversely, it could trigger autoimmune reactions.

Protease specific for human IgA 1 has been reported for *Ureaplasma urealyticum* (human organism) but has not been found in other *Ureaplasma* or *Mycoplasma* species.

Vaccination

It is probably safe to perceive mycoplasmas in general as extracellular parasites, and it should therefore be expected that the humoral immune response plays an important role in resistance. This seems to be the case in all instances except in *M. pulmonis* infection of rats. In mice, adoptive transfer of serum is protective, but in rats, only transferred spleen cells confer protection (Lai et al. 1991a)

Animals convalescing from a mycoplasma infection are generally resistent to rechallenge, although this resistence is of short duration in mastitis. There should, therefore, be at least a theoretical basis for vaccine development. However, it is important to analyse immune responses and antigens and dissect out those with only protective activity, because many mycoplasma diseases may have a hypersensitivity component.

In the case of *M. pneumoniae*, most attention has been paid to the adhesion proteins, in particular the P1 molecule, where oral or respiratory vaccination confers protection in a hamster pneumonia model.

An interesting approach was taken to develop a vaccine for *M. pulmonis* infection in mice and rats. A monoclonal antibody inhibiting mycoplasma growth and attachment to fibroblasts and sheep erythrocytes was selected and shown to also inhibit colonization and disease in the lungs of mice. A 66-kDa protein antigen of *M. pulmonis* with affinity for this monoclonal antibody was subsequently purified and shown to be protective. Furthermore, recombinant protein expressed in *E. coli* was shown to have preserved epitopes and to be protective (Lai et al. 1991b).

Efforts are being made to control the most economically important diseases in food-producing animals by means of vaccination. Various preparations of *M. hyopneumoniae* cells, including formaldehyde-inactivated whole cells, freeze-thaw saline extracts, and culture supernates in Freund's incomplete adjuvant, gave some protection. The protective immunogen appeared stable at 80°C (Ross et al. 1984). There are now several *M. hyopneumoniae* vaccines on the market that reduce the degree of pneumonia in pigs; however, they do not eliminate disease or infection. There is evidence to suggest that the inactivation method affects the protective immunogenicity of these vaccines. Further analyses of virulence and adhesion-associated antigens will pave the way for more refined vaccines. An 85-kDa protein of *M. hyopneumoniae* produced by recombinant DNA technology has been shown to induce significant protection in pigs under laboratory conditions.

Vaccination against pneumonia in calves has shown promise in those outbreaks where *M. bovis* was involved. Likewise, parenteral vaccination with an inactivated suspension of *M. bovis* was efficacious against polyarthritis. Although vaccinated cows cleared mycoplasmas faster from the mammary gland than nonvaccinated animals, the immunization did not prevent the drop in milk yield.

Vaccination of goats against CCPP using inactivated suspensions of Strain F38 appears to be very efficacious. The immunized animals develop antibodies against a carbohydrate moiety of F38 organisms.

Attenuated strains of *M. mycoides*, SC type, have long been in use to control CBPP in countries where test-and-slaughter programs are impractical. Yearly immunizations appear to provide satisfactory protection.

In poultry, an attenuated strain of *M. gallisepticum*, strain F, or a temperature- sensitive mutant, are used to protect against Chronic Respiratory Disease as an alternative to the SPF production strategy. These vaccines are effective when given in the drinking water.

REFERENCES

Almeida, R. A.; Wannemuehler, M. J.; and Rosenbusch, R. F. 1992. Interaction of *Mycoplasma dispar* with bovine alveolar macrophages. Infect Immun 60:2914-19.

Bhandari, S., and Asnani, P. J. 1989. Characterization of phospholipase A_2 of *Mycoplasma* species. Folia Microbiol 34:294-301.

Cole, B. C., and Atkin, C. L. 1991. The *Mycoplasma arthritidis* T-cell mitogen, MAM: A model superantigen. Immunol Today 12:271-76.

Davidson, M. K.; Lindsey, J. R.; Parker, R. F.; Tully, J. G.; and Cassell, G. H. 1988. Differences in virulence for mice among strains of *Mycoplasma pulmonis*. Infect Immun 56:2156-62.

Davidson, M. K.; Davis, J. K.; Simecka, J. W.; Lindsey, J. R. 1990. Pulmonary defense mechanisms against virulent *Mycoplasma pulmonis*. IOM Letters Vol. 1, 8th International Congress of the International Organization for Mycoplasmology, Istanbul, July 1990, pp. 345-46.

Faulkner, C. B.; Davidson, M. K.; Davis, J. K.; and Lindsey, J. R. 1992. Increased levels of tumor necrosis factor-α in the lungs of mice infected with *Mycoplasma pulmonis*. IOM Letters Vol. 2, 9th International Congress of the International Organization for Mycoplasmology, Ames, Iowa, August 1992, p. 125.

Franzoso, G.; Dimitrov, D. S.; Blumenthal, R.; Barile, M. F.; and Rottem, S. 1992. Fusion of *Mycoplasma fermentans* strain incognitus with T-lymphocytes. Fed Eur Biol Soc Lett 303:251-54.

Geary, S. J.; Tourtelotte, M. E.; and Cameron, J. A. 1981. Inflammatory toxin from *Mycoplasma bovis*: Isolation and characterization. Science 212:1032-33.

Lai, W. C.; Bennett, M.; Pakes, S. P.; Kumar, V.; Steutermann, D.; Owusu, I.; and Mikhael, A. 1990. Resistance to *Mycoplasma pulmonis* mediated by activated natural killer cells. J Infect Dis 161:1269-75.

Lai, W. C.; Bennett, M.; Lu, Y-S.; and Pakes, S. P. 1991a. Vaccination of Lewis rats with temperature-sensitive mutants of *Mycoplasma pulmonis*: Adoptive transfer of immunity by spleen cells but not by sera. Infect Immun 59:346-50.

Lai, W. C.; Bennett, M.; Pakes, S. P.; and Murphree, S. S. 1991b. Potential subunit vaccine against *Mycoplasma pulmonis* purified by a protective monoclonal antibody. Vaccine 9:177-84.

Lauerman, L. H., and Reynolds-Vaughn, R. A. 1991. Immunoglobulin G Fc receptors of *Mycoplasma synoviae*. Avian Dis 35:135-38.

Lingwood, C. A.; Quinn, P. A.; Wilansky, S.; Nutikka, A.; Ruhnke, H. L.; and Miller, R. B. 1990. Common sulfoglycolipid receptor for mycoplasmas involved in animals and human infertility. Biol Reprod 43:694-97.

Muhlradt, P. F., and Schade, U. 1991. MDHM, a macrophage-stimulatory product of *Mycoplasma fermentans*, leads to in vitro interleukin-1 (IL-1), IL-6, tumor necrosis factor, and prostaglandin production and is pyrogenic in rabbits. Infect Immun 59:3969-74.

Muhlradt, P. F.; Quentmeier, H.; and Schmitt, E. 1991. Involvement of interleukin-1 (IL-1), IL-6, IL-2, and IL-4 in generation of cytolytic T cells from thymocytes stimulated by a *Mycoplasma fermentans*-derived product. Infect Immun 59:3962-68.

Razin, S., and Freundt, E. A. 1984. Division Tenericutes Div. Nov. (g.v. p. 36). Class I. *Mollicutes*. In Bergey's Manual of Systematic Bacteriology, Vol. 1. Ed. N. R. Krieg and J. G. Holt. Baltimore, London: Williams and Wilkins.

Razin, S., and Jacobs, E. 1992. Mycoplasma adhesion. J Gen Microbiol 138:407-22.

Rosengarten, R., and Wise, K. S. 1991. The Vlp system of *Mycoplasma hyorhinis*: combinatorial expression of distinct size variant lipoproteins generating high-frequency surface antigenic variation. J Bacteriol 173:4782-93.

Ross, R. F.; Zimmermann-Erickson, B. J.; and Young, T. F. 1984. Characteristics of protective activity of *Mycoplasma hyopneumoniae* vaccine. Am J Vet Res 45:1-7.

Simecka, J. W.; Davis, J. K.; Davidson, M. K.; Ross, S. E.; Stadtlander, C. T. K.-H.; Cassell, G. H. 1992. Mycoplasma diseases of animals. In *Mycoplasma*, Molecular Biology and Pathogenesis. Ed. J. Maniloff, R. McElhaney, L. Finch, and J. Baseman. Baltimore: Academic Press.

Stadtlander, C. T. K.-H.; Watson, H. L.; Simecka, J. W.; and Cassell, G. H. 1991. Cytopathic effects of *Mycoplasma pulmonis* in vivo and in vitro. Infect Immun 59:4201-11.

Stuart, P. M.; Cassell, G. H.; and Woodward, J. G. 1989. Induction of class II MHC antigen expression in macrophages by *Mycoplasma* species. J Immunol 142:3392-409.

Thomas, C. B.; Ess, P. van; Wolfgram, L. J.; Riebe, J.; Sharp, P.; and Schultz, R. D. 1991. Adherence to bovine neutrophils and suppression of neutrophil chemiluminescence by *Mycoplasma bovis*. Vet Immun Immunopathol 27:365-81.

Villemot, J. M.; Provost, A.; and Queval, R. 1962. Endotoxin from *Mycoplasma mycoides*. Nature 193:906-7.

Weisburg, W. G.; Tully, J. G.; Rose, D. L.; Petzel, J. P.; Oyaizu, H.; Yang, D.; Mandelco, L.; Sechrest, J.; Lawrence, T. G.; Van Etten, J.; Maniloff, J.; and Woese, C. R. 1989. A phylogenetic analysis of the mycoplasmas: Basis for their classification. J Bacteriol 171:6455-67.

26 / *Chlamydia*

BY A. A. ANDERSEN

CHLAMYDIA **SPP.** are among the most successful of pathogens, being more widespread in nature than most pathogenic organisms. They have evolved to the point where each chlamydial strain is associated with specific diseases in a given host. Our knowledge of their importance as disease agents and the extent of their distribution in nature is increasing rapidly, with improvement in laboratory and diagnostic techniques. In humans, *C. trachomatis* is the leading cause of preventable blindness and the most common agent of sexually transmitted diseases in the United States. *C. pneumoniae* is increasingly being recognized as a significant cause of pneumonia and other acute respiratory tract infections in humans and is now thought to infect most individuals at some time during their life (Grayston 1992). *C. psittaci*, while occasionally infecting humans, is primarily a pathogen of nonhuman species and has been reported in arthropods, mollusks, approximately 130 species of birds, and a wide range of mammals. Pet birds often have clinically inapparent, persistent, *C. psittaci* infections, which constitute a threat to humans. In sheep and goats, chlamydiae are considered the leading cause of reproductive loss in many parts of the world.

The classification of *Chlamydia* is continuing to change as new techniques are exploited to investigate immunologic and genetic relationships. The current system was instituted when Page (1968) proposed that the chlamydiae be divided into two species, *C. trachomatis* and *C. psittaci*, on the basis of susceptibility to sulfadiazine, accumulation of glycogen in inclusions, and inclusion morphology. This effectively classified all the human isolates as *C. trachomatis* and all the animal isolates as *C. psittaci*, with a few exceptions. Currently, *C. trachomatis* includes three biovars: trachoma, lymphogranuloma venereum (LGV), and mouse. Trachoma and LGV together have 18 serovars and the mouse biovar has a single serovar (Moulder et al. 1984; Wang and Grayston 1991). The remaining chlamydial isolates were placed in the *C. psittaci* species and only recently have been subject to division. The first division was made on discovery of a human respiratory isolate (TWAR), that was different from both *C. psittaci* and *C. trachomatis* isolates. This isolate was proposed in 1989 as a third species, *C. pneumoniae*, with one serovar (Grayston et al. 1989). Recently, a fourth species, *C. pecorum*, was proposed to include strains from cattle and sheep that cause polyarthritis, encephalomyelitis, and diarrhea (Fukushi and Hirai 1992). It is expected to include 3 to 4 serovars. The remaining *C. psittaci* isolates are heterogeneous and likely will be subjected to further division as more information becomes available. Currently, 6 avian serovars and 8-10 mammalian serovars have been identified (Perez-Martinez and Storz 1985; Andersen 1991).

Chlamydiae are obligate, intracellular parasites that multiply in the cytoplasm of eukaryotic cells, forming membrane-bound cytoplasmic inclusions. They are dependent on the host cell for energy and a majority of their nucleotide-metabolizing enzymes. The life cycle is unique, having a growth cycle consisting of two major developmental forms. The elementary body (EB) is a condensed form, 200-300 nm in diameter, that is suited to survival outside the cell. The reticulate body (RB), ranging in size from 500 to 1000 nm, is the replicating form found in the cytoplasmic inclusions and predominates throughout most of the developmental cycle. Replication is by binary fission typical of other bacteria, with the exception that chlamydiae rely on the host cell for nutrients. Intermediate forms are usually seen and vary in size from 300 to 500 nm. These often are called dispersing forms or condensing forms, depending on whether they are a transition from an EB to a RB, or vice versa.

For chlamydia to be an effective pathogen it must transverse through five phases: attachment and penetration of the EB into a susceptible host cell; transition of the metabolically inert EB into the metabolically active RB and avoidance of destruction by the host cell; growth and replication of the RB using cellular components without destroying host cellular functions; maturation of the noninfectious RB into an infectious EB; release of the EB from the host cell and the transfer to new cells. What is known about these stages is covered in a number of reviews (Moulder 1984; Ward 1988).

One of the critical stages in the life cycle is the phase following entry into the host cell. During that period, the EB transforms into the replicative RB form and must escape destruction by the host cell. Chlamydiae enter the host cell in endosomes, in which they remain during the complete replicative cycle. The normal cell processes endosomes containing foreign material by acidification of the endosome and subsequent fusing of the endosome with lysosomes. This digestion and processing of antigen by the lysosomes is also critical for the processing of antigen during the immune response. Research shows that live chlamydiae, but not heat-inactivated chlamydiae, prevent acidification of the endosomes and subsequent phagolysosomal fusion. The mechanism by which the organisms prevent acidification of the endosome is not fully understood.

HOST RESPONSE

The role of the immune response to the chlamydial proteins in the pathogenesis of chlamydial disease is not fully understood. However, it is clear in a number of diseases caused by chlamydia that the pathology seen is more severe than would be expected from the infection alone. Models for studying genital and ocular infections have shown that disease severity increases following reinfection or heterotypic infections. Also, early studies of naturally occurring trachoma in humans and studies with experimental vaccines in primates indicate that trachoma is a disease of ocular hypersensitivity.

The molecular mechanism for the increased pathology following reinfection was first elaborated by Watkins et al. (1986). They found that a triton-soluble chlamydial extract would induce conjunctival disease in Guinea pigs that had been previously infected with the Guinea pig inclusion conjunctivitis strain of chlamydia, but not in naive Guinea pigs. Similar results have been demonstrated in primates infected with the trachoma biovar of *C. trachomatis* (Taylor et al. 1987). Recent data suggest that a single 57-kDa protein plays a major part in the immunopathogenesis of chlamydial disease, both in ocular trachoma and in tubal infertility (Morrison et al. 1989; Patton 1985; Wagar et al. 1990). The protein has been determined to be a homolog of the heat-shock protein (hsp) 60 family and shares 48% sequence identity with the human Hu Cha 60 protein (Cerrone et al. 1991). In humans the response to the hsp 60 protein is variable, and infertility due to tubal occlusion and increased ectopic pregnancies has been associated with higher levels of antibody response to that protein (Wagar et al. 1990). Studies in mice indicate that the immune response to the protein is genetically controlled (Tuffrey 1992; Zhong and Brunham 1992).

The role of cytokines in chlamydial infection has been investigated. It has been shown that interferon (IFN), interleukin (IL)-1, IL-6, and tumor necrosis factor α (TNFα) are produced in chlamydial infections (Moulder 1991). IFN has been shown to exert antichlamydial activity in vivo and may be important in controlling early primary chlamydial infections. However, IFN also has been hypothesized to be responsible for inducing a state of latency or persistence in chlamydial infections (Schachter 1992; Treharne 1992). This may be a factor in producing high antibody titers to the heat-shock protein 60 and increased severity of disease. The mechanisms by which gamma interferon inhibits chlamydial replication has been studied in vitro. Byrne et al. (1986) showed that gamma interferon inhibited the replication of *C. psittaci* by inducing indoleamine 2,3-dioxygenase, the enzyme that decycles tryptophan to N-formylkynurenine. This was confirmed in a later study using interferon-treated macrophages, in which tryptophan degradation appeared to be the general mechanism for the inhibition of *C. psittaci* replication (Carlin et al. 1989). A recent study, however, indicated that inhibition of chlamydial growth in murine cells was due to increased production of nitric oxide (Mayer et al. 1993).

Interleukin-1 has been implicated as a mediator of conjunctival inflammation and scarring in ocular trachoma (Rothermel 1989). TNFα has been shown to be produced in vivo during *C. trachomatis* infections and may have a modest protective role in the host defense. The effect of TNFα on chlamydiae resembles that of IFNγ (Moulder 1991).

The neutralization of chlamydial infectivity has been studied both in vivo and in vitro. Studies with monoclonal antibodies have shown both complement-dependent and complement-independent neutralization (Peeling et al. 1984; Andersen and van Deusen 1988). Most studies have placed neutralization by antibody at the stage of blocking the attachment to host cells (Moulder 1991). However, antibodies to the major outer-membrane protein will inhibit inclusion formation by a number of *C. trachomatis* isolates without affecting attachment to the host cells, which indicates that the development cycle is stopped at a point beyond attachment (Peeling et al. 1984). Neutralizing monoclonal antibodies are usually serovar-specific and are thought to react with the major outer-membrane protein. Monoclonal antibodies to the lipopolysaccharide or group reactive component have no neutralizing ability even though they react with the chlamydial elementary bodies.

DISEASES

The severity and type of disease produced by chlamydiae in mammals and birds depends on the strain of chlamydia and on the species of the host. In mammals, chlamydiae have been associated with pneumonia, enteritis, encephalomyelitis, abortion, urogenital infections, polyarthritis, polyserositis, mastitis, hepatitis, and conjunctivitis (Storz 1988). In birds, chlamydia cause pericarditis, airsacculitis, pneumonia, lateral nasal adenitis, peritonitis, hepatitis, and splenitis (Grimes and Wyrick 1991). In both mammals and birds, the severity of the disease may vary from clinically inapparent infections to severe systemic infections. Systemic or generalized infections produce fever, anorexia, lethargy, and, occasionally, shock and death. An asymptomatic carrier state is now thought to be a common sequel, if the infection is not properly treated.

Chlamydial conjunctivitis occurs in humans, other mammals, and birds and the chlamydial strains are often associated with other signs of disease. The eye involvement is usually restricted to the conjunctiva and may last 30-60 days, or longer. The chlamydial strains causing conjunctivitis in humans, cats, guinea pigs, and koalas have also been associated with reproductive tract infections. The conjunctivitis strain in sheep and cattle is associated with polyarthritis. *C. psittaci* has also been isolated from conjunctivitis in swine, but other signs have not been observed.

Enzootic abortion of ewes (EAE) has been recognized for a century, but the causal agent was not identified as chlamydiae until 1950 (Stamp et al. 1950). The ovine abortion strain (immunotype 1) of *C. psittaci* produces abortion or weak young in sheep, goats, and cattle; it has been isolated from cases of human abortion. The agent has been implicated in enteritis and other problems in sheep and cattle, but its role in these syndromes is unclear. Many chlamydial isolates have been recovered from the intestinal tract of sheep and goats; some of these are of immunotype 1. However, recent research indicates that most of the intestinal isolates are of immunotype 2 (polyarthritis-conjunctivitis). Other immunotypes of *C. psittaci* are known to infect sheep and cattle; their roles in disease are not known.

Chlamydial polyarthritis of lambs was first identified in Wisconsin in 1957 as a chlamydial infection of the synovial tissues involving most joints of the limbs. It is now recognized in epizootic proportions in most major sheep-raising areas of the world. Affected lambs have varying degrees of stiffness, lameness, and anorexia. Conjunctivitis is often present. A similar disease caused by the same agent occurs in calves. Polyarthritis is caused by immunotype 2 strains (the newly proposed species *C. pecorum*). It appears that polyarthritis and conjunctivitis are caused by the same agent. Encephalomyelitis may also be caused by this agent or by a serologically similar agent. These organisms are often isolated from fecal specimens and may also be involved in enteritis.

Turkeys infected with strains of high virulence can experience mortality rates of 5-40%, unless early antibiotic treatment is instituted (Grimes and Wyrick 1991). Typical postmortem findings include vasculitis, pericarditis, splenitis, and lateral nasal adenitis. The virulent turkey strains cause little if any disease in chickens, pigeons, or sparrows; however, cockatiels and parakeets succumb rapidly to infections with these agents. The virulent turkey strain is implicated in most human infections from turkeys. Outbreaks of chlamydiosis characterized by low virulence of the strain and little or no human involvement are usually caused by pigeon strains, and mortality usually is less than 5%.

Chlamydiosis is a common, chronic infection of psittacine birds. Many birds show no clinical signs until they are stressed. These birds often shed chlamydiae intermittently and serve as a source of infection for humans and other birds. Infections cause enteritis, airsacculitis, pneumonitis, and hepatosplenomegaly. Chlamydiosis in pigeons is similar to that seen in psittacine birds; infections are usually chronic with survivors becoming asymptomatic carriers. However, signs of disease are more likely to be conjunctivitis, blepharitis, and rhinitis.

Chlamydiosis in ducks is a serious economic and occupational health problem in Europe. Trembling, conjunctivitis, rhinitis, and diarrhea are the most common signs, and mortality can range up to 30%.

PATHOGENESIS

Four specific diseases in mammals and birds will be discussed: placental and fetal infection in ruminants, polyarthritis-polyserositis in ruminants, conjunctivitis (feline pneumonitis) in cats, and chlamydiosis in turkeys. These represent the most thoroughly investigated syndromes caused by chlamydia and are used to illustrate the disease process. Similar diseases occur in other animals and are often caused by distinctly different serovars of chlamydia. It should also be recognized that chlamydia can cause other disease syndromes such as pneumonia, enteritis, meningoencephalitis, and mastitis, and that as better diagnostic techniques are developed, new chlamydial strains and new diseases caused by chlamydiae are being found.

Placental and Fetal Infections in Ruminants

In enzootically infected sheep flocks, abortions occur year after year at a rate of 1-5%, whereas in flocks in which chlamydia has recently been introduced, the abortion rate may reach 30% (Storz 1988). Most abortions occur during the last month of gestation, but some may occur as early as the one-hundredth day. In experimentally infected sheep, fever as high as 40-41°C is seen on day 1 or 2 and lasts for 3-5 days (Stamp et al. 1950). In cattle, chlamydial abortions are usually sporadic but the abortion rate may sometimes reach 20%. Abortions in the natural disease usually occur in the last trimester.

Sheep become infected by ingesting or inhaling *C. psittaci*. It has been suggested that infection is first established in the tonsil (Jones and Anderson 1988), from which it is disseminated by the blood to other organs. The organism then persists in the dam in a latent form and intermittent low-grade chlamydiosis occurs, eventually infecting the placenta (Huang et al. 1990). Placental infections usually become established sometime between 60 and 90 days of gestation, with pathological changes first being detected after 90 days of gestation (Buxton et al. 1990).

The mechanism by which chlamydiae migrate from the maternal side of the placenta to the fetus is still uncertain. After approximately 60 days of gestation, the physiological invasion of the carunculae stroma by chorionic villi coincides with hemorrhage from the maternal vessels and results in the formation of hematomas (Buxton et a. 1990). These hematomas have been suggested as a means for chlamydia in the maternal circulation to make direct contact with the chorionic epithelium. The 30-day interval between the start of formation of hematomas and the first observations of pathological changes in the placenta has raised questions about whether endocrinological and/or related immune-related changes may play a significant role (Buxton et al. 1990). The initial lesions in the placenta involve the limbus of the placentomas in the hilar region, where chlamydial inclusions are seen in the trophoblast (Buxton et al. 1990). Progression of the infection results in a considerable loss of chorionic epithelial cells in both the cotyledonary and intercotyledonary placenta. At this time a mixed-cell inflammatory infiltrate is present. Fibrin deposits and a purulent arteritis are noted in the thickened placental mesenchyme underlying the lesions. Chlamydial inclusions are present in the endometrial epithelium, where affected chorion is in apposition to the maternal tissues. During the later stages, severe necrosis and sloughing of the endometrial epithelium occur.

Infection of the fetus is secondary to placentitis and is not a major factor in the disease. Chlamydial isolations are often made from fetuses; however, titers are of low levels and infectious foci have not been reported. Necrotic foci, sometimes with inflammatory reactions, are frequently found in most fetal organs and tissues and may be embolic in origin (Buxton et al. 1990). Chlamydial antigen, if seen in the foci, is usually present only in small amounts. The popliteal and mesenteric lymph nodes of infected fetuses are usually enlarged, having a demarcated cortex with several follicles and germinal centers. Fetal lambs will respond with specific chlamydial antibodies that can be detected in the indirect fluorescent antibody or immunodiffusion tests.

Polyarthritis-Polyserositis in Ruminants

Polyarthritis in sheep and cattle is caused by immunotype 2 *C. psittaci*. The disease is readily reproduced by inoculating calves or lambs by oral, intramuscular, subcutaneous, intravenous, or intraarticular routes. Under field conditions, the organism is thought to be ingested and to subsequently multiply in the mucosa of the large and small intestines. It may produce a diarrhea during this phase. A chlamydemia follows the multiplication in the intestinal mucosa and distributes the chlamydiae to other parts of the body. Both periarticular and articular tissue changes are seen in the joints. Periarticular changes, including subcutaneous edema and fluid-filled synovial sacs, cause joint enlargement; these enlarged joints will contain excessive greyish-yellow turbid synovial fluid. Fibrin plaques will also be seen in joints with advanced lesions. In severe cases, tendon sheaths may also contain excess fluids. Muscle involvement is limited to the point of tendinous attachment.

Histological changes are primarily an inflammatory reaction in the synovium, tendon sheaths, and subsynovial tissues (Shupe and Storz 1964; Cutlip and Ramsey 1973). In experimentally inoculated joints, granulation tissue replaces much of the fibropurulent exudate, with the formation of large fibrous villi by 21-24 days. By this time the synovial surfaces are again covered by intact lining cells.

Conjunctivitis (Feline Pneumonitis)

C. psittaci infection in cats (feline pneumonitis) is characterized by a severe conjunctivitis with blepharospasm, conjunctival hyperemia, chemosis, and serous and mucopurulent ocular discharges (Wills et al. 1987). Mild respiratory signs with slight nasal discharge, coughing, and sneezing are often seen with the conjunctivitis. Following natural infections, the agent may colonize the gastrointestinal tract (primarily the superficial gastric) and reproductive tract.

The disease is transmitted by direct contact with infected secretions and by droplet infection. Clinical signs usually are seen on day 4 postinoculation (PI) and will last 30 days, after which they gradually subside. Long-term persistent infections are likely responsible for maintaining the agent in the feline population, for the agent has been recovered from the eyes, vagina, rectal swabs, and superficial gastric mucosa for over 150 days after infection (Wills et al. 1987).

Chlamydiosis in Turkeys

Transmission of chlamydiae in turkeys is thought to be primarily through inhalation of the organism, which is excreted in large quantities in both fecal material and in nasal and ocular discharges. In a study using a virulent turkey isolate, only a few of the orally exposed turkeys initially developed a mild subclinical infection. The infection appeared to then spread to the remaining birds by the aerosol route (Page 1958). Chlamydial isolation patterns and serological data support the hypothesis of secondary transmission following infection of only a few of the birds by the oral route. The hypothesis that aerosol exposure is the primary method of exposure is supported by the findings that chlamydia is first isolated from the oral-pharyngeal and nasal secretions, and that titers in fecal material drop rapidly following drying.

Chlamydia is rapidly disseminated throughout the body following aerosol exposure and is recovered from the lung, airsacs, pericardial sac, and mesentery within 4 hours. Consistently detectable levels in the blood are not achieved until 72 hours, at which time chlamydia is found throughout the body and in cloacal materials (Page 1958). Clinical signs vary with the strain of chlamydia (Tappe et al. 1989). A study in which turkeys were given a virulent turkey strain (TT3), a psittacine strain (VS1), or an ovine abortion strain (B577), showed significant differences in clinical signs. The TT3-infected turkeys showed clinical signs on postinoculation (PI) days 3-24 and experienced a 20-40% decrease in body weight compared with birds in other groups. The VS1-infected birds experienced a mild dyspnea on days 4-11 and were slightly lethargic on day 7. Control birds and B577-infected birds remained normal throughout the study.

Gross lesions also varied with the chlamydial strain. Pericarditis was the most severe lesion seen in turkeys infected with TT3. Only mild pericardial lesions were seen in VSI-inoculated birds. However, air sac lesions were more severe in turkeys infected with VS1 than in turkeys infected with TT3. Bronchopneumonia was characteristic of turkeys infected with the psittacine isolate, but not of turkeys infected with the virulent turkey isolate.

Both the VS1 strain and the TT3 strain produced an infection of the lateral nasal glands that was detectable by histological examination through PI day 50. This could be an important source for aerosolization of chlamydiae, since these glands are the main source of moisture for the nasal mucosa.

IMMUNITY

During most chlamydial infections the host develops an immune response that halts the infection and probably eliminates the organism from the body. This immunity provides solid protection from reinfection to the same chlamydial strain for 4-6 months, after which immunity subsides. Immunity likely involves both cell-mediated and antibody-mediated mechanisms. The role of each is not understood and may depend on the location of the parasite in the body, i.e., the eye, the respiratory tract, or the genital tract.

The response to the primary infection includes high levels of antibody to the chlamydial lipopolysaccharide (LPS) or genus-specific epitopes. The response is measured by complement-fixation (CF) and enzyme-linked immunosorbent assay (ELISA) tests; however, antibody levels do not correspond to immunity. In secondary infections, the anamnestic response may or may not give a CF response. The neutralizing or protective antibodies are thought to be primarily to the major outer-membrane protein (MOMP) and are likely serovar specific. With *C. trachomatis*, serovar-specific monoclonal antibodies (MAbs) to the MOMP have been shown to provide neutralization in vitro. Serovar-specific MAbs to the ovine abortion strain provide neutralization in both in vitro tests and by passive immunization in mice (Andersen and van Deusen 1988; Buzoni-Gatel 1990). However, the protein specificity of these MAbs has not been tested, for they fail to react by the Western blot. The MAbs are thought to be to conformational epitopes on the MOMP, as the MOMP is believed to be the site of serovar specificity. High-molecular weight protein may also be a factor in immunity, for MAbs to an 89-kDa protein in the ovine abortion isolate have been shown to provide protection in vitro (Cevenini et al. 1991).

Currently, there are commercial vaccines to the ovine abortion strain and to the feline pneumonitis strain. The ovine abortion vaccine is a formalin-inactivated bacterial preparation, intended to induce production of neutralizing antibody to prevent the spread of the organism through the blood to the placenta. The vaccine has been used extensively in Europe and was initially considered quite effective, but subsequently there have been several reports of vaccine failure. There are indications that the vaccine stocks may have been contaminated with other strains. New serotyping techniques should eliminate this problem.

A subcellular vaccine for ovine abortion has provided good protection (Tan et al. 1990). The vaccine was prepared by procedures that provide a MOMP-enriched preparation while not denaturing the protein. The preparation has the advantage of reducing the response to the genus-specific antigen (LPS), which does not relate to protection. In early trials, the vaccine gave protection comparable to chlamydial bacterins prepared from purified elementary bodies.

Feline pneumonitis is the other chlamydial strain for which commercial vaccines are available. The vaccines include a live attenuated vaccine and a number of inactivated whole-cell preparations. The infection in cats is primarily an infection of the conjunctiva with secondary spread to the gastrointestinal and reproductive tracts. Reports on the specific inactivated products are not available; however, data are available on the use of the live attenuated vaccine and on experimental inactivated vaccines (Shewen et al. 1980; Wills et al. 1987). These vaccines reduce or eliminate clinical signs but do not prevent infection and shedding of the organism.

REFERENCES

Andersen, A. A. 1991. Serotyping of *Chlamydia psittaci* isolates using serovar-specific monoclonal antibodies with the microimmunofluorescence test. J Clin Microbiol 29:707-11.

Andersen, A. A., and van Deusen, R. A. 1988. Production and partial characterization of monoclonal antibodies to four *Chlamydia psittaci* isolates. Infect Immun 56:2075-79.

Byrne, G. I.; Lehmann, L. K.; and Landry, G. J. 1986. Induction of tryptophan catabolism is the mechanism for gamma-interferon-mediated inhibition of intracellular *Chlamydia psittaci* replication in T24 cells. Infect Immun 53:347-351.

Buxton, D.; Barlow, R. M.; Finlayson; J., Anderson, I. E.; and Mackellar, A. 1990. Observations on the pathogenesis of *Chlamydia psittaci* infection of pregnant sheep. J Comp Pathol 102:221-37.

Buzoni-Gatel, D.; Bernard, F.; Andersen, A.; and Rodolakis, A. 1990. Protective effect of polyclonal and monoclonal antibodies against abortion in mice infected by *Chlamydia psittaci*. Vaccine 8:342-46.

Carlin, J. M.; Borden, E. C.; and Byrne, G. I. 1989. Interferon-induced indolamine 2,3-dioxygenase activity inhibits *Chlamydia psittaci* replication in human macrophages. J Interferon Res 9:329-37.

Cerrone, M. C.; Ma, J. J.; and Stephens, R. S. 1991. Cloning and sequence of the gene for heat shock protein 60 from *Chlamydia trachomatis*, and immunological reactivity of the protein. Infect Immun 59:79-90.

Cevenini, R.; Donati, M.; Brocchi, E.; De Simone F.; and La Placa, M. 1991. Partial characterization of an 89-kDa highly immunoreactive protein from *Chlamydia psittaci* A/22 causing ovine abortion. Fed Eur Microbiol Soc 81:111-16.

Cutlip, R. C., and Ramsey, F. K. 1973. Ovine chlamydial polyarthritis: Sequential development of articular lesions in lambs after intraarticular exposure. Am J Vet Res 34:71-75.

Fukushi, H., and Hirai, K. 1992. Proposal of *Chlamydia pecorum* sp. nov. for *Chlamydia* strains derived from ruminants. Int J Syst Bacteriol 42:306-8.

Grayston, J. T. 1992. *Chlamydia pneumoniae*, strain TWAR pneumonia. Annu Rev Med 43:317-23.

Grayston, J. T.; Kuo, C. -C.; Campbell, L. A.; and Wang, S. -P. 1989. *Chlamydia pneumoniae* sp. nov. for *Chlamydia* sp. strain TWAR. Int J Syst Bacteriol 39:88-90.

Grimes, J. E., and Wyrick, P. B. 1991. Chlamydiosis (ornithosis). In Diseases of Poultry, 9th Ed. Ed. B. W. Calnek, H. J. Barnes, C. W. Beard, W. M. Reid, and H. W. Joder, Jr. Ames: Iowa State University Press.

Huang, H. -S.; Buxton, D.; and Anderson, I. E. 1990. The ovine immune response to *Chlamydia psittaci*; histopathology of the lymph node. J Comp Pathol 102:89-97.

Jones, G. E., and Anderson, I. E. 1988. *Chlamydia psittaci*: Is tonsillar tissue the portal of entry in ovine enzootic abortion? Res Vet Sci 44:260-61.

Mayer, J.; Woods, M. L.; Vavrin, Z.; and Hibbs, J. B. Jr. 1993. Gamma interferon-induced nitric oxide production reduces *Chlamydia trachomatis* infectivity in McCoy cells. Infect Immun 61:491-97.

Morrison, R. P.; Belland, R. J.; Lyng, K.; and Caldwell, H. D. 1989. Chlamydial disease pathogenesis: The 57-kDa chlamydial hypersensitivity antigen is a stress response protein. J Exp Med 170:1271-83.

Moulder, J. W. 1984. Looking at chlamydiae without looking at their hosts. Am Soc Microbiol News 50:353-62.

__________. 1991. Interaction of chlamydiae and host cells in vitro. Microbiol Rev 55:143-90.

Moulder, J. W.; Hatch, T. P.; Kuo, C. -C. 1984. Genus *Chlamydia*. In Bergey's Manual of Systematic Bacteriology, Vol. 1. Ed. N. R. Krieg and J. G. Holt. Baltimore: Williams and Wilkins.

Page, L. A. 1958. Experimental ornithosis in turkeys. Avian Dis 3:51-66.

__________. 1968. Proposal for the recognition of two species in the genus *Chlamydia* Jones, Rake, and Stearns, 1945. Int J Syst Bacteriol 18:51-66.

Patton, D. L. 1985. Immunopathology and histopathology of experimental salpingitis. Rev Infect Dis 7:746-53.

Peeling, R. W.; Maclean, I. W.; and Brunham, R. C. 1984. In vitro neutralization of *Chlamydia trachomatis* with monoclonal antibody to an epitope of the major outer membrane protein. Infect Immun 46:484-88.

Perez-Martinez, J. A., and Storz, J. 1985. Antigenic diversity of *Chlamydia psittaci* of mammalian origin determined by micro-immunofluorescence. Infect Immun 50:905-10.

Rothermel, C. D.; Schachter, J.; Lavrich, P.; Lipsitz, E. C.; and Francus, T. 1989. *Chlamydia trachomatis*-induced production of interleukin-1 by human monocytes. Infect Immun 57:2705-2711.

Schachter, J. 1992. The pathogenesis of chlamydial infection. Proc Eur Soc Chlamydia Res 2:67-72.

Shewen, P. E.; Povey, R. C.; and Wilson, M. R. 1980. A comparison of the efficacy of a live and four inactivated vaccine preparations for the protection of cats against experimental challenge with *Chlamydia psittaci*. Can J Comp Med 44:244-51.

Shupe, J. L., and Storz, J. 1964. Pathologic study of psittacosis lymphogranuloma polyarthritis of lambs. Am J Vet Res 25:943-51.

Stamp, J. T.; McEwen, A. D.; Watt, J. A. A.; Nisbet, D. J. 1950. Enzootic abortion in ewes. 1. Transmission of the disease. Vet Rec 62:251-54.

Storz, J. 1988. Overview of animal diseases induced by chlamydial infections. In Microbiology of Chlamydia. Ed. A. L. Barron. Boca Raton, Fla.: CRC Press.

Tan, T.-W.; Herring, A. J.; Anderson, I. E.; and Jones, G. E. 1990. Protection of sheep against *Chlamydia psittaci* infection with a subcellular vaccine containing the major outer membrane protein. Infect Immun 58:3101-8.

Tappe, J. P.; Andersen, A. A.; and Cheville N. F. 1989. Respiratory and pericardial lesions in turkeys infected with avian or mammalian strains of *Chlamydia psittaci*. Vet Pathol 26:386-95.

Taylor, H. R.; Johnson, S. L.; Schachter, J.; Caldwell, H. D.; and Prendergast, R. A. 1987. Pathogenesis of trachoma: the stimulus for inflammation. J Immunol 38:3023-27.

Treharne, J. D. 1992. Chlamydia occular infections. Proc Eur Soc Chlamydia Res 2:129-32.

Tuffrey, M. 1992. Animal models of chlamydial infection. Proc Eur Soc Chlamydia Res 2:99-102.

Wagar, E. A.; Schachter, J.; Bavoil, P.; and Stephens, R. S. 1990. Differential human serologic response to two 60,000 molecular weight *Chlamydia trachomatis* antigens. J Infect Dis 162:922-27.

Wang, S.-P., and Grayston, J. T. 1991. Three new serovars of *Chlamydia trachomatis:* Da, Ia, and L2a. J Infect Dis 163:403-5.

Ward, M. E. 1988. The chlamydial developmental cycle. In Microbiology of Chlamydia. Ed. A. L. Barron. Boca Raton, Fla.: CRC Press.

Watkins, N. D.; Hadlow, W. J.; Moos, A. B.; and Caldwell, H. D. 1986. Ocular delayed hypersensitivity: A pathogenetic mechanism of chlamydial conjunctivitis in guinea pigs. Proc Natl Acad Sci USA 83:74-80.

Wills, J. M.; Gruffydd-Jones, T. J.; Richmond, S. J.; Gaskell, R. M.; and Bourne, F. J. 1987. Effect of vaccination on feline *Chlamydia psittaci* infection. Infect Immun 55:2653-57.

Zhong, G., and Brunham, R. C. 1992. Antibody responses to the chlamydial heat shock proteins hsp60 and hsp70 are *H-2* linked. Infect Immun 60:3143-49.

Index